Feature Papers in Eng 2024—Volume I

Feature Papers in Eng 2024—Volume I

Guest Editor

Antonio Gil Bravo

MDPI

Basel • Beijing • Wuhan • Barcelona • Belgrade • Novi Sad • Cluj • Manchester

Guest Editor
Antonio Gil Bravo
INAMAT2
Universidad Publica de
Navarra
Pamplona
Spain

Editorial Office
MDPI AG
Grosspeteranlage 5
4052 Basel, Switzerland

This is a reprint of the Special Issue, published open access by the journal *Eng* (ISSN 2673-4117), freely accessible at: https://www.mdpi.com/journal/eng/special_issues/NL9E2PWG88.

For citation purposes, cite each article independently as indicated on the article page online and as indicated below:

Lastname, A.A.; Lastname, B.B. Article Title. *Journal Name* **Year**, *Volume Number*, Page Range.

Volume I
ISBN 978-3-7258-3979-7 (Hbk)
ISBN 978-3-7258-3980-3 (PDF)
https://doi.org/10.3390/books978-3-7258-3980-3

Set
ISBN 978-3-7258-3977-3 (Hbk)
ISBN 978-3-7258-3978-0 (PDF)

Contents

eng

MDPI

Editorial

Special Issue: Feature Papers in *Eng* 2024

Antonio Gil Bravo

INAMAT^2, Science Department, Public University of Navarra, Building Los Acebos, Campus of Arrosadia, E-31006 Pamplona, Spain; andoni@unavarra.es

1. Introduction

Similarly to previous Special Issues of this series, the aim of this fourth edition of "Feature Papers in *Eng*" is to collect experimental and theoretical works related to engineering science and technology. The general topics published in *Eng* include electrical, electronic, and information engineering; chemical and materials engineering; energy engineering; mechanical and automotive engineering; industrial and manufacturing engineering; civil and structural engineering; aerospace engineering; biomedical engineering; geotechnical engineering and engineering geology; and ocean and environmental engineering. Many of these topics are included as they contribute to the *circular economy* and *sustainable development*. These aspects are addressed from various points of view and are supported by the fascinating field of engineering and its applications. The following selection of representative works of these topics were published in our journal in 2024–early 2025.

The intersection of *Industry 4.0* and the *circular economy*, considered in a manuscript in this Special Issue [1], offers numerous opportunities for businesses to create more sustainable, efficient, and resilient operations. Some examples include digital product passports, which involve using digital technologies to track product lifecycles, thus enabling more efficient recycling and reuse; predictive maintenance, which involves using Internet of Things (IoT) sensors and machine learning to predict maintenance needs, thus reducing waste and improving resource utilization; and product as a service, which involves offering products as services, encouraging sharing and collaboration, and reducing waste.

In a similar vein, *sustainable charging stations*, alongside *electric vehicles (EVs)* as key components of a low-carbon transportation system [2], can be designed to minimize environmental impacts while supporting the growth of EVs. Key features include charging stations powered by solar, wind, or other renewable energy sources; battery storage systems to reduce peak demand and stabilize the grid; and fast-charging technologies that minimize energy losses and reduce charging times. The benefits reported for *sustainable charging stations* also include reduced greenhouse gas emissions due to the use of renewable energy sources; improved air quality by promoting the adoption of EVs; and enhanced energy security by diversifying energy sources and reducing dependence on fossil fuels [3].

Synthetic nitrogen fertilizers are a widely used agricultural tool that has both benefits and drawbacks for *sustainable development* [4]. Its cited benefits include significant increases in crop yields, helping to meet global food demands; improvements in food security, particularly in regions with limited agricultural resources; and economic benefits for farmers, due to increases in yield and income. However, some of the drawbacks of synthetic nitrogen fertilizers include the significant environmental impacts of its production and use, including water pollution, soil degradation, and greenhouse gas emissions; the significant amounts of energy required, which is often generated from fossil fuels; and the negative health impacts, particularly for farmers and rural communities, due to exposure to toxic chemicals.

Received: 10 April 2025
Accepted: 12 April 2025
Published: 15 April 2025

Citation: Gil Bravo, A. Special Issue: Feature Papers in *Eng* 2024. *Eng* **2025**, 6, 78. https://doi.org/10.3390/eng6040078

Finally, some sustainable alternatives to the use of synthetic nitrogen fertilizers and their development opportunities include organic fertilizers, such as compost and manure, which can provide a more sustainable alternative to synthetic nitrogen fertilizers; crop rotation and intercropping, which can help improve soil fertility and reduce the need for synthetic nitrogen fertilizers; promoting sustainable agriculture practices, such as organic farming and agroecology, which can help reduce the environmental impacts of synthetic nitrogen fertilizers; implementing circular economy approaches, such as recycling and reusing nutrients, which can help reduce waste and minimize environmental impacts; and developing policies and regulatory frameworks that support sustainable agriculture practices and reduce the use of synthetic nitrogen fertilizer, which can help promote sustainable development.

Some of these aspects have been addressed in this Special Issue, allowing for greater discussion among potential readers. For more information, please see the List of Contributions.

2. Overview of the Published Articles

This Special Issue contains 42 papers, including 6 *Reviews* and 1 *Perspective*, published by several authors interested in new cutting-edge developments in the field of engineering. The submitted papers are from authors from 33 countries, including Australia, Bosnia and Herzegovina, Brazil, Brunei, Canada, China, Croatia, the Czech Republic, Estonia, Finland, France, Germany, Greece, Hungary, India, Iran, Italy, Japan, Malaysia, Mexico, Poland, Portugal, the Republic of Korea, Romania, Serbia, Slovakia, Slovenia, Spain, Taiwan, Turkey, Ukraine, the United Kingdom, and the USA.

3. Conclusions

The articles published in this Special Issue present important advancements in engineering. I express my sincere gratitude to all the authors, who have professionally and enthusiastically contributed to this Special Issue, and I thank the managing editors and reviewers who helped to improve the papers and made important contributions to this Special Issue. I hope that the articles showcased in this Special Issue are interesting and inspiring to its readers, especially young scholars who are eager to learn about recent advances in the field and to contribute future research.

Acknowledgments: The author is grateful for the financial support from the Spanish Ministry of Science and Innovation (MCIU/AEI/10.13039/501100011033/FEDER, EU) through project PID2023-146935OB-C21.

Conflicts of Interest: The author declares no conflicts of interest.

List of Contributions

1. Ferreira, V.B.; de Aquino Gomes, R.; Domingos, J.L.; da Fonseca, R.C.B.; Mendes, T.A.; Bouloukakis, G.; da Costa, B.B.F.; Haddad, A.N. Planning Energy-Efficient Smart Industrial Spaces for Industry 4.0. *Eng* **2025**, *6*, 53. https://doi.org/10.3390/eng6030053.
2. Tomanik, E.; Christinelli, W.; Garcia, P.S.; Rajala, S.; Crepaldi, J.; Franzosi, D.; Souza, R.M.; Rovai, F.F. Effect of Graphene Nanoplatelets as Lubricant Additive on Fuel Consumption During Vehicle Emission Tests. *Eng* **2025**, *6*, 18. https://doi.org/10.3390/eng6010018.
3. Lui, M.H.; Liu, H.; Tang, Z.; Yuan, H.; Williams, D.; Lee, D.; Wong, K.C.; Wang, Z. An Adaptive YOLO11 Framework for the Localisation, Tracking, and Imaging of Small Aerial Targets Using a Pan–Tilt–Zoom Camera Network. *Eng* **2024**, *5*, 3488–3516. https://doi.org/10.3390/eng5040182.
4. Rátkai, M.; Géczi, G.; Székely, L. Investigation of the Hottel–Whillier–Bliss Model Applied for an Evacuated Tube Solar Collector. *Eng* **2024**, *5*, 3427–3438. https://doi.org/10.3390/eng5040178.
5. Kobelev, V. On the Game-Based Approach to Optimal Design. *Eng* **2024**, *5*, 3212–3238. https://doi.org/10.3390/eng5040169.

6. Vieira, A.F.C.; Filho, M.R.T.; Eguea, J.P.; Ribeiro, M.L. Optimization of Structures and Composite Materials: A Brief Review. *Eng* **2024**, *5*, 3192–3211. https://doi.org/10.3390/eng5040168.
7. Puška, A.; Nedeljković, M.; Štilić, A.; Božanić, D. Evaluation of Affordable Agricultural Drones for Small and Medium Farms. *Eng* **2024**, *5*, 3161–3173. https://doi.org/10.3390/eng5040166.
8. Armenta-Déu, C.; Sancho, L. Sustainable Charging Stations for Electric Vehicles. *Eng* **2024**, *5*, 3115–3136. https://doi.org/10.3390/eng5040163.
9. Netto, K.; Francis-Pester, G.; Benazic, P.; Edwards, P. Understanding the Musculoskeletal Demand of Ride-On Mowing Using Wearable Technology. *Eng* **2024**, *5*, 3108–3114. https://doi.org/10.3390/eng5040162.
10. Harkonen, J.; Rodriguez, J.M.G.; Mustonen, E. The Role of Productization in End-To-End Traceability. *Eng* **2024**, *5*, 2943–2965. https://doi.org/10.3390/eng5040153.
11. Thakkar, G.; Preradović, N.M.; Tadić, M. Examining Sentiment Analysis for Low-Resource Languages with Data Augmentation Techniques. *Eng* **2024**, *5*, 2920–2942. https://doi.org/10.3390/eng5040152.
12. Thwin, S.M.; Park, H.-S. Enhanced Skin Lesion Segmentation and Classification Through Ensemble Models. *Eng* **2024**, *5*, 2805–2820. https://doi.org/10.3390/eng5040146.
13. Udhwani, L.; Soni, A.; Cuce, E.; Kumarasamy, S. Optical Fiber Technology for Efficient Daylighting and Thermal Control: A Sustainable Approach for Buildings. *Eng* **2024**, *5*, 2680–2694. https://doi.org/10.3390/eng5040140.
14. Sokolov, O.; Iakovets, A.; Andrusyshyn, V.; Trojanowska, J. Development of a Smart Material Resource Planning System in the Context of Warehouse 4.0. *Eng* **2024**, *5*, 2588–2609. https://doi.org/10.3390/eng5040136.
15. Garousi, M.H.; Karimi, M.; Casoli, P.; Rundo, M.; Fallahzadeh, R. Vibration Analysis of a Centrifugal Pump with Healthy and Defective Impellers and Fault Detection Using Multi-Layer Perceptron. *Eng* **2024**, *5*, 2511–2530. https://doi.org/10.3390/eng5040131.
16. Shahriar, A.; Majlesi, A.; Montoya, A. A Computationally Time-Efficient Method for Implementing Pressure Load to FE Models with Lagrangian Elements. *Eng* **2024**, *5*, 2379–2394. https://doi.org/10.3390/eng5030124.
17. Pucci, M.; Zanforlin, S. Improved Lift for Thick Flatback Airfoils in the Inboard Blades of Large Wind Turbines. *Eng* **2024**, *5*, 2345–2361. https://doi.org/10.3390/eng5030122.
18. Paulino, D.; Netto, A.T.; Brito, W.A.T.; Paredes, H. WebTraceSense—A Framework for the Visualization of User Log Interactions. *Eng* **2024**, *5*, 2206–2222. https://doi.org/10.3390/eng5030115.
19. Medvedieva, K.; Tosi, T.; Barbierato, E.; Gatti, A. Balancing the Scale: Data Augmentation Techniques for Improved Supervised Learning in Cyberattack Detection. *Eng* **2024**, *5*, 2170–2205. https://doi.org/10.3390/eng5030114.
20. Suzuki, Y.; Yue, B. Grading Evaluation of Marbling in Wagyu Beef Using Fractal Analysis. *Eng* **2024**, *5*, 2157–2169. https://doi.org/10.3390/eng5030113.
21. Schroeder, B.; Free, M.; Sarswat, P.; Sadler, E.; Burke, J.; Evans, Z. Evaluating Field-Effect Separation on Rare Earth and Critical Metals. *Eng* **2024**, *5*, 2016–2032. https://doi.org/10.3390/eng5030107.
22. Rosset, J.; Olaniyanu, E.; Stein, K.; Almeida, N.D.; França, R. Exploring the Frontier of 3D Bioprinting for Tendon Regeneration: A Review. *Eng* **2024**, *5*, 1838–1849. https://doi.org/10.3390/eng5030098.
23. Ye, X.P. Decarbonizing Nitrogen Fertilizer for Agriculture with Nonthermal Plasma Technology. *Eng* **2024**, *5*, 1823–1837. https://doi.org/10.3390/eng5030097.
24. Panaite, F.A.; Leba, M.; Ionica, A.C. Assessing CNN Architectures for Estimating Correct Posture in Cruise Machinists. *Eng* **2024**, *5*, 1785–1803. https://doi.org/10.3390/eng5030094.
25. Zani, N.; Solazzi, L. Analysis and Prediction of Spring-Back in Cylindrical Helical Springs Using Analytical and Numerical Models. *Eng* **2024**, *5*, 1696–1707. https://doi.org/10.3390/eng5030089.
26. Cascino, A.; Amedei, A.; Meli, E.; Rindi, A. Development, Designing and Testing of a New Test Rig for Studying Innovative Polycrystalline Diamond Bearings. *Eng* **2024**, *5*, 1615–1640. https://doi.org/10.3390/eng5030085.

27. Laganà, F.; Prattico, D.; De Carlo, D.; Oliva, G.; Pullano, S.A.; Calcagno, S. Engineering Biomedical Problems to Detect Carcinomas: A Tomographic Impedance Approach. *Eng* **2024**, *5*, 1594–1614. https://doi.org/10.3390/eng5030084.
28. Caesarendra, W. Bone Drilling: Review with Lab Case Study of Bone Layer Classification Using Vibration Signal and Deep Learning Methods. *Eng* **2024**, *5*, 1566–1593. https://doi.org/10.3390/eng5030083.
29. Kakinuma, T.; Fukuura, Y. Control of Floating Body Waves Due to an Airplane Takeoff from a Very Large Floating Airport. *Eng* **2024**, *5*, 1513–1533. https://doi.org/10.3390/eng5030081.
30. Cascino, A.; Meli, E.; Rindi, A. Development of a Methodology for Railway Bolster Beam Design Enhancement Using Topological Optimization and Manufacturing Constraints. *Eng* **2024**, *5*, 1485–1498. https://doi.org/10.3390/eng5030079.
31. Aviles, M.; Sánchez-Reyes, L.M.; Álvarez-Alvarado, J.M.; Rodríguez-Reséndiz, J. Machine and Deep Learning Trends in EEG-Based Detection and Diagnosis of Alzheimer's Disease: A Systematic Review. *Eng* **2024**, *5*, 1464–1484. https://doi.org/10.3390/eng5030078.
32. Wang, Z.; Qi, X.; Wang, C.; Easa, S.M.; Chen, F.; Cheng, J. Correlation Analysis between Young Driver Characteristics and Visual/Physiological Attributes at Expressway Exit Ramp. *Eng* **2024**, *5*, 1435–1450. https://doi.org/10.3390/eng5030076.
33. Paplomatas, P.; Rigas, D.; Sergounioti, A.; Vrahatis, A. Enhancing Metabolic Syndrome Detection through Blood Tests Using Advanced Machine Learning. *Eng* **2024**, *5*, 1422–1434. https://doi.org/10.3390/eng5030075.
34. Theodorakopoulos, L.; Theodoropoulou, A.; Stamatiou, Y. A State-of-the-Art Review in Big Data Management Engineering: Real-Life Case Studies, Challenges, and Future Research Directions. *Eng* **2024**, *5*, 1266–1297; https://doi.org/10.3390/eng5030068.
35. Falter, J.; Herburger, D.; Binz, H.; Kreimeyer, M. An Investigation of Increased Power Transmission Capabilities of Elastic–Plastic-Designed Press–Fit Connections Using a Detachable Joining Device. *Eng* **2024**, *5*, 1155–1172. https://doi.org/10.3390/eng5030063.
36. Jakschik, M.; Endemann, F.; Adler, P.; Lamers, L.; Kuhlenkötter, B. Assessing the Suitability of Automation Using the Methods–Time–Measurement Basic System. *Eng* **2024**, *5*, 967–982. https://doi.org/10.3390/eng5020053.
37. Chaudhuri, R.A.; Kim, D. Effects of Initial Small-Scale Material Nonlinearity on the Pre-Yield and Pre-Buckling Response of an Externally Pressurized Ring. *Eng* **2024**, *5*, 733–749. https://doi.org/10.3390/eng5020040.
38. Bade, S.O.; Meenakshisundaram, A.; Tomomewo, O.S. Current Status, Sizing Methodologies, Optimization Techniques, and Energy Management and Control Strategies for Co-Located Utility-Scale Wind–Solar-Based Hybrid Power Plants: A Review. *Eng* **2024**, *5*, 677–719. https://doi.org/10.3390/eng5020038.
39. Taghavi, S.A.; Jalali, F.M.; Moezzi, R.; Khaksar, R.Y.; Wacławek, S.; Gheibi, M.; Annuk, A. Numerical Analysis of Bearing Capacity in Deep Excavation Support Structures: A Comparative Study of Nailing Systems and Helical Anchors. *Eng* **2024**, *5*, 657–676. https://doi.org/10.3390/eng5020037.
40. Bajrami, A.; Costa, D.; Palpacelli, M.C.; Emiliani, F. Investigating Collaborative Robotic Assembly: A Case Study of the FANUC CRX-10 iA/L in Industrial Automation at i-Labs. *Eng* **2024**, *5*, 532–543. https://doi.org/10.3390/eng5020029.
41. Rosa, A.; Massaro, A. Process Mining Organization (PMO) Based on Machine Learning Decision Making for Prevention of Chronic Diseases. *Eng* **2024**, *5*, 282–300. https://doi.org/10.3390/eng5010015.
42. Patel, B.; Dubey, V.; Barde, S.; Sharma, N. Optimum Path Planning Using Dragonfly-Fuzzy Hybrid Controller for Autonomous Vehicle. *Eng* **2024**, *5*, 246–265; https://doi.org/10.3390/eng5010013.

References

1. Ferreira, V.B.; de Aquino Gomes, R.; Domingos, J.L.; da Fonseca, R.C.B.; Mendes, T.A.; Bouloukakis, G.; da Costa, B.B.F.; Haddad, A.N. Planning Energy-Efficient Smart Industrial Spaces for Industry 4.0. *Eng* **2025**, *6*, 53. [CrossRef]
2. Armenta-Déu, C.; Sancho, L. Sustainable Charging Stations for Electric Vehicles. *Eng* **2024**, *5*, 3115–3136. [CrossRef]

3. Gil, A. Challenges on Waste-to-Energy for the Valorization of Industrial Wastes: Electricity, Heat and Cold, Bioliquids and Biofuels. *Environ. Nanotechnol. Monit. Manag.* **2022**, *17*, 100615. [CrossRef]
4. Ye, X.P. Decarbonizing Nitrogen Fertilizer for Agriculture with Nonthermal Plasma Technology. *Eng* **2024**, *5*, 1823–1837. [CrossRef]

eng

MDPI

Article

Planning Energy-Efficient Smart Industrial Spaces for Industry 4.0

Viviane Bessa Ferreira [1], Raphael de Aquino Gomes [1,*], José Luis Domingos [1], Regina Célia Bueno da Fonseca [1], Thiago Augusto Mendes [1], Georgios Bouloukakis [2], Bruno Barzellay Ferreira da Costa [3,*] and Assed Naked Haddad [3]

1 Postgraduate Program in Technology, Management, and Sustainability, Instituto Federal de Goiás, Goiânia 74055-110, GO, Brazil; viviane.ferreira@ifg.edu.br (V.B.F.); jose.domingos@ifg.edu.br (J.L.D.); regina.fonseca@ifg.edu.br (R.C.B.d.F.); thiago.mendes@ifg.edu.br (T.A.M.)

2 SAMOVAR, Telécom SudParis, Institut Polytechnique de Paris, 75013 Paris, France; georgios.bouloukakis@telecom-sudparis.eu

3 Environmental Engineering Program, Universidade Federal do Rio de Janeiro, Macaé 21941-630, RJ, Brazil; assed@poli.ufrj.br

* Correspondence: raphael.gomes@ifg.edu.br (R.d.A.G.); bruno.barzellay@macae.ufrj.br (B.B.F.d.C.)

Abstract: Given the significant increase in electricity consumption, especially in the industrial and commercial categories, exploring new energy sources and developing innovative technologies are essential. The fourth industrial revolution (Industry 4.0) and digital transformation are not just buzzwords; they offer real opportunities for energy sustainability, using technologies such as cloud computing, artificial intelligence, and the Internet of Things (IoT). In this context, this study focuses on improving energy efficiency in smart spaces within the context of Industry 4.0 by utilizing the SmartParcels framework. This framework creates a detailed and cost-effective plan for equipping specific areas of smart communities, commonly referred to as parcels. By adapting this framework, we propose an integrated model for planning and implementing IoT applications that optimizes service utilization while adhering to operational and deployment cost constraints. The model considers multiple layers, including sensing, communication, computation, and application, and adopts an optimization approach to meet the needs related to IoT deployment. In simulated industrial environments, it demonstrated scalability and economic viability, achieving high service utility and ensuring broad geographic coverage with minimal redundancy. Furthermore, the use of heuristics for device reuse and geophysical mapping selection promotes cost-effectiveness and energy sustainability, highlighting the framework's potential for large-scale applications in diverse industrial contexts.

Keywords: Industry 4.0; energy efficiency; Internet of Things; smart space planning

Academic Editor: Antonio Gil Bravo

Received: 18 December 2024

Revised: 22 February 2025

Accepted: 13 March 2025

Published: 16 March 2025

Citation: Ferreira, V.B.; Gomes, R.d.A.; Domingos, J.L.; da Fonseca, R.C.B.; Mendes, T.A.; Bouloukakis, G.; da Costa, B.B.F.; Haddad, A.N. Planning Energy-Efficient Smart Industrial Spaces for Industry 4.0. *Eng* **2025**, *6*, 53. https://doi.org/10.3390/eng6030053

1. Introduction

As the world faces increasingly intricate energy production, distribution, and consumption challenges, energy efficiency has emerged as a crucial pillar in tackling these obstacles [1]. In industrial settings, optimizing energy consumption is crucial for competitiveness and cost reduction, a challenge that industries worldwide face. For example, the industrial sector is the largest consumer of energy in Brazil [2].

Based on information from the Brazilian Energy Research Company [3], the industrial class in Brazil leads to an increase in energy consumption, with an annual rate of 3.0% in January 2025. The sector's ongoing growth is evident in this rising pattern, while the population increase and the widespread use of electronic devices contribute to this trend, the primary issues lie in poor device management and the lack of effective strategies for energy conservation.

As growing energy demands push industries toward more efficient solutions, the search for alternative energy sources and technological advancements becomes increasingly valuable [4–6]. Over the recent few decades, industrial sectors have faced mounting pressure to optimize energy use while minimizing the environmental impact. In response, Industry 4.0 [7], also known as the fourth industrial revolution, has emerged as a transformative force, integrating digital technologies into manufacturing and operational processes.

By leveraging artificial intelligence, the Internet of Things (IoT), cloud computing, and big data analytics, Industry 4.0 enables organizations to monitor and optimize energy consumption in real time. For instance, predictive maintenance systems powered by machine learning can identify inefficiencies in machinery before they escalate, thereby preventing unnecessary energy waste [8]. Furthermore, smart grids and automated energy management systems allow industries to dynamically adjust power usage, aligning it with fluctuating production demands [9].

These solutions should be implemented in a manner that ensures satisfactory performance while aiming to reduce costs, such as through resource sharing. Recent studies have emphasized the importance of adopting holistic approaches that integrate advanced technologies and strategic planning to achieve energy efficiency goals [1,10]. For instance, energy-efficient resource allocation strategies and predictive models are essential for optimizing operations in smart industrial environments [11]. These approaches leverage real-time data analytics and adaptive systems to balance the energy supply and demand and enhance operational resilience and sustainability.

Despite efforts to improve energy efficiency, current solutions have limitations in choosing and implementing applications that fulfill these requirements. As a result, this study presents a comprehensive approach that integrates the sensing, communication, computing, and application layers. It utilizes a variety of heterogeneous devices and takes into account the structure and specific needs of different applications. This study outlines the model for the proposed solution and details its implementation. By using the software for planning, we expect that the project's execution will lead to reduced energy consumption through better regulation of connectivity restrictions, resources, and equipment. Additionally, the software shows promise in supporting strategic planning and decision-making processes related to the modeled spaces, with the goals of cutting costs and promoting sustainability.

2. Related Work

Several studies have explored the role of emerging technologies in promoting energy sustainability and the evolution of IoT in manufacturing and service environments. Next, representative works are discussed to contextualize this paper's proposal.

Trstenjak and Cosic [12] discuss how Industry 4.0 redefines process planning through cyber-physical systems (CPSs), IoT, and cloud computing, introducing the concept of a "product planner". The proposal highlights the integration of advanced algorithms to optimize the sequencing of operations and scheduling, and to promote productivity and mass customization. However, since this study is centered exclusively on the manufacturing sector, it does not explore potential challenges or adaptations for applying these technologies in other domains, such as the energy sector.

Oluyisola et al. [13] propose a smart production planning and control (PPC) system using IoT, machine learning, and big data analytics for dynamic and near-real-time actions. Although this study offers an appropriate manufacturing approach, its methodology faces scalability challenges in systems with high device heterogeneity, which is common in complex and multifunctional industrial environments.

Reichardt et al. [14] aims to understand why manufacturing industries adopt IoT energy monitoring systems, what impact these systems have on energy consumption, and how companies can successfully implement them. Therefore, it explores factors like cost reduction, regulatory requirements, and customer expectations as drivers for adoption and examines the role of data collection and organizational capabilities in realizing energy-saving potential.

Chang et al. [15] present QuIC-IoT, a planning platform for temporarily deploying IoT infrastructures in short-term events. The model addresses scenarios, such as controlled fires using mobile and fixed devices, based on physical models for monitoring. Although relevant, the focus on sporadic events limits its applicability in continuous industrial contexts that require integrated solutions for energy management.

Krishna et al. [16] developed IoT Composer, a tool for behavioral modeling and composition of IoT objects, with automated validation and deployment plans. Despite its success in case studies, the tool faces challenges in integrating heterogeneous devices into highly distributed and energy-efficient systems such as those discussed in this paper.

Hu et al. [17] introduce the AIoTML language for modeling artificial intelligence-based cyber-physical systems. The proposed method facilitates platform-independent simulation and control optimization for heterogeneous devices. However, its practical application is limited by the increasing complexity of IoT systems with high device densities and the absence of explicit consideration of energy constraints.

Guan et al. [18] investigate communication scheduling strategies in battery-powered and renewable energy IoT deployments. This study proposes heuristics to minimize energy consumption, but the analysis is restricted to edge computing scenarios without directly addressing sustainability demands in more complex industrial architectures.

Herrera et al. [19,20] explored the quality of service (QoS) and energy consumption optimization in next-generation IoT applications. Although they present significant performance improvements, the proposed frameworks do not directly address the need to balance the energy efficiency and connectivity in industrial contexts with high volumes of data and heterogeneous devices.

Brogi and Forti [21] highlighted the challenges of deploying IoT applications in distributed environments that integrate edge and cloud infrastructures. Although their proposed approach shows promise, it lacks a thorough practical assessment within industrial IoT ecosystems, particularly those characterized by high heterogeneity and strict energy constraints. Similarly, Ghaderi and Movahedi [22] introduced an energy-efficient data management scheme aimed at optimizing power consumption in Industrial Internet of Things (IIoT) networks while also adhering to latency and cache capacity limitations. However, a significant drawback of their approach is its exclusive focus on energy efficiency, which neglects important aspects of network communication performance. In contrast, this study takes a more integrative perspective by incorporating both communication efficiency and cost factors into its optimization framework, ensuring a more balanced and practical solution for IIoT environments.

Although the works discussed offer significant advances in their respective domains, they exhibit limitations that this study aims to address. The majority of existing proposals focus on specific scenarios, such as manufacturing or short-duration events, neglecting the necessity for integrated solutions encompassing sensing, communication, computation, and application in heterogeneous industrial systems. Moreover, few studies have investigated the optimization of connectivity and resource constraints to reduce energy consumption in

IoT architectures. This study addresses this gap and proposes an integrated approach to promote energy sustainability.

3. Proposed Modeling

The proposed modeling is centered around the SmartParcels tool [23,24], a framework designed to develop plans for equipping specific areas within smart communities. The problem is divided into four layers—application, information, infrastructure, and geophysics—as shown in Figure 1.

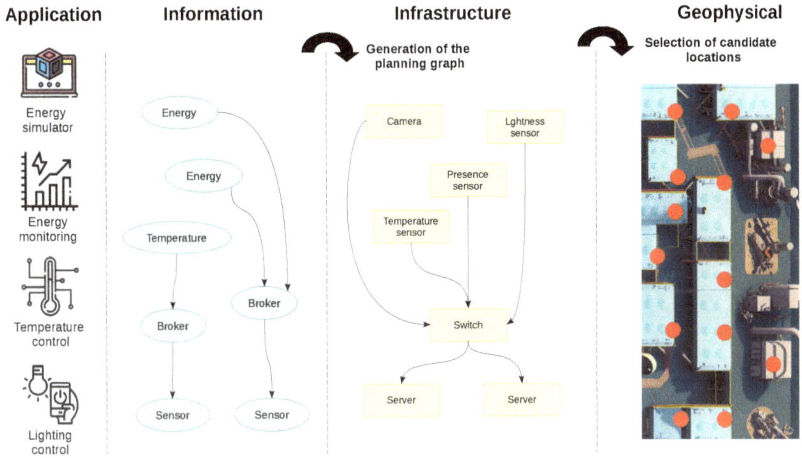

Figure 1. Summary of the proposed modeling.

The scenario unfolds as an optimization challenge, poised to maximize the overall utility of applications following the strategic deployment of IoT devices, edge servers, and network switches. Service utility is quantified through two essential dimensions: (i) coverage, which delineates the geographical scope where application events can be detected, and (ii) accuracy, which gauges the likelihood of accurately identifying those events.

Crafting an effective deployment strategy demands adherence to a series of stringent constraints. These include financial limitations for installation and operation as well as considerations of detection ranges, computing capacity, network bandwidth, and quality-of-service (QoS) expectations. Tackling the intricacies of application planning is seen as a challenging task, given the intricate interdependencies among four vital layers: the application layer, tasked with executing services; the information layer, responsible for managing data processing and transmission; the infrastructure layer, which encompasses vital hardware resources; and the geophysical factors, which significantly impact deployment feasibility and efficiency.

In this complex landscape, the quest for an optimal solution promises enhanced performance and a transformative leap in harnessing technology for better outcomes.

The solution to be developed will serve as a differentiator compared to other energy efficiency promotion solutions. It aims to optimize the quantity of equipment and components needed while considering factors such as functionality and coverage.

The formulated model focuses on a manufacturing facility comprising multiple distinct rooms, denoted as S, where each individual room is represented by $s_i \in S$. Within this industrial environment, a diverse set of applications A_i is required to support various operational tasks, with each specific application being identified as $a_{i,j} \in A_i$.

The industrial environment is formally defined as a tuple $(S, A_i | \forall s_i \in S)$, where S represents the set of rooms, and A_i denotes the set of applications associated with each room s_i. For modeling convenience, each room is represented by the coordinates of its geometric center.

The facility consists of a set L of candidate deployment points, which include conventional infrastructure elements, such as lighting systems, air conditioning units, and computing devices, alongside IoT components like environmental sensors, edge computing nodes, and network communication equipment. These elements serve as potential locations for application execution and data collection.

Since different applications contribute variably to operational efficiency, each room s_i is assigned a weight $\beta_{i,j}$ for each required application $a_{i,j}$, indicating its relative priority within that space. It is assumed—without loss of generality—that the sum of weights across all applications in a given room satisfies $\sum_{\forall a_{i,j} \in Ai} \beta i, j = 1, \forall s_i \in S$. This weighting mechanism facilitates an optimized resource allocation strategy by reflecting the specific functional demands of each industrial zone.

Modeling must be conducted for an industry by considering all its rooms. Assuming the industry has 15 rooms, we can establish a model where $|S|= 15$ and $S = \{s_1, s_2, \cdots, s_{15}\}$. Let us further assume that one of these rooms, specifically room s_1, is an administrative room that requires two applications: temperature control and consumption monitoring. In this case, we can represent these applications as a_{11} and a_{12}. Furthermore, $|A_1|= 2$ and $A_1 = \{a_{11}, a_{12}\}$. This example is intended solely to illustrate the distribution of applications across rooms. The specific weight assigned to each application is detailed in the sections describing information and infrastructure flows, where factors such as coverage, accuracy, and resource allocation are considered. The model ensures that these weights reflect the relative importance of each application based on its functional role in the industrial environment.

The other elements of the model are described below.

3.1. Information Flows for an Application

Each application can be implemented using various combinations of sensor data from IoT devices and analytical algorithms on computing devices, such as edge servers. These devices are connected through directed graphs known as information flows. Different information flows for the same application enable planners to balance quality of service (QoS) with both deployment and operational costs. This allows for the selection of the most appropriate information flow to meet industry requirements.

To implement application $a_{i,j}$, a set of information flows $F_{i,j}^{info}$ can be adopted, with $f_{i,j,k}^{info} \in F_{i,j}^{info}$ being the k-th information flow. More precisely, $f_{i,j,k}^{info} = \left(V^{info}, E^{info}\right)$ is a directed weighted graph where $v \in V^{info}$ denotes a unit of information, which may consist of raw data or components of the communication middleware, and $e(u, v) \in E^{info}$ represents the data flow between these information units. Both the vertices and edges have associated weights. The weight of a vertex $w(v)$ indicates the computing resources consumed by that unit of information, while the weight of an edge $w(e(u, v))$ reflects the bandwidth consumption. Additionally, each information flow specifies the number of sensors needed; for example, three microphones are required for sound source detection using triangulation. Figure 2 illustrates how these information flows can be implemented in each application.

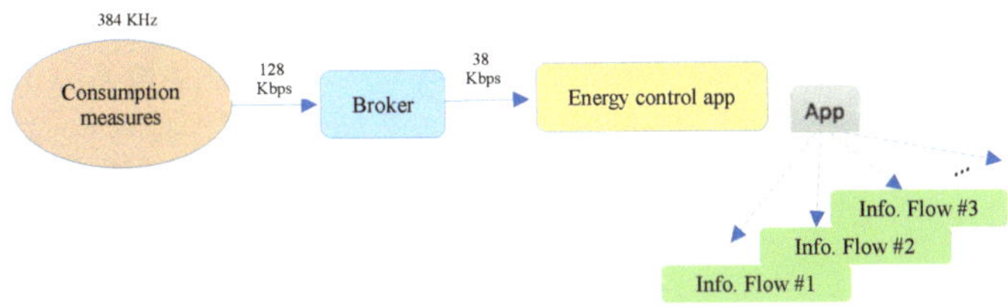

Figure 2. Representation of information flows for an application.

3.2. Infrastructure Flows Implementing an Information Flow

Each information flow can be deployed across various combinations of sensors, edge servers, and network switches, which are referred to as infrastructure flows. Different combinations of infrastructure flows associated with multiple information flows can result in varying levels of resource, or device, sharing. This allows for strategic planning to make use of resource reuse, leading to greater efficiency.

Each information flow $f_{i,j,k}^{info}$ can be represented by a set of infrastructure flows $F_{i,j,k}^{ifr}$, where $f_{i,j,k,m}^{ifr} \in F_{i,j,k}^{ifr}$ denotes the m-th infrastructure flow. Let $f_{i,j,k,m}^{ifr} = \left(V^{ifr}, E^{ifr}\right)$ be a directed weighted graph, where $v \in V^{ifr}$ represents a device and $e(u,v) \in E^{ifr}$ represents the data flow between two devices. For our purposes, we consider various types of devices, such as sensors (for example, power meters or cameras), computing devices (like edge servers), and network switches (including LTE cells or Ethernet switches). The weights assigned to a vertex $w(v)$ and an edge $w(e(u,v))$ reflect the computing resources and network bandwidth they provide, respectively.

A tuple $\left(F_{i,j}^{info}, \left\{F_{i,j,k}^{ifr} \middle| \forall f_{i,j,k}^{info} \in F_{i,j}^{info}\right\}\right)$ encapsulates all information flows and their corresponding infrastructure flows for each application $a_{i,j}$.

Given a $f_{i,j,k}^{info}$ and a $f_{i,j,k,m}^{ifr}$, each processing unit $v \in V^{info}$ is assigned to a device $v' \in V^{ifr}$ through a specific function $R(v) = v'$. Additionally, for a given edge $e(u,v) \in E^{info}$, $\langle R(u), R(v)\rangle$ represents the shortest path in $f_{i,j,k,m}^{ifr}$ that includes the involved devices, reflecting the actual data flow at the infrastructure layer. For the sake of clarity, it is assumed that $\langle R(u), R(v)\rangle$ contains at least one network switch unless devices $R(u)$ and $R(v)$ are identical. If processes u and v operate on the same device, their network bandwidth is significantly higher, leading to the assumption that $w(e(R(u), R(v))) = \infty$.

Figure 3 illustrates how infrastructure flows can facilitate each type of information flow.

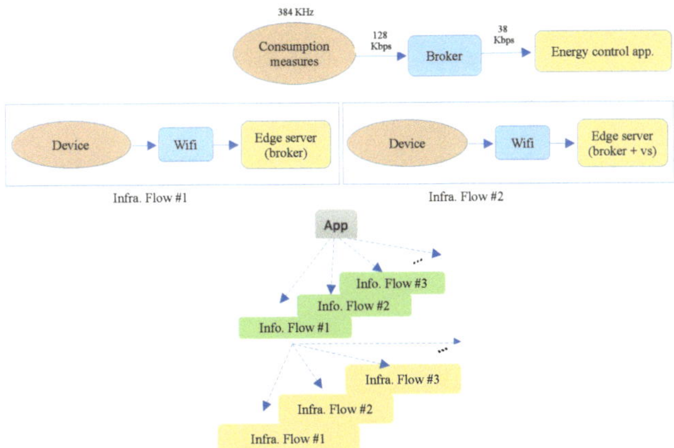

Figure 3. Infrastructure flow representation.

3.3. Planning Graph

Application deployment planning can be carried out by analyzing the flow of information and infrastructure. There are two types of deployments: (i) initial deployment, where no existing IoT infrastructure is in place (such as in a greenfield industry), and (ii) retrofit deployment, where IoT devices, edge servers, and network switches are already integrated (as seen in a growing industry).

An auxiliary structure known as a planning graph is created based on the following concepts. This planning graph is defined as a two-layer graph $G^p = (V^p, E^p)$: the first layer $G_1^p = \left(V_1^p, E_1^p\right)$ consists of a set of information flows, while the second layer $G_2^p = \left(V_2^p, E_2^p\right)$ comprises a set of infrastructure flows. In both layers, flows can share vertices and edges. Additionally, a set of assignment edges E^r is defined, where each edge $e(v, R(v))$ represents the assignment of $v \in V^{info} \subset V_1^p$ to $R(v) \in V^{ifr} \subset V_2^p$ for $f_{i,j,k}^{info}$ and $f_{i,j,k,m}^{ifr}$. This planning graph can be denoted as $V^p = \left\{V_1^p, V_2^p\right\}$ and $E^p = \left\{E_1^p, E_2^p, E^r\right\}$.

3.4. Infrastructure Geophysical Mapping Function

To identify candidate locations, a geophysical mapping function $f(v)$ maps a vertex $v \in V_2^p$ from the infrastructure layer of a planning graph G^p to each candidate location $l \in L$.

A tuple $t_v = \left(r_v^{tr}, r_v^{sin}, \tau_v\right)$ represents each device $v \in V_2^p$. The device's transmission range is denoted by r_v^{tr}, while r_v^{sin} indicates its detection range. τ_v specifies the type of device, which can be either a sensor, a compute unit, or a network device. If $\tau_v \neq$ "network", then r_v^{tr} represents the transmission range of its associated network device u, denoted as $r_v^{tr} = r_u^{tr}, e(v, u) \in E_2^p$. Furthermore, multiple devices can be assigned to the same candidate location. For clarity, $V_{i,j,k,m}^{sin}$ is used to refer to the sensors of $f_{i,j,k,m}^{ifr}$ (that is, $\forall v \in V_{i,j,k,m}^{sin}, \tau_v =$ "sensor"). Figure 4 illustrates the geophysical mapping for an infrastructure flow. To identify candidate locations, a geophysical mapping function $f(v)$ will associate a vertex $v \in V_2^p$ of the infrastructure layer of a planning graph G^p with a potential candidate location $l \in L$.

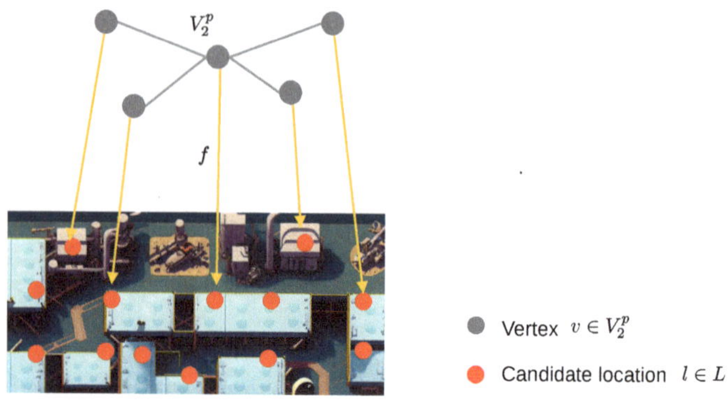

Figure 4. Representation of geophysical mapping.

3.5. Utility of a Service in an Infrastructure Flow

The Euclidean distance between two candidate locations $l_1, l_2 \in L$ is defined as $dist(l_1, l_2)$. By definition, the Euclidean distance between two two-dimensional points (x_1, y_1) and (x_2, y_2) is given by Equation (1):

$$\sqrt{(x_2 - x_1)^2 + (y_2 - y_1)^2} \tag{1}$$

Infrastructure flow is considered connected if all its devices are linked after mapping, i.e., $dist(f(u), f(v)) \leq min(r_u^{tr}, r_v^{tr}), \forall e = (u,v) \in f_{i,j,k,m}^{ifr}$; otherwise, the flow is deemed unconnected. If $f_{i,j,k,m}^{ifr}$ is connected, the service utility in a room s_i is determined by Equation (2):

$$U\left(f_{i,j,k,m}^{ifr}, s_i\right) = A\left(f_{i,j,k,m}^{ifr}\right) \times P\left(V_{i,j,k,m}^{sen}, s_i\right) \tag{2}$$

where $A\left(f_{i,j,k,m}^{ifr}\right)$ represents accuracy, while $P\left(V_{i,j,k,m}^{sin}, s_j\right)$ signifies the probability of detection.

If $f_{i,j,k,m}^{ifr}$ is not connected, then $U\left(f_{i,j,k,m}^{ifr}, s_i\right)$ is defined as 0. Each $f_{i,j,k,m}^{ifr}$ that pertains to the application $a_{i,j}$ includes a precision model $A\left(f_{i,j,k,m}^{ifr}\right)$ that varies based on the method used. For instance, detection based on presence sensors is generally more accurate for identifying presence than detection based on images.

The detection probability models used here, similar to those in SmartParcels, are based on the concept of attenuated truncated mode [25]. This concept indicates that the coverage measure becomes very small when the distance between a spatial point and a sensor is too large. In such cases, the coverage measure can be disregarded, allowing for approximations by truncating the coverage measure at greater distance values. This approach ensures that signal coverage is realistically represented, preventing overestimating utility in areas with limited sensor effectiveness. However, it is important to note that the current implementation focuses on planning rather than real-time adjustments. Future developments could integrate dynamic recalibration mechanisms, adapting detection strategies based on environmental conditions and signal propagation variations in real-time applications.

Initially, for a sensor $v \in V_{i,j,k,m}^{sin}$, the probability is attenuated (or decayed) with increasing distance in s_i, $dist(f(v), s_i)$ and truncated by its detection range r_v^{sin}. Therefore, if $dist(f(v), s_i) \leq r_v^{sin}, \forall v \in V_{i,j,k,m}^{sin}$, the average truncated attenuated detection probability, denoted as Y, is expressed by Equation (3) as follows:

$$\overline{p} = \frac{\sum_{\forall v \in V_{i,j,k,m}^{sen}} e^{-\alpha_v * dist(f(v), s_i)}}{\mid V_{i,j,k,m}^{sen} \mid} \tag{3}$$

where α_v is a parameter that is related to v. If this relationship does not hold, then $\overline{p} = 0$.

The detection probability is limited by the sensors' detection range, as defined in Equation (4):

$$P\left(V_{i,j,k,m}^{sen}, s_i\right) = \begin{cases} \overline{p}, & \text{if } dist(f(v), s_i) \leq r_v^{sen}, \forall v \in V_{i,j,k,m}^{sen}; \\ 0, & \text{otherwise.} \end{cases} \tag{4}$$

The definitions provided complete the concept known as the utility of a service. It is important to note that the proposed algorithms are not dependent on the mathematical properties of this utility. As a result, there is complete freedom to apply various models. For instance, one can enhance the analysis by incorporating non-line-of-sight detection ranges [26]. These scenarios occur when detection is possible despite obstacles blocking the direct line of sight, relying on signal reflections, diffractions, or scattering to reach the receiver.

3.6. Costs

Each device $v \in V_2^p$ in the infrastructure layer is subject to two types of costs:

(i) deployment cost $\delta_{deploy}(v, l)$ due to deploying the device v at the candidate location $l \in L$, and
(ii) operational cost $\delta_{op}(v)$ due to maintaining its operation.

The deployment cost charged once, while the operational cost is recurring. Furthermore, we define B_{dp} and B_{op} as the budgets for deploying and operating the devices. The operational cost refers to the ongoing expenses required to keep the device functioning, such as energy consumption, maintenance, software updates, and other associated costs necessary for its continuous operation.

3.7. Problem Formulation

Given the industry's operational characteristics, the interaction between information and infrastructure flows, and the constraints regarding resource availability, the energy efficiency planning problem seeks to optimize the overall quality of services while adhering to predefined cost budgets. The objective is to derive an optimal planning graph, denoted as G^{p^*}, and determine the corresponding geophysical mapping functions $F^* = f^*(v)|\forall v \in V_2^{p^*}$ that best allocate resources across the industrial environment.

More precisely, the problem involves selecting an optimal set of locations for deploying IoT devices, network infrastructure, and edge computing resources to enhance the energy efficiency without compromising system performance. This requires balancing multiple constraints, including power consumption, latency, computational capacity, and spatial deployment feasibility. The energy efficiency planning problem is formally structured as follows:

$$\text{Maximize} \sum_{\forall s_i \in S} \sum_{\forall f^{ifr^*}_{i,j,k,m} \in G_2^{p^*}} \beta_{i,j} U\left(f^{ifr^*}_{i,j,k,m}, s_i\right), \tag{5}$$

$$\text{subject to}: \sum_{\forall v \in V_2^{p^*}} \delta_{deploy}(v, f^*(v)) \leq B_{dp}, \tag{6}$$

$$\sum_{\forall v \in V_2^{p^*}} \delta_{op}(v) \leq B_{op}, \tag{7}$$

$$\sum_{\forall e(u,v) \in E^{r^*}} w(u) \leq w(v), \forall v \in V_2^{p\ ast}, \tag{8}$$

$$\sum_{\forall e(u,v) \in E_2^{p^*}} w(e(u,v)) \leq \sum_{\forall e(v,u') \in E_2^{p^*}} w(e(v,u')), \forall v \in V_2^{p^*}, \tag{9}$$

$$w(e(u,v)) \leq \min_{\forall e' \in \langle R(u), R(v) \rangle} w(e'), \forall e(u,v) \in E_1^{p^*}. \tag{10}$$

The objective function in Equation (5) is designed to determine the optimal planning graph G^{p^*} and the corresponding geophysical mapping functions F^* that maximize the total utility of the system. This optimization process efficiently balances resource allocation, service quality, and energy consumption.

Budget constraints are enforced through Equations (6) and (7), which, respectively, limit the implementation cost $\left(B_{dp}\right)$ and the operational cost (B_{op}). These constraints ensure that the deployment and ongoing management of the industrial infrastructure remain within financial feasibility.

To maintain computational efficiency, Equation (8) restricts every device $v \in V_2^{p^*}$, ensuring that its available computational resources are sufficient to process all assigned information units u. This constraint means that each device's weight $w(v)$ must be at least equal to the cumulative weight of all associated processing tasks, preventing system overloads and performance degradation.

Furthermore, Equation (9) guarantees that each device's output bandwidth capacity is at least as high as its input bandwidth demand. This condition ensures smooth data transmission across the network, reducing bottlenecks and enhancing system responsiveness.

For every data stream $e(u,v) \in E_1^{p^*}$, Equation (10) enforces that the minimum bandwidth along the allocated communication path $\langle R(u), R(v) \rangle$ meets or exceeds the required bandwidth threshold. This constraint is required to maintain reliable data flow across the industrial IoT network, support real-time applications, and minimize latency issues.

On top of these definitions, the energy efficiency planning problem can be classified as a $\mathcal{NP}$-hard optimization problem, which inherently resists approximation within a factor of $1 - 1/e$. This complexity can be established by demonstrating a polynomial-time reduction from the well-known max K-cover problem [27] to a constrained version of the energy efficiency planning problem.

To achieve this reduction, consider a simplified instance where resource constraints are entirely removed by disregarding Equations (8)–(10). In this special case, each application is assumed to be associated with a single infrastructure flow, while deployment and operational costs are set to a uniform value of one. Additionally, both budget limits are defined as exactly K. Under these conditions, solving the energy efficiency planning problem is equivalent to selecting K infrastructure flows for a set of rooms in a way that maximizes overall service utility.

Since the max K-cover problem is $\mathcal{NP}$-hard and has a polynomial reduction to this special case, the original energy efficiency planning problem retains the same computational complexity. Furthermore, as demonstrated by Feige [28], the max K-cover problem cannot be approximated within a factor better than $1 - 1/e$ unless $\mathcal{P} = \mathcal{NP}$. Consequently, this inapproximability threshold extends to the energy efficiency planning problem, reinforcing the inherent computational challenge associated with optimizing the energy efficiency under deployment and operational constraints.

4. Heuristic Solution

The smart space planning problem is decomposed into two subproblems targeting reducing complexity and avoiding redundant calculations: (i) geophysical mapping selection, which chooses promising mappings among infrastructure flows and candidate locations, and (ii) generation of planning graphs, which calculates mappings between information flows and infrastructure flows that maximize the overall service utility.

Let $G^{p^*} = \left(V^{p^*}, E^{p^*}\right)$ denote the optimized planning graph, where $G_1^{p^*} = \left(V_1^{p^*}, E_1^{p^*}\right)$ represents the set of information flows and $G_2^{p^*} = \left(V_2^{p^*}, E_2^{p^*}\right)$ defines the infrastructure flows. This decomposition is essential because identifying the optimal planning graph requires the continuous generation and assessment of geophysical mappings associated with G^{p^*}. Since these mappings remain largely static, a more efficient strategy involves storing and reusing them rather than recalculating them at each step.

Figure 5 illustrates the interdependence between these subproblems and the computational methods used to solve them. The Selection (SEL) algorithm prioritizes geophysical mappings that maximize service utility and communication reach, aiming to reduce the number of deployed devices. Meanwhile, the Maximum Reusability (MR) algorithm iteratively chooses infrastructure flows that optimize device utilization, minimizing redundancy and enhancing resource efficiency. Integrating these approaches, the proposed model balances service performance, network efficiency, and cost-effectiveness in energy-efficient planning.

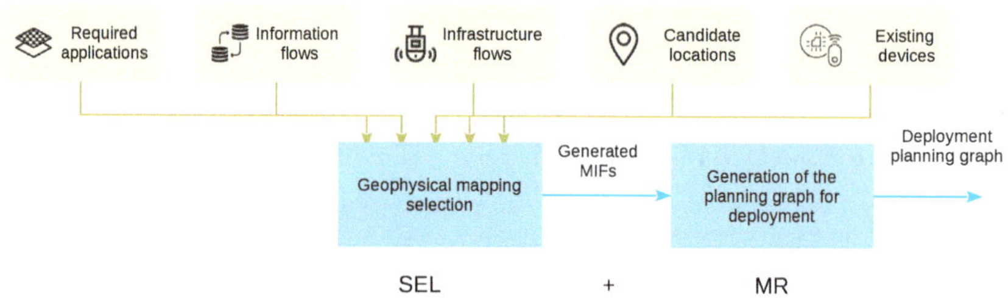

Figure 5. Problem decomposition, main inputs/outputs and proposed algorithms.

Based on the results from [24], it was possible to develop a solution approach based on dynamic programming that generates the planning graph with maximum service utility. However, the execution time increases dramatically when the number of rooms, required applications, implementation methods (information flows and infrastructure), or candidate locations increases. Therefore, the optimal solution is unfeasible due to its extremely long time to complete. As a result, the algorithms presented represent a heuristic solution for each subproblem.

The first heuristic, SEL, is based on selection policies to eliminate less promising mappings. The policies contain the following intuitions: (i) MIFs with more significant utilities should be included earlier and (ii) MIFs with more excellent communication coverage should be included earlier.

For each $f^{ifr}_{i,j,k,m}$, the utility can be estimated by Equation (2) after mapping all equipment (assuming that the graph is connected). Similarly, the communication coverage of a network device can be estimated after mapping. Algorithm 1 presents the adopted heuristic, using M and N to represent the user-specified pruning criteria for utility and communication coverage, respectively.

Algorithm 1: SEL $\left(S, A, F^{info}, F^{ifr}\right)$.

---

input: set of rooms (S), applications (A), information flows (F^{info}), and infrastructure flows (F^{ifr}).
output: promising set of geophysical mappings for possible infrastructure flows.

1 Select all possible geophysical mapping functions $f(v)$ for each sensor $v \in V^{sen}_{i,j,k,m} \subset V^{ifr}$ where s_i is within the sensing radius of v, i.e., $dist(f(v), s_i) \leq r^{sen}_v$
2 Update $\widehat{\mathcal{F}}$ to contain only the M best service utilities
3 **foreach** *possible combination* $\widehat{\mathcal{F}}$ **do**
4 | Select all possible mappings $f(u)$ for each infrastructure u adjacent to the sensors in $f^{ifr}_{i,j,k,m}$, i.e., $dist(f(v), f(u)) \leq min(r^{tr}_v, r^{tr}_u)$
5 | Update $\widehat{\mathcal{F}}$ to contain only the N best communication coverages if the examined infrastructure is a network device
6 **end**
7 Recursively execute Step 3 for infrastructures adjacent to those previously examined until all have been examined

---

In the MR heuristic, instead of examining all possible combinations of MIFs, it iteratively (i) selects an application $a_{i,j} \in \hat{A}$ to implement and (ii) merges an END $\hat{F} \in M^{ifr}_{i,j,k,m}$ in the planning graph $\hat{G}(K)$ according to the reusability of $\hat{F}$, where $M^{ifr}_{i,j,k,m} \in \hat{M}_{i,j}$. A reusability index was defined, considering the investment efficiency and the gain in communication coverage when merging $\hat{F}$.

Investment efficiency is the ratio between the application's utility gain and the cost gain after merging $\hat{F}$ into $\hat{G}(K)$. In other words, the more infrastructures are reused, the lower the costs will be with merging $\hat{F}$. Specifically, whether we choose $\Delta U(\hat{F}, \hat{G}(K))$, and considering the application's utility gain after merging $\hat{F}$ into $\hat{G}(K)$, the cost gain is given by

$$\Delta\delta(\hat{F}, \hat{G}(K)) = \hat{\delta}_{dp}(K) - \hat{\delta}_{dp}(K-1) + \hat{\delta}_{op}(K) - \hat{\delta}_{op}(K-1) \tag{11}$$

Therefore, we have

$$I_{eff}(\hat{F}, \hat{G}(K)) = \frac{\Delta U(\hat{F}, \hat{G}(K))}{\Delta\delta(\hat{F}, \hat{G}(K))} \tag{12}$$

which is the investment efficiency of merging $\hat{F}$ into $\hat{G}(K)$.

The current investment efficiency formula balances utility gain and costs in a structured industrial environment. However, we acknowledge that a more adaptive approach could be beneficial in environments with a highly variable application utility and resource costs. One potential adaptation would be introducing weight adjustments based on real-time demand fluctuations, prioritizing applications with a higher dynamic impact. Additionally, integrating cost prediction models could refine investment decisions by anticipating resource variations. Although this extension is not currently implemented, it represents a promising direction for future work, particularly for applications in heterogeneous or rapidly changing industrial contexts.

Communication coverage gain is determined by the locations of network devices. If a network device has greater communication coverage after deployment, fewer devices will be needed. Let $L_{cov}(K) \subset L$ be the candidate locations in the communication coverage of network devices in $\hat{G}(K)$. The communication coverage gain after merging $\hat{F}$ into $\hat{G}(K)$ is given by

$$I_{cov}(\hat{F}, \hat{G}(K)) = L_{cov}(K) - L_{cov}(K-1) \tag{13}$$

The reusability index is defined as a weighted sum of the investment efficiency and communication coverage gain when merging an FIM $\hat{F}$ into the intermediate planning graph $\hat{G}(K)$. Let α_{eff} and α_{cov} be the weights of the investment efficiency and communication coverage gain, respectively. The reusability index is written as

$$I(\hat{F}, \hat{G}(K)) = \alpha_{eff} I_{eff}(\hat{F}, \hat{G}(K)) + \alpha_{cov} I_{cov}(\hat{F}, \hat{G}(K)) \tag{14}$$

Without the loss of generality, it is assumed that $\alpha_{eff} + \alpha_{cov} = 1$.

With these definitions, MR starts with an empty planning graph $\hat{G}(0)$ and iteratively selects an application $a_{i_j} \in \hat{A}$ to implement by merging an END $\hat{F} \in M^{ifr}_{i,j,k,m}$ in the current intermediate planning graph $\hat{G}(K)$ (as established in Algorithm 2), where $M^{ifr}_{i,j,k,m} \in \hat{M}_{i,j}$.

When viewing Lines 1 to 3 (Algorithm 2), for each application $a_{i_j} \in \hat{A}$, it can be seen that the algorithm examines the reusability index $I(\hat{F}, \hat{G}(K))$ for each END $\hat{F}$ within the set of possible mappings Mi, j, k, m^{ifr}, where $\hat{M}i, j$ represents the set of all mappings for the considered application. In Line 4 (Algorithm 2), the END $\hat{F}$ with the highest reusability index $I(\hat{F}, \hat{G}(K))$ is then incorporated into the planning graph $\hat{G}(K)$. Applications corresponding to this END are excluded from the set A, indicating that their infrastructure needs have already been met.

Algorithm 2: MR $\left(S, A, F^{info}, F^{ifr}\right)$

input: set of rooms (S), applications (A), information flows (F^{info}), and infrastructure flows (F^{ifr}).
output: planning graph.

```
foreach a_{i_j} ∈ Â do
    Examine the reusability index I(F̂, Ĝ(K)) of each MIF
    F̂ ∈ M^{ifr}_{i,j,k,m}, ∀M^{ifr}_{i,j,k,m} ∈ M̂_{i,j}
end
Integrate F̂ with the largest I(F̂, Ĝ(K)) in Ĝ(K) and delete the corresponding
    applications of Â
while at least one of the constraints is not violated for all MIFs of the remaining
    applications or all reusability indices generated are zero do
    Repeat previous steps
end
```

The process repeats the evaluation and selection steps as long as at least one of the constraints is not violated for the MIFs of the remaining applications, or until all calculated reusability indices are zero. This strategy ensures that planning continues to optimize component reuse without violating operational or design constraints. As output, the algorithm produces a planning graph representing the infrastructure and application mapping, prioritizing component reuse.

To establish the convergence of our heuristic methodology, we validate two fundamental characteristics: (i) each successive step enhances the function $U\left(G^{p^*}\right)$, and (ii) these enhancements gradually diminish, yielding convergence.

Initially, observe that the functional expression in the optimization challenge is expressed as

$$U\left(G^{p^*}\right) = \sum_{\forall s_i \in S} \sum_{\forall f^{ifr^*}_{i,j,k,m} \in G_2^{p^*}} \beta_{i,j} U\left(f^{ifr^*}_{i,j,k,m}, s_i\right)$$

where $U\left(f^{ifr^*}_{i,j,k,m}, s_i\right)$ represents the utility of a selected infrastructure flow at room s_i, and $\beta_{i,j}$ is a weighting factor for each application. The MR heuristic iteratively selects a FIM $\hat{F}$ that maximizes the reusability index. Since the heuristic always selects $\hat{F}$ such that it improves utility, it follows that

$$U\left(G^{p^*(t+1)}\right) \geq U\left(G^{p^*(t)}\right), \quad \forall t.$$

This characteristic guarantees that the efficacy function exhibits monotonic growth across iterations.

Next, we demonstrate that these improvements attenuate, culminating in convergence. Let ΔU_t represent the utility gain at iteration t, defined as

$$\Delta U_t = U\left(G^{p^*(t+1)}\right) - U\left(G^{p^*(t)}\right).$$

Initially, when many high-impact infrastructure flows are available, ΔU_t is relatively large. However, as the number of remaining candidate FIMs decreases, additional selections yield redundancies due to overlapping mappings, reducing their incremental contribution to the utility. This behavior can be modeled as an exponential decay in utility gain as follows:

$$\Delta U_t = c \cdot \Delta U_{t-1}, \quad \text{where } 0 < c < 1.$$

Since ΔU_t approaches zero as $t \to \infty$, we conclude that

$$\lim_{t \to \infty} \Delta U_t = 0.$$

Finally, because $U\left(G^{p^*}\right)$ is both monotonically increasing and bounded above by the maximum achievable utility under budgetary and infrastructural constraints, the Monotone Convergence Theorem [29] guarantees that $U\left(G^{p^*}\right)$ converges to a finite value as follows:

$$\lim_{t \to \infty} U\left(G^{p^*(t)}\right) = U^*.$$

Thus, we have established that the heuristic progressively achieves an optimal or near-optimal configuration, ensuring convergence while maximizing energy efficiency throughout industrial environments.

Implementation

The proposed solution was implemented as a prototype that was made available through a Web API. The objective was to develop a flexible application for both local and cloud infrastructure. Based on this, the implemented architecture is presented in Figure 6.

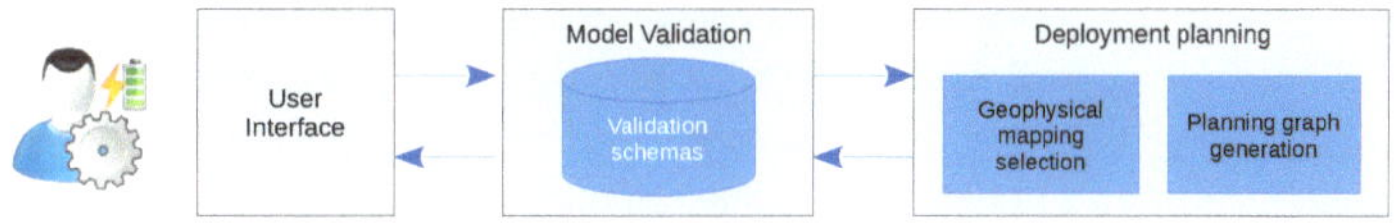

Figure 6. Solution architecture.

The first component is the user interface, which is designed to enable modeling and planning. This interaction occurs through the REST API. User models were validated according to the specifications intended for IoT models. This process is based on standards and specifications developed to facilitate interoperability and simplified data exchange between devices and systems in IoT environments.

Among the standards used, a large part is from [30], which is a collaborative initiative that aims to improve data models aimed at IoT. The data models are compatible with *FIWARE* version 2 [31] and Next-Generation Service Interface-Linked Data (NGSI-LD) [32] specifications, enabling their use by these standards. They encompass detailed definitions of properties, attributes, and relationships between various data entities, ensuring that information is not only accessible but also meaningful and readily usable across diverse systems. In total, 15 application domains were available, and models from 4 of them were used: Smart Energy, Smart Cities, Smart Robotics, and Smart Sensing.

In addition to Smart Data Models, schemas from [33] are used, which is a collaborative initiative led by large search engines aiming to structure information on the Internet through a standardized set of tags via Extensible Markup Language (XML).

To represent information and infrastructure flows, ref. [34] was used, which is a convention used to describe the data structure in graphs using the JSON format, facilitating data storage, manipulation, and transfer. Figure 7 summarizes the models used for data representation.

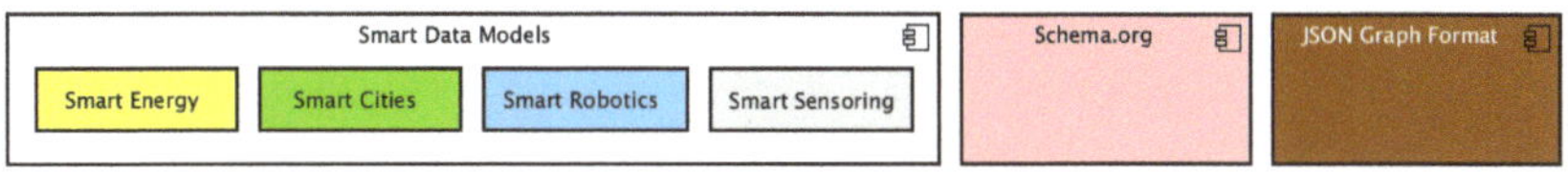

Figure 7. Models used in solution.

After validating the models, planning is performed using the heuristic solution described in the previous section.

The industry and its rooms are represented using GeoJSON [35], an open standard format designed to represent simple geographic features, along with their non-spatial attributes using JSON. Each candidate location is represented by the geographic coordinates of its location and deployment cost for each type of equipment in that location.

In information and infrastructure flows, each node is assigned a role that specifies its role. Vertices can be classified as sensor, representing sensors or any data collection device, network for component parts of the network, or compute, indicating devices with processing capacity. For vertices categorized as sensors, attributes list the data types that a device can collect.

The attributes that were introduced were r_tr and r_sen, which represent the transmission range (r_v^{tr}) and sensing range (r_v^{sin}), respectively. In addition, precision_model was introduced as a representation of the precision model $A\left(f_{i,j,k,m}^{ifr}\right)$. The procedure starts by selecting all possible geophysical mapping functions for each room, considering the sensors whose sensing area includes part of the room. In this analysis, candidate locations in the room and external locations are considered, as long as the room is in the sensing coverage area of a sensor in these locations.

The application was implemented in Python using the Django framework [36]. The choice for these tools was made because Python has a clear and readable syntax and is a robust option with a vast number of libraries available.

In the implementation, Shapely and Geopy libraries were used in Python 3.9. Shapely is used to manipulate and analyze plane geometry. It allows for the creation, manipulation, and analysis of geometries in addition to performing operations such as union, intersection, and difference on geometric objects. On the other hand, Geopy provides a simple and consistent interface for performing various geographic operations, such as geocoding, distance calculation between geographic points, and route calculation between locations. Moreover, it is often used in applications involving geographic data analysis, geolocation, and geocoding.

5. Performance Evaluation

The performance evaluation of our proposed smart space planning framework focuses on assessing the system scalability and cost-effectiveness in industrial environments. Using the simulated scenarios, we evaluated the efficiency of our IoT-based planning model. We employed a multifaceted evaluation that included scalability and cost-effectiveness analysis. By implementing optimized information and infrastructure flows, the system demonstrated consistent efficiency gains even as the number of deployed devices increased.

5.1. Setup

The experiments were implemented using a data generator to simulate data on industries and applications. The goal was to create a structured representation of the industry's physical space and operational processes to simulate possible scenarios for using the proposed solution.

The total area of the specified physical space is divided into functional categories, such as production, offices, utility and storage, ensuring that the spaces are proportional to an industry's typical needs. Each generated room has attributes, such as size, occupancy capacity, type of functional area, and geographic location.

Specific applications were also created for each room by considering the following possibilities: (A) detection of mechanical failures, abnormal vibrations, or compressed air leaks in industrial equipment through acoustic sensing; (B) monitoring of emissions in boilers and thermal processes through smoke sensing; (C) monitoring of the efficiency of ventilation systems through air quality sensing; (D) detection of overheating in machines or industrial processes through temperature sensing; and (E) lighting and cooling control through presence detection.

The experiment was modeled on a typical industrial setup, featuring functional spaces commonly found in industries worldwide. The physical plant was divided into five distinct rooms (s_1 to s_5). The Brazilian case was utilized solely to demonstrate the applicability of the solution. This division aims to simulate the industrial environment in detail, reflecting the typical requirements for evaluating the proposed IoT solution. The rooms were categorized as follows:

- s_1: Production Area: represents the majority of the space, housing essential industry production processes.
- s_2: Storage Area: designated for storing materials and finished products.
- s_3: Utilities' Area: this area is intended for operational support, such as production support machines and general utilities.
- s_4: Office: place dedicated to administrative and management activities.
- s_5: Other Production Area: this area complements the production processes and may include specific production lines.

Specific IoT applications were simulated for each of these areas to meet the functional and operational demands of each space:

- s_1 and s_5 (Production): Applications A (detection of mechanical failures, abnormal vibrations, or compressed air leaks), B (monitoring of emissions in boilers and thermal processes), C (monitoring of the efficiency of ventilation systems), D (detection of overheating), and E (lighting and cooling control).
- s_2 (Storage): Applications C (monitoring of ventilation efficiency) and E (lighting and cooling control).
- s_3 (Utilities): Application E (lighting and cooling control).
- s_4 (Office): Applications C (monitoring of air quality) and E (lighting and cooling control).

The floor plan presented in Figure 8 illustrates the layout of these rooms and their respective applications, simulating real scenarios and allowing for a practical evaluation of the IoT deployment planning tool.

Wi-Fi and Lora were considered for network communication. The equipment parameters used to deploy the applications were the same as those adopted in [24], as listed in Table 1. However, in our representation, we include the memory requirement by considering a buffer of 10 s. Moreover, hertz (Hz) measures the computing requirements by converting bits per second (bps) using

$$f_{\text{Hz}} = \frac{R_{\text{bps}}}{N} \tag{15}$$

where

- f_{Hz} is the frequency in hertz (Hz);
- R_{bps} is the transmission rate in bits per second (bps);
- N is the number of bits transmitted per cycle (we assume it as 64).

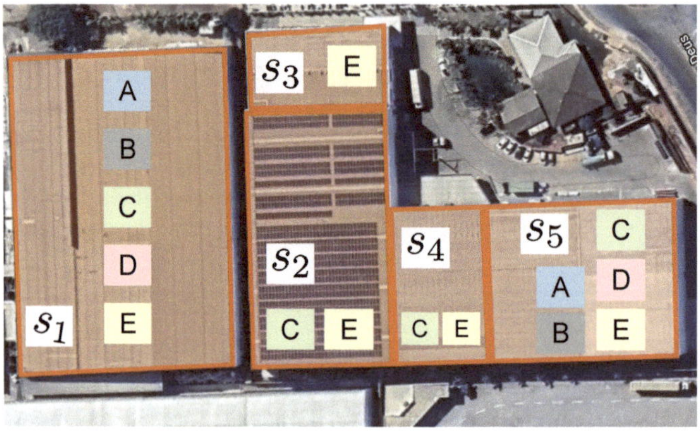

Figure 8. Application deployment plan considered in the experiment.

Table 1. Parameters used in the experiments [24].

<table>
<tr><th colspan="2">Parameters</th><th>Values</th></tr>
<tr><td rowspan="4">Bandwidth consumption</td><td>Image</td><td>10 Mbps</td></tr>
<tr><td>Motion sample</td><td>1.92 Kbps</td></tr>
<tr><td>Emission reading</td><td>0.64 Kbps</td></tr>
<tr><td>Sound</td><td>128 Kbps</td></tr>
<tr><td rowspan="4">Computing resource requirement</td><td>Image</td><td>1.25 MHz</td></tr>
<tr><td>Motion sample</td><td>60 Hz</td></tr>
<tr><td>Emission reading</td><td>20 Hz</td></tr>
<tr><td>Sound</td><td>6 KHz</td></tr>
<tr><td rowspan="4">Memory requirement</td><td>Image</td><td>1.25 GB</td></tr>
<tr><td>Motion sample</td><td>240 B</td></tr>
<tr><td>Emission reading</td><td>80 B</td></tr>
<tr><td>Sound</td><td>16 KB</td></tr>
<tr><td>Computing resource</td><td>Edge server</td><td>6.8 GHz (4 × 1.7 GHz)</td></tr>
<tr><td rowspan="2">Transmission range and bandwidth</td><td>WiFi AP</td><td>50 m/100 Mbps</td></tr>
<tr><td>Lora gateway</td><td>1 km/50 Kbps</td></tr>
<tr><td rowspan="4">Sensing range</td><td>Camera</td><td>15 m</td></tr>
<tr><td>Motion sensor</td><td>10 m</td></tr>
<tr><td>Gas sensor</td><td>600 m</td></tr>
<tr><td>Microphone</td><td>300 m</td></tr>
<tr><td>Sensing parameter α_v in Equation (3)</td><td>All sensors</td><td>Reciprocal of sensing range</td></tr>
</table>

Regarding information flow, two patterns were considered, as illustrated in Figure 9: a single source of sensing data that sends information to an analytical model or multiple sources of sensing data.

Infrastructure flows are generated from each information flow, as specified in the patterns in Figure 10, and a sensor is established for each source of sensing data and an edge server for each analysis model. In addition, one or more network devices were considered.

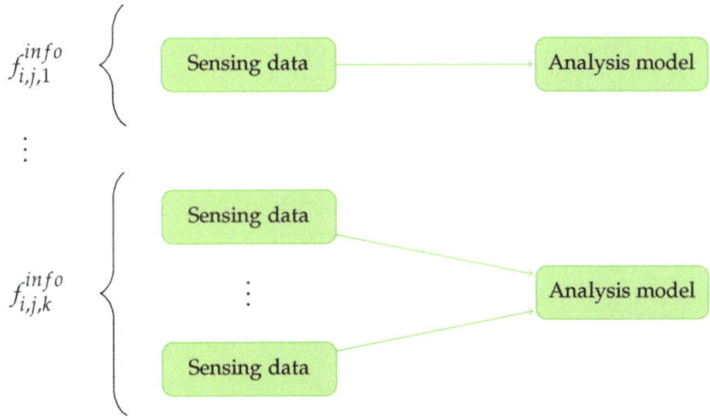

Figure 9. Patterns considered for information flows.

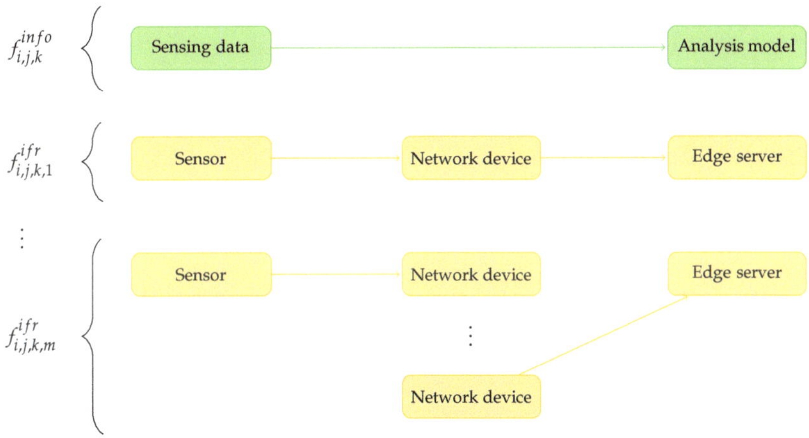

Figure 10. Patterns considered for infrastructure flows.

The values established in [24] were also maintained for the cost of implementing and operating equipment. However, corrections were applied due to inflation, based on what was established by the U.S. Bureau of Labor Statistics [37]. The adjusted values are presented in Table 2.

Table 2. Deployment and operational costs in USD.

Device	Camera	Motion Sensor	Gas Sensor	Mic.	WiFi AP	Lora Gateway	Edge Server
Deployment cost	1613.05	415.08	847.45	790.96	272.11	728.70	433.53
Operational cost	13.86	9.66	6.35	18.16	9.25	23.11	77.28

The number of candidate locations was established using the Brazilian standard ABNT NBR 5410 [38], which defines standards for electrical installations. According to this standard, one power outlet point is sufficient if the area reaches 6 m^2. One outlet point is required for every 5 m, or a fraction thereof, of the perimeter if the area of the room or dependency is greater than 6 m^2; these points should be spaced as evenly as possible.

The proposed solution was developed to be geographically neutral, relying on universal energy efficiency principles and IoT deployment. This neutrality ensures that the solution can seamlessly adapt to industries with varying infrastructural, economic, and regulatory conditions. Therefore, this Brazilian case study served as an illustrative example, and the model can be readily applied to other regions, supported by its modular and flexible architecture.

The experiments were conducted by varying the following parameters: (i) number of applications per room, (ii) number of information flows per application, (iii) number of infrastructure flows per information flow, (iv) deployment and operation budgets, and (v) MR weights to study their implications on various performance metrics.

5.2. Experimental Results

Figure 11 presents the results of running the tool for the considered scenario. When analyzing the implementation cost of the number of flows (Figure 11a), both information and infrastructure, it is possible to observe trends that reveal the interdependence between these parameters. The data suggest that, as the number of flows increases, the cost also tends to increase proportionally. This behavior is expected because more information flows imply an increase in the number of detection and communication devices required to ensure application coverage and reliability.

Similar behavior was observed when analyzing the service utility of the number of flows (Figure 11b). As the number of flows increases, the utility tends to grow. This behavior is justified by the greater density of information collected and transmitted, which expands the application's ability to detect events in a broader area with greater precision. However, utility tends to show decreasing marginal gains as the flows reach a certain level. In other words, after a certain point, the addition of new flows does not generate a significant increase in utility, indicating the presence of redundancies.

Regarding coverage, it is possible to observe a direct and significant relationship in Figure 11c, which reveals how the increase in flows affects the IoT solution's ability to monitor geographic areas and expand event detection. The coverage tends to grow proportionally when the number of flows increases, especially in the initial stages. This behavior occurs because additional flows allow the inclusion of more devices and sensors, increasing network density and expanding the geographic reach of the application. However, as the flow continued to grow, the rate of gain in coverage decreased. This behavior is explained by the fact that after reaching a specific density of sensors, additional flows begin to overlap areas that are already covered, resulting in redundancy and not a significant gain in coverage.

On the other hand, investment efficiency shows a decreasing trend as the number of flows increases (Figure 11d). This behavior reflects a relationship of diminishing returns in which the growth of the utility obtained does not keep pace with the proportional increase in costs. In other words, adding new flows contributes to significant gains; however, these gains become progressively smaller as flows increase, whereas costs continue to grow linearly or exponentially.

When analyzing the relationship between the covered locations (Figure 11e), a particular behavior is observed when $\alpha_{eff} = 1$ and $\alpha_{cov} = 0$. In this configuration, the covered location metric does not reach a value of 5, whereas all other weight configurations manage to reach or exceed this value. This result indicates that the system's ability to expand the served geographic area by exclusively prioritizing efficiency over coverage, which is limited even with increased flows. Furthermore, as the number of flows increased in this specific configuration, the number of covered locations increased at a steadier pace than in the other weight configurations. This behavior suggests that, even under the exclusive prioritization of efficiency, the flow increase still contributes to expanding coverage but is less effective. The absence of weight attributed to coverage limits the impact of new flows on the expansion of the served geographic area. In contrast, in configurations where α_{cov} has positive values, the system can simultaneously optimize efficiency and coverage, achieving higher values for covered locations. This result occurs because the weight attributed to coverage directs resources to expand the system's performance geographically, maximizing utility and the area served.

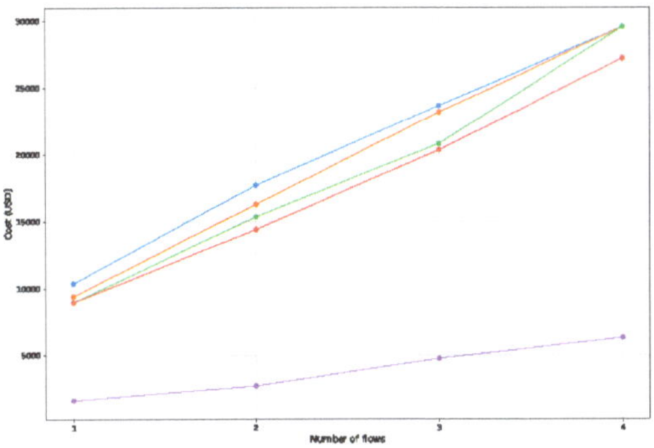

(**a**) Deployment cost.

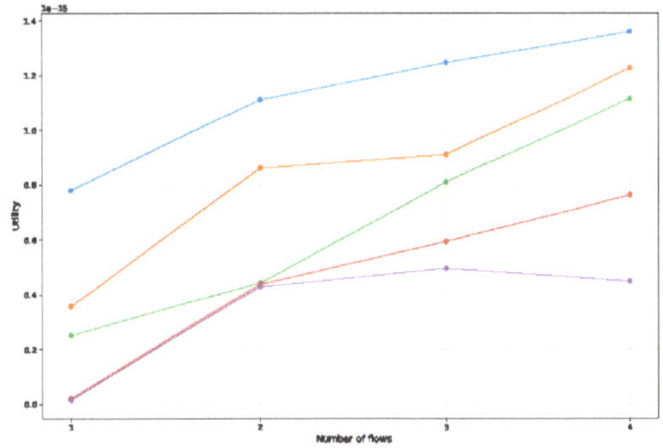

(**b**) Service utility.

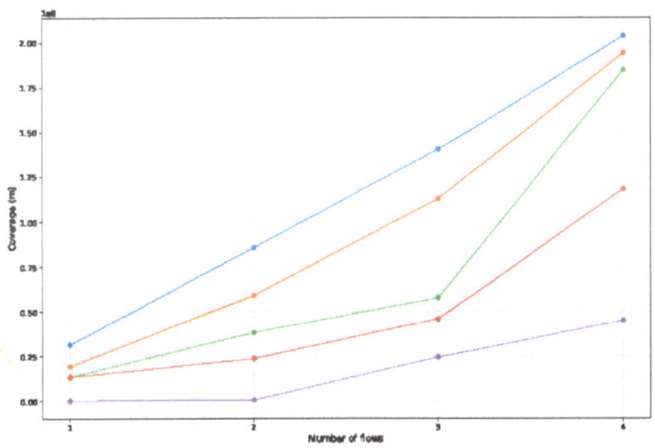

(**c**) Coverage.

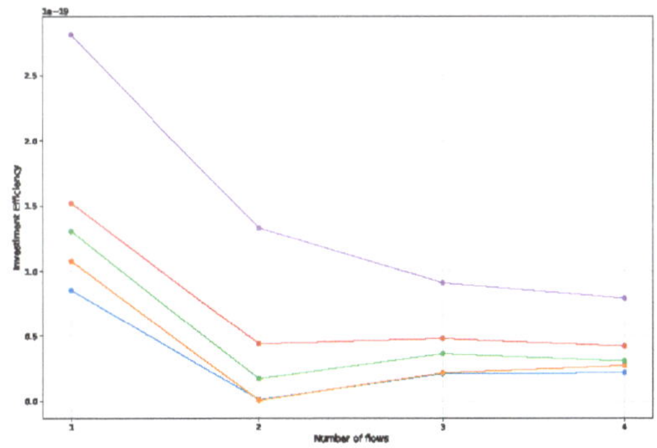

(**d**) Investment efficiency.

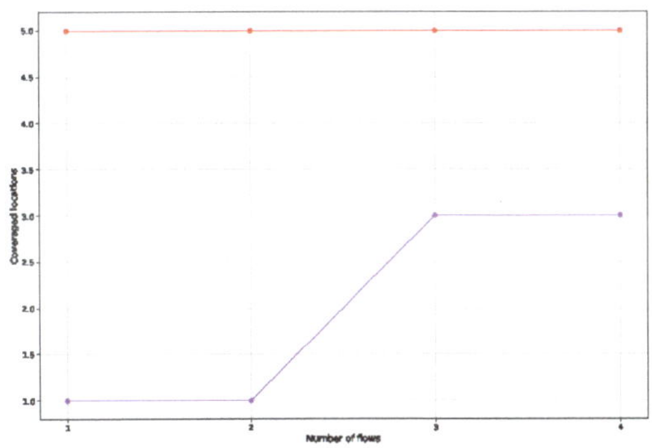

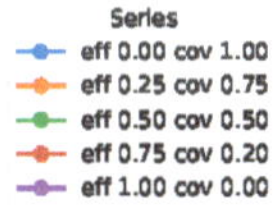

(**e**) Locations covered.

Figure 11. Experimental results.

The experimental results indicate that the proposed model demonstrates scalability to a certain extent, particularly regarding the increasing number of deployed devices and their impact on cost, utility, and coverage. The heuristic-based approach optimizes resource allocation, ensuring efficient device reuse while minimizing redundancy. However, we acknowledge that the scalability assessment did not cover all potential factors, such as extreme increases in the number of rooms or highly heterogeneous industrial environments. Nevertheless, given that the solution is designed for planning within a specific industrial context, we do not anticipate scenarios requiring an exceptionally high number of rooms. The model primarily focuses on optimizing the energy efficiency within predefined industrial spaces rather than handling large-scale deployments across multiple facilities. Future work could explore adaptations to enhance scalability for broader applications beyond the targeted industrial setting.

6. Final Remarks

One of the primary motivations for the solution presented in this work was the increasing complexity of challenges related to energy production, distribution, and consumption. The solution seeks to contribute to the energy efficiency in smart spaces in the context of Industry 4.0. The global demand for solutions that combine sustainability and technological innovation requires models capable of integrating advanced technologies, such as the IoT, artificial intelligence, and cloud computing, to promote more efficient, connected, and resilient industrial operations.

In this scenario, this study presented an adaptation of the SmartParcels framework, which constitutes an integrated model for planning smart spaces, combining sensing, communication, computing, and application layers. The model allows for the optimization of resource utilization and for maximizing the utility of services, respecting cost constraints and ensuring high energy efficiency. Specific heuristics were used to select promising geophysical mappings and maximize the reuse of devices, promoting economic and environmental sustainability. The experimental results demonstrated that the proposed solution is scalable and viable for large-scale applications across diverse industrial contexts. By not assuming region-specific characteristics, the model ensures adaptability to different industries and geographical regions.

Our analysis reveals that implementing a structured methodology for selecting information and infrastructure flows as well as heuristic-based optimization techniques yields substantial efficiency improvements. These insights provide valuable practical guidance for manufacturing, logistics, and industrial automation sectors navigating the transition to Industry 4.0 paradigms. Using our approach, the strategic allocation of IoT devices demonstrates multiple tangible benefits: (i) reduced energy consumption across industrial systems, (ii) minimized infrastructure expenditure, and (iii) enhanced system scalability for future expansion.

The results suggest that organizations can apply these methodologies to achieve optimal resource utilization while maintaining operational effectiveness. This balance between efficiency and performance represents a critical consideration for industries seeking competitive advantages in increasingly digitized operational landscapes.

However, some challenges pave the way for future work. It is important to note that the proposed solution focuses exclusively on the planning phase of smart spaces, specifically the design and deployment of IoT infrastructure to optimize energy efficiency. It does not address tasks or costs associated with the real-time execution of applications within these spaces, which remains outside the scope of this work.

Additionally, the model does not differentiate between constant and variable energy consumption scenarios. For example, energy usage in production areas may vary depending on production volume, while offices and utilities typically exhibit more constant consumption patterns. Future research should incorporate these dynamic factors to improve the model's applicability to real-world industrial contexts. Furthermore, the model could be adapted to various sectors, such as precision agriculture, healthcare, and transportation, expanding its applicability and relevance. Another possibility would be to incorporate algorithms based on machine learning or artificial intelligence to improve the accuracy and adaptability of planning decisions, especially in dynamic scenarios with high device heterogeneity.

Additionally, validating the model in real industrial environments is essential to confirm its robustness and ability to meet complex demands. Such validation would enable a deeper analysis, including statistical hypothesis testing and sensitivity analyses, which are not feasible within the current simulation-based framework. This analysis could involve case studies in diverse industries, providing insights into operational efficiency, sustainability gains, and return on investment while addressing dynamic energy consumption patterns.

Finally, creating more intuitive interfaces and visualization tools could facilitate the model's adoption by multidisciplinary teams, promoting its integration into strategic decision-making processes. Thus, this work lays the foundations for developing even more advanced solutions aligned with contemporary demands for innovative, sustainable, and efficient industrial operations.

Author Contributions: Conceptualization, R.d.A.G. and G.B.; data curation, V.B.F. and R.d.A.G.; formal analysis, R.d.A.G. and B.B.F.d.C.; investigation, V.B.F. and R.d.A.G.; methodology, V.B.F., R.d.A.G. and G.B.; project administration, R.d.A.G.; resources, R.d.A.G.; software, V.B.F. and R.d.A.G.; supervision, J.L.D., R.C.B.d.F., T.A.M., B.B.F.d.C. and A.N.H.; validation, J.L.D., R.C.B.d.F., T.A.M., B.B.F.d.C. and A.N.H.; writing—original draft preparation, V.B.F. and R.d.A.G.; writing—review and editing, T.A.M. and R.d.A.G. All authors have read and agreed to the published version of the manuscript.

Funding: This research received no external funding.

Institutional Review Board Statement: Not applicable.

Informed Consent Statement: Not applicable.

Data Availability Statement: Data are contained within this article.

Conflicts of Interest: The authors declare no conflicts of interest.

Abbreviations

The following abbreviations are used in this manuscript:

ABNT	Brazilian Association of Technical Standards
CPSs	Cyber-physical systems
IoT	Internet of Things
IIoT	Industrial Internet of Things
NGSI-LD	Next-Generation Service Interface-Linked Data
NBR	Brazilian Technical Standard
PPC	Production planning and control
QoS	Quality of service
XML	Extensible Markup Language

References

1. Liu, F.; Sim, J.Y.; Sun, H.; Edziah, B.K.; Adom, P.K.; Song, S. Assessing the role of economic globalization on energy efficiency: Evidence from a global perspective. *China Econ. Rev.* **2023**, 77, 101897. [CrossRef]
2. Empresa de Pesquisa Energética. *Balanço Energético Nacional*; Empresa de Pesquisa Energética: Rio de Janeiro, Brazil, 2023; Technical Report.

3. Empresa de Pesquisa Energ6tica. Resenha Mensal: Ap6s leve Queda em Dezembro, o Consumo Nacional de Eletricidade Volta a Crescer em Janeiro Liderado pela indúStria. Available online: https://www.epe.gov.br/pt/imprensa/noticias/resenha-mensal-apos-leve-queda-em-dezembro-o-consumo-nacional-de-eletricidade-volta-a-crescer-em-janeiro-liderado-pela-industria (accessed on 10 March 2025).
4. Chatterjee, A.; Keyhani, A.; Kapoor, D. Identification of photovoltaic source models. *IEEE Trans. Energy Convers.* **2011**, *26*, 883–889. [CrossRef]
5. Barman, P.; Dutta, L.; Bordoloi, S.; Kalita, A.; Buragohain, P.; Bharali, S.; Azzopardi, B. Renewable energy integration with electric vehicle technology: A review of the existing smart charging approaches. *Renew. Sustain. Energy Rev.* **2023**, *183*, 113518. [CrossRef]
6. Arévalo, P.; Ochoa-Correa, D.; Villa-Ávila, E. Optimizing microgrid operation: Integration of emerging technologies and artificial intelligence for energy efficiency. *Electronics* **2024**, *13*, 3754. [CrossRef]
7. Lasi, H.; Fettke, P.; Kemper, H.G.; Feld, T.; Hoffmann, M. Industry 4.0. *Bus. Inf. Syst. Eng.* **2014**, *6*, 239–242. [CrossRef]
8. Çınar, Z.M.; Abdussalam Nuhu, A.; Zeeshan, Q.; Korhan, O.; Asmael, M.; Safaei, B. Machine learning in predictive maintenance towards sustainable smart manufacturing in industry 4.0. *Sustainability* **2020**, *12*, 8211. [CrossRef]
9. Rathor, S.K.; Saxena, D. Energy management system for smart grid: An overview and key issues. *Int. J. Energy Res.* **2020**, *44*, 4067–4109. [CrossRef]
10. Chen, M.; Sinha, A.; Hu, K.; Shah, M.I. Impact of technological innovation on energy efficiency in industry 4.0 era: Moderation of shadow economy in sustainable development. *Technol. Forecast. Soc. Change* **2021**, *164*, 120521. [CrossRef]
11. Li, S.; Ni, Q.; Sun, Y.; Min, G.; Al-Rubaye, S. Energy-efficient resource allocation for industrial cyber-physical IoT systems in 5G era. *IEEE Trans. Ind. Inform.* **2018**, *14*, 2618–2628. [CrossRef]
12. Trstenjak, M.; Cosic, P. Process planning in Industry 4.0 environment. *Procedia Manuf.* **2017**, *11*, 1744–1750. [CrossRef]
13. Oluyisola, O.E.; Bhalla, S.; Sgarbossa, F.; Strandhagen, J.O. Designing and developing smart production planning and control systems in the industry 4.0 era: A methodology and case study. *J. Intell. Manuf.* **2022**, *33*, 311–332. [CrossRef]
14. Reichardt, A.; Murawski, M.; Bick, M. The use of the Internet of Things to increase energy efficiency in manufacturing industries. *Int. J. Energy Sect. Manag.* **2024**, *19*, 369–388. [CrossRef]
15. Chang, T.C.; Banerjee, T.; Venkatasubramanian, N.; York, R. QuIC-IoT: Model-Driven Short-Term IoT Deployment for Monitoring Physical Phenomena. In Proceedings of the 8th ACM/IEEE Conference on Internet of Things Design and Implementation, San Antonio, TX, USA, 9–12 May 2023; pp. 424–437.
16. Krishna, A.; Le Pallec, M.; Mateescu, R.; Noirie, L.; Salaün, G. IoT Composer: Composition and deployment of IoT applications. In Proceedings of the 2019 IEEE/ACM 41st International Conference on Software Engineering: Companion Proceedings (ICSE-Companion), Montreal, QC, Canada, 25–31 May 2019; pp. 19–22.
17. Hu, M.; Cao, E.; Huang, H.; Zhang, M.; Chen, X.; Chen, M. AIoTML: A Unified Modeling Language for AIoT-Based Cyber–Physical Systems. *IEEE Trans. Comput.-Aided Des. Integr. Circuits Syst.* **2023**, *42*, 3545–3558. [CrossRef]
18. Guan, P.; Dangwal, A.; Taherkordi, A.; Wolski, R.; Krintz, C. Energy-Aware IoT Deployment Planning. In Proceedings of the 21st ACM International Conference on Computing Frontiers, Ischia, Italy, 7–9 May 2024; pp. 61–70.
19. Herrera, J.L.; Galán-Jimnez, J.; Garcia-Alonso, J.; Berrocal, J.; Murillo, J.M. Joint optimization of response time and deployment cost in next-gen iot applications. *IEEE Internet Things J.* **2022**, *10*, 3968–3981. [CrossRef]
20. Herrera, J.L.; Galán-Jiménez, J.; Berrocal, J.; Bellavista, P.; Foschini, L. Energy-Efficient QoS-Aware Application and Network Configuration for Next-Gen IoT. In Proceedings of the 2023 IEEE 28th International Workshop on Computer Aided Modeling and Design of Communication Links and Networks (CAMAD), Edinburgh, UK, 6–8 November 2023; pp. 105–110.
21. Brogi, A.; Forti, S. QoS-aware deployment of IoT applications through the fog. *IEEE Internet Things J.* **2017**, *4*, 1185–1192. [CrossRef]
22. Ghaderi, A.; Movahedi, Z. An energy-efficient data management scheme for industrial IoT. *Int. J. Commun. Syst.* **2025**, *38*, e5167. [CrossRef]
23. Chang, T.C.; Bouloukakis, G.; Hsieh, C.Y.; Hsu, C.H.; Venkatasubramanian, N. Demo abstract: SmartParcels: A what-if analysis and planning tool for IoT-enabled smart communities. In Proceedings of the IoTDI 2021: 6th ACM/IEEE International Conference on Internet-of-Things Design and Implementation, Nashville, TN, USA, 18–21 May 2021; pp. 267–268.
24. Chang, T.C.; Bouloukakis, G.; Hsieh, C.Y.; Hsu, C.H.; Venkatasubramanian, N. Smartparcels: Cross-layer iot planning for smart communities. In Proceedings of the International Conference on Internet-of-Things Design and Implementation, Charlottesvle, VA, USA, 18–21 May 2021; pp. 195–207.
25. Wang, B. Coverage problems in sensor networks: A survey. *ACM Comput. Surv. (CSUR)* **2011**, *43*, 1–53. [CrossRef]
26. Adeyeye, A.O.; Lynch, C.; Eid, A.; Hester, J.G.; Tentzeris, M.M. Energy autonomous two-way repeater system for non-line-of-sight interrogation in next generation wireless sensor networks. *IEEE Trans. Microw. Theory Tech.* **2022**, *70*, 1779–1788. [CrossRef]
27. Chierichetti, F.; Kumar, R.; Tomkins, A. Max-cover in map-reduce. In Proceedings of the 19th international conference on World Wide Web, Raleigh, NC, USA, 26–30 April 2010; pp. 231–240.
28. Feige, U. A threshold of ln n for approximating set cover. *J. ACM (JACM)* **1998**, *45*, 634–652. [CrossRef]

29. Asplund, E. Asplund, E. A monotone convergence theorem for sequences of nonlinear mappings. In *Nonlinear Functional Analysis, Proceedings of Symposia in Pure Mathematics*; American Mathematical Society: Providence, RI, USA, 1970; Volume 18, pp. 1–9.
30. Smart Data Models. Smart Data Models program. Available online: https://smartdatamodels.org (accessed on 7 March 2022).
31. Cirillo, F.; Solmaz, G.; Berz, E.L.; Bauer, M.; Cheng, B.; Kovacs, E. A standard-based open source IoT platform: FIWARE. *IEEE Internet Things Mag.* **2019**, *2*, 12–18. [CrossRef]
32. Privat, G.; Medvedev, A. Guidelines for Modelling with NGSI-LD. *ETSI White Pap.* **2021**, *42*. Available online: https://www.etsi.org/images/files/ETSIWhitePapers/etsi_wp_42_NGSI_LD.pdf (accessed on 21 October 2024).
33. Schema.org. Schemas. Available online: https://schema.org/ (accessed on 21 October 2024).
34. Bargnesi, A.; DiFabio, A.; Hayes, W.; Shibaev, G.; Benz, C.; Pyle, H.; Giggy, T. JSON Graph Format (JGF). Available online: https://jsongraphformat.info/ (accessed on 21 October 2024).
35. Butler, H.; Daly, M.; Doyle, A.; Gillies, S.; Schaub, T.; Schmidt, C. GeoJSON. Available online: http://geojson.org (accessed on 21 October 2024).
36. Foundation, D.S. django—The Web Framework for Perfectionists with Deadlines. Available online: https://www.djangoproject.com/ (accessed on 5 July 2023).
37. Consumer Price Index. Available online: https://www.bls.gov/cpi/ (accessed on 23 November 2024).
38. Associação Brasileira de Normas Técnicas. Catálogo. Available online: https://abnt.org.br (accessed on 23 November 2024).

eng

MDPI

Article

Effect of Graphene Nanoplatelets as Lubricant Additive on Fuel Consumption During Vehicle Emission Tests

Eduardo Tomanik [1,2,*], Wania Christinelli [1], Pamela Sierra Garcia [1], Scott Rajala [3], Jesuel Crepaldi [4], Davi Franzosi [2], Roberto Martins Souza [2] and Fernando Fusco Rovai [5,6]

1 Gerdau Graphene Brazil, São Paulo 05402-918, Brazil; wania.christinelli@gerdaugraphene.com (W.C.); pamela.garcia@gerdaugraphene.com (P.S.G.)
2 Surface Phenomena Laboratory, Polytechnic School, University of São Paulo, São Paulo 05508-030, Brazil; davi.franzosi.oliveira@usp.br (D.F.); robertosouza@usp.br (R.M.S.)
3 Idemitsu Lubricants America Corp., Jeffersonville, IN 47130, USA; srajala.1460@idemitsu.com
4 Idemitsu Lube South America Ltda, São Paulo 01419-001, Brazil; jcrepaldi.0450@idemitsu.com
5 VW do Brasil—Way to Zero Center, São Paulo 09850-901, Brazil; fernando.rovai@volkswagen.com.br or ffusco@fei.edu.br
6 Department of Mechanical Engineering, Centro Universitário FEI, São Paulo 09850-901, Brazil
* Correspondence: eduardo.tomanik@gmail.com or eduardo.tomanik@ex.gerdaugraphene.com

Abstract: Lubricant friction modifier additives are used on lower viscosity engine oils to mitigate boundary friction. This work presents the development of a graphene-based material as an oil friction modifier additive, from formulation to actual vehicle tests. The graphene material was initially characterized using scanning electron microscopy (SEM) and Raman spectroscopy, which revealed the predominance of graphene nanoplatelets (GNPs) with an average of nine layers. After functionalization, two GNP additive variants were initially mixed with a fully formulated SAE 0W-20 engine oil and tribologically evaluated using reciprocating sliding tests at 40 and 120 °C and Hertzian pressure up to 1.2 GPa when both variants presented friction reduction. Then, the GNP additive variant with better performance was evaluated in a vehicle emission test using a fully formulated 5W-20 SAE oil as a reference. The addition of 0.1% of GNPs reduced fuel consumption by 2.6% in urban conditions and 0.8% in highway ones. The urban test cycle was FTP75 and higher benefits of the GNP additive occurred especially on the test start, when the engine and oil were still cold and on test portions where the vehicle speed was lower.

Keywords: graphene; friction; fuel economy; viscosity

Academic Editor: Antonio Gil Bravo

Received: 8 December 2024
Revised: 9 January 2025
Accepted: 11 January 2025
Published: 16 January 2025

Citation: Tomanik, E.; Christinelli, W.; Garcia, P.S.; Rajala, S.; Crepaldi, J.; Franzosi, D.; Souza, R.M.; Rovai, F.F. Effect of Graphene Nanoplatelets as Lubricant Additive on Fuel Consumption During Vehicle Emission Tests. *Eng* **2025**, *6*, 18. https://doi.org/10.3390/eng6010018

1. Introduction

The search for reducing fuel consumption and CO_2 emission has led to continuous efforts for reducing the mechanical losses caused by friction. For combustion engines, lower viscosity oils are being introduced to reduce engine-dominant hydrodynamic friction losses but with the risk of increasing boundary friction [1,2]. To protect engine parts against potential damage related to metal-to-metal contact, due to the increasing trend in reducing oil viscosity, some lubricant formulation strategies are applied. These initiatives include introducing the right type of friction modifier, choosing high-viscosity index base oils and selecting efficient viscosity index improvers based on olefin copolymers to minimize shear thinning, as demonstrated in previous works [3–6].

In [5], different types of FM additives (MoDTC, three variants based on ester polymer and an Amine-based one) were tested in TE-77 rotational and floating liner reciprocating testers. The tested MoDTC additive provided the highest reductions in friction force and friction losses. Combining those efforts with engine adaptations to operate in the presence of ultra-low viscosity oils, the results of the fuel economy in homologation test cycles may reach values up to 5.5% depending on the baseline oil of a given engine, as shown in [7,8].

More recently, nanoparticles are being investigated as lubricant additives. Spikes [9] has mentioned five potential advantages of using nanoparticles as lubricant additives: (1) insolubility in nonpolar base oils, (2) low reactivity with other additives in the lubricant, (3) high possibility of film formation on many different types of surfaces, (4) more durability and (5) high nonvolatility to withstand high temperatures. Different elements have been investigated as nanoparticles, including metals, ceramics, chalcogenides, MXenes and carbon-based materials [9–16]. Lubricants with nanoparticles are also called "Nanolubricants", with the nanoparticles acting as an anti-wear or friction modifier. However, these solutions usually require the use of dispersants and surfactants to functionalize the nanoparticles as an oil additive [11,13,14]. Carbon-based nano materials include carbon dots, Carbon Nanotubes (CNTs), graphene, Graphene Oxide and others [13,14]. Graphene-based materials, due to unique properties such as low shear resistance, high stiffness and thermal conductivity, are attractive materials for tribological applications, including improvement on the properties of lubricants [13,14]. The exact mechanism that improves tribological performance is still being investigated and probably more than one occur on actual applications. Figure 1 summarizes the main potential tribological mechanisms of nanoparticles. Graphene and other nanosheets may also work as a viscous modifier (see discussion in [13]).

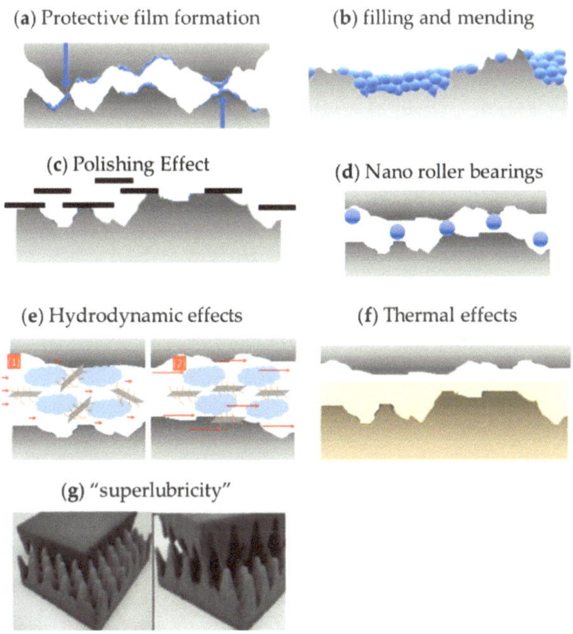

Figure 1. Graphene's tribological mechanisms. (**a**) Typical FM tribofilm; (**b**) surface filling and mending; (**c**) polishing effect; (**d**) nano roller bearings; (**e**) hydrodynamics at (**1**) low shear rate, (**2**) high shear rate; (**f**) thermal effects; (**g**) superlubricity, incommensurable contact. Reproduced from [9].

Synergic or antagonistic mechanisms with other oil additives as well as with the materials in contact may occur [17]. In summary, despite all the research already conducted, there are several knowledge gaps about the use of nanoparticles as lubricant additives. Especially for engine oils, it appears mandatory to test fully formulated oils and test on actual applications. Following the steps of the formulation and the initial characterization of a given engine oil, the evaluation of tribological behavior usually starts with laboratory tests [5,6]. Despite the importance that these initial tests have, it is frequently difficult to evaluate how much a difference in the coefficient of friction in a laboratory represents in terms of the overall performance of real engines, for example in terms of fuel consumption. Factors that contribute to these difficulties include the following: (i) the relatively large fuel consumption dispersion on actual vehicle tests; (ii) the impact that other energy losses (e.g., thermal) have on efficiency [8]; and (iii) the diversity of tribological systems inside an engine, each one with specific tribological conditions including lubrication regimes that can vary from boundary to hydrodynamic ones [2]. The use of laboratory data as inputs to the numerical simulations of an engine, or of a specific system within, may help to bridge the gap between laboratory and engine tribological results [1,2].

This work aims to cover all the main steps mentioned in the paragraphs above. Two variations of graphene-based additives of engine oil were developed and characterized. Laboratory lubricated reciprocating sliding tests were then conducted with these oils, to evaluate the friction reduction potential. Finally, the investigation was completed by vehicle tests to compare fuel consumption when oils with or without the additive were used.

2. Materials and Methods

2.1. Graphene Characterization

Graphene samples, after deposition as a powder over a conductive carbon adhesive tape, were characterized by scanning electron microscopy (SEM) using a Hitachi SU5000 model. Raman spectroscopy was used to characterize the structure of the samples using Oxford Instruments (Ulm, Germany) Witec Alpha 300 RA equipment with a 532 nm laser. A typical Raman spectrum of graphene has three main bands that describe the crystalline quality of the material and stacking characteristics, such as the number of coupled interlayers. The D band, located at 1350 cm^{-1}, is activated by the disorder generated at 1580 cm^{-1}, caused by stretching the C-C covalent bonds common in all carbon systems with sp2 hybridization. The 2D band, located at approximately 2700 cm^{-1}, is the overtone of the D band, with two transverse optical phonons.

Raman spectra were the inputs of an improved version of the protocol described in [18] to quantify crystalline defects and the number of graphene-coupled interlayers (see Table 1 and Figures 2 and 3). The GNP used had on average 9 layers and a lateral size of 71.1 nm. The diagram (b) in Figure 2, proposed by Silva et al. [18], relates point defects and linear defects based on the ratio between the areas of the G and D bands and the G bandwidth, thereby analyzing the structure of graphene. In this diagram, values corresponding to larger crystallite sizes and greater distances between point defects (indicating fewer defective samples) are located in the lower left corner. As the defects in the structure increase, the position in the diagram shifts upward and to the right. Thus, it can be observed that the graphene used in this work has a preserved crystalline structure with a low density of defects. Under higher magnification SEM, it is possible to notice that its sheets are aggregated in a spherical manner (Figure 4).

Table 1. GNP characterization.

Characteristic	Unit	Mean	Q90
Number of layers—<N>2D (nm)	-	9	11
Surface density of point defects—nD	10^{10} cm^{-2}	2.8	4.3
Lateral size—L_a	nm	71.1	99.4
D to G peak intensity ratio (I_D/I_G)		0.28	0.44
Percentile of volume-based particle size distribution		9.6 (D_{50})	19.3 (D_{90})

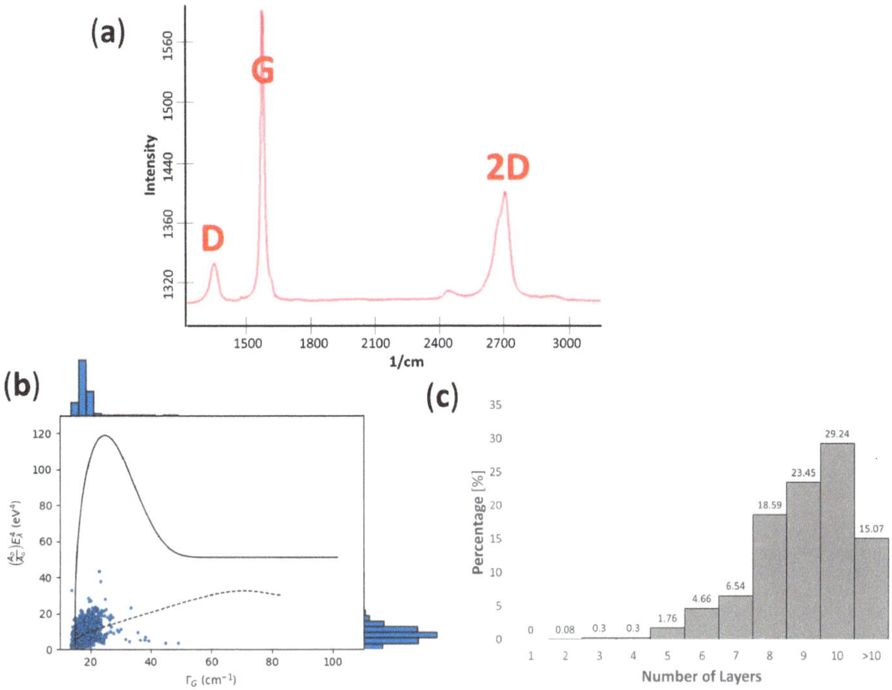

Figure 2. (**a**) Characteristic Raman spectrum. (**b**) Scatter plot with the frequency distributions of the G-band that is full-width at half max (ΓG) and the ratio between the integrated areas of the D and G bands (AD/AG) multiplied by the fourth power. (**c**) Layer distribution.

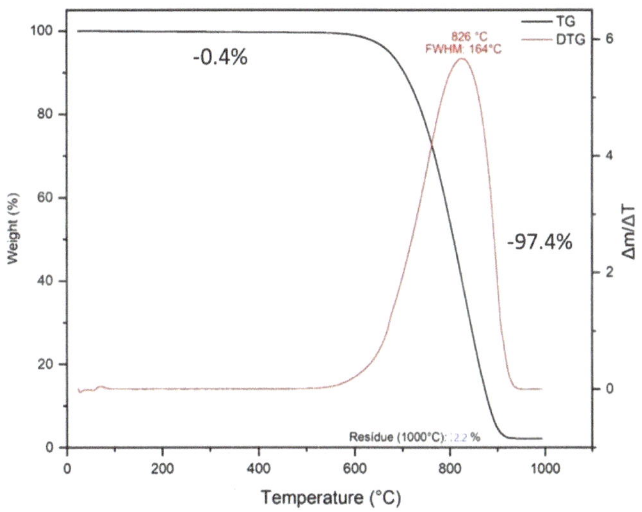

Figure 3. Thermogravimetric and derivative curves obtained under air gas flow of 100 mL/min and at a heating rate of 10 °C/min.

Figure 4. GNP scanning electron microscopy—SEM photos with increasing magnifications.

2.2. *Booster with Graphene*

To ensure effective interaction between graphene and lubricants, a molecule featuring a highly reactive cyclic group and an oxygen functional group was used to functionalize the graphene powders (samples L66_1 and L66_2). This functionalization process was followed by treatment with an organic long-chain compound to enhance compatibility with the lubricant matrix. Both samples underwent advanced preparation methods tailored to optimize their performance in lubrication systems.

L66_1 was produced on an industrial scale using a high-energy mixing process, yielding a concentrated formulation with approximately 38% graphene nanoplatelets (GNPs). L66_2, on the other hand, followed the same preparation method as L66_1 but incorporated an additional exfoliation step purely through shear mixing. This extra step was introduced to further reduce graphene aggregation, resulting in a more homogenous dispersion. The effectiveness of this modification was evidenced by a notable decrease in the viscosity of L66_2 (see Figure 5).

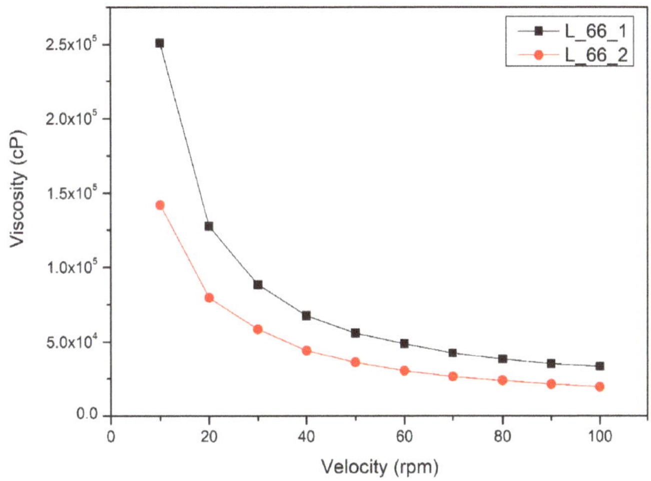

Figure 5. Viscosity of the additives L66_1 and L66_2.

The specific substances and techniques employed for graphene functionalization and mixing are proprietary and cannot be disclosed.

2.3. Tribological Tests

For the tribological tests, the additives L66_1 and L66_2 were mixed with a fully formulated oil, SAE 0W-20 SN. To mix the GNP additives, the oil was heated to 40 °C and the GNP mass required to achieve a 0.2 w/w% concentration was weighed on an analytical scale. The additive was then added to the warmed oil. The mixture was first stirred manually with a glass rod and then placed in an ultrasonic bath for 45 min. After this period, no sediment was observed. The mixture was stored, and photographs were taken as a function of time, as presented in Figure 6. Immediately after mixing (as new), and after 10 days, the dispersion remained visually stable. However, after 20 days, some sedimentation of the additive was observed at the bottom. The dispersion could easily be restored by gentle shaking and a brief ultrasonic bath treatment.

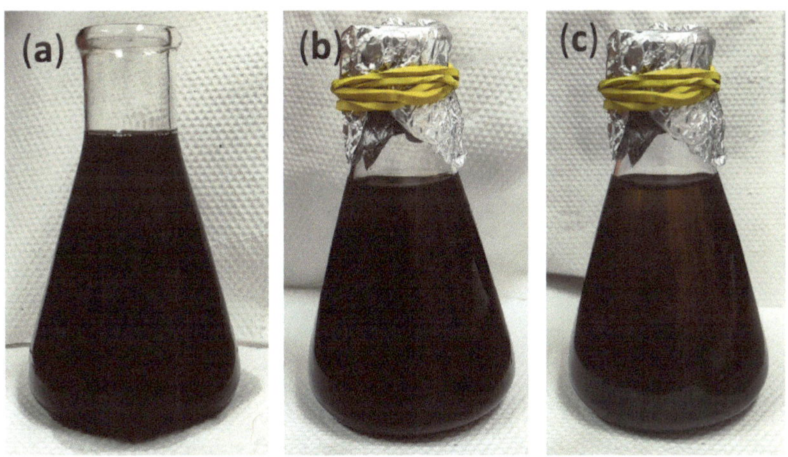

Figure 6. L66_1 on 0W20 (**a**) as new, (**b**) after 10 days, (**c**) after 20 days.

Tribological tests were conducted using an SRV tribometer (Optimol, München, Germany). This test involved the reciprocating sliding of a ball against the flat surface of a 24 mm diameter AISI H13 steel disc specimen. To ensure consistent roughness across all tests, the disc was polished, with the final polishing stage performed using a paste containing 1 μm diamond particles. After polishing, the surface roughness (Sa) was measured using a 3D laser interferometer. The ball was made of AISI 52100 steel, presenting a diameter of 10 mm. See Table 2.

Table 2. SRV sample parameters.

Sample	Diameter (mm)	Material	Hardness (Hv)	Young Modulus (GPa)	Poisson Ratio	Roughness Sa (μm)
Ball	10	AISI 52100	813 ± 6	210	0.3	0.042 ± 0.004
Disc	24	AISI H13	615	210	0.3	0.012 ± 0.002

Tribological tests were conducted in triplicate. Each repetition followed the procedure detailed in Table 3. Each repetition lasted 105 min, divided into steps of 15 min each. Five drops of oil were applied at each test start, covering the entire disc surface. After each test, residual oil was observed on the surface, indicating consistent lubrication throughout the test.

Table 3. Tribological test procedure.

Parameter	Unit					
Temperature	°C		40		120	
Load	N	20	5	5	20	5
Max. Hertzian Pressure	GPa	1.2	0.5	0.5	1.1	0.5
Stroke	mm			5		
Frequency	Hz			5		
Duration	min/per step			15		

2.4. Vehicle Emission Tests

Vehicle tests were part of a larger test program comparing different lubricants, and the GNPs were added to a fully formulated 5W-20 oil. After the tests with SAE 5W-20 reference oil for the vehicle emission test, the engine was started and ran until the oil temperature reached the operation value, 90 °C. The engine was stopped and 500 mL of oil was removed. From these, 250 mL was kept as a sample after the test and the other 250 mL, while still hot, was used to disperse the graphene additive. The mix was conducted only manually with the help of a "spoon" (Such a simple mixing method was conducted to somehow mimic the expected application of the additive as a booster, on a common workshop). Then, the 250 mL plus additive was returned to the engine. The engine was completed with a volume of new oil considering the small amount of additive to ensure that the test sequence started with the same volume as with the reference oil. The engine was again restarted and run for a few minutes before being conditioned ("soaking period") according to the test procedure standard. Brazil uses the so-called flex fuel and Brazilian standards (NBR) define fuel consumption in liters per 100 km, calculated from the balance of carbon in the emissions to calculate the fuel mass converted to volume using the density of the test fuel.

The experimental emissions tests were performed with a large sport utility vehicle in an emissions laboratory following a combined cycle over a chassis dyno, according to NBR7024 [19], composed of 55% in an urban cycle (FTP75) and 45% in a highway cycle. To better investigate the influence of the GNP additive, the urban FTP75 cycle was divided into three phases: Ph1, Ph2 and Ph3. See Figure 7 and Table 4. Ph3 has an identical vehicle speed profile as Ph1, but as the engine and oil are already hot, fuel consumption is significantly lower than in Ph1. Due to the temperature on the aftertreatment 3-way catalytic converter, almost all pollutant emissions occur in Ph1. See discussion in [20].

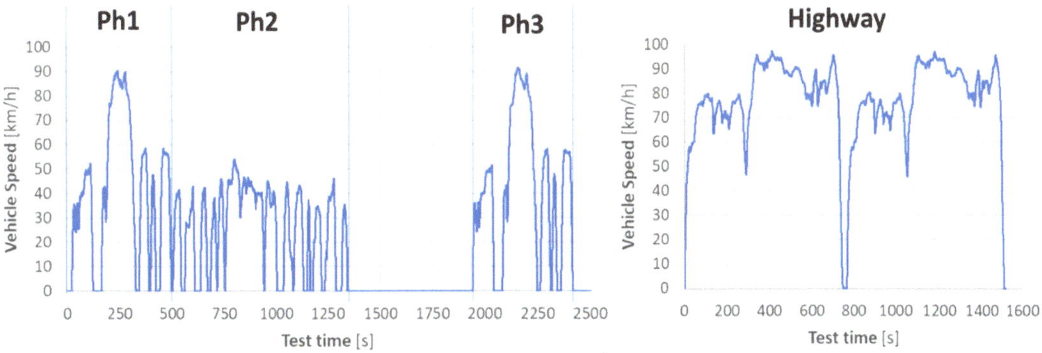

Figure 7. NBR7024 chassis emissions test. Ph1 to Ph3 are identical to the FTP75 emission test.

Table 4. Emission test cycle details.

	Units	Ph1 and 3	Ph2	Highway
Duration	s	505	864	765
Distance	km	5.78	6.21	16.45
Mean Velocity	km/h	41.20	25.88	77.73
Max. Velocity	km/h	91.2	55.2	96.56
Stops		6	13	None

The vehicle is equipped with a 4-cylinder, spark ignition, direct injection, turbocharged engine coupled to a 6-gear automatic transmission by a torque converter. At least two tests were performed with each lubricant version. The test uncertainties were compensated in terms of vehicle speed profile and battery voltage based on ECU data measurements by ETAS Inca. The test compensation factors were determined by 1D numerical simulation with a vehicle mathematical model in a GT-Suite v.2024 from Gamma Technologies. The test compensation is detailed in [21].

3. Results

3.1. Reciprocating Friction Tests

The data were analyzed using the "all data" file generated by the SRV acquisition system. This file records CoF (coefficient of friction) data for 1 minute at intervals of 5 min, with measurements taken every 1.9×10^{-5} s during the recording minute. In this setup, each step consists of three such 5 min intervals, totaling 15 min of testing under specific conditions (e.g., load or temperature). For each step, an average COF is calculated for the three individual measurements, and the overall CoF for the step is determined as the average of these three values. Figure 8 presents the CoF results for each step. Here, L66_1 and L66_2 refer to the dispersion of SAE 0W-20 with the respective additive variant.

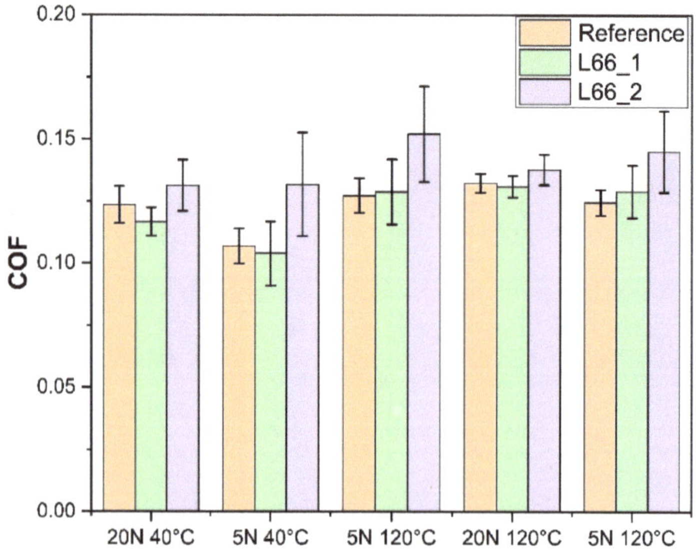

Figure 8. Cycle average CoF.

Adding L66_1 and L66_2 decreased the CoF compared to the reference oil, SAE 0W-20, under all test conditions. The CoF values for L66_1 and L66_2 were similar; L66_2 presented a lower CoF than L66_1 in the first test steps, with the difference between the two additive variants reducing along the test. It can be speculated that along the test the GNPs exfoliated in fewer layers and created a tribofilm on the surface. Such processes reduced the advantages of the more exfoliated and dispersed L66_2 in comparison to the 66_1, while in others L66_1 shows a slight advantage. The largest difference between the L66 additives and the reference oil was observed in the test steps with 5N. The error bars in Figure 8 reflect the variation in the COF during a single stroke, superimposed on the variations in the COF throughout the strokes during the 15 min evaluation period. Of these two causes, variations during the strokes may be significant, as presented in Figure 9, which reflects specific conditions at each contact point and the effect of the varying velocity during reciprocating motion. The lowest values of the COF were obtained near mid-stroke, where higher velocities lead towards the hydrodynamic lubrication regime. Figure 7 also indicates a trend for larger error bars for the tests conducted with the lower 5 N, which impacts the precision of the measurement of friction load.

Another way to analyze the tribological results is in terms of friction losses [22]. The energy dissipation due to the friction force is calculated through the force–displacement amplitude (F–D) hysteresis loops for each cycle during the test. Figure 9 shows one typical example of each lubricant variant and test step. Figure 10 shows the average results. As for the CoF, L66_2 presented a slight advantage compared to L66_1 in most of the test steps.

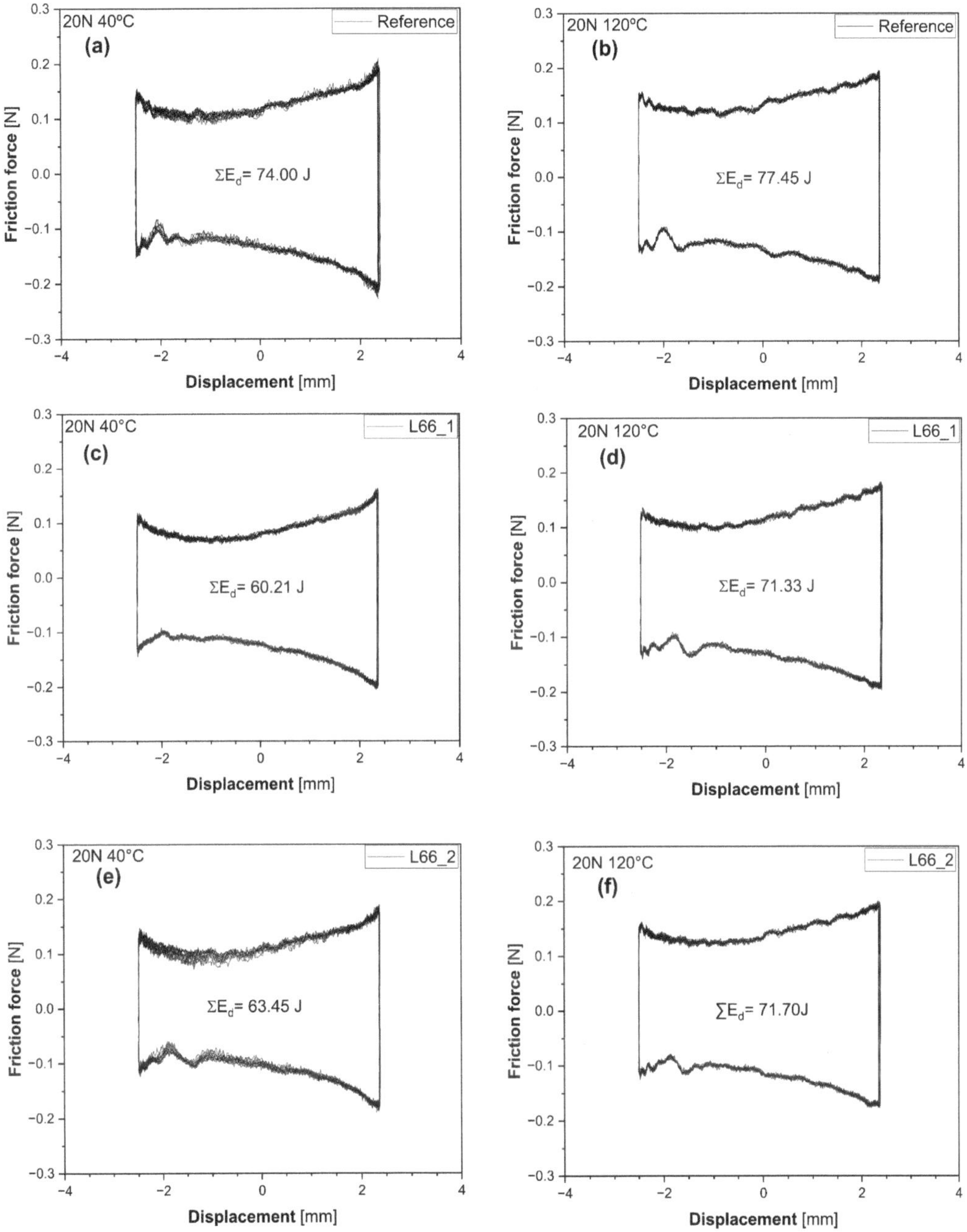

Figure 9. Friction force–displacement curves for the steps with 20 N of applied load. (**a**,**b**) Reference oil 0W-20m (**c**,**d**) with L66_1, (**e**,**f**) with L66_2.

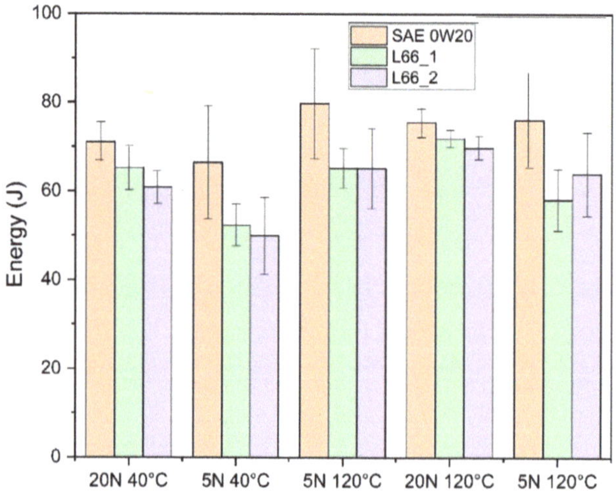

Figure 10. Dissipated energy calculated based on force–displacement amplitude (F–D) hysteresis loops during the reciprocating tests.

3.2. *Fuel Consumption*

As mentioned before, fuel consumption was measured by using the carbon balance converted to liters per 100 km. To allow more detailed analysis, the FTP75 cycle was divided into three phases. Ph1 and Ph3 have identical speed profiles, but Ph1 starts with the engine at room temperature, so oil viscosity is significantly higher than in Ph3. For this reason and for normal combustion issues, fuel consumption is also significantly higher in Ph1 than in Ph3. Figure 11 shows the delta fuel consumption (difference with respect to the consumption using the reference oil) in each of the three FTP75 phases, as well as: the accumulated one, the one in the highway cycle and the NBR7024 one (indicated as "combined"), which is composed of 55% of the FTP75 values and 45% of the highway values. Compared with the 5W-20 reference oil, tests with graphene additive presented a fuel saving of 2.6% in the FTP75 cycle and 0.8% in the highway cycle, providing combined NBR7024 standard of 1.9% fuel saving. Figure 10 shows the range obtained with the minimum and average compensations described in [21]. The values in the plot refer to the ones with the average compensation.

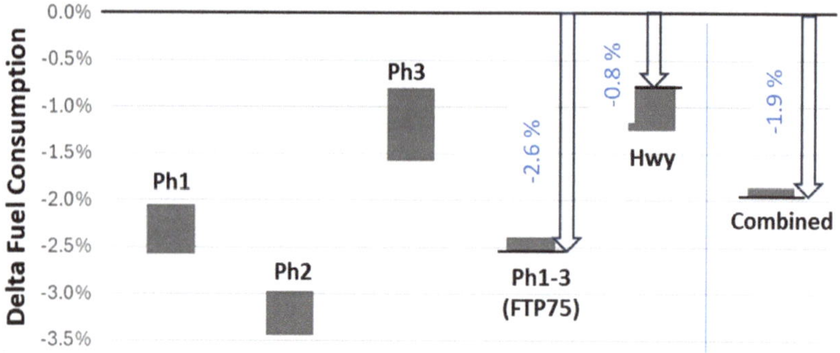

Figure 11. Delta fuel consumption with the L66_2, 0.1% GNPs in the different test steps.

4. Discussion

Paying attention to Figure 11, it can be observed that, as expected, fuel-saving reductions with the GNP additive were more significant in the Ph1 and Ph2 phases, where friction losses have more impact on fuel consumption. Internal combustion engines present several lubricated systems, which vary in terms of the predominant lubrication regime. In part of the systems, such as in cam-follower, the boundary lubrication may prevail, while in journal bearings hydrodynamic lubrication is expected to be the most important. Thus, the decrease in fuel consumption with the use of the GNP additives can be due to both boundary and hydrodynamic effects. In addition, graphene additives have shown the potential of increasing the lubricant conductivity. In the conducted tests, the oil with the GNP additive showed a slighter quicker temperature drop during the vehicle stop interval between the cold and hot phases. See Figure 12. Such behavior suggests that the addition of graphene increased the oil thermal conductivity as seen by other authors. Alqahtani [23] obtained a 20% increase in the thermal conductivity ofSAE 5W-30 when this oil presented a concentration of 0.09 wt% of graphene. A similar increase in thermal conductivity was seen in [24,25]. In the vehicle test described in this work, the oil with GNPs started the hot phase approximately 2 °C cooler. The impact on viscosity is negligible, but such an increase in thermal conductivity could be beneficial in terms of wear and on applications such as Electrical Vehicles [26] and rolling bearings [24,27].

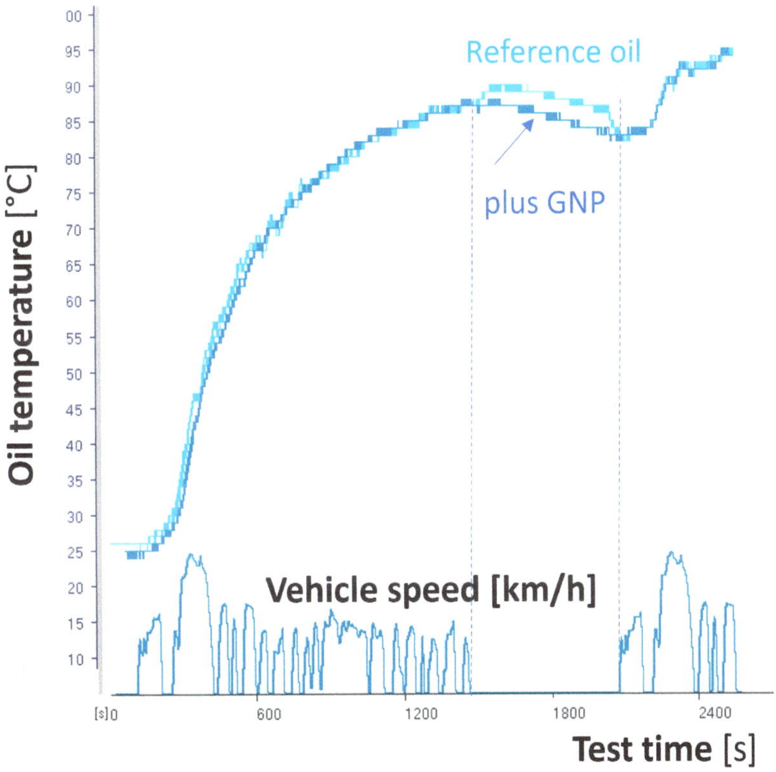

Figure 12. Oil temperature along the test.

5. Conclusions

The use of graphene nanoplatelets, with an average of nine layers, after functionalization to work as a lubricant additive reduced both the CoF and friction losses in a reciprocating test. Specifically, at the more severe test condition of 40 N and 120 °C, the L66_2 additive reduced the CoF and energy losses in 5% and 8%, respectively, in comparison with the reference oil, a fully formulated SAE 0W-20.

In vehicle emission tests, adding 0.1% w/w of GNPs on a fully formulated 5W-20 SAE oil reduced fuel consumption by 2.6% in the FTP-75 cycle and 0.8% in the highway one, resulting in 1.9% in the combined cycle.

The conducted work showed promising fuel savings for SI engines under vehicle emission tests. This work is part of a larger project where durability and additive degradation are key factors. Additive selection is especially important for Diesel engines, where MoDTC is not used because it causes clogging in the Diesel Particulate Filter (DPF). As with other friction modifier additives, the benefits of using GNPs are expected to be higher on oils with ultra-low viscosity.

Author Contributions: Conceptualization, E.T. and F.F.R.; investigation, W.C., P.S.G., D.F., R.M.S. and F.F.R.; resources, W.C., P.S.G., S.R., J.C., R.M.S. and F.F.R.; writing—original draft preparation, E.T. and F.F.R.; writing—review and editing, all authors. All authors have read and agreed to the published version of the manuscript.

Funding: Part of this work was funded by the Fundação de Desenvolvimento da Pesquisa—Fundep—Rota 2030 with the project 27192.24.

Institutional Review Board Statement: Not applicable.

Informed Consent Statement: Not applicable.

Data Availability Statement: The datasets presented in this article are not readily available because they contain company-proprietary information about the vehicle test and the GNP additive. In the paper, authors tried to share as many details as possible, including detailed characterization of the GNP and partial results from the emission tests. Requests to access the datasets should be directed to the correspondence author, Eduardo Tomanik.

Conflicts of Interest: Authors Eduardo Tomanik, Wania Christinelli and Pamela Sierra Garcia were employed by the Gerdau Graphene Brazil; Scott Rajala was employed by the Idemitsu Lubricants America Corp.; Jesuel Crepaldi was employed by the Idemitsu Lube South America Ltda; Fernando Fusco Rovai was employed by the VW do Brasil. The remaining authors declare that the research was conducted in the absence of any commercial or financial relationships that could be construed as a potential conflict of interest.

References

1. Tormos, B.; Pla, B.; Bastidas, S.; Ramírez, L.; Pérez, T. Fuel economy optimization from the interaction between engine oil and driving conditions. *Tribol. Int.* **2019**, *138*, 263–270. [CrossRef]
2. Taylor, R.; Morgan, N.; Mainwaring, R.; Davenport, T. How much mixed/boundary friction is there in an engine—And where is it? *Proc. Inst. Mech. Eng. Part J J. Eng. Tribol.* **2020**, *234*, 1563–1579. [CrossRef]
3. Kennedy, M.; Hoppe, S.; Esser, J. Weniger Reibleistung Durch Neue Kolbenringbeschichtung. *MTZ-Mot. Z.* **2014**, *75*, 48–51. [CrossRef]
4. Schommers, J.; Scheib, H.; Hartweg, M.; Bosler, A. Reibungsminimierung Bei Verbrennungsmotoren. *MTZ-Mot. Z.* **2013**, *74*, 566–573. [CrossRef]
5. Tamura, K.; Kasai, M.; Nakamura, Y.; Enomoto, T. Impact of Boundary Lubrication Performance of Engine Oils on Friction at Piston Ring-Cylinder Liner Interface. *SAE Int. J. Fuels Lubr.* **2014**, *7*, 875–881. [CrossRef]
6. Tamura, K.; Kasai, M.; Nakamura, Y.; Enomoto, T. Influence of Shear-Thinning of Polymer-Containing Engine Oils on Friction at the Piston Ring-Cylinder Liner Interface. In Proceedings of the SAE/KSAE 2013 International Powertrains, Fuels & Lubricants Meeting, Seoul, Republic of Korea, 21–23 October 2013; SAE International: Warrendale, PA, USA, 2013; Volume 11.

7. Michlberger, A.; Morgan, P.; Delbridge, E.E.; Gieselman, M.D.; Kocsis, M. Engine Oil Fuel Economy Testing—A Tale of Two Tests. *SAE Int. J. Fuels Lubr.* **2017**, *10*, 478–486. [CrossRef]
8. Holmberg, K.; Andersson, P.; Erdemir, A. Global Energy Consumption Due to Friction in Passenger Cars. *Tribol. Int.* **2012**, *47*, 221–234. [CrossRef]
9. Spikes, H. Friction modifier additives. *Tribol. Lett.* **2015**, *60*, 1–26. [CrossRef]
10. Wu, Y.Y.; Tsui, W.C.; Liu, T.C. Experimental analysis of tribological properties of lubricating oils with nanoparticle additives. *Wear* **2007**, *262*, 819–825. [CrossRef]
11. Gulzar, M.; Masjuki, H.H.; Kalam, M.A.; Varman, M.; Zulkifli, N.W.M.; Mufti, R.A.; Zahid, R. Tribological performance of nanoparticles as lubricating oil additives. *J. Nanopart. Res.* **2016**, *18*, 223. [CrossRef]
12. Shahnazar, S.; Bagheri, S.; Hamid, S. Enhancing lubricant properties by nanoparticle additives. *Int. J. Hydrogen Energy* **2016**, *41*, 3153–3170. [CrossRef]
13. Tomanik, E.; Christinelli, W.; Souza, R.M.; Oliveira, V.L.; Ferreira, F.; Zhmud, B. Review of Graphene-Based Materials for Tribological Engineering Applications. *Eng* **2023**, *4*, 2764–2811. [CrossRef]
14. Nyholm, N.; Espallargas, N. Functionalized carbon nanostructures as lubricant additives—A review. *Carbon* **2023**, *201*, 1200–1228. [CrossRef]
15. Gao, J.; Du, C.F.; Zhang, T.; Zhang, X.; Ye, Q.; Liu, S.; Liu, W. Dialkyl Dithiophosphate-Functionalized $Ti_3C_2T_x$ MXene Nanosheets as Effective Lubricant Additives for Antiwear and Friction Reduction. *ACS Appl. Nano Mater.* **2021**, *4*, 11080–11087. [CrossRef]
16. Boidi, G.; de Queiróz, J.C.F.; Profito, F.J.; Rosenkranz, A. $Ti_3C_2T_x$ MXene Nanosheets as Lubricant Additives to Lower Friction under High Loads, Sliding Ratios, and Elevated Temperatures. *ACS Appl. Nano Mater.* **2023**, *6*, 729–737. [CrossRef]
17. Singh, A.; Dwivedi, R.; Suhane, A. In the Context of Nano Lubrication, Do Nanoparticles Exhibit Favourable Impacts on All Tribo Surfaces? A Review. *Prot. Met. Phys. Chem. Surf.* **2022**, *58*, 325–338. [CrossRef]
18. Silva, D.L.; Campos, J.L.E.; Fernandes, T.F.; Rocha, J.N.; Machado, L.R.; Soares, E.M.; Miquita, D.R.; Miranda, H.; Rabelo, C.; Neto, O.P.V.; et al. Raman spectroscopy analysis of number of layers in mass-produced graphene flakes. *Carbon* **2020**, *161*, 181–189. [CrossRef]
19. *NBR7024*; NBR7024: Veículos Rodoviários Automotores Leves—Medição do Consumo de Combustível—Método de Ensaio. Associação Brasileira de Normas Técnicas (ABNT): Rio de Janeiro, Brazil, 2010.
20. Tomanik, E.; Miedviedieva, N.; Tomanik, V.; Miedviediev, B. Use of digital twins to analyze and predict CO_2 and emissions on Hybrid vehicles. In Proceedings of the International Scientific Conference Intelligent Transport Systems: Ecology, Safety, Quality, Comfort, Kiev, Ukraine, 26–27 November 2024.
21. Rovai, F.F.; Tomanik, E. Lubricant viscosity impact in fuel economy: Experimental uncertainties compensation. *Lubricants* **2025**. *to be published*.
22. Rustamov, I.; Xiang, L.; Xia, Y.; Peng, W. Tribological and mechanical endowments of polyoxymethylene by liquid-phase exfoliated graphene nanofiller. *Polym. Int.* **2024**. [CrossRef]
23. Alqahtani, B.; Hoziefa, W.; Abdel Moneam, H.M.; Hamoud, M.; Salunkhe, S.; Elshalakany, A.B.; Abdel-Mottaleb, M.; Davim, J.P. Tribological Performance and Rheological Properties of Engine Oil with Graphene Nano-Additives. *Lubricants* **2022**, *10*, 137. [CrossRef]
24. Ota, J.; Hait, S.; Sastry, M.; Ramakumar, S. Graphene dispersion in hydrocarbon medium and its application in lubricant technology. *RSC Adv.* **2015**, *5*, 53326. [CrossRef]
25. Contreras, E.M.C.; Oliveira, G.A.; Filho, E.P.B. Experimental analysis of the thermohydraulic performance of graphene and silver nanofluids in automotive cooling systems. *Int. J. Heat Mass Transf.* **2019**, *132*, 375–387. [CrossRef]
26. Canter, N. *Tribology and Lubrication for E-Mobility: Findings from the Inaugural STLE Conference on Electric Vehicles*; Society of Tribologists and Lubrication Engineers: Park Ridge, IL, USA, 2022.
27. Nassef, M.; Soliman, M.; Nassef, B.; Daha, M.; Nassef, G. Impact of Graphene Nano-Additives to Lithium Grease on the Dynamic and Tribological Behavior of Rolling Bearings. *Lubricants* **2022**, *10*, 29. [CrossRef]

eng

MDPI

Article

An Adaptive YOLO11 Framework for the Localisation, Tracking, and Imaging of Small Aerial Targets Using a Pan–Tilt–Zoom Camera Network

Ming Him Lui [1], Haixu Liu [1], Zhuochen Tang [1], Hang Yuan [2], David Williams [3], Dongjin Lee [4], K. C. Wong [1,*] and Zihao Wang [1]

1 School of Aerospace, Mechanical and Mechatronic Engineering, The University of Sydney, Sydney, NSW 2006, Australia; mlui0805@uni.sydney.edu.au (M.H.L.); hliu2490@uni.sydney.edu.au (H.L.); ztan6808@uni.sydney.edu.au (Z.T.); zihao.wang@sydney.edu.au (Z.W.)
2 School of Engineering, Australian National University, Canberra, ACT 2601, Australia; hyua0366@uni.sydney.edu.au
3 SiNAB Pty Ltd., Sydney, NSW 2229, Australia; dwilliams@sinab.com
4 Department of Unmanned Aircraft Systems, Hanseo University, Seosan 31963, Republic of Korea; djlee@hanseo.ac.kr
* Correspondence: kc.wong@sydney.edu.au

Abstract: This article presents a cost-effective camera network system that employs neural network-based object detection and stereo vision to assist a pan–tilt–zoom camera in imaging fast, erratically moving small aerial targets. Compared to traditional radar systems, this approach offers advantages in supporting real-time target differentiation and ease of deployment. Based on the principle of knowledge distillation, a novel data augmentation method is proposed to coordinate the latest open-source pre-trained large models in semantic segmentation, text generation, and image generation tasks to train a BicycleGAN for image enhancement. The resulting dataset is tested on various model structures and backbone sizes of two mainstream object detection frameworks, Ultralytics' YOLO and MMDetection. Additionally, the algorithm implements and compares two popular object trackers, Bot-SORT and ByteTrack. The experimental proof-of-concept deploys the YOLOv8n model, which achieves an average precision of 82.2% and an inference time of 0.6 ms. Alternatively, the YOLO11x model maximises average precision at 86.7% while maintaining an inference time of 9.3 ms without bottlenecking subsequent processes. Stereo vision achieves accuracy within a median error of 90 mm following a drone flying over 1 m/s in an 8 m × 4 m area of interest. Stable single-object tracking with the PTZ camera is successful at 15 fps with an accuracy of 92.58%.

Keywords: object detection; object tracking; data augmentation; Stable Diffusion; pan–tilt–zoom; camera calibration; YOLO11x; YOLOv8-Nano; Segment Anything Model; BicycleGAN

Citation: Lui, M.H.; Liu, H.; Tang, Z.; Yuan, H.; Williams, D.; Lee, D.; Wong, K.C.; Wang, Z. An Adaptive YOLO11 Framework for the Localisation, Tracking, and Imaging of Small Aerial Targets Using a Pan–Tilt–Zoom Camera Network. *Eng* **2024**, *5*, 3488–3516. https://doi.org/10.3390/eng5040182

Academic Editor: Antonio Gil Bravo

Received: 16 October 2024
Revised: 9 December 2024
Accepted: 10 December 2024
Published: 20 December 2024

1. Introduction

The widespread use of small drones has created opportunities for improving efficiency and technology advancements in many applications, such as precision agriculture, wildlife monitoring, and urban logistics. However, the booming drone usage also increases drone incidents and threats in both the commercial and defence sectors. Unauthorised drone flights pose a significant safety threat to the public as they may potentially collide with airliners and buildings. One of the most notorious recent examples is the 2018 incident at London's Gatwick Airport, where multiple unregistered drones resulted in the cancellation of approximately 800 flights [1]. In addition, large outdoor gatherings such as open-air concerts and carnivals attract unauthorised drone flights, which pose a serious threat to public safety. Yet, a suitable system that can detect and track these illegal drones is almost non-existent. This paper presents the development and experimental validation of a portable and field-deployable PTZ camera system that can detect, track and classify small aerial targets, with the object detection computer vision algorithm being a key component of the study. To demonstrate the versatility of the method, the cameras used in this study are unmodified commercial off-the-shelf products controlled by the original firmware.

Detecting and tracking aerial targets has always been highly challenging in object detection. The cost and accessibility appeal of unmanned aerial vehicles (UAVs), more commonly known as drones, will only raise the possibility of airspace obstacles and covert attacks. Wildlife also falls under this category of aerial targets, with frequent bird strikes below altitudes of 1000 m threatening aircraft take-off and approach [2]. In these cases, vision-based autonomous airspace monitoring holds advantages over existing radar and surveillance systems, which struggle to detect small targets.

Small targets may be classified as taking up less than 0.1% of the frame, adapted from the definition from the International Society for Optics and Photonics [3]. Fewer pixels—and, thus, a narrower feature space—create challenges when identifying small objects in varying environmental conditions. Preliminary studies have shown that targets such as birds were only a few pixels in size, far below the standard definition of small objects in typical detection tasks. The size, movement speed, and trajectory of aerial targets add complexity to environmental clutter and occlusions. Fast-moving objects become blurred or distorted in camera frames, making them difficult to identify compared to when stationary. Furthermore, an overlooked concern arises when annotating many small targets. While a significant burden, segments of objects are more prone to being excluded from their bounding boxes as they more easily blend into their environments. Existing datasets that are not sufficiently expansive may also encounter targets that cannot be identified. Finally, balancing model performance given fixed computational resources tends to be neglected in existing literature. Current object detection frameworks such as Ultralytics and MMDetection [4] support hundreds of detection algorithms, though they cannot assess algorithm suitability for system-specific deployment.

The proposed detection and tracking framework first investigates the effectiveness of two mainstream object detector frameworks: You Only Look Once (YOLO) and MMDetection. The former is known for its accurate real-time capabilities, while the latter provides modularity depending on the required task. Both are actively developed in the computer vision community. To construct a dataset, semi-supervised techniques assist in labelling small targets. Two dataset augmentation methods are proposed: one is based on the Segment Anything Model (SAM) [5] and Stable Diffusion [6] to replace image backgrounds; the other uses Llama 3 [7] and Stable Diffusion to generate image pairs for training generative adversarial networks (GANs) [8] to simulate environmental variations. Target ID tracking between frames is tested and implemented between two accessible and state-of-the-art object trackers: BoT-SORT and ByteTrack. While object detection effectiveness is determined on the mean average precision (mAP) and inference time, object tracking is independently measured based on a set of proposed metrics.

The complete framework is deployed into a pan–tilt–zoom (PTZ) camera network for detailed imaging of small aerial objects. The network comprises two low-cost cameras for target triangulation and a PTZ camera for high-definition visualisation and tracking. Such a system offers advantages over standalone PTZ trackers by enabling target localisation and depth perception over a significantly larger monitoring area. To ensure real-time capabilities, the YOLO model is deployed through TensorRT to dramatically decrease inference time. Following camera calibration, triangulation accuracy is validated using a motion capture system. The resulting PTZ frame balances between keeping the target within view and having it fill a notable proportion of the frame.

The primary contribution of this study is the development and experimental validation of a networked camera system for ground-to-air surveillance. The computer vision techniques used are specifically optimised for the detection and classification of small flying targets. Furthermore, the proposed PTZ camera network implementation offers the advantage of depth perception and, thus, target localisation over single PTZ camera tracking.

2. Related Work

2.1. Data Augmentation

Data augmentation is a common training strategy used to enhance the performance and robustness of object detection models. It is commonly integrated into existing image detection frameworks. Traditional image augmentation typically includes synchronously rotating and flipping images and bounding boxes, altering brightness and contrast, and randomly adding motion blur and noise.

Modern augmentation methods have appeared through CutMix [9] and Mixup [10], proposed in ICLR 2018 and ICCV 2019, respectively. These methods improve a model's generalisation ability by combining two images and their labels. YOLO adopts the mosaic augmentation method, which is an extension of CutMix. This process randomly crops multiple images to combine into a single image. From YOLOv4 [11], each version has improved mosaic augmentation up to YOLOv8 [12], which introduces negative samples of background frames into mosaic augmentation. This enhancement method significantly improves training efficiency and effectiveness.

Yanming et al. [13] proposed a particular data augmentation method against small targets. It divides the image into nine equal parts and randomly changes the scale of the targets in each part, replicating them within that section. The advantage of this approach is that it ensures the semantics of the augmented targets in the overall image remain unchanged. Thus, the augmented results become more suitable for backbone networks like Swin Transformer [14] and ViT [15] that tend to extract global features. However, these methods do not resolve interference from background pixels of irrelevant objects within the bounding boxes. This leads to poor model generalisation and an over-sensitivity to backgrounds during training.

With the introduction of GANs, pixel-level augmentation schemes for datasets have been proposed where generative tasks assist object detection tasks. However, training GANs often requires paired datasets or two sets of datasets with sufficiently different styles, which are almost nonexistent in aviation.

The advent of the Stable Diffusion model [6] allows for generating high-quality images based on prompts and original images. However, there are two issues with directly using Stable Diffusion in practice. First, the inference time is excessively long, leading to low augmentation efficiency. Second, some small targets may not be preserved during generation, or details may be severely distorted. Therefore, adapted from the methods in InstructPix2Pix [16], this article proposes a new augmentation scheme.

2.2. Object Detection

The fundamental goal of object detection tasks is to classify relevant targets in an image and to locate the dimensions of bounding boxes that most tightly enclose these targets. In practice, when the quality of dataset annotations meets the requirements of semantic segmentation tasks, object detection models can be extended to instance segmentation tasks. This is attributed to the principle of instance segmentation, which identifies the region where the instance is located and classifies the pixels within that region based on whether they belong to such an instance. Since it can more precisely distinguish between the target and the background, it is more accurate than general object detection models. However, inference speed is increased as this process requires pixel-level processing.

This article collectively defines object detection and instance segmentation as generalised object detection tasks, typically implemented using three independent sequential network structures: backbone, neck, and head. These are used, respectively, for feature extraction, feature fusion, and outputting results.

Currently, the mainstream object detection frameworks are Ultralytics and MMDetection. Ultralytics primarily supports algorithms based on one-stage object detection (without region proposal networks), covering the YOLO series and Real-Time Detection Transformer (RT-DETR) [17] algorithms. It allows developers to freely integrate their modules into

YAML files to improve existing YOLO models. Due to its flexibility, ease of deployment, and rapid inference, its primary users are from industry.

MMDetection integrates state-of-the-art (SOTA) object detection algorithms proposed at top computer vision conferences up to 2022. It supports users in integrating their developed backbone, neck, and head structures and combining them with existing structures. As it is highly suitable for comparative experiments with existing SOTA algorithms and routinely maintained by the Multimedia Laboratory (MMLab), its primary users are from academia.

Since its inception, the YOLO model has evolved to its 11th generation. The original YOLO simplified object detection into an end-to-end regression problem, utilising a single neural network to simultaneously predict bounding boxes and categories. Starting from the second generation, YOLO drew inspiration from another well-known single-stage object detection algorithm of the same era, SSD [18], incorporating prior anchor boxes and detection based on multi-scale feature maps. From the third generation onward, YOLO introduced the iconic CSPDarknet53 backbone network. Since then, each generation of YOLO has made improvements to its signature one-stage detection head and backbone network, reducing the number of parameters and inference latency.

The region-based convolutional neural network (R-CNN) series is the most iconic two-stage object detection framework. The original R-CNN [19] was proposed very early and used CNNs for feature extraction, SVMs for object classification, and linear regression for bounding box adjustment - typical model combinations for SOTA algorithms of that era. Subsequently, Fast R-CNN [20] and Faster R-CNN [21] gradually replaced all non-neural network components with neural networks that can be graphics processing unit (GPU)-accelerated. Mask R-CNN, proposed by Kaiming [22], extended the R-CNN series to instance segmentation tasks. Currently, the highest-scoring R-CNN algorithm in the MMDetection leaderboard is Cascade Mask R-CNN [23], an improved method proposed by Zhaowei and Nuno. It enhances the model's performance in detecting difficult (low Intersection over Union) targets by adding a cascade matching mechanism to the detection head.

In recent years, Transformer-based object detection algorithms in the DETR [24] series have been proposed. These solutions can better handle occlusion and densely packed objects in complex scenes through global features, which aligns well with scenarios such as detecting flocks of birds. However, DETR faces challenges such as difficulty in convergence, slow inference, and poor performance in small object detection. To address these issues, DINO [25], RT-DETR [17], and Deformable DETR [26] have been proposed. The improvements in the aforementioned algorithms focus on the Head component. Standard optimisations are usually based on the classic feature pyramid network (FPN), such as PAFPN [27] and BiFPN [28]. Additionally, the SOTA solution on MMDetection's leaderboard, DyHead [29], is an improvement based on the neck. It adaptively extracts features through dynamic convolution and enhances feature representation and fusion by introducing spatial, attention, and task attention. The process balances classification and regression tasks during training, thus achieving significant performance improvements.

The backbone network is the core of object detection tasks and is responsible for extracting features from images. It largely determines the algorithm's inference time and accuracy. Common backbone network baselines include Darknet from the YOLO series and ResNet proposed by Kaiming. With the advent of neural architecture search (NAS) technology, EfficientNet [30] has become the SOTA solution for lightweight convolutional backbone networks. With the application of Transformers in visual tasks, using Transformer-based SOTA feature extractors as backbone networks, such as Swin Transformer, have become common due to their ability to capture global features. Recent SOTA backbone networks, such as ConvNeXt [31], have combined the advantages of both CNNs and Transformers. It has been implemented on Cascade Mask R-CNN to achieve results surpassing ResNet [32], but lacks comparative experiments. This study aims to fill this gap in the literature.

2.3. Multi-Object Tracking

Both Ultralytics and MMDetection support the integration of object trackers into existing object detection models. In recent years, mainstream multi-object tracking (MOT) tasks mostly use tracking-by-detection methods based on Kalman filter motion state estimation. End-to-end methods have also been explored to integrate detection and tracking into a single network.

The most classic MOT algorithm is the simple, online, and real-time (SORT) [33] algorithm based on Kalman filters and the Hungarian matching algorithm. The DeepSORT [34] algorithm builds upon this by adding a Re-ID component based on CNNs to extract target features and calculate similarity, assisting in matching trajectories with low reliability. The ByteTrack [35] algorithm proposed in 2022, as an improvement over SORT, does not use Re-ID to enhance computational efficiency. Instead, it re-matches low-confidence detections with the trajectories of high-confidence targets that were mismatched in previous frames, thereby reducing the mismatch rate of small and occluded targets and improving trajectory continuity. In the same year, the BoT-SORT [36] algorithm was proposed to improve the ByteTrack model. It uses an enhanced Kalman filter and camera motion compensation while reintegrating Re-ID to assist in trajectory matching.

2.4. PTZ Tracking Applications

Most literature on object tracking using a PTZ camera integrates object detection within itself rather than supplementary cameras. This method relies on the target remaining visible over consecutive frames to apply the relevant PTZ commands. This imposes a limitation on smaller targets that move quickly. Tracking aerial targets such as first-person view drones thus becomes difficult, balancing between observing the target with sufficient zoom at the risk of losing sight.

PTZ tracking is more popular in fixed surveillance systems, simplifying object detection processes due to greater background consistency. Kang et al. [37] leveraged this observation using the concept of background subtraction. While the PTZ camera remains at a single view, a corresponding background frame is generated on observations over 20–40 s. Their results were some of the earliest examples showing the viability of real-time PTZ tracking. Caterina et al. [38] provided a benchmark of comparison for real-time capability by achieving an execution time of under 38ms with an internet protocol (IP) PTZ camera.

Unlu et al. [39] presented the closest literature to this study, exploring a UAV-based PTZ tracking system based on deep learning techniques. Their work involved K-nearest neighbours' background subtraction to segment moving objects and validate the presence of a UAV using a modified ResNet identifier. This object detector was trained on a 55:45 ratio using a mix of open-source datasets and in-house images versus negatives. The UAV target was successfully tracked for 71.2% and 60.8% of frames over indoor and outdoor locations separately, estimating the 3D position to be within 0.67 m of the ground truth within a $2 \times 2 \times 2$ m space.

2.5. Camera Modelling and Calibration

Conventional optical cameras are modelled through the pinhole projection model, which relates 3D world coordinates to its projected point on a camera frame. This model requires knowledge of characteristics including its intrinsic parameters, extrinsic parameters and distortion coefficients. Such values are estimated through camera calibration. This process has become more accessible following the input of calibration patterns of varying orientations, introduced by Zhang [40]. Camera calibration is currently freely available via MATLAB's Camera Calibration Toolbox or OpenCV's calibrateCamera.

With PTZ cameras, the calibration approach remains largely similar. Varying zoom changes lens properties and, thus, the camera's intrinsic parameters. Hence, the most conventional method for PTZ camera calibration is calibrating at discrete zoom values. Calibration results from Sinha and Pollefeys [41] have shown multiple possible trends between PTZ camera models. Focal length is expected to increase with a linear or parabolic relationship to zoom level, while the principle point deviates more severely from the frame centre with increasing zoom. Distortion effects are less critical for this study as the PTZ camera will only be used for visual identification. However, it is noted that radial distortion effects are less prominent with increasing zoom.

Additional errors in a PTZ camera may arise from mechanical variation depending on manufacturing quality. Wu and Radke [42] proposed a method to determine such errors by capturing a calibration pattern before and after a pan–tilt command. A homography matrix may be calculated to reflect the true amount of rotation. The results showed that angular error increased linearly as the camera rotated further from its directional origin, as expected of the stepper motor used in the construction. Other errors that arose included random, accumulated, and power-cycling errors. The same azimuth and elevation command was observed to have varied by 5–7 pixels after 30 min of movement, increasing to 38 pixels after 200 h.

2.6. Triangulation

Triangulation describes extracting 3D information based on 2D information from two or more scene images. Apart from stereo vision, it is also commonly used for tasks such as structure from motion and simultaneous localisation and mapping. After detecting a target's bounding box, a line of sight (LOS) may be determined from the camera's optical centre through the bounding box centroid. These LOS rays ideally intersect a single point representing the object's 3D world coordinates. In reality, these lines are always skew due to measurement noise, uncertainties of intrinsic or extrinsic parameters, and subpixel 2D inaccuracies [43].

How optimal a triangulation algorithm is deemed is most widely determined by how well it minimises reprojection error [44], known as 2D error. This describes how well the calculated 3D point projects back onto the observed 2D points of each frame. A 3D error is also a valuable metric, taken from a comparison to the true 3D position in world space. Additionally, the robustness of an algorithm is determined by its invariance to affine transformations (scaling, rotation, etc.), projective transformations (changes in perspective), and Euclidean (position and orientation) reconstruction.

An elementary form of triangulation is the midpoint method, which takes the midpoint where two rays approach each other the closest. Hartley and Sturm [45] criticise this method for lacking affine and projective invariance despite the ease of computation. Lee and Civera [44] proposed a generalised weighted midpoint method that extracts the depths from each frame to a single 3D point, locates a point on each ray of the corresponding depth, and computes a weighted average of such points. More recently, Nasiri et al. [43] claimed that the midpoint method was advantageous over alternative methods due to having less sensitivity to error when there is uncertainty in extrinsic camera parameters.

Another widely used form of triangulation is Direct Linear Transformation (DLT). With a known camera projection matrix $\mathbf{P}$ and set of homogenous frame coordinates $\vec{\mathbf{x}}$, the corresponding homogenous 3D coordinates $\vec{\mathbf{X}}$ may be estimated with (x_W, y_W, z_W) representing the world frame. This system may be solved through a least-squares solution using pseudo-inverses, singular value decomposition, or the solution corresponding to the smallest eigenvector of the matrix $\mathbf{A}^T\mathbf{A}$ [45]. Triangulation via DLT is used by OpenCV's triangulatePoints.

Additionally, Hartley and Sturm [45] proposed a notable alternative to the midpoint and DLT methods by solving a sixth-degree polynomial to minimise a chosen cost function. While computational time is of a higher relative order to alternatives, it was shown to have affine and projective invariance.

3. Methodology

3.1. Dataset Collection and Annotation

This study obtained 49 videos containing birds, drones, aircraft, ships, and unidentified flying objects through on-site filming. Five of these videos were reserved for a test set, with the remaining videos used as the training set. From the training videos, one frame out of every five frames was extracted for annotation.

Subsequently, manual annotations on 18,205 images, including images without targets, were performed using LabelImg in the VOC format. Some birds occupied too few pixels in the images. However, birds usually appear in groups. In such cases, movement patterns of dense bird groups were observed in adjacent frames and annotated as several bird flocks.

Finally, a semi-supervised annotation tool was trained using YOLOv7x to label the remaining images, with manual checks and adjustments conducted on these annotations. The manual and semi-supervised annotations amounted to 199,141 labels. Images without targets were deleted, resulting in a total of 21,217 images. The cleaned dataset was divided into an 80:20 split for training and validation, then annotations were converted into COCO and YOLO formats using X-Anylabeling.

3.2. Dataset Augmentation

This article proposes a novel data augmentation method to overcome the over-sensitivity of object detection models to background pixels within bounding boxes. It relies on the coordinated work among the latest open-source pre-trained large models in segmentation, text generation, and image generation tasks.

First, there must be considerations towards segmenting the target from the original image background rather than cropping the entire bounding box. This article leverages SAM to accomplish this task. Compared to traditional segmentation models, SAM has the advantage of accepting bounding boxes and coordinate points as input prompts, accurately extracting the masks of targets within the bounding boxes or associated with the coordinate points. Based on this feature, an algorithm filters bounding boxes of birds, aircraft and drones occupying more than 30 pixels. These bounding boxes are passed to SAM to complete instance segmentation and obtain object masks.

Next, a method is required to generate backgrounds appropriate to deployment environments. Limitations arise from directly using existing background image libraries or generating images via text prompts using Stable Diffusion, as the scene composition is not guaranteed to remain consistent between the original and augmented background. For aerial object detection, unrealistic scenes can be generated, such as an aircraft in front of a forest background. These combinations stray the model away from real-world conditions while increasing the training burden, thus cannot effectively improve the model's generalisation ability.

This study's dataset includes environments of farmland, forests, oceans, courtyards, and indoor laboratories. Based on detailed text prompts and ControlNet, which can use the original image as a control condition, this article refines and guides the generation process of Stable Diffusion. This process generates frames with new scenes, such as parks, roads, cities, hills, lakes, warehouses, ports, and beaches. The targets segmented by SAM are randomly copied within preset pixel coordinate ranges to obtain new images with enhanced backgrounds, demonstrated in Figure 1. Simultaneously, this process records the coordinates of the new target bounding boxes to generate new labels.

(**a**) (**b**) (**c**)

Figure 1. SAM and Stable Diffusion Augmentation Process. (**a**) Original Frame. (**b**) Object Mask. (**c**) Augmented Frame.

A procedure was adopted to ensure bounding box locations were consistent when converted into augmented images. First, the segmentation mask of the object was extracted, where a value of 1 represents the segmented pixels of the target and a value of 0 represents the background. Element-wise multiplication of the mask with the original image was performed to isolate the segmented target, resulting in an image where all pixels outside the target are set to 0. Subsequently, the new background, resized to match the dimensions of the original image, was multiplied element-wise with the binary inverse of the segmentation mask, effectively removing the regions corresponding to the embedded object. Finally, the isolated target and the background image with the target region removed were combined through element-wise addition to generate the final augmented image.

Apart from background sensitivity, aerial object detectors are also highly sensitivity to lighting conditions affected by the time of day, weather, and seasons. Hence, the following process will also enhance images of the same background under these conditions while preserving existing details. Therefore, the computational cost and volatility of Stable Diffusion are too high. GANs can overcome these two issues. This step is motivated by the CVPR highlight article "InstructPix2Pix" to obtain a high-quality training set for the GAN. Suitable images from the LAION-Aesthetics dataset are filtered based on captions containing words representing seasons, weather, day and night, and outdoor scenes.

Next, using Llama 3 and prompt templates, the captions are batch-modified into Stable Diffusion prompts. For example, if the original image's caption is "a snowy lake in winter", the generated prompt is "Please strictly base the generated image on the scene provided in my image and generate a lake in sunny winter". The generated prompts and filtered original images are fed into a Stable Diffusion model guided by ControlNet to obtain a large-scale, high-quality set of paired images. Large-scale generation is conducted on 16 NVIDIA cloud GPUs operating for over 168 h. The paired image dataset trains BicycleGAN [46], which augments these images while retaining their core features. If particular image pairs have notably different details, CycleGAN [47] is trained instead. The GAN weights are then used to obtain augmented images based on the required changes in time of day, weather and season, such as those in Figure 2.

(a) (b) (c)

Figure 2. LLM and Stable Diffusion augmentation processes. (**a**) Hilly terrain—raining. (**b**) Hilly terrain—evening. (**c**) Hilly terrain—winter.

This approach consumes a large amount of time in the data preparation stage for training the GANs. However, it achieves a once-and-for-all solution. Once the GANs are trained, images can be rendered quickly without repeatedly recalling the Stable Diffusion model. The efficiency of these large models is improved through inference acceleration methods provided by Hugging Face. The lightweight version of SAM, MobileSAM, was evaluated to be less precise in segmentation performance at the benefit of efficiency. For Stable Diffusion, optimised samplers were utilised to reduce the number of sampling steps. Model deployment techniques were also used to accelerate the inference time of Stable Diffusion and Llama 3, including mixed precision (FP16) to load pre-trained models and multi-GPU pipeline parallelism.

3.3. Object Detection

The presented framework supports both the Ultralytics and MMDetection libraries. The Ultralytics library was further developed to integrate enhancement modules open-sourced by YOLO-related developer communities and ensure compatibility with YOLOv7, the only version from 5 to 11 omitted from the native library. Most importantly, this modification allows the framework to fetch pre-trained models from the Hugging Face timm library as backbone networks.

This study selects mAP50 and mAP50-95 as the primary evaluation metrics while also considering floating point operations per second (FLOPS), parameter count, and inference time to ensure that the model size and computational requirements remain within reasonable limits for real-time inference. Confusion matrices are generated to determine class-specific detection capabilities.

Hyperparameter selection is a further consideration. Based on the champion of the 2023 ICCV UAV Detection Challenge, increasing the input size significantly enhances the performance of detection models. Therefore, the input size was set to 1080p for all models. Additionally, the number of epochs is set to 30 as most models converge at this value during preliminary experiments. Other parameters, such as training strategies and augmentation methods, use the models' default settings.

Initially, experiments are conducted on 38 models from the Ultralytics library, encompassing various network architectures from the YOLOv5 to YOLO11 generations. Subsequently, the following algorithms are chosen from the MMDetection framework: the SOTA DyHead algorithm, the best-performing Cascade Mask R-CNN from the R-CNN series, and the highest-scoring Transformer-based representative algorithm Deformable DETR. These algorithms are paired with backbone networks, including ResNet, Swin Transformer, ConvNeXt, and EfficientNet, resulting in 12 comparative experiments. An example of a standard network architecture layout is visualised in Figure 3, showing that of YOLO11x.

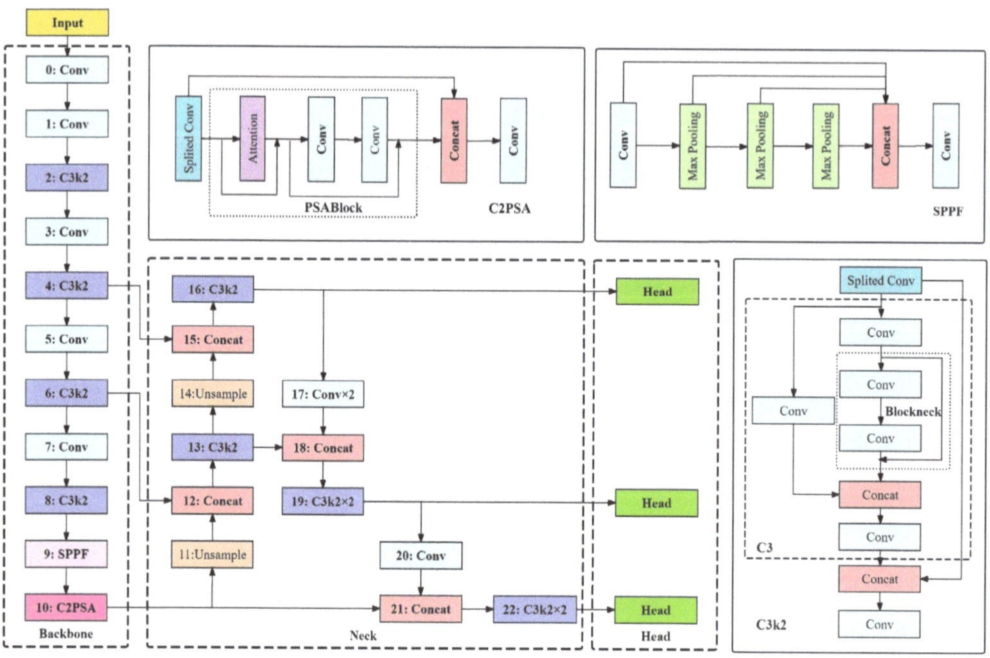

Figure 3. YOLO11x network architecture diagram.

3.4. Object Tracking

The chosen object tracking methods employed the representative tracking-by-detection algorithms ByteTrack and BoT-SORT. They determine whether the targets identified in consecutive frames belong to the same object and assign IDs accordingly. BoT-SORT is derived from ByteTrack, though the former possesses more complex hyperparameters and distance calculation logic than the latter. The robustness of these algorithms is evaluated by running them on test videos containing only a single drone. The drone performs steady flight at constant speeds, then flight with complex trajectories at varying speeds.

Ultralytics does not provide evaluation methods to determine MOT metrics. Therefore, assuming that only one target in the video exists, an appropriate set of metrics is proposed, as follows:

1. **Number of tracks (NOT):** Since it is known in advance that only one target appears in the video from start to finish, the optimal result is that all detected targets are assigned a single track ID. Therefore, a larger number of track IDs corresponds to lesser model stability. It is noted that track IDs are not sequential in ByteTrack. When the target temporarily mismatches, the algorithm assumes a new target has appeared and assigns a new track ID. Once these targets are successfully re-matched with the previous target, the previous track ID is reused, and the newly assigned ID is removed. Therefore, track IDs are not equivalent to the number of tracks.
2. **Tracking length (TL)**: The number of consecutive frames in which the algorithm can continuously track the target in the longest identified trajectory in the video.
3. **Average tracking length (ATL)**: The mean tracking length of all trajectories. A higher value indicates a more robust algorithm.
4. **Matching rate (MR)**: The percentage of frames where targets are assigned IDs to all frames where targets are detected. A higher value indicates a more robust algorithm.
5. **Long-term matching rate (LTMR)**: The percentage of the total number of frames from all trajectories with tracking lengths exceeding a set threshold to all frames where targets are detected. A higher value indicates a more robust algorithm.

3.5. Camera Calibration

Camera calibration will be conducted using MATLAB 2024a's stereo camera calibrator. Intrinsic parameters are calibrated by placing two cameras sufficiently close so that a checkerboard pattern fills up most of both frames. Outlier frame pairs are removed until the reprojection error is sufficiently low. Schmalz et al. suggested aiming for a reprojection error of less than 0.3 for "day-to-day calibrations" [48]. The calculated intrinsic parameters are then fixed when calculating extrinsic parameters, as it becomes much more difficult to determine distortion effects, with the calibration pattern taking up much less of the screen. A comparison of pattern distances to cameras is visualised in Figure 4.

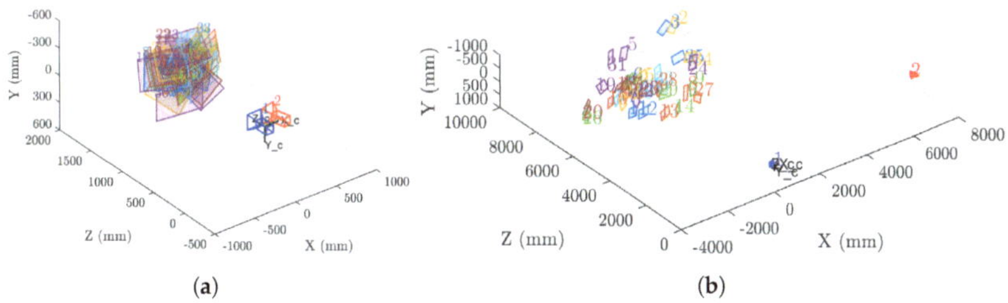

Figure 4. Calibration pattern visualisation. (**a**) Intrinsic parameters. (**b**) Extrinsic parameters.

MATLAB's single-camera stereo calibrator is then used to determine the intrinsic parameters of the PTZ camera. Frames are taken at appropriate increments of zoom values, with smaller increments taken closer to 100% zoom due to the dramatic increase in focal length from initial observations. The scale factor (S) of a frame will then vary with respect to the corresponding focal length (f) and the focal length at 0% zoom (f_0):

$$S = \frac{f}{f_0} \tag{1}$$

A relationship between the zoom (Z) and focal length will be curve-fitted depending on the appropriate shape as a function $f = g(Z)$.

3.6. Stereo Triangulation

Object detector models are deployed on two optical cameras facing an area of interest. As variations in the target pose are unknown, the bounding box centroid is extracted from each frame for triangulation. Following results from Nasiri et al. on the midpoint method's advantages under uncertainty of intrinsic parameters, this method was chosen to conduct indoor testing and validation [43]. The deployment pipeline is largely conducted through Python 3.11. Firstly, lens distortion is corrected through OpenCV's undistortPointswhich iterates through normalised distorted pixels to reach the corresponding undistorted pixels.

When introducing equations for the midpoint method, the notation is equivalent to that used in Section 2.6. First, the unit vector of each line of sight from frame coordinates (u_j, v_j) to world coordinates $\mathbf{v}_j$ are determined, derived from the pinhole projection model on the concept of similar triangles:

$$\mathbf{v}_j = \mathbf{R}^T \left[\frac{u_j - u_0}{f_x}, \frac{v_j - v_0}{f_y}, 1 \right]^T \quad \text{for } j = 1, 2 \tag{2}$$

The shortest distance between these skew rays is determined, corresponding to the distance between two points on each ray $\mathbf{p}_1, \mathbf{p}_2$. The distances from the optical centre to each of these points d_1, d_2 are such, with $\mathbf{R}_j, \mathbf{t}_j$ representing the corresponding rotational and translational matrices of Camera j:

$$d_1 = \frac{((\mathbf{R}_1^T \mathbf{t}_1 - \mathbf{R}_2^T \mathbf{t}_2) \times \mathbf{v}_2) \cdot (\mathbf{v}_1 \times \mathbf{v}_2)}{|\mathbf{v}_1 \times \mathbf{v}_2|^2} \tag{3}$$

$$d_2 = \frac{((\mathbf{R}_2^T \mathbf{t}_2 - \mathbf{R}_1^T \mathbf{t}_1) \times \mathbf{v}_1) \cdot (\mathbf{v}_2 \times \mathbf{v}_1)}{|\mathbf{v}_2 \times \mathbf{v}_1|^2} \tag{4}$$

The coordinates of the calculated point on each ray are such:

$$\mathbf{p}_j = -\mathbf{R}_j^T \mathbf{t}_j + d_j \cdot \mathbf{v}_j \quad \text{for } j = 1, 2 \tag{5}$$

From this step, the midpoint method takes the average of both points. When multiple objects are detected in the frame, a distance threshold between points is imposed. The calculated midpoint for a certain frame is only passed if the two predicted points are under such a threshold. These equations are extendable to take the point averages across n cameras.

3.7. Coordinate Transformation

The direction vector from the PTZ camera position to the target may be converted to azimuth θ, elevation ϕ, and distance $|\vec{v}|$:

$$\theta = \text{atan2}(y, x) \tag{6}$$

$$\phi = \tan^{-1}\left(\frac{z}{\sqrt{x^2 + y^2}}\right) \tag{7}$$

$$|\vec{v}| = \sqrt{x^2 + y^2 + z^2} \tag{8}$$

The pan and tilt commands correspond to azimuth and elevation, respectively. To determine an appropriate zoom command, an object's size is assumed to be inversely proportional to its distance from the camera. This consideration is used to determine the required scale factor S for the frame. For Camera j, we have the following:

$$S_j = \frac{|\vec{v}|}{\text{Cam to Point Dist.}} \times \min\left(\frac{\text{Frame Width}}{\text{BBox Width}}, \frac{\text{Frame Height}}{\text{BBox Height}}\right) \quad (9)$$

The minimum value between S_1 and S_2 is taken after calculating for instances of each camera to ensure the whole target is within the frame. A zoom undershoot factor $0 < UF < 1$ is also applied to account for PTZ camera errors addressed by Wu and Radke [42] and display a complete view of the surrounding environment:

$$S = UF \times \min(S_1, S_2) \quad (10)$$

3.8. Model Deployment and Experimental Setup

The chosen YOLO models are trained in Python to output a weighted format. This format generally has a longer inference time and becomes less advantageous for real-time deployment. TensorRT is widely accepted to most effectively minimise inference speed while sacrificing negligible mAP results, while having been tested with YOLOv5 and announced on the Ultralytics forum.

Stereo vision is conducted over an area of interest measuring 8 m × 4 m, visualised in Figures 5 and 6. The experimental hardware is listed in Table 1. Before conducting extrinsic calibration, the webcams are positioned at appropriately high parallax to minimise triangulation error. Notably, tracking performance is independent of any obstructions to the PTZ camera, such as the glass wall. Such a configuration is only reliant on the visibility of the fixed cameras.

Figure 5. Experimental setup.

Inferences are conducted on a Jetson TX2 for portability, which returns results to a Ryzen 9 5900HX laptop for localisation and PTZ control. If real-time capabilities cannot be achieved, inference shall be conducted solely on the laptop, which supports TensorRT through a mobile RTX3050 GPU. The motion capture system is positioned overhead. The asynchronous framework in Python will allow the task of dual object detection and localisation to occur simultaneously with the task of sending PTZ inputs through cURL commands.

Table 1. Experimental Hardware.

Component	Manufacturer	Details	Manufacturer Country
GPU	NVIDIA	Geforce RTX3050	Santa Clara, CA, USA
CPU	AMD	Ryzen 9 5900HX	Santa Clara, CA, USA
Webcam	Logitech	C922	Lausanne, Switzerland
PTZ Camera	FLIR	M300C	Washington, DC, USA
Embedded System Module	NVIDIA	Jetson TX2	Santa Clara, CA, USA

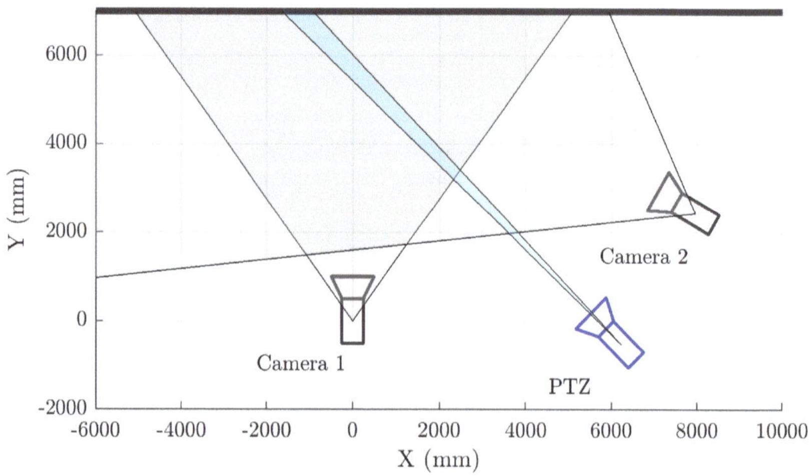

Figure 6. Top view—environment visualisation.

A motion capture system is used to validate the accuracy of triangulation and determine the 3D error. The effectiveness of the PTZ camera in tracking a moving aerial target is determined by its computational efficiency and the frequency with which it maintains the target within the frame. Hence, PTZ tracking performance is evaluated as follows:

- **E2E time**: Median end-to-end processing time from the webcam frame capture to PTZ command.
- **Success rate:** Percentage of frames where the target is completely in the PTZ frame.

4. Results and Discussion

4.1. Detection Performance Metrics

Appendix A Table A1 provides model comparisons over the YOLO, cascade R-CNN, DyHead, and deformable DETR algorithms. The accuracy rankings of the models implemented using MMDetection align with their corresponding official accuracy rankings. However, the best model configuration, DyHead combined with Swin Transformer, only achieved performance comparable to the early lightweight YOLO models. Notably, MMDetection models typically converge faster than the YOLO series, which may be due to differences in training strategies. A significant difference in inference time is noted between the YOLO and MMDetection series. YOLO indeed has an advantage in inference time, though this is likely related to the implementation of the frameworks. The Ultralytics framework, geared towards industrial applications, likely incorporates more inference optimisations.

Most of the MMDetection models are from work published two years ago, so it is reasonable that they have been surpassed in performance by the more recent YOLO versions. Furthermore, most models exhibited abnormal convergence when using EfficientNet as the backbone, so it was not included in the comparison.

Due to its notable performance advantages, focus shall be placed on comparing the YOLO version models. A performance comparison between YOLO model formats from v5 to v11 is shown in Figures 7 and 8, with extensive metrics in Appendix A Table A1. The best-performing model based on test set precision is YOLO11x at a mAP50 of 86.7%, with a mAP50-95 of 60.9%. However, the YOLOv8n model, among the lightweight variants, holds the highest mAP50 at 82.2% while maintaining the advantage of the lowest inference time at 0.6 ms. Older versions of v5 and v7 show abnormal results that do not follow the trend where the mAP and inference time increase with larger models. YOLOv7 likely lacks streamlined integration as Ultralytics do not natively support it. The larger YOLOv6 models likely overfitted during training to cause lower mAP even with more parameters. At IoU thresholds of 50–95%, overfitting is also observed with YOLOv6, though YOLOv7 holds results that are more consistent with expected trends. The mAP degrades from v8 to v10 with smaller models, suggesting unsuitability in model architecture or hyperparameters to the existing dataset.

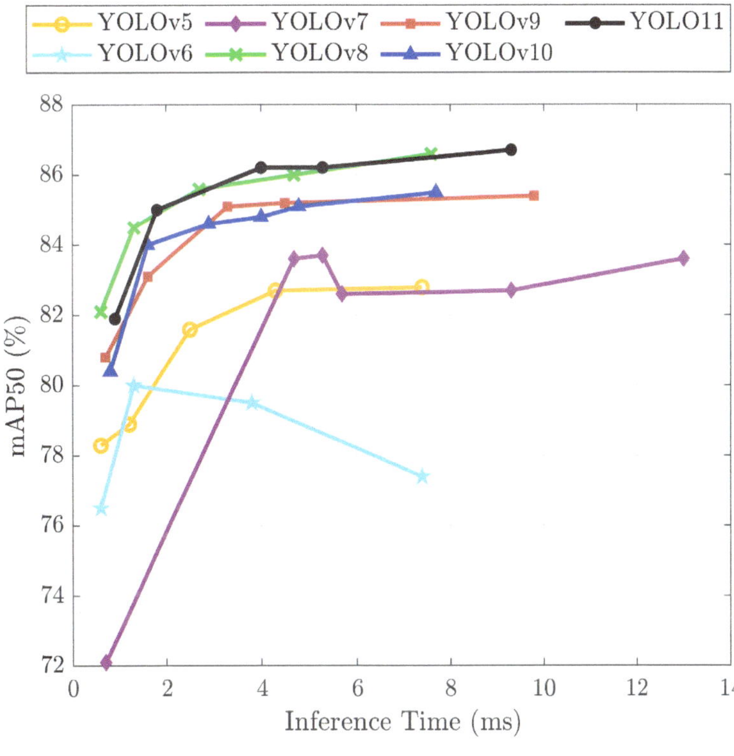

Figure 7. mAP50 vs. inference time.

Figure 9 presents performance variation of key metrics during the training process to demonstrate stability. Notably, there was a significant spike in training loss around the 20th epoch due to the training process disabling YOLO's built-in mosaic augmentation. While the training distribution focal loss (DFL) increased, the validation metrics did not deteriorate, validating the rationale behind this proposed training strategy. Precision plateaus at 30 epochs while recall, and thus mAP, retains a positive gradient. There is potential for improvement in results by increasing the number of epochs, though it comes at the risk of overfitting to the test set.

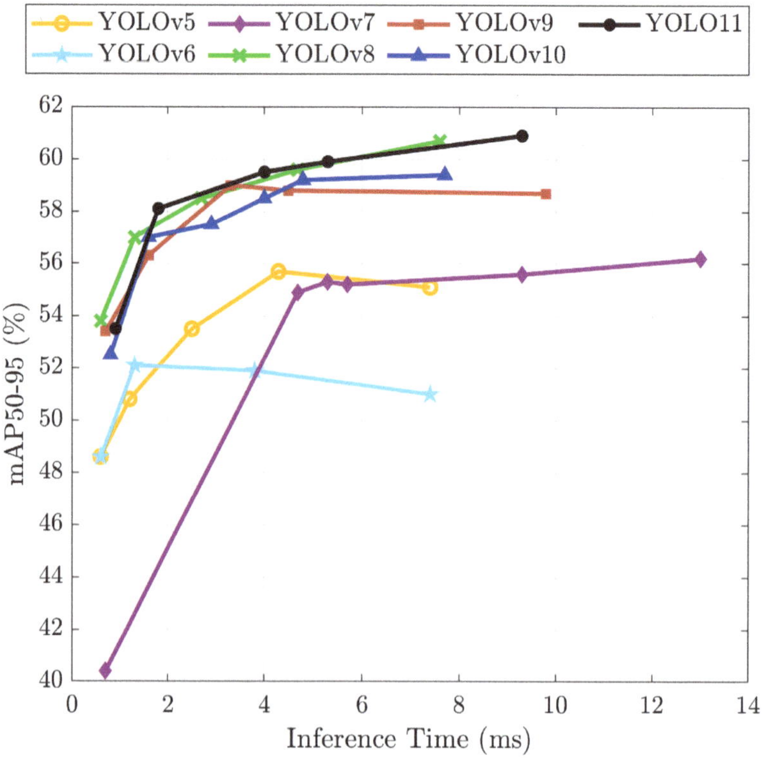

Figure 8. mAP50-95 vs. inference time.

The confusion matrices for the highlighted YOLOv8n and YOLO11x models are shown in Figure 10. Both models achieved high detection accuracy for the critical targets of drones, aircraft, and flocks of birds. YOLO11x showed notable improvements in detecting larger objects of aircraft and ships, as more parameters were available for feature extraction. Both models struggled with the classes of birds and bird flocks, commonly declaring false positives to background objects. The contrary of this observation was also true, with birds often failing to be detected. Annotations on the bird class likely lacked distinctive features, such as being predominantly composed of one colour for cases of crows and ravens at a distance. The proposed data augmentation strategy also could not be applied to objects under 30 pixels due to the limitations of SAM. Class-based F1 score metrics are provided in Appendix A Figure A1, to visualise the balance of precision and recall with varying confidence.

Table 2 compares BoT-SORT and ByteTrack through the proposed object tracking metrics. ByteTrack holds higher matching and long-term matching rates, showing higher confidence in linking an existing track to the corresponding ID. However, the higher NOT suggests that ByteTrack is more susceptible to hastily assigning new IDs when a track disappears and is thus less stable. A higher average tracking length also validates that BoT-SORT holds higher stability than ByteTrack and will thus be chosen for subsequent experiments. However, both algorithms are seamlessly interchangeable within Ultralytics, and ByteTrack can be substituted if experiments miss too many existing IDs.

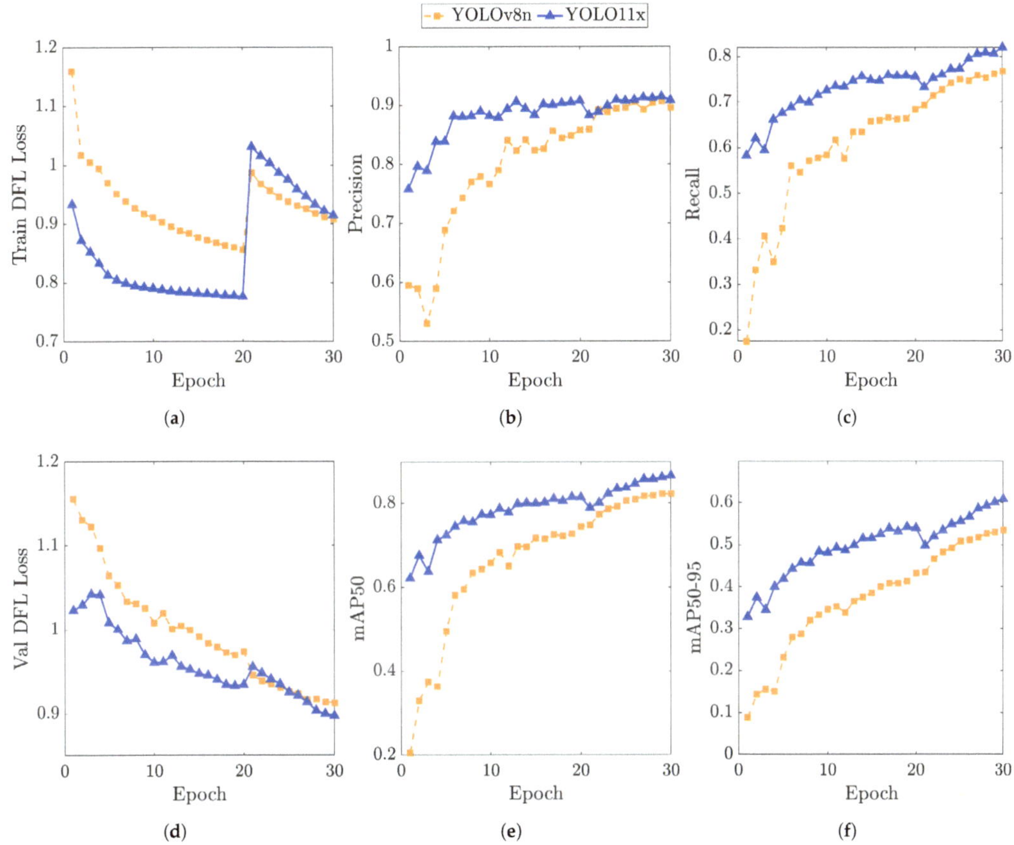

Figure 9. YOLO training metrics for 30 epochs. (**a**) Train DFL. (**b**) Precision. (**c**) Recall. (**d**) Val DFL. (**e**) mAP50. (**f**) mAP50-95.

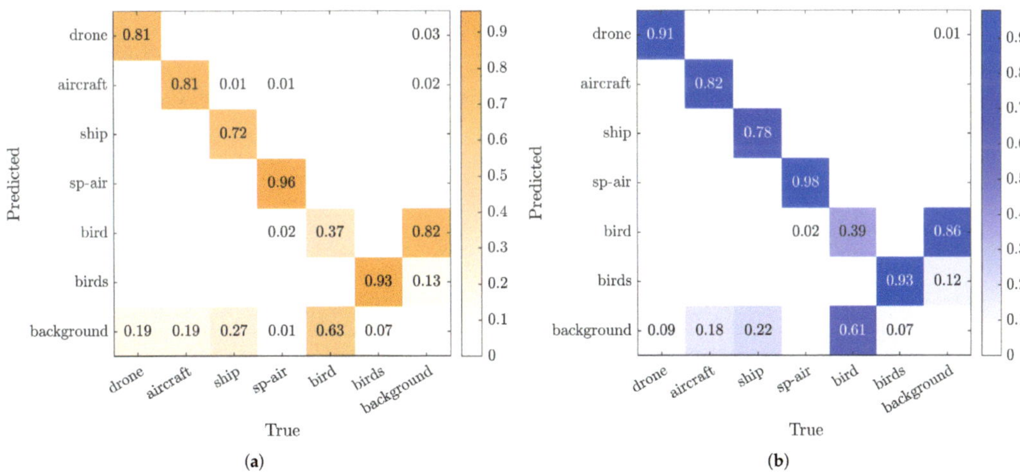

Figure 10. YOLO Confusion matrices on test datasets. (**a**) YOLOv8n. (**b**) YOLO11x.

Table 2. Object tracking metrics.

Tracker	NOT	TL	ATL	MR (%)	LTMR (%)
BoT-SORT	45	262	79.91	87.79	67.92
ByteTrack	51	276	72.37	96.64	68.48

Deployment with the TensorRT model is tested on a Jetson TX2 due to its small form factor, achieving an inference time of 105.08 ms at 9.38 frames per second (FPS). Applied to two cameras, the FPS will halve. While there will be difficulty in implementing real-time localisation and tracking on this device, the feasibility of this pipeline is proven. Improvements may be made through network pruning, further inference optimisation, and switching to higher-performing devices. Hence, deployment is conducted on a laptop (the specifications are proposed in Section 3.8).

4.2. Stereo Validation

Outlier image pairs of high mean error were filtered until the overall mean error reached around 1 pixel or less. Image pairs and overall mean reprojection error are shown in Table 3 with individual errors shown in Appendix A Figure A2. Image pairs with higher mean errors, such as pairs 3 and 14, had their calibration planes at a greater angle to the camera field of view. Further uncertainties arise from calibrating extrinsic parameters using an A1-sized checkerboard pattern, as the grid corners are no longer in focus at greater distances. Without fixing the intrinsic parameters before extrinsic calibration, earlier calibration experiments have also shown overfitting on a local cluster of patterns. In these cases, the mean reprojection error becomes misleadingly low. Overfitting can be noticed by observing discrepancies between the estimated camera extrinsic characteristics and pattern locations.

Table 3. Calibration metrics.

	Number of Image Pairs	Overall Mean Error (px)
Intrinsic	33	0.40
Extrinsic	35	0.10

Intrinsic parameters were generated with 33 image pairs to a mean reprojection error of 0.40 pixels. These parameters were set as defaults when calculating extrinsic parameters. A mean reprojection error of 0.10 pixels is achieved, validating against overfitting.

Triangulation 3D error was verified through controlled drone flying experiments in fixed and randomised paths. Three flight tests were conducted for each scenario to be compared with the motion capture system. Figures 11–14 provide a visualisation of the travelled paths and 1D results between the ground truth and predicted coordinates. Pre-processing was conducted to correct any constant offsets in either axes. The results for the highest drone target velocity and, thus, the highest median Cartesian error are shown. Additional results are shown in Appendix A Figures A3–A6.

Predictions do not return a positive result at around 10 s, consistent across all three experiments. At this position, the drone is furthest from both cameras and appears much smaller in frame. Hence, the detection framework may struggle more to return positive detections. The prediction error is most significant when approaching closest to Camera 1, around 25 to 30 s. Camera calibration accuracy may become less reliable as the target strays from the cameras' working distances. In this case, the object is so close to Camera 1 that it almost leaves the field of view of Camera 2. Both observations expose the shortcomings of the asymmetrical camera positioning, which are limited by device connection distances and laboratory space.

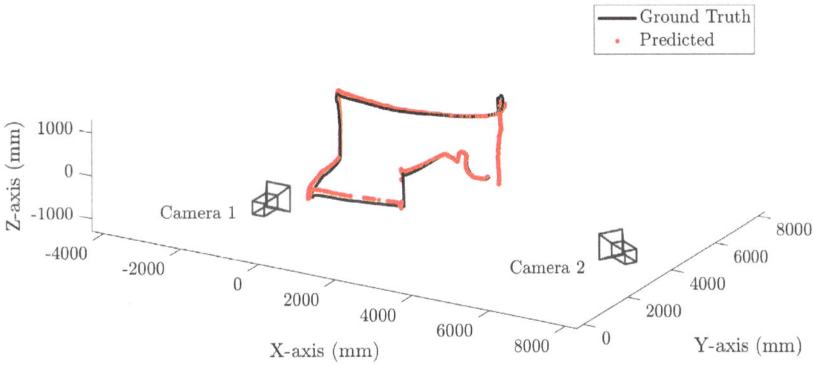

Figure 11. Experiment 3—path visualisation.

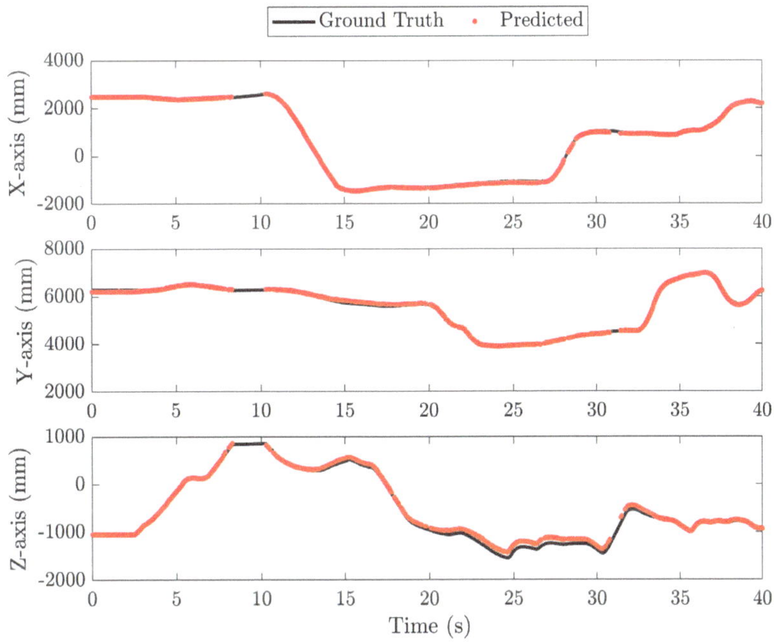

Figure 12. Experiment 3—predicted vs. ground truth coordinates.

Results from the random flight test expose more significant errors in the y and z axes. The geometry of the camera setup is separated by a greater x-distance, resulting in a more considerable disparity along the corresponding x-axis. Otherwise, the magnitude of error remains proportional to the target velocity. Less significant predictive discrepancies also remain in the experimental setup. Algorithm errors may arise from skewed camera lines of sight or mean projection errors during calibration, especially as only one pair of sight lines is used for localisation. Up to two tracking markers may also disappear due to the inherent motion capture configuration and propeller inference. In these cases, one marker is used to estimate ground truth, which will have a discrepancy from the true centroid. Outside this article's scope of investigation, temperature variation and mechanical changes over time will also influence camera accuracy. Table 4 displays the 1D axes and Cartesian errors of all experiments.

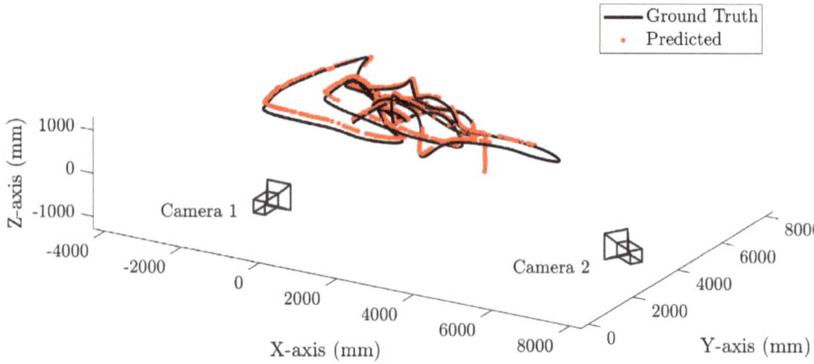

Figure 13. Experiment 6—Path Visualisation.

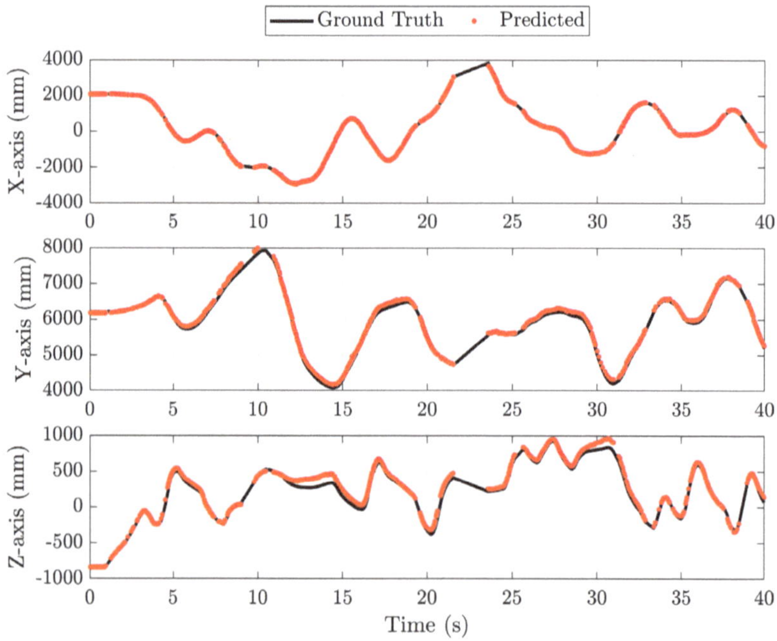

Figure 14. Experiment 6—predicted vs. ground truth coordinates.

Table 4. Predicted vs. ground truth metrics.

Experiment	Mean Target Velocity (mm/s)	Median X Error (mm)	Median Y Error (mm)	Median Z Error (mm)	Recall (%)
Fixed: 1	345	10	36	33	74.96
Fixed: 2	357	12	40	44	66.78
Fixed: 3	363	30	29	28	72.78
Random: 4	449	16	38	39	68.30
Random: 5	684	16	42	39	77.68
Random: 6	1026	14	72	34	85.71

Table 4 displays the 1D axes and Cartesian errors compared to the drone velocity, reflecting reliable accuracy in estimating the 3D position of the target. The median errors for the x- and y-axes fall within half the dimensions of the DJI Spark, which are $289 \times 245 \times 56$ mm. However, the z-axis error may localise the drone's position away from the ground truth within an order of 10^{-2} mm. It should be noted that these errors may compound as the viewable area of interest increases. The recall metric is determined by a true positive from both cameras. Greater variation was present in random experiments, with a higher value corresponding to when the drone flew more often towards Camera 2. The drone class within the dataset likely holds fewer annotations of such scale viewed by Camera 2.

Figure 15 visualises the 3D error distributions combined over the six experiments over 5123 data points, taken as the norm of all 1D errors. Similar trends are observable, with the x-direction having a smaller error than the other directions at a median of less than 20 mm. Considering outliers, the upper bound of errors approaches equal magnitudes of up to 180 mm. The results remain positively skewed, supporting prediction reliability.

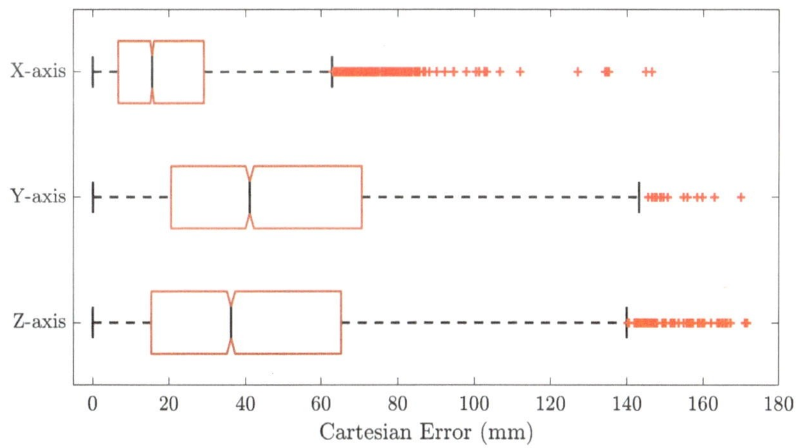

Figure 15. Error distribution across axes for all experiments. Each box plot displays the distribution of each quartile alongside the measured outliers, which are shown as red crosses towards the right.

Single camera calibration at zoom segments verified that zoom and focal length did not follow a linear relationship. Thus, calibration points were curve-fit using an exponential relationship. Figure 16 displays results for the x-axis focal length, as the difference in x- and y-directions was negligible.

A relationship between the zoom and principle point was also determined to ensure that corresponding PTZ commands would keep the target close to centred. The linear fit on u mostly aligns with the ideal optical centre of $u_0 = 540$ for a 1080p display, while variation in v holds uncertainty due to deviations at 100% zoom. A horizontal correction shall be included during deployment. Calibration at each zoom was repeated for over 50 image pairs at the same requirement of under a 0.50 mean reprojection error. Thus, errors may have arisen from limited pattern angles or motion blur as frame pairs were adjusted.

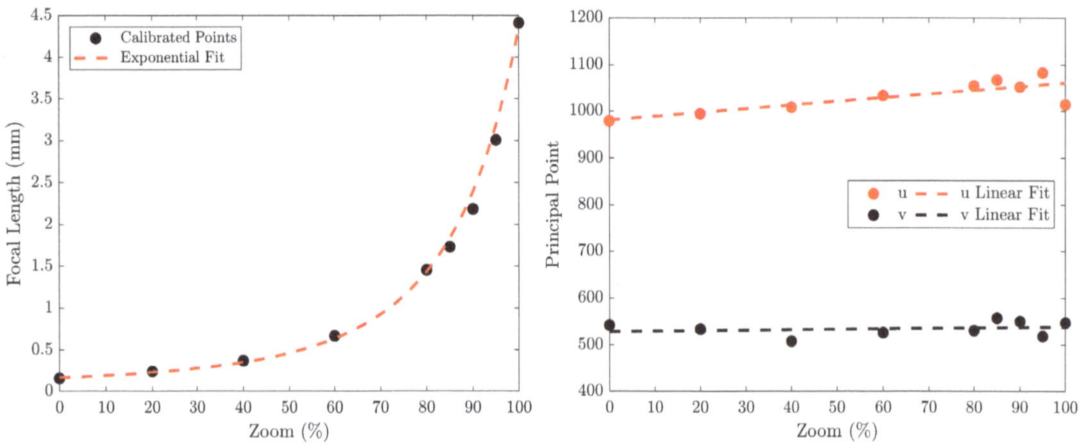

Figure 16. Zoom vs. focal length (**left**). Zoom vs. principal point (**right**).

4.3. PTZ Tracking Accuracy

Results are extracted using the fixed path from earlier experiments over 40 s. For this case, an undershoot factor of 0.2 is applied to show the surrounding environment adequately, as seen in Figure 17. Table 5 quantifies the final performance of the end-to-end system holding real-time performance at 15 FPS to a high tracking efficacy of 92.58%.

Table 5. PTZ tracking performance metrics.

E2E Time (ms)	Success Rate (%)
64.23	92.58

Since asynchronous processes are used to send a PTZ command and calculate the subsequent 3D coordinates, the end-to-end processing time entirely depends on the PTZ IP delay. The final deployment runs at an equivalent of 15 FPS. The results have a notably higher success rate than existing literature, such as that of Unlu et al. [39] at 71.2% while displaying the target at higher resolution. Figures 17 and 18 display frame segments where the PTZ camera undergoes a pan–tilt–zoom combination.

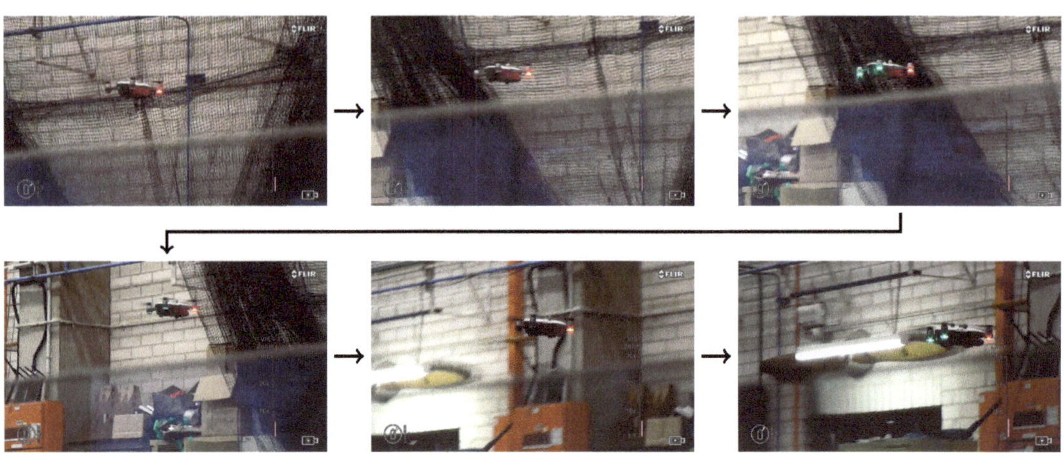

Figure 17. Pan, tilt, and zoom feed (UF = 0.2).

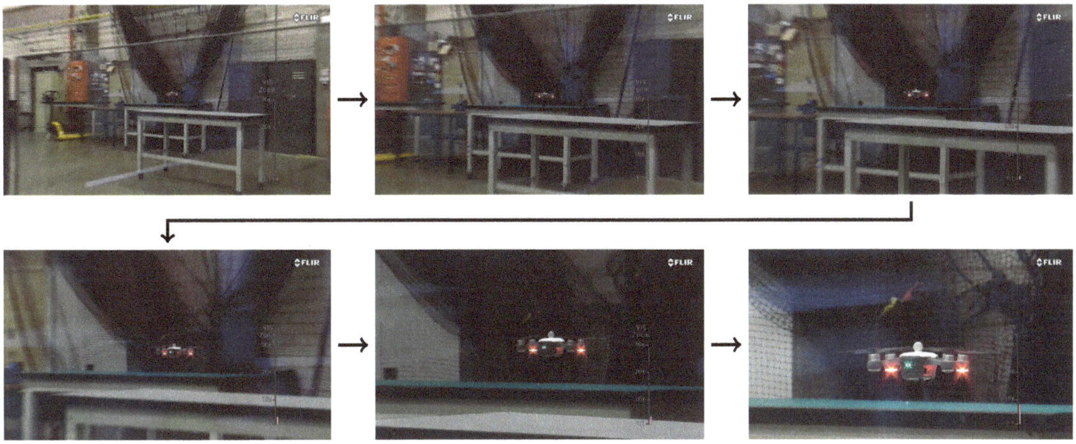

Figure 18. Zoom-only feed (UF = 0.4).

The drone veering from the frame centre reflects the input lag contributed by the inherent delay of IP cameras, distinguishable in Figure 17. This deviation increases as the drone's velocity increases, likely due to negative detections from motion blur. When stationary, the drone exhibits the same central offset that varies depending on its position within the area of interest. A coordinate offset is likely present between the world frame and the PTZ coordinate frame. Figure 18 shows that zoom estimation is capable within this small-scale environment. An undershoot factor of 0.4 corresponds to the drone width, taking up approximately 40% of the PTZ camera frame. Frames that lose sight of the drone are almost always a result of the detection framework failing to detect it.

5. Conclusions

The proposed camera network successfully detected, localised, and tracked a single aerial target in real-time. The pipeline was operational at 15 FPS, with the target visible in 92.58% of the PTZ camera frames. The object detection framework successfully augmented a dataset of aerial targets to output YOLO models, achieving an accuracy of at least 82.2% even in its most lightweight version through YOLOv8n. Maximising accuracy through YOLO11 at 86.7% retained real-time capabilities.

5.1. Contributions

This article provides the following contributions:

- Collection and annotation of a large dataset of videos for object detection in flight under various backgrounds, seasons, and weather conditions.
- A data augmentation pipeline that utilises knowledge distillation based on the collaboration of open-source pre-trained large models for different tasks.
- Comparative experiments to analyse how detection metrics vary by combining representative model structures (head, neck, and backbone) and backbone network sizes. In particular, this study incorporates the latest object detection model at the time of publication: YOLO11.
- Metrics to aid the evaluation of single object tracking performance and optimise hyperparameters in the absence of video annotations.
- A solution towards the absence of depth perception for PTZ-based imaging systems in the literature, improving the means of characterising aerial targets through localisation.

5.2. Limitations and Future Work

In real-world scenarios, aerial objects may remain unrecognised or sparsely annotated in existing datasets. Techniques such as out-of-distribution detection or zero-shot learning can be introduced to improve the detector's performance in handling such cases. The linear Kalman filter may perform poorly in motion estimation for targets with complex trajectories. Hence, tracking can extend to non-linear motion estimation algorithms or adopt end-to-end trained trackers to achieve more accurate motion predictions. This also resolves the cases where the PTZ camera loses track of the target by filling in between available predictions.

Providing rectangular boxes to SAM can generate masks for the instances within the boxes, yielding segmentation labels. This approach allows for extending object detection datasets into instance segmentation datasets, enabling the training of more confident detectors. Furthermore, many studies have proposed neural network modules specifically designed to enhance small object detection and moving object detection, such as SPD convolution and DyHead. These modules can be integrated into YOLO to improve inference accuracy.

Furthermore, the object detection dataset augmentation method proposed in this study holds significant potential for further research. In our future work, we will explore how to better optimise the coordination between these large models, which involves engineering optimisations, such as employing inference acceleration techniques. It may also involve structural and learning strategy optimisations, such as improving the network architecture at the task interface of the collaborative large models and leveraging fine-tuning techniques to make the proposed workflow as end-to-end as possible. Alternatively, we may explore knowledge distillation to learn more lightweight networks as substitutes for large models.

Deployment shows functionality within a small-scale environment of an 8 m $\times$ 4 m area of interest. However, accuracy will deteriorate with increasing surveillance area, given that object scale and camera hardware remain identical. Deployment should be considered for a large-scale environment where cameras can be connected wirelessly, such as through 900 MHz LoRa radio modules. In the case of monitoring very large airspace, the system is envisioned to be a modular setup where each computational node only communicates with a few cameras that cover a smaller part of the airspace of interest. The computer vision is done locally at each node. The global position of the tracked target is then computed by a central computer once the data from each individual node is collected. In this case, the accuracy and coverage may improve by including more modules without compromising the stability of the whole system.

An extrinsic calibration process suggested in research by Wu and Radke [42] may be explored to improve accuracy in keeping the drone precisely in the PTZ frame centre. Alternative PTZ camera models should also be tested to minimise end-to-end processing and reduce latency, thus improving the operational FPS.

Author Contributions: Conceptualisation, Z.W., D.W. and K.C.W.; methodology, M.H.L., H.L., Z.T., H.Y., Z.W. and D.W.; software, M.H.L., H.L., Z.T., H.Y., Z.W. and D.W.; validation, M.H.L., H.L. and Z.T.; formal analysis: M.H.L. and H.L.; investigation: M.H.L., H.L., Z.T., H.Y., Z.W. and D.L.; resources: Z.W., D.W. and K.C.W.; data curation: M.H.L., H.L., Z.T. and H.Y. All authors have read and agreed to the published version of the manuscript.

Funding: This project is supported by the Vacation Research Scholarship from the Faculty of Engineering at the University of Sydney. The PTZ camera is provided by SiNAB Pty Ltd.

Institutional Review Board Statement: Not applicable.

Informed Consent Statement: Not applicable.

Data Availability Statement: The original contributions presented in the study are included in the article, further inquiries can be directed to the corresponding author.

Acknowledgments: The dataset was provided by Zihao Wang, and annotated by Haixu Liu and Hang Yuan. The use of this dataset requires citation of this article.

Conflicts of Interest: Author David Williams was employed by the company SiNAB Pty Ltd. The remaining authors declare that the research was conducted in the absence of any commercial or financial relationships that could be construed as a potential conflict of interest.

Abbreviations

The following abbreviations are used in this manuscript:

ATL	average tracking length
DFL	distribution focal loss
DLT	direct linear transformation
E2E	end-to-end
FLOPS	floating point operations per second
FPN	feature pyramid network
FPS	frames per second
GAN	generative adversarial network
GPU	graphics processing unit
IP	Internet Protocol
LOS	line of sight
LTMR	long-term matching rate
mAP	mean average precision
MMLab	Multimedia Laboratory
MOT	multi-object tracking
MR	matching rate
NOT	number of tracks
PTZ	pan–tilt–zoom
R-CNN	region-based convolutional neural network
RT-DETR	Real-Time Detection Transformer
SAM	Segment Anything Model
SOTA	state-of-the-art
SORT	simple, online, and real-time
TL	tracking length
UAV	unmanned aerial vehicle
UF	undershoot factor
YOLO	You Only Look Once

Appendix A

Table A1. Performance metrics of tested models. Bolded data relate to chosen models of the main body.

Model	Scale	mAP50	mAP50:100	Parameters	Flops (G)	Inference (ms)
YOLOv5	n	0.783	0.486	2,509,634	7.2	0.6
	s	0.789	0.508	9,124,514	24.1	1.2
	m	0.816	0.535	25,068,610	64.4	2.5
	l	0.827	0.557	53,167,970	135.3	4.3
	x	0.828	0.551	97,205,186	246.9	7.4
YOLOv6	n	0.765	0.486	4,238,738	11.9	0.6
	s	0.800	0.521	16,307,010	44.2	1.3
	m	0.795	0.519	51,998,962	161.6	3.8
	l	0.774	0.510	110,897,826	391.9	7.4
	x	0.789	0.515	173,025,874	611.2	8.3
YOLOv7	tiny	0.721	0.404	8,116,226	21.3	0.7
	vanilla	0.836	0.549	44,224,385	132.2	4.7
	x	0.837	0.553	44,224,386	132.2	5.3
	w6	0.826	0.552	102,496,192	–	5.7
	e6	0.827	0.556	141,203,328	–	9.3
	e6e	0.836	0.562	195,713,904	–	13
	d6	0.835	0.562	197,285,568	–	10

Table A1. *Cont.*

Model	Scale	mAP50	mAP50:100	Parameters	Flops (G)	Inference (ms)
YOLOv8	n	**0.822**	0.538	3,012,018	8.2	**0.6**
	s	0.845	0.570	11,137,922	28.7	1.3
	m	0.856	0.585	25,859,794	79.1	2.7
	l	0.860	0.596	43,634,466	165.4	4.7
	x	0.866	0.607	68,158,386	258.1	7.6
YOLOv9	t	0.808	0.534	2,006,578	7.9	0.7
	s	0.831	0.563	7,289,730	27.4	1.6
	m	0.851	0.590	20,162,658	77.6	3.3
	c	0.852	0.588	25,533,858	103.7	4.5
	e	0.854	0.587	58,149,538	192.7	9.8
YOLOv10	n	0.804	0.525	2,709,380	8.4	0.8
	s	0.840	0.570	8,070,980	24.8	1.6
	m	0.846	0.575	16,491,076	64.0	2.9
	b	0.848	0.585	20,460,276	98.7	4.0
	l	0.851	0.592	25,774,580	127.2	4.8
	x	0.855	0.594	31,666,420	171.1	7.7
YOLO11	n	0.819	0.535	2,591,010	6.4	0.9
	s	0.85	0.581	9,430,098	21.6	1.8
	m	0.862	0.595	20,057,618	68.2	4
	l	0.862	0.599	25,315,090	87.3	5.3
	x	**0.867**	0.609	56,880,690	195.5	**9.3**
Cascade R-CNN	ResNet	0.504	0.316	69,167,000	166	21.4
	Swin Transformer	0.618	0.417	93,883,000	229	36.1
	ConvNeXt	0.704	0.522	94,501,000	224	29.9
DyHead	ResNet	0.739	0.506	38,901,000	70.5	59.5
	Swin Transformer	0.785	0.545	210,000,000	569	78.7
	ConvNeXt	0.75	0.522	64,276,000	130	61.4
Deformable DETR	ResNet	0.573	0.245	40,100,000	127	28.5
	Swin Transformer	0.111	0.04	61,908,000	191	42.6
	ConvNeXt	0.626	0.31	62,525,000	184	36.7

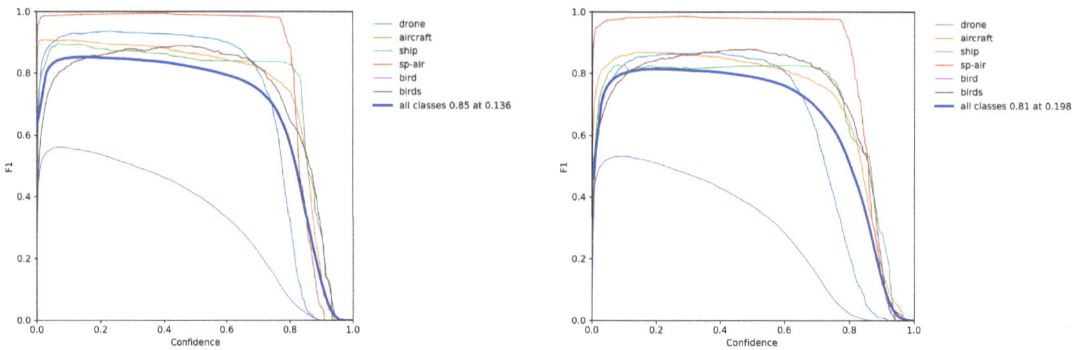

(**a**) (**b**)

Figure A1. YOLO F1-confidence curves. (**a**) YOLO11x. (**b**) YOLOv8n.

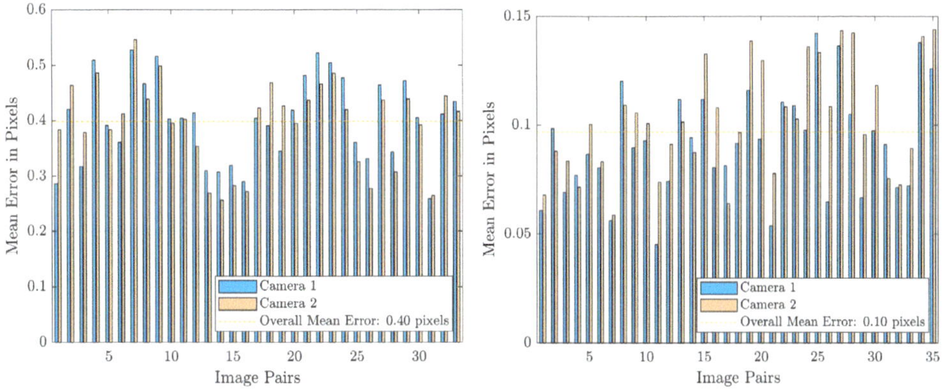

(**a**) (**b**)

Figure A2. Mean reprojection error per image pair. (**a**) Intrinsic parameters. (**b**) Extrinsic parameters.

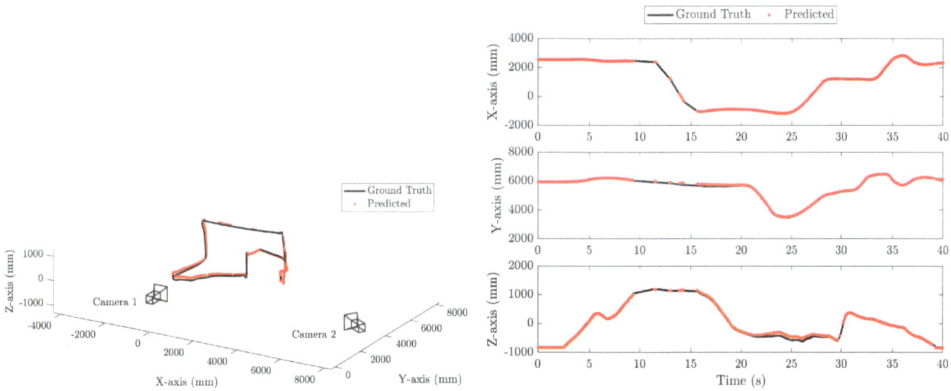

(**a**) (**b**)

Figure A3. Experiment 1—fixed flight. (**a**) Target path visualisation. (**b**) Predicted vs. ground truth coordinates.

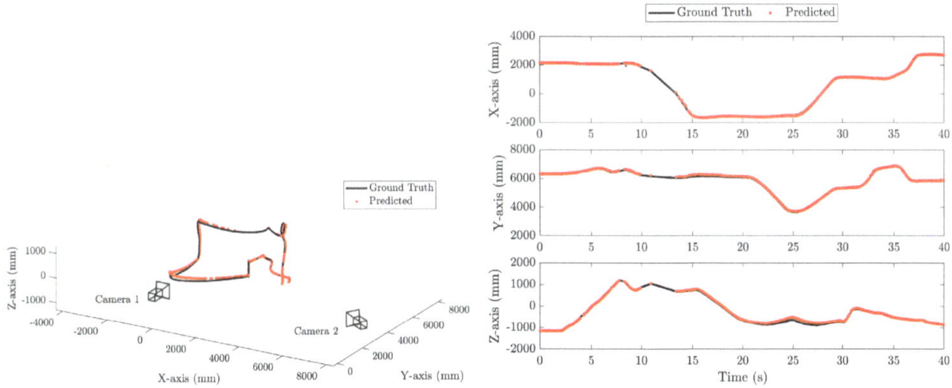

(**a**) (**b**)

Figure A4. Experiment 2—fixed flight. (**a**) Target path visualisation. (**b**) Predicted vs. ground truth coordinates.

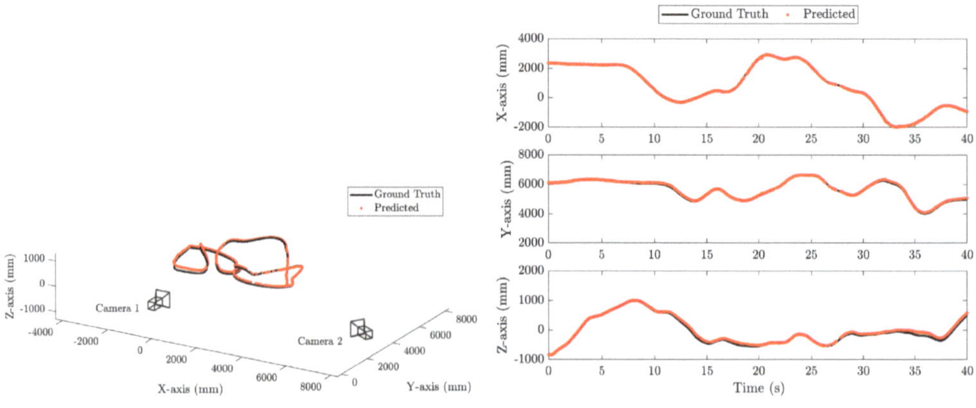

(a) (b)

Figure A5. Experiment 4—random flight. (**a**) Target path visualisation. (**b**) Predicted vs. ground truth coordinates.

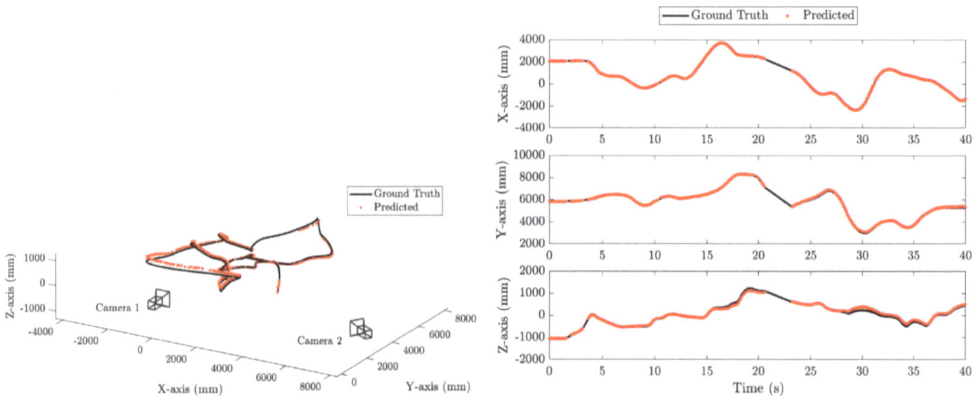

(a) (b)

Figure A6. Experiment 5—Random Flight. (**a**) Target Path Visualisation. (**b**) Predicted vs. Ground Truth Coordinates.

References

1. O'Malley, J. The no drone zone. *Eng. Technol.* **2019**, *14*, 34–38. [CrossRef]
2. Metz, I.C.; Ellerbroek, J.; Mühlhausen, T.; Kügler, D.; Hoekstra, J.M. Analysis of risk-based operational bird strike prevention. *Aerospace* **2021**, *8*, 32. [CrossRef]
3. Zhang, W.; Cong, M.; Wang, L. Algorithms for optical weak small targets detection and tracking. In Proceedings of the International Conference on Neural Networks and Signal Processing, 2003, Proceedings of the 2003, Nanjing, China, 14–17 December 2003; IEEE: New York, NY, USA, 2003; Volume 1, pp. 643–647.
4. Chen, K.; Wang, J.; Pang, J.; Cao, Y.; Xiong, Y.; Li, X.; Sun, S.; Feng, W.; Liu, Z.; Xu, J.; et al. MMDetection: Open mmlab detection toolbox and benchmark. *arXiv* **2019**, arXiv:1906.07155.
5. Kirillov, A.; Mintun, E.; Ravi, N.; Mao, H.; Rolland, C.; Gustafson, L.; Xiao, T.; Whitehead, S.; Berg, A.C.; Lo, W.Y.; et al. Segment anything. In Proceedings of the IEEE/CVF International Conference on Computer Vision, Paris, France, 1–6 October 2023; pp. 4015–4026.
6. Rombach, R.; Blattmann, A.; Lorenz, D.; Esser, P.; Ommer, B. High-Resolution Image Synthesis with Latent Diffusion Models. *arXiv* **2021**, arXiv:2112.10752.
7. Dubey, A.; Jauhri, A.; Pandey, A.; Kadian, A.; Al-Dahle, A.; Letman, A.; Mathur, A.; Schelten, A.; Yang, A.; Fan, A.; et al. The llama 3 herd of models. *arXiv* **2024**, arXiv:2407.21783.
8. Goodfellow, I.; Pouget-Abadie, J.; Mirza, M.; Xu, B.; Warde-Farley, D.; Ozair, S.; Courville, A.; Bengio, Y. Generative adversarial networks. *Commun. ACM* **2020**, *63*, 139–144. [CrossRef]

9. Yun, S.; Han, D.; Oh, S.J.; Chun, S.; Choe, J.; Yoo, Y. Cutmix: Regularization strategy to train strong classifiers with localizable features. In Proceedings of the IEEE/CVF International Conference on Computer Vision, Seoul, Republic of Korea, 27–28 October 2019; pp. 6023–6032.
10. Zhang, H.; Cisse, M.; Dauphin, Y.N.; Lopez-Paz, D. mixup: Beyond Empirical Risk Minimization. *arXiv* **2018**, arXiv:1710.09412.
11. Bochkovskiy, A.; Wang, C.Y.; Liao, H.Y.M. YOLOv4: Optimal Speed and Accuracy of Object Detection. *arXiv* **2020**, arXiv:2004.10934.
12. Jocher, G.; Chaurasia, A.; Qiu, J. Ultralytics YOLOv8. 2023. Available online: https://www.scirp.org/reference/referencespapers?referenceid=3532980 (accessed on 15 October 2024).
13. Hui, Y.; Wang, J.; Li, B. STF-YOLO: A small target detection algorithm for UAV remote sensing images based on improved SwinTransformer and class weighted classification decoupling head. *Measurement* **2024**, *224*, 113936. [CrossRef]
14. Liu, Z.; Lin, Y.; Cao, Y.; Hu, H.; Wei, Y.; Zhang, Z.; Lin, S.; Guo, B. Swin transformer: Hierarchical vision transformer using shifted windows. In Proceedings of the IEEE/CVF International Conference on Computer Vision, Montreal, BC, Canada, 11–17 October 2021; pp. 10012–10022.
15. Alexey, D. An image is worth 16x16 words: Transformers for image recognition at scale. *arXiv* **2020**, arXiv:2010.11929.
16. Brooks, T.; Holynski, A.; Efros, A.A. InstructPix2Pix: Learning to Follow Image Editing Instructions. *arXiv* **2022**, arXiv:2211.09800.
17. Wang, S.; Xia, C.; Lv, F.; Shi, Y. RT-DETRv3: Real-time End-to-End Object Detection with Hierarchical Dense Positive Supervision. *arXiv* **2024**, arXiv:2409.08475.
18. Liu, W.; Anguelov, D.; Erhan, D.; Szegedy, C.; Reed, S.; Fu, C.Y.; Berg, A.C. SSD: Single Shot MultiBox Detector. In Proceedings of the Computer Vision—ECCV 2016: 14th European Conference, Amsterdam, The Netherlands, 11–14 October 2016; Proceedings, Part I; Springer International Publishing: Cham, Switzerland, 2016; pp. 21–37.
19. Girshick, R.; Donahue, J.; Darrell, T.; Malik, J. Rich feature hierarchies for accurate object detection and semantic segmentation. In Proceedings of the Computer Vision and Pattern Recognition, Columbus, OH, USA, 23–28 June 2014.
20. Girshick, R. Fast R-CNN. In Proceedings of the 2015 IEEE International Conference on Computer Vision (ICCV), Santiago, Chile, 7–13 December 2015; pp. 1440–1448. [CrossRef]
21. Ren, S.; He, K.; Girshick, R.; Sun, J. Faster R-CNN: Towards Real-Time Object Detection with Region Proposal Networks. *IEEE Trans. Pattern Anal. Mach. Intell.* **2017**, *39*, 1137–1149. [CrossRef]
22. He, K.; Gkioxari, G.; Dollár, P.; Girshick, R. Mask R-CNN. In *IEEE Transactions on Pattern Analysis and Machine Intelligence*; IEEE: New York, NY, USA, 2017.
23. Cai, Z.; Vasconcelos, N. Cascade R-CNN: High quality object detection and instance segmentation. *IEEE Trans. Pattern Anal. Mach. Intell.* **2019**, *43*, 1483–1498. [CrossRef]
24. Carion, N.; Massa, F.; Synnaeve, G.; Usunier, N.; Kirillov, A.; Zagoruyko, S. End-to-end object detection with transformers. In *European Conference on Computer Vision*; Springer: Berlin/Heidelberg, Germany, 2020; pp. 213–229.
25. Zhang, H.; Li, F.; Liu, S.; Zhang, L.; Su, H.; Zhu, J.; Ni, L.M.; Shum, H.Y. Dino: Detr with improved denoising anchor boxes for end-to-end object detection. *arXiv* **2022**, arXiv:2203.03605.
26. Zhu, X.; Su, W.; Lu, L.; Li, B.; Wang, X.; Dai, J. Deformable detr: Deformable transformers for end-to-end object detection. *arXiv* **2020**, arXiv:2010.04159.
27. Liu, S.; Qi, L.; Qin, H.; Shi, J.; Jia, J. Path aggregation network for instance segmentation. In Proceedings of the IEEE Conference on Computer Vision and Pattern Recognition, Salt Lake City, UT, USA, 18–23 June 2018; pp. 8759–8768.
28. Tan, M.; Pang, R.; Le, Q.V. Efficientdet: Scalable and efficient object detection. In Proceedings of the IEEE/CVF Conference on Computer Vision and Pattern Recognition, Seattle, WA, USA, 13–19 June 2020; pp. 10781–10790.
29. Dai, X.; Chen, Y.; Xiao, B.; Chen, D.; Liu, M.; Yuan, L.; Zhang, L. Dynamic head: Unifying object detection heads with attentions. In Proceedings of the IEEE/CVF Conference on Computer Vision and Pattern Recognition, Nashville, TN, USA, 20–25 June 2021; pp. 7373–7382.
30. Tan, M. Efficientnet: Rethinking model scaling for convolutional neural networks. *arXiv* **2019**, arXiv:1905.11946.
31. Liu, Z.; Mao, H.; Wu, C.Y.; Feichtenhofer, C.; Darrell, T.; Xie, S. A convnet for the 2020s. In Proceedings of the IEEE/CVF Conference on Computer Vision and Pattern Recognition, New Orleans, LA, USA, 18–24 June 2022; pp. 11976–11986.
32. He, K.; Zhang, X.; Ren, S.; Sun, J. Deep residual learning for image recognition. In Proceedings of the IEEE Conference on Computer Vision and Pattern Recognition, Las Vegas, NV, USA, 27–30 June 2016; pp. 770–778.
33. Bewley, A.; Ge, Z.; Ott, L.; Ramos, F.; Upcroft, B. Simple online and realtime tracking. In Proceedings of the 2016 IEEE International Conference on Image Processing (ICIP), Phoenix, AZ, USA, 25–28 September 2016; IEEE: New York, NY, USA, 2016; Volume 9. [CrossRef]
34. Wojke, N.; Bewley, A.; Paulus, D. Simple Online and Realtime Tracking with a Deep Association Metric. *arXiv* **2017**, arXiv:1703.07402.
35. Zhang, Y.; Sun, P.; Jiang, Y.; Yu, D.; Weng, F.; Yuan, Z.; Luo, P.; Liu, W.; Wang, X. ByteTrack: Multi-Object Tracking by Associating Every Detection Box. In *European Conference on Computer Vision*; Springer Nature: Cham, Switzerland, 2022.
36. Aharon, N.; Orfaig, R.; Bobrovsky, B.Z. BoT-SORT: Robust associations multi-pedestrian tracking. *arXiv* **2022**, arXiv:2206.14651.
37. Kang, S.; Paik, J.K.; Koschan, A.; Abidi, B.R.; Abidi, M.A. Real-time video tracking using PTZ cameras. In *Proceedings of the Sixth International Conference on Quality Control by Artificial Vision*; SPIE: Bellingham, WA, USA, 2003; Volume 5132, pp. 103–111.

38. Di Caterina, G.; Hunter, I.; Soraghan, J.J. An embedded smart surveillance system for target tracking using a PTZ camera. In Proceedings of the 4th European Education and Research Conference (EDERC 2010), Nice, France, 1–2 December 2010; pp. 165–169.
39. Unlu, H.U.; Niehaus, P.S.; Chirita, D.; Evangeliou, N.; Tzes, A. Deep learning-based visual tracking of UAVs using a PTZ camera system. In Proceedings of the IECON 2019–2045th Annual Conference of the IEEE Industrial Electronics Society, Lisbon, Portugal, 14–17 October 2019; IEEE: New York, NY, USA, 2019; Volume 1, pp. 638–644.
40. Zhang, Z. A flexible new technique for camera calibration. *IEEE Trans. Pattern Anal. Mach. Intell.* **2000**, *22*, 1330–1334. [CrossRef]
41. Sinha, S.N.; Pollefeys, M. Towards calibrating a pan-tilt-zoom camera network. In *Proceedings of the 5th Workshop Omnidirectional Vision, Camera Networks and Non-Classical Cameras*; Citeseer: University Park, PA, USA, 2004; pp. 42–54.
42. Wu, Z.; Radke, R.J. Keeping a pan-tilt-zoom camera calibrated. *IEEE Trans. Pattern Anal. Mach. Intell.* **2012**, *35*, 1994–2007. [CrossRef] [PubMed]
43. Nasiri, S.M.; Hosseini, R.; Moradi, H. The optimal triangulation method is not really optimal. *IET Image Process.* **2023**, *17*, 2855–2865. [CrossRef]
44. Lee, S.H.; Civera, J. Triangulation: Why optimize? *arXiv* **2019**, arXiv:1907.11917.
45. Sturm, P. Richard I. Hartley and Peter Sturm. *Computer Vision and Image Understanding*. Available online: https://citeseerx.ist.psu.edu/document?repid=rep1&type=pdf&doi=970be04e6e469d841cb8a214f3ad95e5c659cc2f (accessed on 15 October 2024).
46. Zhu, J.Y.; Zhang, R.; Pathak, D.; Darrell, T.; Efros, A.A.; Wang, O.; Shechtman, E. Toward multimodal image-to-image translation. In Proceedings of the Advances in Neural Information Processing Systems, Long Beach, CA, USA, 4–9 December 2017.
47. Zhu, J.Y.; Park, T.; Isola, P.; Efros, A.A. Unpaired Image-to-Image Translation using Cycle-Consistent Adversarial Networks. In Proceedings of the 2017 IEEE International Conference on Computer Vision (ICCV), Venice, Italy, 22–29 October 2017.
48. Schmalz, C.; Forster, F.; Angelopoulou, E. Camera calibration: Active versus passive targets. *Opt. Eng.* **2011**, *50*, 113601.

eng

MDPI

Article

Investigation of the Hottel–Whillier–Bliss Model Applied for an Evacuated Tube Solar Collector

Márton Rátkai [1,*], Gábor Géczi [2] and László Székely [3]

1 Mechanical Engineering Doctoral School, Hungarian University of Agriculture and Life Sciences, Páter Károly utca 1., 2100 Gödöllő, Hungary

2 Department of Environmental Analysis and Technologies, Institute of Environmental Sciences, Hungarian University of Agriculture and Life Sciences, Páter Károly utca 1., 2100 Gödöllő, Hungary; geczi.gabor@uni-mate.hu

3 Department of Mathematics and Modelling, Institute of Mathematics and Basic Science, Hungarian University of Agriculture and Life Sciences, Páter Károly utca 1., 2100 Gödöllő, Hungary; szekely.laszlo@uni-mate.hu

* Correspondence: ratkai.marton.3@phd.uni-mate.hu

Abstract: The goal of this research is to examine the applicability of the Hottel–Whillier–Bliss model, developed for flat-plate collectors, to evacuated tube solar collectors. During this study, the model is applied to an evacuated tube collector, and then the identification and validation of the model are made with the help of measurements performed on the collector. This research also includes the application, identification and validation of the energy balance model for the investigated solar collector. This model works for both flat-plate and evacuated tube collectors. The results obtained with the two different models are then compared. By comparing the modelled results with the measured values, the accuracy and applicability of the models can be determined. Based on the results, the Hottel–Whillier–Bliss model works excellently with evacuated tube solar collectors for predicting the outlet temperature of the medium from the solar collector. It is important to note that the identification gives negative heat transfer parameter values. According to the validation, the average absolute error is 0.8 °C, and the average relative error is 1%. For the energy balance model, these values are 0.87 °C and 1.1% respectively, indicating that the accuracy of the Hottel–Whillier–Bliss model is very similar, and even slightly higher. Additionally, the research provides further proof of the applicability of the energy balance model to evacuated tube collectors.

Keywords: solar energy; solar collector; mathematical modelling; simulation results; measurements

Citation: Rátkai, M.; Géczi, G.; Székely, L. Investigation of the Hottel–Whillier–Bliss Model Applied for an Evacuated Tube Solar Collector. *Eng* **2024**, *5*, 3427–3438. https://doi.org/10.3390/eng5040178

Academic Editors: Antonio Gil Bravo and Maria Founti

Received: 20 November 2024
Revised: 15 December 2024
Accepted: 17 December 2024
Published: 18 December 2024

1. Introduction

Modelling methods make the examination of different systems significantly easier. Their fundamental advantage is that they enable testing the operation of a specific construction without having to physically build it [1], so costs and time can be saved with their help. In the field of mathematical modelling, physically based, or white-box, and black-box models are the most common ones. The grey-box model also occurs as a combination of the previous ones.

Physically based modelling assumes precise knowledge and understanding of the system's operation and physical principles. In this method, the structure of the system, its components and the relationships between them are known in detail. The modelling process is characterized by transparency and is based on explicit knowledge of the structure of the system [2]. For physically based models, it can be stated that the more complex the modelled system is, the more challenging it is to prepare the model. However, with its help, we can gain more accurate understanding of the operation of the system.

Black-box modelling is a more result-oriented approach. In this type, the internal operation of the system is treated as a so-called black box, and the analysis focuses on the observed inputs and outputs. The internal physical structure and processes are not known or are not particularly taken into account during modelling. Black-box modelling is typically used in cases where, due to the complexity of the system, it is difficult to fully understand its internal operation, or the focus is on the system's behaviour [3,4]. Based on this, this modelling method is particularly useful in cases with many influencing factors, as it enables the preparation of predictive models based on observed behaviour. These are often probabilistic, statistical models or such methods are used in their examination [5].

In practice, a hybrid approach known as grey-box modelling is used. Grey-box modelling combines the components of the physically based (white-box) and black-box methods, thus enabling a balance between transparency and practicality. The approach is especially useful when we have partial knowledge of the internal operation of a system, and the focus is on general understanding, taking feasibility into account as well [6].

Both physically based and black-box modelling have advantages and disadvantages. Physically based models provide deep understanding of the underlying physical content, thus enabling the mapping of the operation of the system. In addition, however, they can be computationally demanding, require precise knowledge of many parameters, especially for complex systems, and may not be practical in cases where a complete understanding of the internal operation is too challenging. Black-box models are often simpler to implement, more flexible and easier to adapt to real conditions. At the same time, it is not certain that their accuracy provides sufficient and adequate depth of insight into the internal functioning of the system [7].

The decision between physically based and black-box modelling depends on the exact goals of the analysis and the nature and complexity of the system to be investigated. Both methods are important tools that contribute to a holistic understanding of different systems, effective problem solving and optimization.

Models for solar collector systems can be found in the literature. In [8], a system of ordinary differential equations is presented, which can describe the transient processes in a solar collector–solar storage system with adequate accuracy. For the model in [8], an analytical solution that enables the quick and easy determination of the collector and solar storage temperatures can be found in [9]. In [10], a mathematical model for parabolic trough collector systems is proposed to perform heat transfer analysis. In [11], a model was developed for an evacuated tube collector-thermal storage system to analyze the performance of a solar cabinet dryer equipped with the collector. In [12], the use of physics-informed neural networks is proposed for optimizing solar collector design. There are models which also include the length and other physical properties of pipes, too. The physically based model in [13] is a system of differential equations enabling the calculation of the inlet temperature, and Ref. [14] contains a grey-box model describing the temperature of pipes conveying liquid. There are also models describing other system elements like solar storages and heat exchangers. The study in [15] presents complete analytical models for fully mixed and stratified seasonal solar storage tanks. In [16], a model was used to determine optimal flow control in a solar collector system with fully mixed storage tanks. In [17], a black-box type model for solar storage tanks is presented. The model is based on multiple linear regression. Practical examples are provided for different solar storage modelling strategies in [18]. In [19], a one-dimensional model is proposed for stratified storage tanks. The model considers buoyancy and mixing effects. A dynamic, transient simulation model is presented in [20] for solar heating and cooling systems including several storage tanks. Several one-dimensional models for stratified solar storage tanks are compared in [21] and performance studies of such models can be found in [22]. In [23], a multi-node dynamic model for stratified storage tanks is developed and analyzed. A heat transfer model for crossflow heat exchangers is presented in [24] and thermal hydraulic models can be found in [25]. In [26], three-dimensional modelling of compact heat exchangers is presented. Mathematical models for heat exchangers considering heat gain and loss to the environment are proposed in [27]. In [28], second-order concentrated parameter models based on the logarithmic mean temperature difference approach are discussed. The results showed that these models can describe the dynamics of heat exchangers reliably.

There are also solar collector models in the literature. In [29], several mathematical models describing the dynamic behaviour of different flat-plate collectors without glazing can be found. The most complex of these is the physically based model marked D2. By validating the D2 model with measurement data, it was proven that it can be well used for modelling the temperature distribution of the solar collector even in cases where the change in volumetric flow rate is significant.

In [30], a device used for domestic hot water production and heating was investigated, in which a solar collector was combined with a mirror. A physically based mathematical model was made for evacuated tube collectors, which can be used to estimate the total solar irradiance reaching the collector. The isotropic diffuse sky model, with certain simplifications, was also used. The model was validated with data from a meteorological database for one year. The modelled results followed the measured values well. The validated model made it possible to run the simulations that were used to determine the optimal angle of inclination of the solar collector and the mirror. By this, it was possible to alleviate the two typical problems of solar collectors, which are the reduced energy production in the heating season and the overheating occurring in the summer season. The optimal angle of the mirror and the collector, which differs in the two seasons, enables the maximization of energy production during the heating season, and the maximum possible energy production in the summer season without overheating.

In [31], a physically based mathematical model for an evacuated tube collector is made. Using the model, it is possible to predict the thermal efficiency of the collector and the temperature of the exiting water through simulation.

The Hottel–Whillier–Bliss physically based mathematical model [32,33], specially created for flat collectors, can be used to describe the temperature distribution of the working medium along the length of the collector as a function of time and place.

Among black-box models, the so-called neural network models can be found. During the design of these models, probabilistic and statistical methods are used, even though their primary functionality is typically driven by optimization techniques. The goal in [34] was to create a neural network that can be used to describe the thermal behaviour of a flat-plate collector. Knowing the intensity of solar radiation, the ambient temperature and the temperature of the heat transfer medium entering the collector, with a constant mass flow rate, the temperature of the exiting working medium can be predicted with the model. The neural network was created using measured data and calculated data using the validated Hottel–Whillier–Bliss model. The latter were necessary because training the model requires a lot of data in order to operate with sufficient accuracy in different operating conditions, but the measurement possibilities were limited. The Levenberg–Marquardt algorithm [35,36] was also used. The model proved to be suitable for learning the thermal behaviour of the investigated solar collectors. The temperatures measured and calculated with the Hottel–Whillier–Bliss model were very close to the values calculated with the neural network model.

In [37], a physically based mathematical model that can be used to describe the behaviour of a solar collector is created. The model describes the temperature of the working medium leaving the collector as a function of the variables and parameters. The model is based on the energy balance of the collector, which is why it is also called the energy balance model.

In [38], a black-box model based on multiple linear regression for solar collectors is described. The inputs of the model are the corresponding values of solar irradiance, ambient temperature and temperature of the entering and exiting working medium. The appropriate temperature of the exiting working medium can be calculated as an output. The model takes the volumetric flow rate of the collector circuit pump to be zero or constant corresponding to the on–off control, which is one of the most common controlling methods. The effect of the inputs on the output is not immediate (due to the finite propagation speed of effect), which is taken into account with a time delay. The identified and validated model was compared to the energy balance model, and it was able to predict the temperature of the exiting working medium more accurately. The importance of the model, in addition to the appropriate accuracy (average accuracy better than 5%), is its simplicity. It has few calculation demands and can be used for any type of collector.

The goal of this research is to examine the applicability of the Hottel–Whillier–Bliss (hereinafter HWB) model, originally created for flat-plate collectors, to evacuated tube collectors. Although the HWB model was extended to different types of collectors [39,40], it has never been examined before with evacuated tube collectors.

The contribution of the present paper is as follows:

1. An experimental evacuated tube solar collector system is assembled. Different variables of the system are measured.
2. Application of the Hottel–Whillier–Bliss and the energy balance model for the examined collector, identification and validation of the two models, and performing simulations with the validated models.
3. It is proven with measurement and simulation data that the Hottel–Whillier–Bliss model, originally created for flat-plate solar collectors, works excellently for evacuated tube solar collectors as well.
4. The applicability of the energy balance model for evacuated tube collectors is further confirmed with measurement and simulation data.

2. Materials and Methods

During the research, the HWB model is applied to an evacuated tube solar collector, and then the model is identified and validated with the help of measurements made on the collector. The accuracy and applicability of the model can be determined by comparing the modelled results with the measured values. The research also includes the application, identification and validation of the energy balance (hereafter EE) model for the investigated solar collector, which is suitable for both flat-plate and evacuated tube collectors, and then the comparison of the results obtained with the EE model with the results of the HWB model. All this makes it possible to draw further conclusions regarding the applicability of the HWB model to evacuated tube solar collectors.

2.1. Description of the Measurement System

The tested solar collector is an evacuated tube collector (Figure 1), the surface of which is 2.2 m^2. It is oriented to the south; its inclination angle is 40°. The volume of the working medium of the collector (including the connecting pipes) is 16 litres. The working medium is water, its density is 1000 kg/m^3 and its specific heat is 4200 J/kgK. The volumetric flow rate of the pump is a constant 100 litres/hour. The equipment is located on the Szent István Campus of the Hungarian University of Agriculture and Life Sciences in Gödöllő, Hungary.

Figure 1. The examined solar collector.

2.2. Measurements

A total of three temperature values were measured using type K thermocouples, which have an average measurement uncertainty of 1 °C. These are the temperature of the working medium entering the collector, the temperature of the working medium leaving the collector and the temperature of the solar collector's environment. The flow rate was measured using a Kobold Unirota URM-33 33H G5 0 type rotameter (Nyíregyháza, Hungary). The uncertainty of the instrument is 0.1 litres/min. A Theodor Friedrichs 6003.3000 BG global irradiance meter (Elmshorn, Germany) placed in the plane of the collector was used to measure solar irradiance (measurement uncertainty 30 W/m^2). The frequency of measurements was 30 s. The sensors were connected to an Almemo 2590 data logger manufactured by Ahlborn (Ilmenau, Germany), which was connected to a laptop via a data cable to read the measured data. The data were read out on the computer using the Ahlborn AMR Win Control V5 software. The measurement data stored in the program's memory can be saved in a text document.

2.3. Mathematical Modelling

The modelling and simulation tasks were performed using Matlab and Simulink (https://www.mathworks.com/products/simulink.html, accessed on 16 December 2024). The HWB model, if we are only interested in the temperature of the medium leaving the collector, takes the following form, depending only on time:

$$T_{c,out}(t) = T_{c,e}(t) + \frac{I(t)}{k_{ae}} + \left[T_{c,in}(t) - T_{c,e}(t) - \frac{I(t)}{k_{ae}}\right] \cdot e^{\frac{-k_{we} \cdot A_c}{c \cdot \dot{m}}}, \tag{1}$$

where the ambient temperature of the collector is $T_{c,e}$ [K], the collector inlet water temperature is $T_{c,in}$ [K], the collector outlet water temperature is $T_{c,out}$ [K], the global solar irradiance in the plane of the collector is I [W/m^2], the mass flow rate in the collector circuit is $\dot{m}$ [kg/s], the specific heat capacity of water is c [J/kgK], the surface of the collector is A_c [m^2], the heat transfer coefficient between the absorber and the environment is k_{ae} [W/m^2K], the heat transfer coefficient between the water and the environment is k_{we} [W/m^2K] and time is denoted by t [s]. The HWB model constructed in Simulink can be seen in Figure 2.

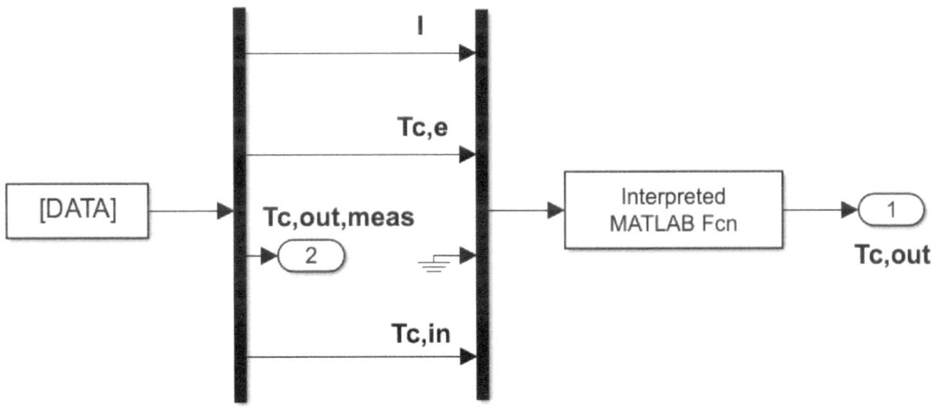

Figure 2. The HWB model in Simulink.

The EE model, written for the derivative of the outlet water temperature of the evacuated tube collector, is as follows:

$$\frac{dT_{c,out}}{dt} = \frac{A_c \eta_o}{\rho \cdot c \cdot V_c} I - \frac{UA_c}{\rho \cdot c \cdot V_c}(T_{c,av} - T_{c,e}) + \frac{\dot{V}_c}{V_c}(T_{c,in} - T_{c,out}), \tag{2}$$

where the average collector temperature is $T_{c,av}$ [K] (arithmetic mean of $T_{c,in}$ and $T_{c,out}$), the volumetric flow rate in the collector circuit is $\dot{v}$ [m^3/s], the optical efficiency is η_o [-], the overall heat loss coefficient is U [W/m^2K], the volume of the collector is V_c [m^3], the density of water is ρ and the specific heat capacity of water is c [J/kgK]. Figure 3 shows the model realized in Simulink. Since the solution for the differential equation of the EE model is similar to Equation (1), it is expected that the HWB model should work quite well for evacuated tube collectors, too. The solution for Equation (2) is as follows:

$$T_{c,out}(t) = \left(T_{c,out,0} - \frac{(2\rho c\dot{v}_c - UA_c)T_{c,in} + 2A_c(UT_{c,e} + \eta_o I)}{2\rho c\dot{v}_c + UA_c}\right)e^{\frac{-(2\cdot\rho\cdot c\cdot\dot{v}_c + UA_c)t}{2\cdot\rho\cdot c\cdot V_c}} + \frac{(2\rho c\dot{v}_c - UA_c)T_{c,in} + 2A_c(UT_{c,e} + \eta_o I)}{2\rho c\dot{v}_c + UA_c}, \quad (3)$$

where $T_{c,out,o}$ is the initial value.

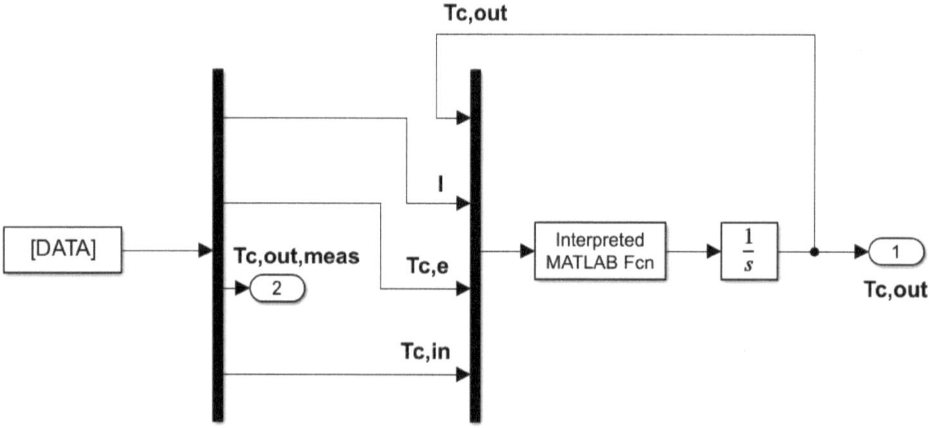

Figure 3. The EE model in Simulink.

The values of the terms in the equations are either known or determined by measurement. However, the values of certain parameters are obtained using parameter identification. Both models were identified based on the data of the same two measurement days (24 May and 11 June 2024). One day was cloudier, the other one was clear and sunny (Figure 4). In the case of the HWB model, the two parameters determined by identification are the heat transfer coefficient between the absorber and the environment and the coefficient between the working medium and the environment. For the EE model, these are the optical efficiency and the overall heat loss coefficient. For the identification, we need an objective function. During the measurements, the collector outlet temperature was also measured, so the modelled values can be compared with them. Since the measured values are discrete, the objective function is also used in a discrete form. The value of the objective function is the absolute error calculated from the difference between the measured and modelled temperatures, summed up in all measurement points. This value depends on the two parameters, so the parameter values of the investigated collector can be determined by minimum search; thus, the model can be made more precise. The objective function J can be written in the following form:

$$J[p_1, p_2] = \sum_{i=1}^{n} |T_{c,out,meas} - T_{c,out}|, \quad (4)$$

where p_1 and p_2 are the two parameters and n is the number of measurement points.

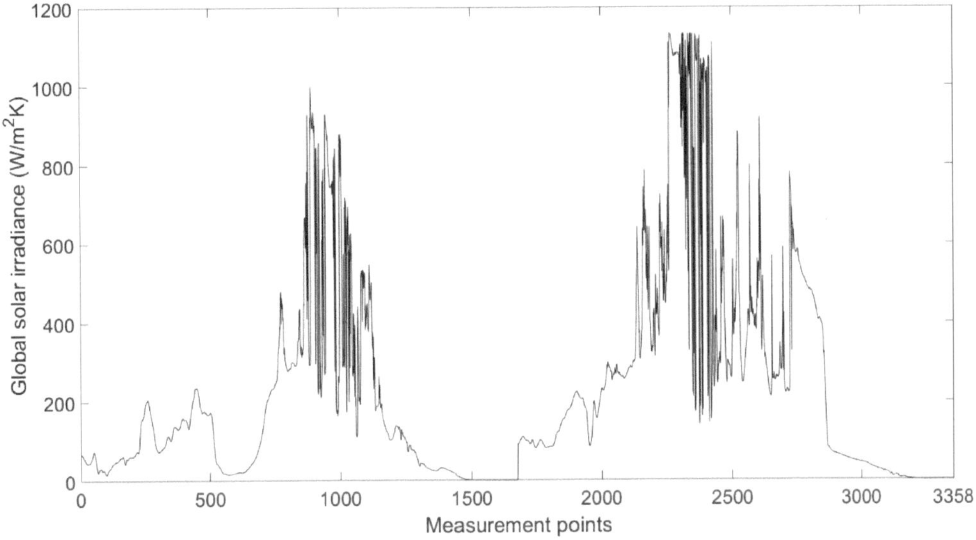

Figure 4. Global irradiance on the days used during the identification.

The validation of the models was carried out using the obtained parameter values and the data of a total of nine days at the end of May and June, which were not used for identification. Figure 5 shows the global solar irradiance for these measurement days. The results were evaluated with the average absolute error and the average absolute error divided by the difference between the largest and smallest measured outlet temperatures, i.e., the average relative error.

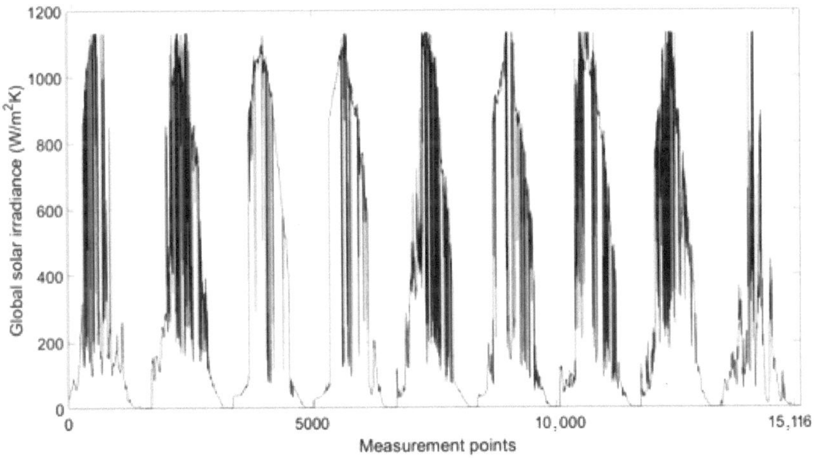

Figure 5. Global irradiance on the days used during the validation.

3. Results

3.1. Identification

The values of heat transfer coefficients after the identification of the HWB model are $k_{ae} = -7.14$ [W/m^{2}K] and $k_{we} = -1.96$ [W/m^{2}K]. These values are negative despite the fact that they should be positive according to the physical content of the model. However, when the model is run for the days of identification, it works excellently with the obtained negative parameter values. This can be seen in Figure 6, and the average absolute error of 0.58 °C and the average relative error of 1% also confirms this.

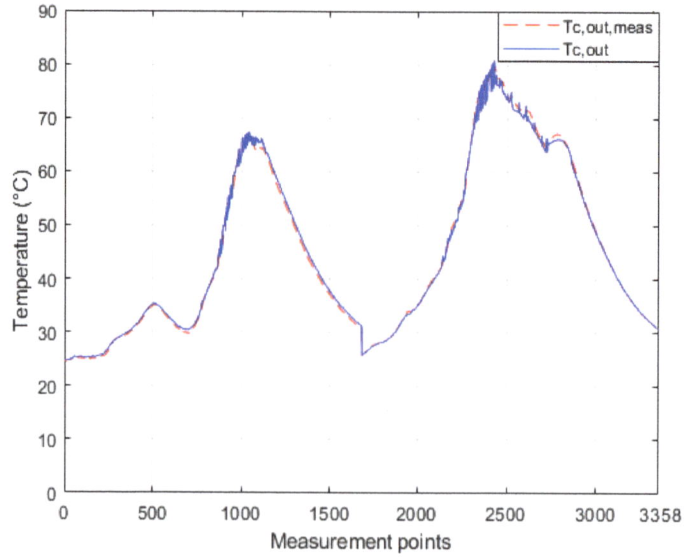

Figure 6. Measured and modelled collector outlet temperatures for the days of identification (HWB model).

The value of the optical efficiency after the identification of the EE model is $\eta_o = 0.68$ [-], and the overall heat transfer coefficient is $U = 1.54$ [W/m^{2}K]. The model works excellently with the obtained parameter values when it is run for the days of identification (Figure 7). The average absolute error is 0.41 °C and the average relative error is 0.7%.

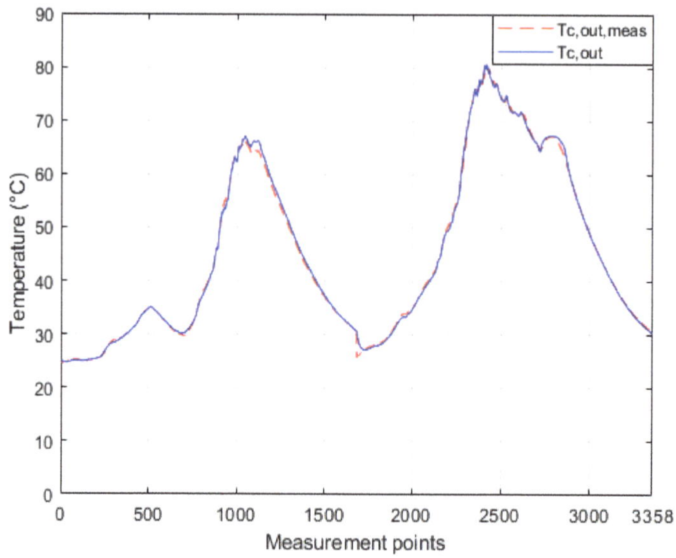

Figure 7. Measured and modelled collector outlet temperatures for the days of identification (EE model).

3.2. Validation

Despite the negative parameter values, the HWB model works with high accuracy even with the independent data, so the validation was successful. Figure 8 shows the measured and modelled collector outlet temperatures. The average absolute error is 0.8 °C and the average relative error is 1%.

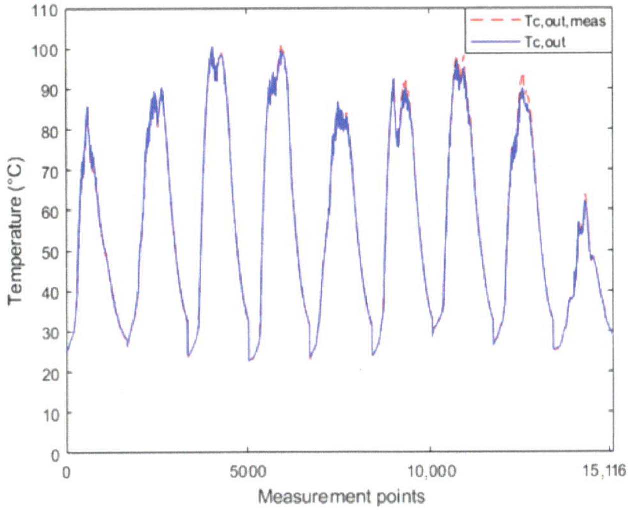

Figure 8. Measured and modelled collector outlet temperatures for the days of validation (HWB model).

The EE model also works with high accuracy with the independent data, so the validation proved to be successful in this case as well. The measured and modelled outlet temperatures are shown in Figure 9. The average absolute error is 0.87 °C, and the average relative error is 1.1%.

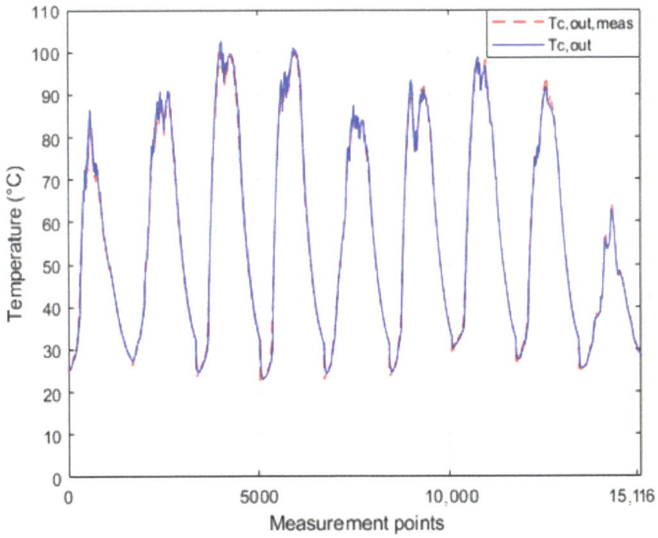

Figure 9. Measured and modelled collector outlet temperatures for the days of validation (EE model).

4. Conclusions

The aim of this research was to investigate the applicability of the HWB model, created for flat collectors, to evacuated tube solar collectors. During this research, the model was applied on an evacuated tube collector, and then the model was identified and validated with the help of measurements made on the collector. This research also included the application, identification and validation of the EE model for the tested solar collector. This model is applicable to both flat-plate and evacuated tube collectors. Then, a comparison of the results obtained with the two different models was made. The accuracy and applicability of the model can be determined by comparing the modelled results with the measured values. The identification of the HWB model resulted in negative heat transfer coefficients, even though they should be positive according to the model's physical content. The reason for this is most likely the operating principle of the evacuated tube solar collector, which is different from the one of flat-plate collectors. Evacuated tube collectors involve more complex heat transfer processes, including boiling and condensation, which are not adequately captured by the simpler HWB model designed for flat-plate collectors. The geometry of evacuated tube collectors, such as the presence of absorber fins and the vacuum between the absorber and the glass tube, affects the heat transfer differently compared to flat-plate collectors. These structural differences are not accounted for in the HWB model. The mentioned discrepancies can lead to negative parameters when the model is applied to evacuated tube collectors. By extending the HWB model (which was originally created for flat-plate collectors) to evacuated tube collectors, these parameters, instead of being the usual heat transfer coefficients, serve as their generalization and thus lose their precise physical meaning. This explains why the model works with high accuracy during validation with the obtained negative parameter values. Based on the results, the HWB model can also be used excellently in the case of evacuated tube collectors for predicting the outlet temperature of the solar collector's working medium. Based on the validation, the average absolute error was 0.8 °C, and the average relative error was 1%. The same values for the EE model were 0.87 °C and 1.1%, so the accuracy of the HWB model created for flat collectors was very similar, and even slightly higher. During the research, the applicability of the EE model to evacuated tube collectors was further proven.

Author Contributions: Conceptualization, M.R.; methodology, M.R. and L.S.; software, M.R.; investigation, M.R. and G.G.; resources, M.R., G.G. and L.S.; data curation, M.R. and G.G.; writing—original draft preparation, M.R.; writing—review and editing, M.R., G.G. and L.S.; visualization, M.R. and G.G.; supervision, L.S.; project administration, M.R., G.G. and L.S. All authors have read and agreed to the published version of the manuscript.

Funding: This research received no external funding.

Institutional Review Board Statement: Not applicable.

Informed Consent Statement: Not applicable.

Data Availability Statement: The data used to support findings of this study are available from the corresponding author upon request.

Conflicts of Interest: The authors declare no conflicts of interest.

References

1. Hachicha, A.A.; Yousef, B.A.A.; Said, Z.; Rodríguez, I. A review study on the modeling of high-temperature solar thermal collector systems. *Renew. Sustain. Energy Rev.* **2019**, *112*, 280–298. [CrossRef]
2. Kalogirou, S.A.; Panteliou, S.; Dentsoras, A. Modeling of Solar Domestic Water Heating Systems Using Artificial Neural Networks. *Sol. Energy* **1999**, *65*, 335–342. [CrossRef]
3. Brus, L.; Zambrano, D. Black-box identification of solar collector dynamics with variant time delay. *Cont. Eng. Prac.* **2010**, *18*, 1133–1146. [CrossRef]
4. Zheng, J.; Febrer, R.; Castro, J.; Kizildag, D.; Rigola, J. A new high-performance flat-plate solar collector. Numerical modelling and experimental validation. *Appl. Energy* **2023**, *355*, 1–14. [CrossRef]

5. Castillo Alvarez, Y.; González González, Y.; Jiménez Borges, R.; Iturralde Carrera, L.A.; Álvarez-Alvarado, J.M.; Rodríguez-Reséndiz, J. Energy Efficiency and Mathematical Modeling of Shrimp Pond Oxygenation: A Multiple Regression Experimental Study. *Eng* **2024**, *5*, 2862–2885. [CrossRef]
6. Sánchez-Mora, H.; Quezada-García, S.; Polo-Labarrios, M.A.; Cázares-Ramírez, R.I.; Torres-Aldaco, A. Dynamic mathematical heat transfer model for two-phase flow in solar collectors. *Case Stud. Therm. Eng.* **2022**, *40*, 1–13. [CrossRef]
7. Amrizal, N.; Chemisana, D.; Rosell, J.I.; Barrau, J. A dynamic model based on the piston flow concept for the thermal characterization of solar collectors. *Appl. Energy* **2012**, *94*, 244–250. [CrossRef]
8. Buzás, J.; Farkas, I. Solar domestic hot water system simulation using blockoriented software. In Proceedings of the 3rd ISES-europe Solar World Congress, CD-ROM Proceedings, København, Denmark, 19–22 June 2000; pp. 1–9.
9. Székely, L.; Kicsiny, R.; Hermanucz, P.; Géczi, G. Explicit analytical solution of a differential equation model for solar heating systems. *Sol. Energy* **2021**, *222*, 219–229. [CrossRef]
10. Castellanos, L.S.M.; Noguera, A.L.G.; Velásquez, E.I.G.; Caballero, G.E.C.; Lora, E.E.S.; Cobas, V.R.M. Mathematical modeling of a system composed of parabolic trough solar collectors integrated with a hydraulic energy storage system. *Energy* **2020**, *208*, 1–16. [CrossRef]
11. Iranmanesh, M.; Akhijahani, H.S.; Jahromi, M.S.B. CFD modeling and evaluation the performance of a solar cabinet dryer equipped with evacuated tube solar collector and thermal storage system. *Renew. Energy* **2020**, *145*, 1192–1213. [CrossRef]
12. Cáceres, M.; Avila, C.; Rivera, E. Thermodynamics-Informed Neural Networks for the Design of Solar Collectors: An Application on Water Heating in the Highland Areas of the Andes. *Energies* **2024**, *17*, 4978. [CrossRef]
13. Chen, C.; Wu, L.; Yang, Y. Estimation of Time-Varying Inlet Temperature and Heat Flux in Turbulent Circular Pipe Flow. *J. Heat Trans.* **2006**, *128*, 44–52. [CrossRef]
14. Kicsiny, R. Grey-box model for pipe temperature based on linear regression. *Int. J. Heat Mass Trans.* **2017**, *107*, 13–20. [CrossRef]
15. Hiris, D.P.; Pop, O.G.; Balan, M.C. Analytical modeling and validation of the thermal behavior of seasonal storage tanks for solar district heating. *Energy Rep.* **2022**, *8*, 741–755. [CrossRef]
16. Badescu, V. Optimal control of flow in solar collector systems with fully mixed water storage tanks. *Energy Conv. Man.* **2008**, *49*, 169–184. [CrossRef]
17. Kicsiny, R. Black-box model for solar storage tanks based on multiple linear regression. *Renew. Energy* **2018**, *125*, 857–865. [CrossRef]
18. Eicker, U. *Solar Technologies for Buildings*; John Wiley & Son, Ltd.: Stuttgart, Germany, 2003. [CrossRef]
19. Lago, J.; De Ridder, F.; Mazairac, W.; De Schutter, B. A 1-dimensional continuous and smooth model for thermally stratified storage tanks including mixing and buoyancy. *Appl. Energy* **2019**, *248*, 640–655. [CrossRef]
20. Buonomano, A.; Calise, F.; Palombo, A. Solar heating and cooling systems by CPVT and ET solar collectors: A novel transient simulation model. *Appl. Energy* **2013**, *103*, 588–606. [CrossRef]
21. Zurigat, Y.H.; Maloney, K.J.; Ghajar, A.J. A Comparison Study of One-Dimensional Models for Stratified Thermal Storage Tanks. *J. Sol. Energy Eng.* **1989**, *111*, 204–210. [CrossRef]
22. Kleinbach, E.M.; Beckmann, W.A.; Klein, S.A. Performance study of one-dimensional models for stratified thermal storage tanks. *Sol. Energy* **1993**, *50*, 155–166. [CrossRef]
23. Cadau, N.; De Lorenzi, A.; Gambarotta, A.; Morini, M.; Rossi, M. Development and Analysis of a Multi-Node Dynamic Model for the Simulation of Stratified Thermal Energy Storage. *Energies* **2019**, *12*, 4275. [CrossRef]
24. Bradley, J. Counterflow, crossflow and cocurrent flow heat transfer in heat exchangers: Analytical solution based on transfer units. *Heat Mass Trans.* **2010**, *46*, 381–394. [CrossRef]
25. Pacio, J.C.; Dorao, C.A. A review on heat exchanger thermal hydraulic models for cryogenic applications. *Cryogenics* **2011**, *51*, 366–379. [CrossRef]
26. Zohuri, B. *Compact Heat Exchangers, Selection, Application, Design and Evaluation*; Springer International Publishing: Cham, Switzerland, 2017. [CrossRef]
27. Géczi, G.; Kicsiny, R.; Korzenszky, P. Modified effectiveness and linear regression based models for heat exchangers under heat gain/loss to the environment. *Heat Mass Trans.* **2019**, *55*, 1167–1179. [CrossRef]
28. Zavala-Río, A.; Santiesteban-Cos, R. Reliable compartmental models for double-pipe heat exchangers: An analytical study. *Appl. Math. Mod.* **2007**, *31*, 1739–1752. [CrossRef]
29. Hilmer, F.; Vajen, K.; Ratka, A.; Ackermann, H.; Fuhs, W.; Melsheimer, O. Numerical solution and validation of a dynamic model of solar collectors working with varying fluid flow rate. *Sol. Energy* **1999**, *65*, 305–321. [CrossRef]
30. Mao, C.; Li, M.; Li, N.; Shan, M.; Yang, X. Mathematical model development and optimal design of the horizontal all-glass evacuated tube solar collectors integrated with bottom mirror reflectors for solar energy harvesting. *Appl. Energy* **2019**, *238*, 54–68. [CrossRef]
31. Korres, D.N.; Tzivanidis, C.; Koronaki, I.P.; Nitsas, M.T. Experimental, numerical and analytical investigation of a U-type evacuated tube collectors' array. *Renew. Energy* **2019**, *135*, 218–231. [CrossRef]
32. Hottel, H.C.; Woertz, B.B. The performance of flat-plate solar-heat collectors. *Trans. Am. Soc. Mech. Eng.* **1942**, *64*, 91–104.
33. Hottel, H.C.; Whillier, A. Evaluation of flat-plate collector performance. *Trans. Conf. Use Sol. Energy* **1955**, *3*, 74–104.
34. Géczyné Víg, P. Modelling of Solar Collector Systems with Neural Network. Ph.D. Thesis, Szent István University, Gödöllő, Hungary, 2007.

35. Levenberg, K. A method for the solution of certain non-linear problems in least squares. *Quart. Appl. Math.* **1944**, *2*, 164–168. [CrossRef]
36. Marquardt, D.W. An Algorithm for Least-Squares Estimation of Nonlinear Parameters. *J. Soc. Ind. Appl. Math.* **1963**, *11*, 431–441. [CrossRef]
37. Buzás, J.; Farkas, I.; Biró, A.; Németh, R. Modelling and simulation aspects of a solar hot water system. *Math. Comp. Sim.* **1998**, *48*, 33–46. [CrossRef]
38. Kicsiny, R. Multiple linear regression based model for solar collectors. *Sol. Energy* **2014**, *110*, 496–506. [CrossRef]
39. Florschuetz, L.W. Extension of the Hottel-Whillier model to the analysis of combined photovoltaic/thermal flat plate collectors. *Sol. Energy* **1979**, *22*, 361–366. [CrossRef]
40. Tiwari, G.N.; Meraj, M.; Khan, M.E.; Mishra, R.K.; Garg, G. Improved Hottel-Whillier-Bliss equation for N-photovoltaic thermal-compound parabolic concentrator (N-PVT-CPC) collector. *Sol. Energy* **2018**, *166*, 203–212. [CrossRef]

eng

Article

On the Game-Based Approach to Optimal Design

Vladimir Kobelev

Faculty IV, School of Science and Technology, University of Siegen, Paul-Bonatz-Str. 9–11, D-57076 Siegen, Germany; vladimir.kobelev@uni-siegen.de

Abstract: A game problem of structural design is defined as a problem of playing against external circumstances. There are two classes of players, namely the "ordinal" and "cardinal" players. The ordinal players, designated as the "operator" and "nature", endeavor to, respectively, minimize or maximize the payoff function, operating within the constraints of limited resources. The fundamental premise of this study is that the action of player "nature" is a priori unknown. Statistical decision theory addresses decision-making scenarios where these probabilities, whether or not they are known, must be considered. The solution to the substratum game is expressed as a value of the game "against nature". The structural optimization extension of the game considers the value of the game "against nature" as the function of certain parameters. Thus, the value of the game is contingent upon the design parameters. The cardinal players, "designers", choose the design parameters. There are two formulations of optimization. For the single cardinal player, the pursuit of the maximum and minimum values of the game reduces the problem of optimal design. In the second formulation, there are multiple cardinal players with conflicting objectives. Accordingly, the superstratum game emerges, which addresses the interests of the superstratum players. Finally, the optimal design problems for games with closed forms are presented. The game formulations could be applied for optimal design with uncertain loading, considering "nature" as the source of uncertainty.

Keywords: structural optimization; game theory; design under uncertainty

Citation: Kobelev, V. On the Game-Based Approach to Optimal Design. *Eng* **2024**, *5*, 3212–3238. https://doi.org/10.3390/eng5040169

Academic Editor: Antonio Gil Bravo

Received: 10 October 2024
Revised: 19 November 2024
Accepted: 28 November 2024
Published: 4 December 2024

1. Introduction

1.1. The overarching objective of optimal structural design theory is to identify a structure that optimizes a specific mechanical characteristic while adhering to the prescribed constraints. In the classical optimal design problem, there is a single decision-maker, or "designer", who must consider the shapes, sizes, material properties, and mutual positions of the structural members. These parameters are referred to as design parameters. Additionally, the conditions of exploitation and external actions on the structure must be defined.

The scalar and vector problems are commonly studied in structural optimization. The goal of scalar design optimization is to select the values of the design variables, given various constraints, so that a single objective function reaches an extreme value. A characteristic feature of such multi-criteria optimization problems is the occurrence of objective conflict, i.e., none of the feasible solutions allows for the simultaneous minimization of all objectives or the individual solutions of each individual objective function differ. Consequently, multiobjective optimization deals with all kinds of conflicting problems, (Section 1.2 in [1]). In the current context, the scalar optimization formulation is represented by an antagonistic game formulation. Analogously, the vector optimization formulation corresponds to the bi-matrix game. Each of the two players has its own payoff matrix. These concepts will be discussed in detail later.

1.2. Game theory is a framework for understanding situations of conflict and cooperation between rational decision-makers. It builds on the ideas of mainstream decision theory and economics, which say that people act rationally when they choose actions that maximize their payoff, given the constraints they face. The field of game theory is primarily concerned with the logical foundations of decision-making processes in situations where the outcome is contingent upon the actions of two or more autonomous agents. A crucial aspect of such scenarios is that each decision-maker possesses only partial control over the resulting outcomes. The phrase "the theory of interdependent decision-making" more accurately encapsulates the core tenets of the theory. Game theory pertains to situations wherein the options at the disposal of each decision-maker and their potential consequences are clearly delineated, and each decision-maker exhibits consistent preferences regarding the prospective outcomes. The primary objective of game theory is to identify solutions to games. Matrix games are two-person games with finite strategy sets. A matrix game with the matrix $A = \{a_{ij}\}$ of size $m \times n$ is a zero-sum game with two players. In classical game theory, the coefficients of the payoff matrix are the given fixed values. The first player has m strategies. This player is referred to as "nature" in the current context. The opposite player has n strategies. In this context, the second player is called "operator". The payoff function of the first player is defined by $a_{i,j}$. The payoff function of the second player is defined by $-a_{i,j}$. The first player selects a row i of the matrix and wants to maximize $a_{i,j}$, while the opponent second player selects a column j of the matrix and wants to minimize $a_{i,j}$. A matrix game is a zero-sum game, which means that, when the strategies of row and column players are fixed, the sum of the payoffs for the two players is zero.

The most important theorem in matrix games is Neumann's minimax theorem [2]. Neumann's theorem was proved using Brouwer's fixed point theorem. Another proof from 1944 was based on dual linear programming [3]. A solution is defined as a set of criteria that delineates the decision to be made and the subsequent outcome that will be reached if the decision-makers adhere to the established rationality criteria. One potential approach to structural optimal design can be formulated as follows. External actions are suboptimal in that they result in the greatest stress intensity, maximal deflection, or highest level of fracture. In game theory, the term "payoff function" is typically used in place of the "aim function", which is the goal of the optimal design problem. Game theory is concerned with analyzing the conflict situations in which the participants are conscious and rational beings trying to achieve a certain goal. However, in many cases, one of the participants cannot be considered a conscious individual with preferences and goals. Consequently, the other players cannot assume that this participant will behave rationally.

1.3. The application of game theory to optimal design is a long-standing field of study. It has been demonstrated [4] that the role of the payoff function in ensuring integral compliance can be expressed in terms of the minimal eigenvalue of the inverse operator of the system (continual case) or the response matrix (discrete case). This is a fundamental characteristic of the game, and it is, therefore, referred to as the upper game value. A game-theoretic approach to robust topological optimization with uncertain loading is illustrated using three different games for the design of both two-dimensional and three-dimensional structures.

1.4. For optimization purposes, the payoff matrix is modified by the player "designer" acting according to the "cardinal" strategy. The alteration is made in favor of the player who receives the winnings. Once the "substratum" game has commenced, it is no longer possible to modify the payout matrix. The degrees of freedom of the "cardinal" players are known under the same names as those used in classical structural optimization. The degrees of freedom of the "cardinal" players are identified as "control functions" or "design parameters". In the event that there is only one "cardinal" player, no conflict of interest arises. In the event that there are multiple "cardinal" players with disparate interests, a situation of conflict may arise. The aforementioned conflict gives rise to the "superstratum" game on the upper level. This scenario is typical in the field of engineering. For example, the objective of the system designer is to achieve the optimal performance of the system. The production designer seeks to implement the system in the most effective and cost-efficient manner. The operating ecology manager aims to ensure the lowest emission level during the system's operational phase. The production ecology manager recognizes the necessity of reducing emissions during the manufacturing stage. Ultimately, the customer strives to minimize the system's overall cost and operational expenses. However, these objectives are not always aligned, leading to potential conflicts among the "cardinal players".

1.5. The term "stratified game" can be more easily defined by reference to the concept of matrix games. The typical game formulations utilize a fixed game matrix. The objective of the game is to identify the optimal strategies for the players, given the predefined and unchanging game matrix. In light of the aforementioned statements, the term "stratified game" is defined as a game formulation with a controllable value. In a stratified game, the elements of the matrix are predefined functions of the design parameters. Accordingly, the value of the game is a function of the design parameters. The value of the game is the result of the lower-level (substratum) game task for each fixed set of design parameters. The determination of the maximum value of the game value as a function of the design parameters is the upper-level (superstratum) task. For an upper-level task, the methods of optimization or control theories are applied. The occurrence of these two levels leads to the term "stratified game".

The mentioned line of reasoning can be applied to diverse optimization problems that involve multiple levels of decision-making. To illustrate, lawmakers can be considered the "superstratum" of a social game. The actions of the superstratum players shape the governing equations of the game, such as those related to taxation or ecological laws. Meanwhile, the players in the "substratum" adhere to the legislation that defines their objectives and the associated modus vivendi.

1.6. There is a fundamental difference between the concepts of game theory in social sciences and operations theory on the one hand, and in engineering and natural sciences on the other. The difference lies in the understanding of the resource limits of the game players. It is possible to examine the issue of resource constraints from the perspective of functional analysis. In functional analysis and related areas of mathematics, a sequence space is a vector space whose elements are infinite sequences of real or complex numbers. L_p-norm is a norm on the space of p-integrable functions [5]. These are special cases of L_p spaces for the counting measure on the set of natural numbers. The most important sequence spaces in the analysis are the l_p spaces, which consist of sequences summable to the p-th power with the p-norm. l_p-norm denotes the norm in the space of p-summable sequences.

The total resources in social sciences and operations theory are the linear sum of the partial resources of each player. For example, the bank account represents the total amount of money of the player. Each element of the sequence is the transaction to or from the account. That is, the actual state of the account is the linear sum of sequences. Each operation is the linear addition of some amount of money. The power p in the sequences that arise in the usual game theory of social sciences is one. The summation of probabilities in the decision-making theory is linear per definition as well. The corresponding game theory with linearly summed resources (i.e., $p = 1$) is fully established in well-known works [2,3] and does not require revision or extension.

The application of game-theoretic methods to problems of a physical nature presents a different situation. The total resources in engineering and science are the quadratic sum of the partial resources of each player (i.e., $p = 2$). The Euclidean norm, standard norm, or 2-norm is a vector norm commonly used in science and engineering. In two- and three-dimensional Euclidean space, the Euclidean norm corresponds to the visual length or magnitude of a vector and can be calculated using the Pythagorean theorem. More generally, the Euclidean norm is also defined for real and complex vector spaces of arbitrary finite dimension and is then the norm derived from the standard scalar product. For example, the total energy of vibration is composed of a sequence of vibrational modes. From the viewpoint of the sequence space concept, each element of the sequence is the energy of a vibrational mode. The total energy is the quadratic sum of the partial energies of each mode. That is, the total energetic resource is the quadratic sum of sequences. The power p in the sequences arising in the common game theory for engineering and science is two. The possible applications of quadratic sequences to the quadratic estimation of energy are numerous. The Poynting vector represents the directional energy flux (the energy transfer per unit area, per unit time) or power flux of an electromagnetic field. The intensity of an electromagnetic wave is the average of the squares of the sinusoidal functions for each mode. The quadratic summation of sequences is widely used in the natural sciences and is used to evaluate the energy of seismic waves and the energy of wind and ocean currents and to calculate the elastic deformation energy of external excitations. In light of the application of game theory in engineering, it is necessary to revisit the concept of resource limitation, which should be revised from a linear summation to a quadratic one.

1.7. In summary, the tasks of the study and the central ideas for their solution are the following.

- The fundamental premise of this study is that the probability distribution governing nature's "choice" of states remains unknown. Statistical decision theory addresses decision-making scenarios where these probabilities, whether objective or subjective, are pertinent factors;
- In classical structural optimization, the "cardinal" players are responsible for assuming the role of "design parameters" or "control functions". If the number of design freedoms is finite, one speaks about "design parameters". For the continuous design freedom, the "control functions" are involved according to Pontryagin's maximum principle [6]. In an actual context, the "cardinal" players modify the governing equations and payoff functions in the game formulations [1]. In structural optimization, the "cardinal" players are responsible for determining the coefficients of the governing equations. In essence, their function is to establish the rules of engagement, which may result in conflict between certain players. In the context of the stratified game approach, the "cardinal" players form the "superstratum", or the upper level of the game. The "cardinal payers" are designated as "designers" due to their capacity to alter, for instance, the coefficients of matrices in matrix games with the objective of attaining a superior value for the game;
- Furthermore, other participants act in accordance with the governing equations, which are determined by the "cardinal" players. In the context of the stratified game approach, the "ordinal" players represent the "substratum". "Ordinal" players are permitted to make decisions within their respective stratum, but they are unable to impact the governing equations. The "ordinal" payers could be referred to as "nature" or "operators". Certain "ordinal" participants represent the external forces. These external factors are typically referred to as "nature". For the sake of clarity, it may be more appropriate to refer to nature's strategies as "states" rather than strategies. Such games are, therefore, classified as "games against nature". The remaining "ordinal" participants aim to offset the impact of "nature" in order to mitigate potential risks or to achieve the most favorable outcome [4]. For the sake of clarity, these participants will henceforth be referred to as "operators" and "nature". In the case of an antagonistic matrix game, the payout matrix is only relevant for "operators". The role of "cardinal players" is inherited from the common optimization formulations;
- The conflict between the two "ordinal" players, namely "nature" and "operators", is studied for the linearly summed resources using the common principles of game theory. This case is typical for interdependent decision-making in operational research and social games;
- The application of game formulations for problems of engineering and natural sciences assumes the quadratically summed resources of the ordinal players. The solutions of matrix and bi-matrix games with quadratically summed resources are essential for the application of a game-theoretical approach to engineering and physical tasks.

1.8. The following summarizes the results of the current study.

- In the matrix game, there is a single payoff matrix. From the perspective of optimization theory, there is a single objective function. In consequence, the optimization problem is of a scalar nature. This value represents the goal function of the matrix game. At the low-level (substratum level), the win of "nature" is the loss of the "operator" This game is antagonistic, with a single goal function. The theory of quadratically constrained matrix and bi-matrix games is presented in Sections 2 and 3;

- The solution of the stratified matrix game is presented in Section 2. In a bi-matrix game, payoff matrices are employed. From the perspective of optimization theory, there are two distinct goal functions. The optimization problem is of a vector type. These goal functions are derived from the solution of a bi-matrix game. The equilibrium between the interests of "nature" and "operator" results in a solution to the game at the lower level. The designer determines the optimal value for the goal function in accordance with the methods of vector optimization. The solution of the stratified bi-matrix game is presented in Section 3;
- In Section 4, the results of the aforementioned sections are generalized for self-adjoint positive definite differential operators. The optimization of elastic energy was selected as an example of an optimization task with a closed form of solution. The action of "Nature" is the external load. The load that results in the greatest structural response among all permissible loads and represents the strategy of "nature" in low-level tasks. The upper value of the game could be determined in all cases. The third player, the "designer", acts on the upper level. The "Designer" determines the optimal value for the aforementioned goal function, manipulating the "design variables". The designer optimizes the upper value, creating a stiffer structural element for the most dangerous action of "nature". The corresponding examples are given in Sections 4–7;
- For a special type of game with stored energy, the value of the game is shown to be equal to the eigenvalue of the structural matrix. The optimal strategies of the "operator" and "nature" are uniquely determined. In the simplest formulation, the resources of the "operator" are limited to zero. In this case, there is no opposite reaction to all possible actions of "nature" on the lower level. In this case, the role of the opposite player takes the cardinal player, the "designer". The "designer" modifies the coefficients of the payoff matrix, while "nature" alters both the left and right vectors of the payoff function;
- The application of the aforementioned considerations to structural optimization problems will be discussed in the following sections. Sections 5 and 6 demonstrate the applications of the developed technique for structural optimization problems with continuous control functions. In Section 5, the beam is subjected to arbitrary bending efforts, and the designer determines the optimal beam shape for resisting these forces. In Section 6, the rod is subjected to arbitrary twist moments, and the designer determines the optimal design of the twisted bar, ensuring the greatest stiffness;
- Section 7 presents a mathematical analysis of the solutions presented in Sections 5 and 6, employing the tools of inequalities theory to elucidate the underlying mathematical principles of the PARETO and NASH fronts.

2. Antagonistic Matrix Stratified Games

2.1. This section contains basic information from the theory of finite antagonistic (matrix) games. The existence theorem of the equilibrium situation in the class of mixed strategies, the properties of optimal mixed strategies, and methods for solving matrix games are well-established areas of research.

The study of game theory commences with the most basic static model, namely a matrix game in which two players engage, the set of strategies available to each player is finite, and the gain of one player is equal to the loss of the other:

$$\Gamma_{K[\boldsymbol{a}]} = (X, Y, K[\boldsymbol{a}]), \tag{1}$$

where X and Y are nonempty sets, and the function $K[\boldsymbol{a}] : X \times Y \to R^1$ is called an antagonistic stratified game in normal form. The elements $x \in X$ and $y \in Y$ are called the strategies of ordinal players 1 and 2, respectively, in the game $\Gamma_{K[\boldsymbol{a}]}$. The elements of the Cartesian product $X \times Y$ (i.e., the pair of strategies (x, y) where $x \in X$ and $y \in Y$) are situations, and the function $K[\boldsymbol{a}]$ is the win function of player 1. The payoff of player 2 in situation (x, y) is assumed to be equal to $[-K[\boldsymbol{a}](x, y)]$. Therefore, the function $K[\boldsymbol{a}]$ is also called the win function of game $\Gamma_{K[\boldsymbol{a}]}$ itself, and game $\Gamma_{K[\boldsymbol{a}]}$ is called a zero-sum game.

Thus, using the accepted terminology to define game $\Gamma_{K[\boldsymbol{a}]}$, it is necessary to define the sets of strategies X, Y of ordinal players 1 and 2 and also the winning function $K[\boldsymbol{a}]$, defined on the set of all situations $X \times Y$.

The stratified game $\Gamma_{K[\boldsymbol{a}]}$ is interpreted as follows. Ordinal players simultaneously and independently choose strategies $x \in X$, $y \in Y$. In the substratum game, ordinal player 1 then receives a payoff equal to $K[\boldsymbol{a}](x, y)$, and ordinal player 2 receives $(-K[\boldsymbol{a}](x, y))$. The elements $\boldsymbol{a} \in$ A are called the strategies of cardinal players in the stratified game $\Gamma_{K[\boldsymbol{a}]}$.

In a stratified game, the elements $\boldsymbol{a} \in$ A are referred to as the strategies of cardinal players. The superstratum game pertains to the strategies for cardinal players $\boldsymbol{a}$ that ensure the maximal and minimal values of the substratum payoff function. When there is only one cardinal player, the superstratum game reduces to the optimization problem. In contrast, when there are two or more cardinal players, their interests may be in opposition, resulting in what is known as an antagonistic game. The total payoff for both ordinal players on the lower level is zero. The goal function on the upper level is the payoff of the oppositely acting ordinal players, e.g., the "nature" and "operator". This objective function is modified by the "designer" at the upper level.

2.2. The following definition is proposed: antagonistic games in which both ordinal players possess finite strategy sets are designated as substratum matrix games.

In the matrix game, ordinal player 1 is assumed to have only m strategies. The set of strategies available to the first ordinal player, X, must be ordered [7,8]. That is, a one-to-one correspondence must be established between X and $M = \{1, 2, \ldots, m\}$. The same process must be repeated for the second ordinal player, with $N = \{1, 2, \ldots, n\}$ and Y. The sets M and N are then ordered in a one-to-one correspondence with X and Y, respectively, where M and N are finite sets of the cardinalities m and n, respectively.

The substratum matrix game Γ is thus completely defined by the matrix $\boldsymbol{A} = \boldsymbol{A}[\boldsymbol{a}]$, where $\boldsymbol{A}$ is defined as follows:

$$\boldsymbol{A} = \{a_{ij}\}, \tag{2}$$

where:

$$\begin{gathered} a_{ij} = K(x_i, y_j) \\ (x_i, y_j) \in X \times Y, \quad (i, j) \in M \times N, \\ i \in M, \quad j \in N. \end{gathered}$$

This is the rationale behind the name of the game, which is derived from the aforementioned matrix. In this instance, the game Γ is realized as follows. Player 1 selects a row, $i \in M$, and player 2 (simultaneously with player 1 and independently of him) chooses a column, $j \in N$. Ordinal player 1 then receives a payoff, a_{ij}, and ordinal player 2 receives $-a_{ij}$. In the event that the payoff is a negative number, it constitutes an actual loss for the ordinal player.

We denote the substratum game Γ with win matrix $\boldsymbol{A}[\boldsymbol{a}]$ by $\Gamma_{A[\boldsymbol{a}]}$ and refer to it as an $(m \times n)-$ game, in accordance with the dimensions of matrix $\boldsymbol{A}[\boldsymbol{a}]$ with the fixed values of strategies for cardinal players $\boldsymbol{a}$.

2.3. The question of the optimal behavior of players in an antagonistic game is worthy of consideration. It is reasonable to conclude that a situation (x^*, y^*) in the game $\Gamma_{K[\boldsymbol{a}]} = (X, Y, K[\boldsymbol{a}])$ is optimal if deviating from it is not favorable for any of the players. Such a situation (x^*, y^*) is referred to as an equilibrium, and the optimality principle based on the construction of an equilibrium situation is known as the equilibrium principle.

For an equilibrium situation to exist in substratum game $\Gamma_{K[\boldsymbol{a}]} = (X, Y, K[\boldsymbol{a}])$, it is necessary and sufficient that there exist a minimax and a maximin:

$$\min_y \sup_x K[\boldsymbol{a}](x, y), \ \max_x \inf_y K[\boldsymbol{a}](x, y),$$

and the equality is satisfied:

$$v^*[\boldsymbol{a}] = \min_y \sup_x K[\boldsymbol{a}](x,y) = \max_x \inf_y K[\boldsymbol{a}](x,y) = v^{**}[\boldsymbol{a}]. \quad (3)$$

Equation (3) establishes a connection between the equilibrium principle and the minimax and maximin principles in an antagonistic game. Games in which equilibrium situations exist are called well-defined games. Therefore, this theorem establishes a criterion for a well-defined game and can be reformulated as follows. For a game to be well-defined, it is necessary and sufficient that there exist min sup and max inf in Equation (3), and the equality in minimax is satisfied.

If there exists an equilibrium situation, then the minimax is equal to the maximin. According to the definition of the equilibrium situation, each player can communicate his optimal (maximin) strategy to the opponent, and neither player can obtain an additional benefit from it.

2.4. Now, suppose that there is no equilibrium situation in the substratum game $\Gamma_{A[\boldsymbol{a}]}$. Since a random variable is characterized by its distribution, we will further identify a mixed strategy with a probability distribution on the set of pure strategies of ordinal players. Thus, the mixed strategy x of ordinal player 1 in the substratum game is an m-dimensional vector, which is constrained by the following equation:

$$\boldsymbol{x} = (\xi_1, \ldots, \xi_m), \qquad \|\boldsymbol{x}\|_{p,m} = 1. \quad (4)$$

where the L^p norm in the real vector space $\mathbb{R}^q$ is defined as [5]:

$$\|\boldsymbol{x}\|_{p,q} \stackrel{\text{def}}{=} \left(\sum_{i=1}^{q} \xi_i{}^p\right)^{1/p}. \quad (5)$$

Similarly, the mixed strategy y of player 2 is an n-dimensional vector:

$$\boldsymbol{y} = (\eta_1, \ldots, \eta_n), \qquad \|\boldsymbol{y}\|_{p,n} = 1. \quad (6)$$

The positive natural number p determines the class of the game. Note that the numbers n and m must not equal.

2.5. If $p = 1$, the values $\xi_i \geq 0$ and $\eta_i \geq 0$ are the probabilities of choosing pure strategies $i \in M$ and $j \in N$, respectively, when ordinal players use mixed strategies x and y. Let us denote by X and Y the sets of mixed strategies of the first and second players, respectively. It is easy to see that the set of mixed strategies of each player is a compact in the corresponding finite-dimensional Euclidean space (a closed, bounded set). A mixed set represents an extension of the pure strategy space available to the player. An arbitrary matrix game is well-defined within the class of mixed random strategies. The von Neumann theorem of matrix games states that in the case $p = 1$, every matrix game has an equilibrium situation within the context of mixed strategies [7]. The cited literature provides an overview of the methods used to evaluate game values. The setting $p = 1$ is typical for the application of game theory fields of economics, political science, and the social sciences. This setting reflects the fact that the mixed strategy for ordinal players is simply a probability distribution over their pure strategies. The probability of any event must be positive, and the total probability of all events must be one. Consequently, any mixed strategy must adhere to the following conditions:

$$\|\boldsymbol{x}\|_{1,m} = 1, \quad \|\boldsymbol{y}\|_{1,m} = 1.$$

2.6. If $p = 2$, the values ξ_i and η_i are the Euclidian coordinates of vector strategies $\boldsymbol{x}, \boldsymbol{y}$ of the ordinal players. The Euclidean length of a vector $\boldsymbol{x}, \boldsymbol{y}$ in the real vector spaces $\mathbb{R}^n$ and $\mathbb{R}^m$ are given by their Euclidean norms:

$$\|\boldsymbol{x}\|_{2,m} \equiv \left(\sum_{i=1}^{m} \xi_i{}^2\right)^{\frac{1}{2}}, \qquad \|\boldsymbol{y}\|_{2,n} \equiv \left(\sum_{i=1}^{n} \eta_i{}^2\right)^{1/2}. \quad (7)$$

As illustrated in the aforementioned examples, this scenario is typical in the game formulations of engineering and physical applications. In such applications, the module of actions for the ordinal players is restricted. The modules of strategy vectors are less than or equal to one:

$$\|x\|_{2,m} \leq 1, \quad \|y\|_{2,n} \leq 1. \tag{8}$$

With definitions (4) and (5), the payoff function of the matrix game on the lower substratum level reads:

$$\mathcal{I}[x, y, a] = \frac{1}{2}\langle A[a]x, y\rangle. \tag{9}$$

The Lagrangian combines the payoff function (9) with the constraints (8), taken with the non-negative multipliers λ, μ:

$$\mathcal{L}[x, y, a] \stackrel{\text{def}}{=\!=} \mathcal{I}[x, y, a] + \lambda\Big(\|x\|_{2,m} - 1\Big) + \mu\Big(\|y\|_{2,n} - 1\Big). \tag{10}$$

For the payoff function (9) with the conditions (8), the equilibrium state (x^*, y^*) satisfies the equations:

$$\tfrac{\partial \mathcal{L}}{\partial x} \equiv A^T[a]y + \lambda x = 0, \quad \tfrac{\partial \mathcal{L}}{\partial y} \equiv A[a]x + \mu y = 0. \tag{11}$$

The resolution of Equation (11) reads:

$$A[a]A^T[a]y = \lambda\mu y, \quad A^T[a]A[a]x = \lambda\mu x. \tag{12}$$

The left sides of Equation (12) contain two auxiliary matrices:

$$\mathcal{K}_A \stackrel{\text{def}}{=\!=} A^T[a]A[a], \quad \mathcal{K}_B \stackrel{\text{def}}{=\!=} A[a]A^T[a].$$

The matrix $\mathcal{K}_A$ is an m square Hermitian matrix. The matrix $\mathcal{K}_B$ is an n square Hermitian matrix. A real symmetric matrix is Hermitian. A Hermitian matrix is always self-adjoint. It follows from (Theorem 2.8, Section 2.4 in [9]), that both matrices $\mathcal{K}_A$ and $\mathcal{K}_B$ have the same nonzero eigenvalues, counting multiplicity. The matrices $\mathcal{K}_A$ and $\mathcal{K}_B$ are positive-semidefinite (Theorem 7.3, Section 7.1 in [9]). The number of zero eigenvalues of $\mathcal{K}_A$ and $\mathcal{K}_B$ is at least $|\mathrm{m} - \mathrm{n}|$. Let $\mathcal{K}_1$ be the matrix with the smallest dimensions of $\mathcal{K}_A$ and $\mathcal{K}_B$. In other words,

$$\mathcal{K}_1 \stackrel{\text{def}}{=\!=} \begin{cases} \mathcal{K}_A & \text{if } m \leq n, \\ \mathcal{K}_B & \text{if } m > n. \end{cases}$$

Generally saying, the matrix $\mathcal{K}_1$ is positive-semidefinite. The eigenvalues of matrix $\mathcal{K}_1$ are:

$$\begin{gathered} \mathbb{L}_i = \text{eigenvalues}(\mathcal{K}_2) \geq 0, \\ i = 1, \ldots, j, \\ j \stackrel{\text{def}}{=\!=} \min(m, n). \end{gathered} \tag{13}$$

If $m = n$, then the matrices are equal $\mathcal{K}_A = \mathcal{K}_B$ and have the same set of eigenvalues.

If $\mathcal{K}_1$ is positive definite, the number of its zero eigenvalues is exactly $|\mathrm{m} - \mathrm{n}|$. The eigenvalues of the positive definite matrix $\mathcal{K}_S$ are:

$$\begin{gathered} \mathbb{L}_i = \text{eigenvalues}(\mathcal{K}_1) > 0, \\ i = 1, \ldots, j, \\ j \stackrel{\text{def}}{=\!=} \min(m, n). \end{gathered}$$

From Equation (12), it follows that $\lambda_i\mu_i = \mathbb{L}_i$. Finally,

$$\lambda_i = \mu_i = \sqrt{\mathbb{L}_i}, \;\; i = 1, \ldots, j. \tag{14}$$

Consequently, every matrix quadratic game possesses the equilibrium situation within the context of mixed strategies.

2.7. In this section, the solution (14) to the substratum matrix game with constrained actions (Equations (4), (6) and (8)) is identified. The objective of the superstratum task is to determine the optimal solution for the "designer" by searching for the extremal value of (14) through a manipulation of the design variables [10].

If the resources available to both players are identical, as illustrated in Equation (8), the resulting total payoff is zero. In the event that player $\boldsymbol{x}$ possesses a greater quantity of resources than player $\boldsymbol{y}$, the possibilities available to them are subject to different constraints: $\|\boldsymbol{x}\|_{2,m} > \|\boldsymbol{y}\|_{2,n}$. In this case, the "designer", who shares similar interests with the "operator", attempts to minimize the maximum potential damage. The maximal potential damage is represented by the maximal eigenvalue $\mathbb{L}_j$ with a negative sign. Consequently, the objective is to minimize the maximal eigenvalue, $\mathbb{L}_j[\boldsymbol{a}]$.

In the case that player $\boldsymbol{x}$ holds fewer resources, than player $\boldsymbol{y}$, we get the following inequality: $\|\boldsymbol{x}\|_{2,m} < \|\boldsymbol{y}\|_{2,n}$. The designer will ensure that the payoff is guaranteed to be certain outcome, regardless of the potential actions of the first player $\boldsymbol{x}$, designated as "nature". The equilibrium point is identified as the minimal positive eigenvalue $\mathbb{L}_1$. In this case the objective of "designer" is to maximize the minimal eigenvalue, $\mathbb{L}_1[\boldsymbol{a}]$. This latter scenario will be the focus of subsequent analysis.

3. Bi-Matrix Stratified Games

3.1. In game theory, a bi-matrix game is defined as a simultaneous game for two players, each of whom has a finite set of possible actions. Such a game can be represented by two matrices: matrix $\boldsymbol{A}$, which outlines the payoffs for ordinal player 1, and matrix $\boldsymbol{B}$, which outlines the payoffs for ordinal player 2. The substratum bi-matrix game deals with two $m \times n$ matrices, $\boldsymbol{A}[\boldsymbol{a}]$ and $\boldsymbol{B}[\boldsymbol{a}]$, whose elements depend parametrically upon the design vector $\boldsymbol{a}$ of the cardinal players. For the fixed values of strategies for cardinal players $\boldsymbol{a}$, the wins of the first and second ordinal players are, correspondingly:

$$\mathcal{I}_1[\boldsymbol{x},\boldsymbol{y},\boldsymbol{a}] = \tfrac{1}{2}\langle \boldsymbol{A}[\boldsymbol{a}]\,\boldsymbol{x},\boldsymbol{y}\rangle,\quad \mathcal{I}_2[\boldsymbol{x},\boldsymbol{y},\boldsymbol{a}] = \tfrac{1}{2}\langle \boldsymbol{B}[\boldsymbol{a}]\,\boldsymbol{x},\boldsymbol{y}\rangle. \tag{15}$$

The mixed strategies of both ordinal players at the substratum level of the game are vectors, which are constrained by the following equations:

$$\begin{aligned} \boldsymbol{x} &= (\xi_1,\ldots,\xi_m), \quad \|\boldsymbol{x}\|_{p,m} = 1, \\ \boldsymbol{y} &= (\eta_1,\ldots,\eta_n), \quad \|\boldsymbol{y}\|_{p,n} = 1. \end{aligned} \tag{16}$$

3.2. It is well known that, for $p = 1$, the bi-matrix game has an equilibrium in randomized strategy [7,8,11]. Thus, the linear case does not require further investigation.

3.3. In the vector case $p = 2$, the substratum bi-matrix game has an equilibrium solution, which reduces to the generalized eigenvalue problem [12,13]. We study the following bi-matrix game:

$$\max_{x} \mathcal{I}_1[\boldsymbol{x},\boldsymbol{y},\boldsymbol{a}] \text{ subject to } \mathcal{I}_2[\boldsymbol{x},\boldsymbol{y},\boldsymbol{a}] \geq 1 \text{ for all } \boldsymbol{y}. \tag{17}$$

According to the generalized Rayleigh–Ritz quotient method [14], this optimization problem can be restated as:

$$\min_{x}\max_{y}\frac{\mathcal{I}_1[\boldsymbol{x},\boldsymbol{y},\boldsymbol{a}]}{\mathcal{I}_2[\boldsymbol{x},\boldsymbol{y},\boldsymbol{a}]}. \tag{18}$$

The Lagrangian for (17) reads:

$$\mathcal{L}[\boldsymbol{x},\boldsymbol{y},\boldsymbol{a}] \stackrel{\text{def}}{=} \mathcal{I}_1[\boldsymbol{x},\boldsymbol{y},\boldsymbol{a}] + \lambda\Big(\|\boldsymbol{x}\|_{2,m} - 1\Big) + \mu\Big(\|\boldsymbol{y}\|_{2,n} - 1\Big) + \Lambda(\mathcal{I}_2[\boldsymbol{x},\boldsymbol{y},\boldsymbol{a}] - 1), \tag{19}$$

where λ, μ, Λ are the Lagrange multipliers. Equating the derivatives of $\mathcal{L}$ to zero gives:

$$\begin{cases} \frac{\partial \mathcal{L}}{\partial x} \equiv \boldsymbol{C}^T[\boldsymbol{a}]\boldsymbol{y} + \lambda \boldsymbol{x} = \boldsymbol{0}, \\ \frac{\partial \mathcal{L}}{\partial y} \equiv \boldsymbol{C}[\boldsymbol{a}]\boldsymbol{x} + \mu \boldsymbol{y} = \boldsymbol{0}, \\ \boldsymbol{C}[\boldsymbol{a}, \rho] = (1-\Lambda)\boldsymbol{A}[\boldsymbol{a}] + \Lambda \boldsymbol{B}[\boldsymbol{a}]. \end{cases} \tag{20}$$

The resolution of Equation (20) reads:

$$\begin{cases} \mathcal{K}_C \boldsymbol{y} - \lambda \mu \boldsymbol{y} = \boldsymbol{0}, \\ \mathcal{K}_D \boldsymbol{x} - \lambda \mu \boldsymbol{x} = \boldsymbol{0}. \end{cases} \tag{21}$$

The symmetric semi-positive matrices in (21) are the following:

$$\begin{cases} \mathcal{K}_C[\boldsymbol{a}, \Lambda] \stackrel{\text{def}}{=\!=} \boldsymbol{C}\,\boldsymbol{C}^T \equiv (1-\Lambda)^2 \boldsymbol{A}\boldsymbol{A}^T + \Lambda \cdot (1-\Lambda)\left(\boldsymbol{A}\boldsymbol{B}^T + \boldsymbol{B}\boldsymbol{A}^T\right) + \Lambda^2 \boldsymbol{B}\boldsymbol{B}^T, \\ \mathcal{K}_D[\boldsymbol{a}, \Lambda] \stackrel{\text{def}}{=\!=} \boldsymbol{C}^T \boldsymbol{C} \equiv (1-\Lambda)^2 \boldsymbol{A}^T\boldsymbol{A} + \Lambda \cdot (1-\Lambda)\left(\boldsymbol{B}^T\boldsymbol{A} + \boldsymbol{A}^T\boldsymbol{B}\right) + \Lambda^2 \boldsymbol{B}^T\boldsymbol{B}. \end{cases}$$

For briefness, the auxiliary matrix $\mathcal{K}_2$ will be defined. This matrix represents the matrix with the smallest dimensions of $\mathcal{K}_C$ and $\mathcal{K}_D$. In other words,

$$\mathcal{K}_2[\boldsymbol{a}, \rho] \stackrel{\text{def}}{=\!=} \begin{cases} \mathcal{K}_C & \text{if } m \le n, \\ \mathcal{K}_D & \text{if } m > n. \end{cases}$$

The $\boldsymbol{x}, \boldsymbol{y}$ are the eigenvectors, and the $\lambda = \mu$ are the non-zero eigenvalues of the symmetric semi-positive matrix $\mathcal{K}_2$:

$$\begin{gathered} \widetilde{\mathbb{L}}_i \stackrel{\text{def}}{=\!=} \text{eigenvalues}(\mathcal{K}_2) \ge 0, \\ 0 \le \widetilde{\mathbb{L}}_1 \le \widetilde{\mathbb{L}}_2 \le \ldots \le \widetilde{\mathbb{L}}_j, \quad \widetilde{\mathbb{L}}_{\min} \stackrel{\text{def}}{=\!=} \widetilde{\mathbb{L}}_1, \quad \widetilde{\mathbb{L}}_{\max} \stackrel{\text{def}}{=\!=} \widetilde{\mathbb{L}}_j, \\ i = 1, \ldots, j, \quad j \stackrel{\text{def}}{=\!=} \min(m, n). \end{gathered} \tag{22}$$

The eigenvalues depend parametrically upon the strategies for cardinal players $\boldsymbol{a}$ and the parameter Λ of Equation (20):

$$\widetilde{\mathbb{L}}_i = \widetilde{\mathbb{L}}_i(\boldsymbol{a}, \Lambda). \tag{23}$$

The eigenvectors $\boldsymbol{x}, \boldsymbol{y}$ are the functions of these parameters as well. The parameter Λ plays the role of Lagrange multiplier. It parametrizes the front of the bi-matrix game. If the second player fixes its payoff in Equation (17) to $\mathcal{I}_2[\boldsymbol{x}, \boldsymbol{y}, \boldsymbol{a}] = 1$, the value of the Lagrange multiplier Λ^* follows from this constraint. If Equation (17) displays a maximization problem, the eigenvector is the one with the largest eigenvalue of matrix $\mathcal{K}_2$:

$$\widetilde{\mathbb{L}}_{\max}[\boldsymbol{a}, \Lambda^*] = \max_{u} \frac{\langle \mathcal{K}_2[\boldsymbol{a}, \Lambda^*]\boldsymbol{u}, \boldsymbol{u} \rangle}{\langle \boldsymbol{u}, \boldsymbol{u} \rangle}.$$

Alternatively, if Equation (17) is a minimization problem, the eigenvector is the one with the smallest eigenvalue:

$$\widetilde{\mathbb{L}}_{min}[\boldsymbol{a}, \Lambda^*] = \min_{u} \frac{\langle \mathcal{K}_2[\boldsymbol{a}, \Lambda^*]\boldsymbol{u}, \boldsymbol{u} \rangle}{\langle \boldsymbol{u}, \boldsymbol{u} \rangle}.$$

The payoff of the first player satisfies the inequalities:

$$\begin{cases} \lambda_{min} \le \mathcal{I}_2 \le \lambda_{max}. \\ \lambda_{min}[\boldsymbol{a}, \Lambda^*] \stackrel{\text{def}}{=\!=} \sqrt{\widetilde{\mathbb{L}}_{min}[\boldsymbol{a}, \Lambda^*]}, \\ \lambda_{max}[\boldsymbol{a}, \Lambda^*] \stackrel{\text{def}}{=\!=} \sqrt{\widetilde{\mathbb{L}}_{max}[\boldsymbol{a}, \Lambda^*]}. \end{cases} \tag{24}$$

This solves the substratum bi-matrix game problem for the ordinal players on the lower, substratum level in the vector case $p = 2$.

Based on the substratum game value, the superstratum problem optimizes the eigenvalue in accordance with the goals of the cardinal players:

$$\lambda[\boldsymbol{a}] \underset{\boldsymbol{a}}{\rightarrow} \text{extremum}.$$

If there is only one cardinal player, namely a "designer", finding the extremum in the above equation is a common optimization task. In the actual context, there is only one cardinal player involved, which reduces the extremum search to a common task of mathematical programming.

In general, finding the extremum in the above equation can be a game optimization task, depending on the number of cardinal players involved. If there is no conflict between these players, finding the extremum is a standard optimization. Studying this is beyond the scope of the current manuscript.

3.4. This section presents the solution (24) to the substratum bi-matrix game with constrained actions (Equation (16)). The objective of the superstratum task is to identify the optimal solution for the "designer" by searching for the extremal value of (24) through variation of the design variables.

4. Optimization Games with One "Cardinal Player"

4.1. The results of Section 3 can be generalized for self-adjoint positive definite differential operators using the technique of control theory [15]. A comparable interpretation will be made from the perspective of game theory, with regard to the optimization tasks involving an infinite number of design parameters for each player. In lieu of the payoff function, the payoff functional emerge. The system described by the equilibrium equations is to be considered in the following form:

$$\mathcal{K}u = m. \tag{25}$$

The self-adjoint positive definite operator $\mathcal{K}$ describes the state of the system. In structural optimization, there are definite "ordinal players". In the most basic formulation of the self-adjoint operator of the continuous system, the actions of both ordinal "nature" and "operator" match. In the matrix formulation, it was shown in Section 2, Equation (12). The "designer" modifies the coefficients of the payoff matrix, while both ordinal players simultaneously alter symmetrically both the left and right vectors of the payoff function. In the self-adjoint formulation, the solution of the game problem for the continuous system is essentially simplified. In Equation (25), u is the function of the state variables and m is the function of the external loads of an ordinal player, "nature" or "operator". All values are determined in some domain Ω. In the one-dimensional case, the domain Ω could be thought of as an open interval.

The vector of "nature" and "operator" actions should belong to the set of admissible external loads $m \in \boldsymbol{P}$, $\tilde{m} \in \boldsymbol{P}$, which L_p-norm on the space of p-integrable functions [5] is constrained. We consider the set of real measurable functions whose absolute value raised to the p-th power has a constraint integral:

$$\|m\|_p \leq 1, \ \left\|\tilde{m}\right\|_p \leq 1. \tag{26}$$

As discussed in Section 2, in addition to the "ordinal players" there is the "cardinal player". The "cardinal player" owns the "design variables" $\boldsymbol{a}$. For example, the function $\boldsymbol{a}$ describes the mechanical properties along the length of the element. In essence, this function can be conceived of as a vector function. The coefficients of the operator depend upon $\boldsymbol{a}$:

$$\mathcal{K} = \mathcal{K}[\boldsymbol{a}]. \tag{27}$$

As usual, certain isoperimetric conditions restrict the possible designs. For example, the total volume of the element could be restricted:

$$\mathcal{V}[\boldsymbol{a}] \leq \mathcal{V}_0. \tag{28}$$

The payoff functional is the functional of the design and loads of both players:

$$\mathcal{I} = \mathcal{I}\left[m, \widetilde{m}, \boldsymbol{a}\right]. \tag{29}$$

This functional characterizes the essential mechanical characteristic of the structure, for example, compliance, period of vibration, maximal stress, etc. Putting it roughly, the natural aim of the "operator" $\widetilde{m}$ is to minimize the functional $\mathcal{I}$ for all possible actions of "nature".

The upper and lower game values are defined as follows:

$$\begin{cases} \mathcal{P}^*[\boldsymbol{a}] = \min_{\widetilde{m}\in \boldsymbol{P}} \max_{f\in \boldsymbol{P}} \mathcal{I}\left[m, \widetilde{m}, \boldsymbol{a}\right], \\ \mathcal{P}^{**}[\boldsymbol{a}] = \max_{m\in \boldsymbol{P}} \min_{\widetilde{m}\in \boldsymbol{P}} \mathcal{I}\left[m, \widetilde{m}, \boldsymbol{a}\right]. \end{cases} \tag{30}$$

The minimax theorem states that, in general,

$$\mathcal{P}^{**}[\boldsymbol{a}] \leq \mathcal{P}^*[\boldsymbol{a}] \tag{31}$$

If the upper value of the game is equal to the lower value, the common value of minimaxes $\mathcal{P}^*$ and $\mathcal{P}^{**}$ is called the value of the game:

$$\mathcal{G}[\boldsymbol{a}] = \mathcal{P}^{**}[\boldsymbol{a}] = \mathcal{P}^*[\boldsymbol{a}] \tag{32}$$

4.2. The case $p = 1$ was considered in (Ch. 16, [8]). In classical game theory, the games played over the unit square are considered as a generalization of the matrix games. The payoff function in "game played over the unit square" is thus defined for the non-negative functions. In this case, a single continuous variable was retained for each individual due to the limitations imposed upon the strategies of each player:

$$0 \leq a \leq 1, \|m\|_1 \leq 1, \left\|\widetilde{m}\right\|_1 \leq 1, m \geq 0, \widetilde{m} \geq 0.$$

4.3. In the current contest of application of game theory to structural optimization the value $p = 2$. The payoff functional will be the stored elastic energy of the structural element [16]. This functional has the physical meaning of integral compliance. Thus, this is the game with the payoff functional (29) with the scalar product:

$$\langle m, u \rangle \stackrel{\text{def}}{=} \int_\Omega m(x)u(x)\mathrm{d}x \tag{33}$$

In Equation (33), the symbol $\langle m, u \rangle$ stays for the bilinear form, or scalar product, satisfying:

$$\langle u, m \rangle = \langle m, u \rangle, \langle m, m \rangle > 0 \text{ for } m \neq 0. \tag{34}$$

The solution to Equation (25) may be expressed in the following form:

$$u = \mathcal{L}[\boldsymbol{a}]m. \tag{35}$$

In Equation (35), the operator

$$\mathcal{L}[\boldsymbol{a}] = \mathcal{K}^{-1}[\boldsymbol{a}] \tag{36}$$

is the symbolic inverse of the $\mathcal{K}[\boldsymbol{a}]$ operator.

As shown above, the restriction is assumed in quadratic form:

$$\|m\|_2 \equiv \int_\Omega m(x)\cdot m(x)\,\mathrm{d}x \le 1. \tag{37}$$

4.4. Following the substitution of (35) into Equation (33), the bi-linear payoff functional is expressed as follows:

$$\mathcal{I} = \left\langle \mathcal{L}[a]m, \tilde{m} \right\rangle. \tag{38}$$

The objective is to reduce the stored energy (38), given the restricted resources (37). The expression

$$\mathcal{G}[a] \stackrel{\text{def}}{=} \left\langle \mathcal{L}[a]\mathfrak{m}, \tilde{\mathfrak{m}} \right\rangle = \min_{\tilde{m}\in P}\max_{m\in P}\left\langle \mathcal{L}[a]m, \tilde{m} \right\rangle \tag{39}$$

stays for the average elastic energy, which cause the stochastic actions of "nature" under the stochastic compensating action of the "operator", $m = -\tilde{m}$. The determination of the minimal value of the game $\mathcal{G}[a]$ is reduced to an ordinary optimization problem with the unknown design function a. The optimization game with the stored elastic energy as the payoff function could be referred to as a "compliance game".

Using the variational property of eigenvalues [17], one can obtain the equivalent expression:

$$\mathcal{G}[a] = \frac{\mathbb{P}}{\mathbb{L}[a]}.$$

In the above Equation, $\mathbb{L}[a]$ is the minimal eigenvalue of the operator $\mathcal{K}[a]$. Correspondingly, $\mathbb{L}^{-1}[a]$ is the maximal eigenvalue of the operator $\mathcal{L}[a]$:

$$\begin{cases} \mathbb{L}[a] = \min\limits_{u} \frac{\langle \mathcal{K}[a]u,u\rangle}{\langle u,u\rangle} \equiv \frac{\langle \mathcal{K}[a]\mathrm{u},\mathrm{u}\rangle}{\langle \mathrm{u},\mathrm{u}\rangle}, \\ \mathbb{L}^{-1}[a] = \min\limits_{m\in P} \frac{\langle m,m\rangle}{\langle \mathcal{L}[a]m,m\rangle} \equiv \frac{\left\langle \mathfrak{m},\tilde{\mathfrak{m}}\right\rangle}{\left\langle \mathcal{L}[a]\tilde{\mathfrak{m}},\tilde{\mathfrak{m}}\right\rangle}. \end{cases} \tag{40}$$

Formulas (40) manifest in the sense of equilibrium strategies $\mathfrak{m}$ of "nature". The strategy of both concurrent players is given by the eigenvector of operator $\mathcal{L}[a]$, which corresponds to its eigenvalue $\mathbb{L}^{-1}[a]$:

$$\mathcal{L}[a]\mathfrak{m} = \frac{1}{\mathbb{L}[a]}\cdot\mathfrak{m}.$$

In the case, a single continuous variable was selected for each individual, given the limitations imposed by the strategies of each player:

$$0 \le a \le 1, \|m\|_2 \le 1, \left\|\tilde{m}\right\|_2 \le 1.$$

The equilibrium loads represent in the saddle point of structural response among all permissible loads. At least the upper value of the game could be determined in all cases. The application of the aforementioned considerations to structural optimization problems will be discussed in the following sections.

5. Game Formulation for the Beam Subjected to Arbitrary Bending Moments

5.1. Consider the beam of a certain cross-section. The beam, or rod, is placed horizontally along the x axis. The beam is subjected to an external moment $m(x)$ distributed perpendicularly to a longitudinal axis of the element. The applied compensating moment $\tilde{m}(x)$ is a priori unknown. When a transverse load is applied on it, the beam deforms, and stresses develop inside it. According to the Euler–Bernoulli theory of slender beams, the equation-describing beam deflection can be presented as:

$$E\,J\,u'' = m.$$

E is the Young's modulus, J is the area moment of inertia of the cross-section, and m is the internal bending moment in the beam. The quantity $E\,J$ in the above equation stays for the bending stiffness of the beam. The bending moment m is a two-times continuously differentiable function on $0 \le x \le l$. The area moment of inertia of the cross-section is given by the relation:

$$J = \mathcal{C}\,a^k,$$

where k is the shape exponent and $\mathcal{C} = \mathcal{C}(k)$ is the shape factor. The shape factor depends on the cross-sectional shape. The admissible cross-sectional area of the rod is the scalar function $a = a(x)$. The function is two times continuously differentiable. The cross-sectional area a plays the role of the strategy of "designer". The shape exponent k takes the values of one, two, and three [12]. In all of these cases, the game will be convex, and both players have pure strategies. The case $k = 2$ corresponds to a similar variation of the form of the cross-section. The areas, area moment of inertia, and shape factors for the similar variation of the form of the cross-section $k = 2$ are shown in Table 1. These formulas are valid for both a horizontal and a vertical axis through the centroid and, therefore, are also valid for an axis with an arbitrary direction that passes through the origin of the regular cross-sections. The a priori unknown cross-sectional area a is the variable function along the span of the beam.

Table 1. Areas, area moment of inertia, and shape factors for the similar variation of the cross-section $k = 2$.

Solid or (Filled) Cross Section	Equilateral Triangle	Regular Hexagon	Square	Circular Area
		side length of s		radius s
area a	$\frac{\sqrt{3}s^2}{4}$	$\frac{3\sqrt{3}s^2}{2}$	s^2	πs^2
area moment of inertia J	$\frac{s^4}{32\sqrt{3}}$	$\frac{5\sqrt{3}s^4}{16}$	$\frac{s^4}{12}$	$\frac{\pi s^4}{4}$
shape factor $\mathcal{C}$	$\frac{\sqrt{3}}{18}$	$\frac{5\sqrt{3}}{108}$	$\frac{1}{12}$	$\frac{1}{4\pi}$

For brevity, the authors use the integral of an arbitrary function $\varphi(x)$, $0 \le x \le l$ for the symbolization:

$$\langle\varphi\rangle \stackrel{\text{def}}{=\!=} \int_0^l \varphi(\xi)\mathrm{d}\xi$$

The boundary conditions must be prescribed as well. For easiness, the rod is hinged at the end $x = -l$ and clamped at the end $x = 0$. The boundary conditions are [4]:

$$m'\big|_{x=0} = 0,\ u'\big|_{x=0} = 0,\ m\big|_{x=l} = 0,\ u\big|_{x=l} = 0. \tag{41}$$

5.2. The payoff functional $\mathcal{P}$ is the integral distortion of the structural element. It is reasonable to express the integral distortion as the stored elastic energy $\mathcal{W}$:

$$\mathcal{I}\Big[m, \tilde{m}, a\Big] \stackrel{\text{def}}{=\!=} \mathcal{W} = \left\langle \frac{m\,\tilde{m}}{EJ} \right\rangle. \tag{42}$$

The game problem is defined exactly as already presented for the general status quo. There are two active ordinal players, "nature" and "operator" who behave recurrently and stochastically. The player ("nature") can apply an arbitrary admissible external load $m = m(x)$ in order to affect the maximal elastic energy $\mathcal{W}$. Due to the symmetry of the game, the "operator" applies the opposite moments $\tilde{m} = \tilde{m}(x)$. These moments compensate for the deformation caused by "nature" m. The total efforts of the "nature" and "operator" are restricted:

$$\left\langle m'^2 \right\rangle = \left\langle \tilde{m}'^2 \right\rangle = 1. \tag{43}$$

Admissible is any effort of "nature" and "operator", which is a continuously differentiable function, that satisfies (43) and certain boundary conditions. Because the quadratic norm (43) is restricted, the sign of the moment statistically plays no role. The game will be symmetric in the sense of Nash [18].

The third player, "designer", attempts to select the most appropriate shape $J = \mathcal{C}\, a^k(x)$, which will guarantee the smallest deformation energy during the exploitation of the structural element. The most appropriate shape is generally referred to as $\boldsymbol{a}$. Once completed, the design remains unaltered over the period of exploitation. That is why the "designer" is considered as the "passive" player, contrary to both of the other opponent players, "nature" and "operator".

The stiffest design corresponds to the minimal integral measure of deformation, which is presented by the stored energy $\mathcal{I} = \mathcal{W}$. The attempts of the "designer" to withstand the deformation are also limited. Namely, the material volumes of all of the competing designs of the rods are less than the certain, fixed volume of material $\mathcal{V}$:

$$\langle a \rangle \leq \mathcal{V}. \tag{44}$$

The conflict leads to the antagonistic game formulation of the optimization task. This game is referred to as the functional game because the values of the game depend on function a. According to Equation (40), the optimal value of the game in mixed strategies is equal to:

$$\mathcal{G}_{opt} = \min_{\langle a \rangle \leq \mathcal{V}} \mathcal{G}[a], \quad \mathcal{G}[a] = \frac{\mathbb{P}}{EC\mathbb{L}[a]}. \tag{45}$$

The symbol $\mathbb{L}$ in (45) signifies the minimal eigenvalue of the game of the ordinary differential equation with the boundary conditions:

$$\begin{cases} m'' + \mathbb{L}[a]\cdot a^{-k} m = 0 \\ m'|_{x=0} = 0, \ m|_{x=l} = 0. \end{cases} \tag{46}$$

The optimization game with the minimal eigenvalue as the payoff function could be referred to as the "eigenvalue game". From (45) and (46), it follows that the maximization of the upper value of the game reduces to the maximization of the minimal eigenvalue:

$$\frac{1}{\mathcal{G}_{opt}} = \frac{1}{\min\limits_{\langle a \rangle \leq \mathcal{V}} \mathcal{G}[a]} = \max_{\langle a \rangle \leq \mathcal{V}} \frac{1}{\mathcal{G}[a]} = \frac{EC}{\mathbb{P}} \max_{\langle a \rangle \leq \mathcal{V}} \mathbb{L}[a]. \tag{47}$$

The cross-section for the Nash equilibrium state will be designated with the capital letter $\mathcal{A}$, leaving the small letter a for any admissible cross-section function. The optimal cross-section presents the strategy for the active players. In other words, the beam with the thickness distribution $\mathcal{A}$ possesses the guaranteed stiffness:

$$\frac{1}{\mathcal{G}_{opt}} = \frac{EC}{\mathbb{P}} \cdot \max_{\langle a \rangle \leq \mathcal{V}} \mathbb{L}[a] \stackrel{\text{def}}{=\!=} \frac{EC}{\mathbb{P}} \mathbb{L}[\mathcal{A}]. \tag{48}$$

The eigenvalue problem (46) is self-adjoint. The boundary conditions (46) are of the Sturm type, see [19]. There exists an infinite set of eigenvalues. All eigenvalues are real and positive and can be arranged as a monotonic sequence, and each eigenvalue is simple:

$$\mathbb{L}[a] \equiv \lambda_1[a] < \ldots < \lambda_{2k-1}[a] < \lambda_{2k}[a] < \ldots. \tag{49}$$

Rayleigh's quotient for the eigenvalue problem (46) reads as:

$$\mathcal{R}[\mathfrak{m}, a] = \frac{\left\langle \mathfrak{m}'^2 \right\rangle}{\left\langle a^{-k} \mathfrak{m}^2 \right\rangle}.$$

In the Rayleigh's quotient, the admissible functions $\mathfrak{m}(x)$ are all functions, having piecewise continuous first derivatives, satisfying the boundary conditions (46). The admissible moment of both active players: $\mathfrak{m} = \mathfrak{m}(x)$ stays as the argument in the Rayleigh's quotient. The Rayleigh's quotient plays the role of the payoff functional of "nature". The player "nature" wants to maximize $-\mathcal{R}[\mathfrak{m}, a]$, while the opposite player "operator" wants to minimize $\mathcal{R}[\mathfrak{m}, a]$. The game is a zero-sum game, which means that, when the strategies of both players are fixed, the sum of the payoffs for the two players is zero. The self-adjointness of the eigenvalue problem implies that the action of the "operator" is opposite, but proportional to the action of the physical process, which we may refer to as "nature".

Among all of the admissible strategies of "nature", the most favorable strategy set for "nature" is $m = m(x)$. This choice delivers the minimal value for the Rayleigh's quotient:

$$\mathbb{L}[a] = \min_{\mathfrak{m}} \mathcal{R}[\mathfrak{m}, a], \ \mathbb{L}[a] \equiv \lambda_1[a]. \tag{50}$$

The favorable strategy set that minimizes the Rayleigh's quotient increases, according to (45), the energy of deformation $\mathcal{W}$ from (42) under its condition (43).

5.3. The "designer" has the opposite task. The necessity of the "designer" is to minimize the energy of deformation for all admissible a under his condition (44). This task is equivalent to the maximization of the Rayleigh's quotient with the restriction (43), see [20]. The Lagrangian functional is the sum of (50) and (44):

$$\mathcal{I} = \langle l \rangle, \ l = a^{-k} m^2 + \Lambda a.$$

Here, Λ represents the Lagrange multiplier of the variational calculus problem. The variation of the Hamiltonian functional reads:

$$\delta \mathcal{I} = \left\langle \frac{\partial l}{\partial a} \right\rangle \delta a, \frac{\partial l}{\partial a} = -k a^{-k-1} m^2 + \Lambda.$$

The nullification of the derivative $\partial l / \partial a = 0$ leads to the necessary optimality condition. The strategies in the state of Hash equilibrium are $m \to \mathcal{M}$, $a \to \mathcal{A}$, and consequently:

$$\mathcal{M}^2 = \frac{\Lambda}{k} \mathcal{A}^{k+1} \tag{51}$$

From the viewpoint of the "designer", the thickness distribution $\mathcal{A}$ must satisfy the necessary optimality condition (51).

From the viewpoint of the "designer", the beam with the thickness distribution $\mathcal{A}$ guarantees the highest effectiveness for the most unfavorable effort of the player ("nature"). Again, due to the self-adjoint nature of differential equations, the answer of the "operator" in equilibrium is exactly the opposite of the action of "nature".

5.4. The next task is to determine the Nash equilibrium strategies for the "designer" $\mathcal{A}$ and for the "nature" $\mathcal{M}$. The governing equation is the nonlinear ordinary differential equations of the second order:

$$\mathcal{M}'' + \mathbb{L} \cdot \mathcal{A}^{-k} \mathcal{M} = 0 \qquad \mathcal{A} = \mathcal{M}^{\frac{2}{k+1}}. \tag{52}$$

In the Equations of this Section, $\mathbb{L}$ signifies the minimal eigenvalue (49) in the Nash equilibrium point:

$$\mathbb{L} = \mathbb{L}[\mathcal{A}] \equiv \lambda_1[\mathcal{A}].$$

The boundary conditions (41) could be proven to be optimal for the boundary conditions of the Sturm type. The equilibrium conditions of the game task are equivalent to the necessary optimality condition for a column's Euler buckling load. The substitution of Equation (51) into (52) gives:

$$\mathcal{M}'' + \mathbb{L} \cdot \mathcal{M}^{\frac{1-k}{1+k}} = 0. \tag{53}$$

The solution of the boundary value problem (41), and (53) determines the strategy of "nature" $\mathcal{M}$. With this solution, the highest possible eigenvalue (52) and, consequently, the upper value of the game have to be evaluated:

$$\mathcal{P}^* = \frac{\mathbb{P}}{EC\mathbb{L}}, \quad \mathbb{L} = \mathbb{L}[\mathcal{A}].$$

The dependent variable $\mathcal{M}$ and independent variable x of Equation (53) are to be exchanged. In the new variables $x(\mathcal{M})$, Equation (53) turns into the ordinary differential equation of the second order of Emden–Fowler type (Section 14.3 of [21]):

$$\frac{d^2x}{d\mathcal{M}^2} = \mathbb{L}\cdot\left(\frac{dx}{d\mathcal{M}}\right)^3 \mathcal{M}^R, \qquad R \stackrel{\text{def}}{=} \frac{1-k}{1+k}., \tag{54}$$

The Emden–Fowler Equation (54) has the following parameters:

$$\frac{d^2x}{d\mathcal{M}^2} = \mathbb{L}\cdot\left(\frac{dx}{d\mathcal{M}}\right)^{1-S} \mathcal{M}^{P-1}, \qquad S = -2, \quad R = P-1, \quad P = \frac{2}{k+1}. \tag{55}$$

The closed-form solution results in an arbitrary acceptable value of the shape exponent k. The general solution of Equation (54) for $k > 1$ is:

$$x = i\sqrt{P}\cdot\int_n^{\mathcal{M}} \frac{\mathrm{d}t}{\sqrt{2\mathbb{L}t^P - \mathcal{N}}} \tag{56}$$

The symbols n and $\mathcal{N}$ stay for the integration constants. The integration constants are to evaluated from the solution of the nonlinear transcendental equations. To avoid the solutions of the nonlinear equations, the authors prefer the symmetrical sights. Specifically, the sense of the constant n is the moment on the end $x = 0$:

$$n \stackrel{\text{def}}{=} \mathcal{M}(x=0).$$

Due to the symmetry of the equations, with respect to the point $x = 0$, the function $\mathcal{M}(x)$ must be an even function of the variable x. This condition fixes the relation between the integration constants: $\mathcal{N} = 2\mathbb{L}n^P$.

With this value, integral (56) evaluates in the closed form. For the shape exponent $k \geq 1$ and $K = 0$, the solution reads as:

$$\begin{cases} x(\mathcal{M}) = c_L\cdot\int_0^{\mathcal{M}} \frac{\mathrm{d}s}{\sqrt{n^P - s^P}}, \\ x(\mathcal{M}) = n^{-\frac{1}{k+1}}\cdot\mathcal{M}\cdot{}_2\mathbf{F}_2\left(\left[\frac{1}{2}, \frac{k+1}{2}\right], \left[\frac{k+3}{2}\right], \left(\frac{\mathcal{M}}{n}\right)^{\frac{2}{k+1}}\right), \\ x(n) = \sqrt{\pi}n^{-\frac{k}{k+1}}\cdot\frac{\Gamma\left(\frac{k+3}{2}\right)}{\Gamma\left(\frac{k+1}{2}\right)}. \end{cases} \tag{57}$$

Equation (57) presents the axial coordinate x as the function of the new independent parameter $\mathcal{M}$. For $k = 1, 2, 3$, the hypergeometric function from Equation (57) expresses in terms of elementary functions (Table 2). The graphs of the solutions are displayed for different values of parameters k on Figure 1.

Table 2. Expressions for axial coordinate x as the function of $\mathcal{M}$ for $k = 1$ and $k = 2$.

k	$x(\mathcal{M})$
1	$x = \frac{\mathcal{M}}{1+\sqrt{1-4\mathcal{M}}}$
2	$x = \frac{2}{\pi}\cdot\arcsin\left(\frac{\sqrt{3\pi}}{2}\sqrt[3]{\mathcal{M}}\right) - \frac{\sqrt{3}}{2\sqrt{\pi}}\cdot\sqrt[3]{\mathcal{M}}\cdot\sqrt{4-3\pi\sqrt[3]{\mathcal{M}^2}}$

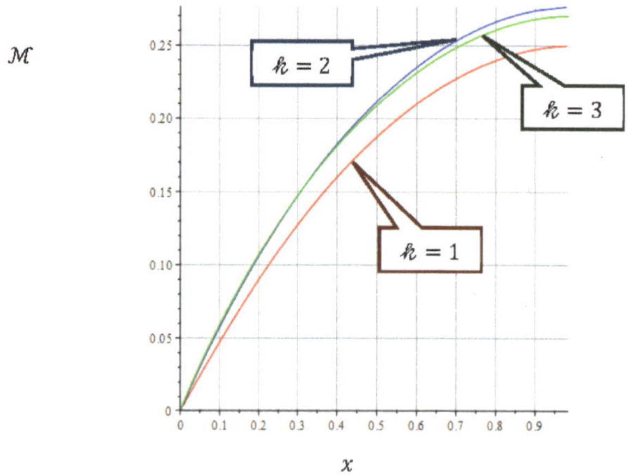

Figure 1. Shapes of beams for the Nash equilibrium states from Equation (57).

According to the boundary conditions (41), the dimensionless moment m vanishes on the hinged end: $\mathcal{M}|_{x=l} = 0$. From this condition, the length of the rod could be determined as the function of an unknown integration constant n. Because the length l of the rod is known, the unknown constant n evaluates from the solution of the equation:

$$x(n) = l.$$

From its solution, the integration constant n follows as:

$$n = \left(\frac{\Gamma\left(\frac{k}{2}+1\right)}{\Gamma\left(\frac{k+3}{2}\right)} \frac{l}{\sqrt{\pi}} \right)^{\frac{k+1}{k}}. \tag{58}$$

To find the volume of the element, the authors evaluate the proper integral of the cross-section area $\mathcal{A}(\mathcal{M})$. For all values of k that are higher that one, the volume of the optimal structural element will be:

$$\mathcal{V} = \frac{1+k}{2+k} \cdot \sqrt[k]{\frac{k^2}{\pi} \frac{l^{2+k}}{(1+k)^2} \frac{\Gamma^2\left(\frac{k}{2}\right)}{\Gamma^2\left(\frac{k+1}{2}\right)}}. \tag{59}$$

The authors define one other constant that was referred to above as the total stiffness $\mathcal{E}$ of the structural element. The total stiffness $\mathcal{E}$ of the beam expresses as an integral of the moment of inertia of the cross-sections along the length l of the beam. To find the total stiffness of element in the Nash equilibrium state, the authors evaluate another proper integral. The integral follows of the energy density $\mathcal{M}^2/EC\mathcal{A}^k$ as:

$$\frac{\mathcal{W}}{EC} = \frac{1+k}{2+k} \cdot \sqrt[k]{\frac{4}{\pi} \frac{l^{2+k}}{(2+k)^2} \frac{\Gamma^2\left(2+\frac{k}{2}\right)}{\Gamma^2\left(\frac{k+3}{2}\right)}}. \tag{60}$$

According to Equation (60), the elastic energy density is constant over the length of the structural element.

Finally, the eigenvalue $\mathbb{L}[\mathcal{A}] \equiv \lambda_1[\mathcal{A}]$ equals the Nash equilibrium point to:

$$\mathbb{L} \equiv \frac{2k^2 l^2}{9\pi} \cdot \frac{k+2}{(k+\frac{2}{3})(k+1)^3} \cdot \frac{\Gamma\left(\frac{3k+3}{2}\right)}{\Gamma\left(\frac{3k}{2}\right)} \cdot \frac{\Gamma^3\left(\frac{k}{2}\right)}{\Gamma^3\left(\frac{k+1}{2}\right)} \tag{61}$$

Equations (59)–(61) characterize the principal mechanical properties in the Nash equilibrium state. The stored elastic energy and volume for Nash equilibrium $\mathbb{L} = \mathbb{L}[\boldsymbol{\mathcal{A}}]$ is presented for different values of the constant k in Table 3.

Table 3. Stored elastic energy and volume for Nash equilibrium $\mathbb{L} = \mathbb{L}[\boldsymbol{\mathcal{A}}]$.

k	$\mathbb{P}$	$\frac{\mathcal{W}}{EC} \sim \mathcal{V}$	$\boldsymbol{\Psi}(k)$
1	$\frac{l^5}{30}$	$\frac{l^3}{6}$	$\frac{1}{5}$
2	$\frac{35 l^4}{27\pi^3}$	$\frac{l^2}{\pi}$	$\frac{35}{27\pi^2}$
3	$\frac{1}{154}\sqrt[3]{\frac{3^5}{l^{11}}}$	$\frac{1}{5}\sqrt[3]{\frac{3^2}{l^5}}$	$\frac{15}{154}$

For the state of Nash equilibrium, the eigenvalue depends on the known constants:

$$\begin{gathered} \Psi(k) \overset{\text{def}}{=\!=} EC\frac{\mathbb{P}}{\mathcal{E}} = \frac{\mathbb{L}}{l^2}, \\ \Psi(k) = k^3 \cdot \frac{\Gamma\left(\frac{3k+3}{2}\right)}{\Gamma\left(\frac{3k+4}{2}\right)} \cdot \frac{\Gamma^3\left(\frac{k}{2}\right)}{\Gamma^3\left(\frac{3+k}{2}\right)} \cdot \frac{2+k}{16\pi}. \end{gathered} \tag{62}$$

The dimensionless $\Psi(k)$ function is displayed on Figure 2.

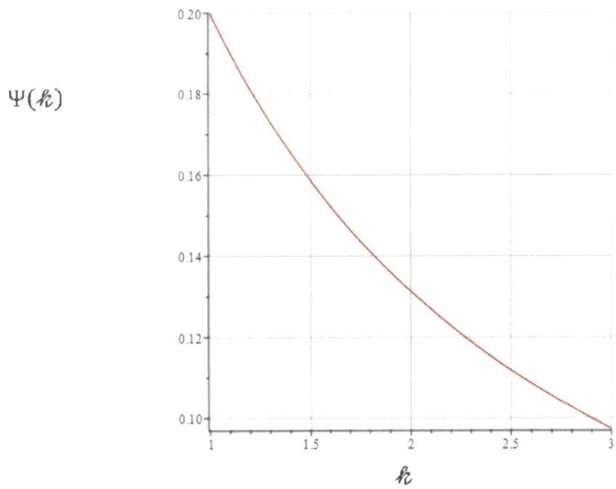

Figure 2. Dimensionless function $\Psi(k)$ from Equation (62).

5.5. Our next task is to establish the relation between the eigenvalues $\mathbb{L}[\mathcal{A}]$ and $\mathbb{L}[a]$. In the convex case $k \geq 1$, the relation could be rigorously proven as the certain isoperimetric inequality. For this purpose, the lowest eigenvalues for two different thickness functions a and $\mathcal{A}$ are the minimal values of the two corresponding Rayleigh's quotients [17]:

$$\mathcal{R}[\mathfrak{m}, a] = \frac{\langle \mathfrak{m}'^2 \rangle}{\langle a^{-k}\mathfrak{m}^2 \rangle}, \qquad \mathcal{R}[\mathfrak{m}, \mathcal{A}] = \frac{\langle \mathfrak{m}'^2 \rangle}{\langle \mathcal{A}^{-k}\mathfrak{m}^2 \rangle}.$$

The Rayleigh's quotients are the functionals of functions $\mathfrak{m}$ and a or $\mathcal{A}$, correspondingly. The functions a or $\mathcal{A}$ are assumed in this section to be fixed. The function $\tilde{\mathfrak{m}}$ is an arbitrary admissible function, which is differentiable and satisfies the boundary conditions (41). The critical point of a functional is that point where the functional attains a minimum (or maximum) in the presence of constraints [17]. We, therefore, examine the conditions when a functional attains a minimum. For the thickness function a, the critical point of the Rayleigh's quotient $\mathcal{R}[\mathfrak{m}, a]$ is the eigenfunction m. To the eigenfunction m corresponds the lowest eigenvalue $\mathbb{L}[a]$. This guarantees that:

$$\mathbb{L}[a] = \frac{\langle m'^2 \rangle}{\langle a^{-k} m^2 \rangle} = \min_{\mathfrak{m}} \mathcal{R}[\mathfrak{m}, a],$$

Analogously, for the thickness function $\mathcal{A}$, the critical point of the Rayleigh's quotient $\mathcal{R}[\mathfrak{m}, \mathcal{A}]$ is the eigenfunction $\mathcal{M}$. To this function corresponds the lowest eigenvalue $\mathbb{L}[\mathcal{A}]$:

$$\mathbb{L}[\mathcal{A}] = \frac{\langle \mathcal{M}'^2 \rangle}{\langle \mathcal{A}^{-k} \mathcal{M}^2 \rangle} = \min_{\mathfrak{m}} \mathcal{R}[\mathfrak{m}, \mathcal{A}]\,.$$

The right sides of the two above Equations must be compared. In order to state the desired isoperimetric inequality, the following auxiliary inequalities have to be approved:

$$\begin{cases} \frac{\langle m'^2 \rangle}{\langle a^{-k} m^2 \rangle} \leq \frac{\langle \mathcal{M}'^2 \rangle}{\langle \mathcal{A}^{-k} \mathcal{M}^2 \rangle}, \\ \mathcal{M}'' + \mathbb{L}[\mathcal{A}]\, \mathcal{A}^{-k} \mathcal{M} = 0, \\ m'' + \mathbb{L}[a]\, a^{-k} m = 0. \end{cases} \tag{63}$$

The numerators of the fractions to the left and right of the auxiliary inequality (63) are identical. The denominators in the first Equation (63) are, however, different. Thus, the denominators should be compared. The inequality for the denominators that has to be proven reads:

$$\left\langle \mathcal{A}^{-k} \mathcal{M}^2 \right\rangle \leq \left\langle a^{-k} \mathcal{M}^2 \right\rangle \leq \left\langle a^{-k} m^2 \right\rangle. \tag{64}$$

At this point, the optimality condition (51) with $\lambda = k$ will be used:

$$\mathcal{M}^2 = \mathcal{A}^{k+1}.$$

Namely, the substitution of the optimality condition (51) into (64) delivers the inequality for $k \geq 1$:

$$\left\langle \frac{\mathcal{A}^{k+1}}{\mathcal{A}^k} \right\rangle \leq \left\langle \frac{\mathcal{A}^{k+1}}{a^k} \right\rangle. \tag{65}$$

The equality in (65) takes place only for $\mathcal{A} = a$. The validity of the as yet suspected inequality (65) follows directly from the Hölder's inequality (Ch. III of [22]). Consequently, from Hölder's inequality follows the inequality for denominators (65) and, finally, the desired inequality (63).

Combining (63) and (65) delivers:

$$\mathbb{L}[a] = \frac{\langle m'^2 \rangle}{\langle a^{-k} m^2 \rangle} \leq \frac{\langle \mathcal{M}'^2 \rangle}{\langle a^{-k} \mathcal{M}^2 \rangle} \equiv \frac{\langle \mathcal{M}'^2 \rangle}{\left\langle \frac{\mathcal{A}^{k+1}}{a^k} \right\rangle} \leq \frac{\langle \mathcal{M}'^2 \rangle}{\left\langle \frac{\mathcal{A}^{k+1}}{\mathcal{A}^k} \right\rangle} = \frac{\langle \mathcal{M}'^2 \rangle}{\langle \mathcal{A}^{-k} \mathcal{M}^2 \rangle} \equiv \mathbb{L}[\mathcal{A}].$$

Consequently, it was proven that, for all arbitrary a, the eigenvalue is less than $\mathbb{L}[\mathcal{A}]$:

$$\mathbb{L}[a] \leq \mathbb{L}[\mathcal{A}]. \tag{66}$$

The equality in Equation (66) was attained only for the optimal beam, which has the optimal shape $\mathcal{A}$ and the maximal possible volume of material $\mathcal{V}$. For the shape exponent $k \geq 1$, the game is convex. Generally, the game will be convex for any convex function $J = J\,(a)$.

Finally, the beam, which obeys the necessary optimality conditions (51), delivers the lowest possible upper value of the game:

$$\mathcal{G}[a] = \frac{\mathbb{P}}{EC\;\mathbb{L}[a]} \geq \frac{\mathbb{P}}{EC\;\mathbb{L}[\mathcal{A}]} \stackrel{\text{def}}{=\!=} \mathcal{G}_{opt}. \tag{67}$$

Consequently, the value of functional game $\mathcal{G}_{opt} = \mathcal{W}[\mathcal{A}]$ for the optimal distribution of thickness $\mathcal{A}(\xi)$ is lower than the payoff functional (distortion energy) for an arbitrary distribution of thickness $a(\xi)$ of the same volume:

$$\langle a \rangle = \langle \mathcal{A} \rangle = \mathcal{V}. \tag{68}$$

If the volume of an arbitrary thickness distribution is lower, the stronger inequality follows:

$$\langle a \rangle < \langle \mathcal{A} \rangle \Longrightarrow \mathbb{L}[a] < \mathbb{L}[\mathcal{A}]. \tag{69}$$

The relations (67)–(69) solve the "superstratum" game with the undefined, but constrained, bending moment function. From the viewpoint of the "designer", the optimal distribution of thickness $\mathcal{A}(\xi)$ guarantees the best compromise for the most unfavorable option for the "designer" action of "nature". Because of the self-adjoint nature of the eigenvalue problem, the action of the "operator" is proportional to the action of "nature". This compromise is the equilibrium in the functional game. The eigenvalue $\mathbb{L}[\mathcal{A}]$ in the equilibrium point is shown on Figure 3 as the function of the length l and the shape factor k.

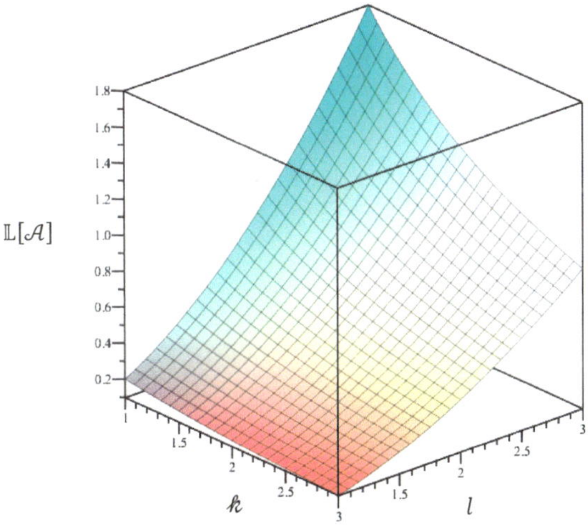

Figure 3. Eigenvalue $\mathbb{L}$ as the function of length l and shape factor k.

5.6. In Section 5, the beam is subjected to arbitrary bending efforts, and the designer determines the optimal beam shape for resisting these forces.

6. Game Formulation for the Rod Subjected to Arbitrary Torque

6.1. Consider the rod with the circular cross section, subjected to the positive distributed torque along its axis. The torque distribution is arbitrary, but the quadratic integral of the torque is fixed:

$$\left\langle u^2 \right\rangle = \left\langle u^2 \right\rangle = 1.$$

The governing equation for this problem is similar to Equation (46), but the order is twice lower [2]. In this case, the symbol $\mathbb{L}$ signifies the minimal eigenvalue of the game of the ordinary differential equation:

$$\left(ECa^{-k}u'\right)' + \mathbb{L}[a]u = 0, \tag{70}$$

with the boundary conditions:

$$u|_{x=0} = 0, \; u'|_{x=l} = 0. \tag{71}$$

The optimal cross-section will be designated with the capital letter $\mathcal{A}$, leaving the small letter a for any admissible cross-section function. The optimal cross-section presents the strategy for the "nature". In other words, the rod with the thickness distribution $\mathcal{A}$ possesses a guaranteed stiffness:

$$\frac{1}{\mathcal{P}^*} = \frac{EC}{\mathbb{P}} \max_{\langle a \rangle \leq \mathcal{V}} \mathbb{L}[a] \overset{\text{def}}{=\!=} \frac{EC}{\mathbb{P}} \mathbb{L}[\mathcal{A}]. \tag{72}$$

The eigenvalue problem (70) and (71) is self-adjoint. The conditions (71) are again of the Sturm type. Once again, there is an infinite set of simple eigenvalues, and all eigenvalues are real and positive and can be arranged as a monotonic sequence:

$$\mathbb{L}[a] \equiv \lambda_1 < \lambda_2 < \ldots < \lambda_{2k-1} < \lambda_{2k} < \ldots. \tag{73}$$

The Rayleigh's quotient is:

$$\widetilde{\mathcal{R}}[\mathrm{u}, a] = \frac{\left\langle a^k \mathrm{u}'^2 \right\rangle}{\langle \mathrm{u}^2 \rangle}. \tag{74}$$

In Equation (74), the admissible functions $\mathrm{u}(x)$ are all functions, having piecewise continuous first derivatives and satisfying the boundary conditions (71). In the Rayleigh's quotient stays the admissible twist of "nature": $\mathrm{u} = \mathrm{u}(x)$.

6.2. Among all admissible strategies of "nature", the most favorable strategy set for "nature" is $u = u(x)$. This choice delivers the minimal value for the Rayleigh's quotient:

$$\mathbb{L}[a] = \min_{\mathrm{u}} \widetilde{\mathcal{R}}[\mathrm{u}, a], \qquad \mathbb{L}[a] \equiv \Lambda_1[a]. \tag{75}$$

The favorable strategy set minimizes the Rayleigh's quotient and increases the energy of deformation $\left\langle a^k \mathrm{u}'^2 \right\rangle$ under the condition $\langle \mathrm{u}^2 \rangle = 1$ and (71).

The "designer" has the opposite task. The necessity of "designer" is to minimize the energy of deformation for all admissible a under the condition. This task is equivalent to the maximization of the Rayleigh's quotient with the restriction (44). The Lagrangian functional is the sum of (74) and (44):

$$\mathcal{I} = \langle l \rangle, \; l = a^k u'^2 + \Lambda a.$$

Here, Λ represents the Lagrange multiplier of the variational calculus problem. The variation of the Lagrangian functional reads:

$$\delta \mathcal{I} = \left\langle \frac{\partial l}{\partial a} \right\rangle \delta a, \qquad \frac{\partial l}{\partial a} = k a^{k-1} u'^2 - \Lambda.$$

The nullification of the derivative $\partial l / \partial a = 0$ leads to the necessary optimality condition. The strategies in the state of Hash equilibrium are $u \to \mathcal{U}$, $a \to \mathcal{A}$, and consequently:

$$\mathcal{A} = \sqrt[k-1]{\frac{\Lambda}{k \mathcal{U}'^2}}. \tag{76}$$

From the viewpoint of the "designer", the thickness distribution $\mathcal{A}$ must satisfy the necessary optimality condition (76). Remarkably, that the equilibrium conditions of the game task are equivalent to the necessary optimality condition for the twist divergence of a wing [23,24].

6.3. The next task is to determine the Nash equilibrium strategies for the "designer" $\mathcal{A}$ and for the "nature" $\mathcal{M}$. For the briefness of formulas, we temporarily put $\lambda = k$. The governing equation is the nonlinear ordinary differential equations of the second order:

$$\left(\mathcal{A}^k \mathcal{U}' \right)' + \mathbb{L} \mathcal{U} = 0 \qquad \mathcal{A} = \mathcal{U}'^{\frac{2}{1-k}}. \tag{77}$$

In the above Equations, $\mathbb{L}$ signifies the minimal eigenvalue in the Nash equilibrium point:

$$\mathbb{L} = \mathbb{L}[\mathcal{A}] \equiv \lambda_1[\mathcal{A}].$$

The substitution of Equation (77) into (70) gives:

$$\left((\mathcal{U}')^{1/R}\right)' + \mathbb{L}\mathcal{U} = 0, \qquad R \overset{\text{def}}{=\!=} \frac{1-k}{1+k}. \tag{78}$$

For the solution of the boundary value problem (77), (78) determines the strategy of "nature" $\mathcal{U}$. The dependent and independent variables of Equation (78) are to be exchanged. In the new variables, Equation (78) turns into the Emden–Fowler equation [21]:

$$\frac{d^2x}{d\mathcal{U}^2} = \mathbb{L}\cdot R\cdot \mathcal{U}\cdot \left(\frac{dx}{d\mathcal{U}}\right)^{\frac{k-3}{k-1}}, \qquad x(\mathcal{U}=0)=0. \tag{79}$$

The Emden–Fowler Equation (79) has the following parameters:

$$S = \frac{2}{k-1}, \quad P = 2.$$

The closed-form solution results for an arbitrary value of the shape exponent are $k \geq 0$. The solution of (79) reads as:

$$\begin{cases} x(\mathcal{U}) = \left(\frac{\mathbb{L}}{1+k}\right)^{\frac{k-1}{2}} \mathcal{U}\, {}_2\mathbf{F}_2 2\left(\left[\frac{1}{2}, \frac{1-k}{2}\right], \left[\frac{3}{2}\right], \mathcal{U}^2\right), \\ x(\mathcal{U}=1) = \frac{\sqrt{\pi}}{2}\left(\frac{\mathbb{L}}{1+k}\right)^{\frac{k-1}{2}} \frac{\Gamma\left(\frac{k+1}{2}\right)}{\Gamma\left(\frac{k+2}{2}\right)}. \end{cases}, \tag{80}$$

Equation (80) presents the axial coordinate x as the function of the independent parameter $\mathcal{U}$. For $k = 1, 2, 3$, the hypergeometric function from Equation (80) expresses in terms of elementary functions (Table 4). Optimal shapes of twisted rods for the Nash equilibrium states are shown on Figure 4.

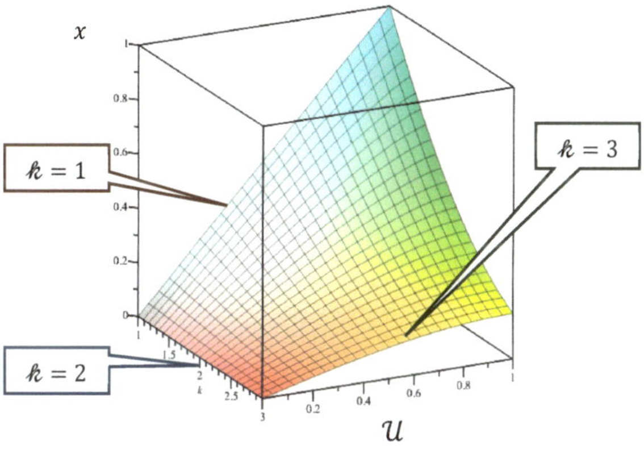

Figure 4. Optimal shapes of twisted rods for the Nash equilibrium states.

Table 4. Expressions for axial coordinate x as the function of $\mathcal{U}$ for $k = 1, 2, 3$.

k	$x(\mathcal{U})$	$x(1)$
1	$\mathcal{U}$	1
2	$\frac{1}{2}\sqrt{\frac{\mathbb{L}}{3}}\left(\mathcal{U}\sqrt{1-\mathcal{U}^2} + \arcsin(\mathcal{U})\right)$	$\sqrt{\frac{\mathbb{L}}{12}}$
3	$\frac{\mathbb{L}\mathcal{U}}{4}\left(1 - \frac{\mathcal{U}^2}{3}\right)$	$\frac{\mathbb{L}}{6}$

6.4. The next task is to establish the relation between the eigenvalues $\mathbb{L}[\mathcal{A}]$ and $\mathbb{L}[a]$. In the convex case, where $k \geq 1$, the relation could be rigorously proven as the certain isoperimetric inequality. For this purpose, the lowest eigenvalues for two different thickness functions a and $\mathcal{A}$ are the minimal values of the two corresponding Rayleigh's quotients:

$$\widetilde{\mathcal{R}}[\mathrm{u},a] = \frac{\left\langle a^k \mathrm{u}'^2 \right\rangle}{\langle \mathrm{u}^2 \rangle}, \qquad \widetilde{\mathcal{R}}[\mathrm{u},\mathcal{A}] = \frac{\left\langle \mathcal{A}^k \mathrm{u}'^2 \right\rangle}{\langle \mathrm{u}^2 \rangle}.$$

The Rayleigh's quotients are the functionals of functions $\widetilde{\mathrm{m}}$ and a or $\mathcal{A}$, correspondingly. The functions a or $\mathcal{A}$ are assumed in this section to be fixed. The function $\widetilde{\mathrm{m}}$ is an arbitrary admissible function, which is differentiable and satisfies boundary conditions. The critical point of a functional is that point where the functional attains a minimum (or maximum) in the presence of constraints. We, therefore, examine the conditions when a functional attains a minimum. For the thickness function a, the critical point of the Rayleigh's quotient $\widetilde{\mathcal{R}}[\mathrm{u},a]$ is the eigenfunction u. To the eigenfunction u corresponds the lowest eigenvalue $\mathbb{L}[a]$. The guarantees that:

$$\mathbb{L}[a] = \frac{\left\langle a^k u'^2 \right\rangle}{\langle u^2 \rangle} = \min_{\mathrm{u}} \widetilde{\mathcal{R}}[\mathrm{u},a]. \tag{81}$$

Analogously, for the thickness function $\mathcal{A}$, the critical point of the Rayleigh's quotient $\widetilde{\mathcal{R}}[\mathrm{u},\mathcal{A}]$ is the eigenfunction $\mathcal{U}$. To this function corresponds the lowest eigenvalue $\mathbb{L}[\mathcal{A}]$:

$$\mathbb{L}[\mathcal{A}] = \frac{\left\langle \mathcal{A}^k \mathcal{U}'^2 \right\rangle}{\langle \mathcal{U}^2 \rangle} = \min_{\mathrm{u}} \widetilde{\mathcal{R}}[\mathrm{u},\mathcal{A}]. \tag{82}$$

The right sides of Equations (81) and (82) must be compared. In order to state the desired isoperimetric inequality, the following auxiliary inequality has to be approved:

$$\begin{gathered} \frac{\left\langle a^k u'^2 \right\rangle}{\langle u^2 \rangle} \leq \frac{\left\langle \mathcal{A}^k \mathcal{U}'^2 \right\rangle}{\langle \mathcal{U}^2 \rangle} \text{ for } \langle \mathcal{U}^2 \rangle = \langle u^2 \rangle, \\ \left(\mathcal{A}^k \mathcal{U}'\right)' + \mathbb{L}[\mathcal{A}]\mathcal{U} = 0, \qquad \left(a^k u'\right)' + \mathbb{L}[a]u = 0. \end{gathered} \tag{83}$$

The denominators of the fractions to the left and right of the auxiliary inequality (83) are identical. The nominators in the inequality (83) are, however, different. Thus, the nominators should be compared. The inequality for the nominators, which has to be proven, reads:

$$\left\langle a^k u'^2 \right\rangle \leq \left\langle a^k \mathcal{U}'^2 \right\rangle \leq \left\langle \mathcal{A}^k \mathcal{U}'^2 \right\rangle. \tag{84}$$

At this point, the optimality condition (71) with $\lambda = k$ will be used:

$$\mathcal{A} = \mathcal{U}'^{\frac{2}{1-k}}.$$

Namely, the substitution of the optimality condition (71) into (84) delivers the inequality for $k \geq 1$:

$$\left\langle a^k \mathcal{A}^{1-k} \right\rangle \leq \left\langle \mathcal{A}^k \mathcal{A}^{1-k} \right\rangle, \qquad \langle \mathcal{A} \rangle = \langle a \rangle. \tag{85}$$

The equality in (85) takes place only for $\mathcal{A} = a$. The validity of the yet suspected inequality (85) follows directly from the Hölder's inequality [22]. Consequently, it follows the inequality for nominators (85) and, finally, the desired inequality (83).

Combining (83) and (85) delivers:

$$\mathbb{L}[a] = \left\langle a^k u'^2 \right\rangle \leq \left\langle a^k \mathcal{U}'^2 \right\rangle \equiv \left\langle a^k \mathcal{A}^{1-k} \right\rangle \leq \left\langle \mathcal{A}^k \mathcal{A}^{1-k} \right\rangle = \left\langle \mathcal{A}^k \mathcal{U}'^2 \right\rangle \equiv \mathbb{L}[\mathcal{A}]$$

It was thus demonstrated that, for any arbitrary value of a, the associated eigenvalue is less than

$$\mathbb{L}[a] \leq \mathbb{L}[\mathcal{A}]. \tag{86}$$

The equality in Equation (86) is only achieved for the optimal configuration of twisted rod, characterized by its optimal shape, $\mathcal{A}$, and the limited volume of material, $\mathcal{V}$. When the shape exponent, $k \geq 1$, is employed, the game is convex. In general, the game will be convex if the function $J = J\,(a)$ is convex with respect to its argument.

In accordance with the necessary optimality conditions (71), the rod is capable of delivering the lowest possible upper value of the game:

$$\mathcal{G}[a] = \frac{\mathbb{P}}{EC\,\mathbb{L}[a]} \geq \frac{\mathbb{P}}{EC\,\mathbb{L}[\mathcal{A}]} \stackrel{\text{def}}{=\!=} \mathcal{G}_{opt}. \tag{87}$$

It follows from (87), that the value of functional game $\mathcal{G}_{opt} = \mathcal{W}[\mathcal{A}]$ for the optimal distribution of thickness $\mathcal{A}(\xi)$ is less than the payoff functional (distortion energy $\mathcal{G}[a]$) for an arbitrary distribution of thickness $a(\xi)$ of the same volume:

$$\langle a \rangle = \langle \mathcal{A} \rangle = \mathcal{V}. \tag{88}$$

If the volume of an arbitrary thickness distribution is inferior to the volume of the optimal thickness distribution, the stronger inequality follows.

$$\langle a \rangle < \langle \mathcal{A} \rangle \Rightarrow \mathbb{L}[a] < \mathbb{L}[\mathcal{A}]. \tag{89}$$

The relations (87)–(89) solve the "superstratum" game for the twisted rod with a torque distribution that is not uniquely defined.

6.5. In Section 6, the rod is subjected to arbitrary twist moments, and the designer determines the optimal design of the twisted bar, ensuring the greatest stiffness.

7. Nash and Pareto Fronts

7.1. Pareto efficiency is a specific characteristic of multifunctional equilibrium. A situation is defined as efficient, or Pareto-optimal, if no entity can improve its satisfaction of needs through further activities without threatening any other entities. The central prerequisite for this state is perfect competition, without which the optimum cannot be realized in principle, as only this leads to the necessary equilibrium efforts [25]. Simply put, Pareto efficiency improvement makes at least one player better off but no one worse off. Consequently, a strategy combination is called Pareto-efficient if no Pareto efficiency improvement is possible.

We examine the above solution from the viewpoint of inequalities theory and reveal the mathematical sense of the PARETO and NASH fronts [26,27]. There exists the relation between objective functionals, which expresses as the isoperimetric inequality:

$$ECl^2\frac{\mathbb{P}}{\mathcal{W}} = \mathbb{L}[a] \leq \mathbb{L}[\mathcal{A}] = ECl^2\frac{\mathbb{P}}{\mathcal{G}}. \tag{90}$$

7.2. The Pareto and Nash fronts were coincided in this case [28]. The fronts emerge as the equality case $\mathcal{G} = \mathcal{W}[\mathcal{A}]$. Pareto fronts for the states of Nash equilibrium for different shape factors are displayed on Figure 5.

Section 7 examines the solution of Sections 5 and 6 from the perspective of inequalities theory, elucidating the mathematical significance of the Pareto and Nash fronts [29].

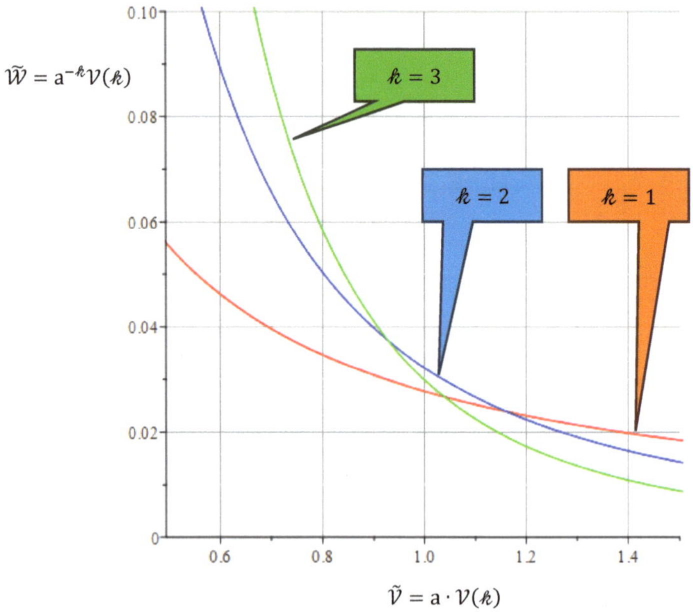

Figure 5. Pareto fronts for the states of Nash equilibrium for different shape factors.

8. Conclusions

The premise of this study is that there is currently no definitive understanding of the factors that influence the states of nature. The game method for structural optimization problems can be applied in situations when the external loads are not definitively prescribed [30–32]. Statistical decision theory is concerned with making decisions in the absence of precise probability information. The conflict between nature and operators ("ordinal players") is examined through the lens of game theory ("substratum" game). The payout matrix is relevant for one ordinal player ("operator") and the profit he makes if he employs his strategy while the second natural player ("nature") is in a specific state. For greater accuracy, we propose referring to nature's strategies as states. The content of the consequent theory could be summarized as follows. In the matrix game, there is a single payoff matrix. From the viewpoint of the optimization theory, there is one goal function. Thus, the optimization problem is of a scalar type. This goal function is the value of the matrix game. The win of "nature" is the loss of the "operator" as both players on the low-level. This game is antagonistic with only one goal function. The designer achieves the most appropriate value for this goal function.

In the bi-matrix game, there are two payoff matrices. From the viewpoint of optimization theory, there are two goal functions. The optimization problem is of a vector type. These goal functions come from the solution of bi-matrix game. The equilibrium state between the interests of "nature" and "operator" results in the solution of the game on the low-level. The designer achieves the most appropriate value for this goal function in accordance with the methods of vector optimization

The essence of the method is to identify the structure that optimally resists the worst external load. In the context of classical structural optimization, the "cardinal" players are tasked with assuming the role of "control functions". Similarly, the "cardinal" players are responsible for modifying the governing equations and payoff functions in the game formulations ("superstratum" game). If a relation between objective functionals, which could be expressed in terms of an isoperimetric inequality, exists, then the Pareto and Nash fronts coincide and emerge as the equality case of the isoperimetric inequality.

Funding: This research received no external funding.

Data Availability Statement: The numerical results presented in this document can be replicated using the methodology and formulations described here. The analytical expressions used in the examples are available upon request to the authors.

Conflicts of Interest: The author declares no conflicts of interest.

References

1. Eschenauer, H.; Koski, J.; Osyczka, A. *Multicriteria Design Optimization: Procedures and Applications*; Springer: Berlin/Heidelberg, Germany, 1990.
2. von Neumann, J. Zur Theorie der Gesellschaftsspiele. *Math. Ann.* **1928**, *100*, 295–320. Available online: https://link.springer.com/content/pdf/10.1007/BF01448847.pdf (accessed on 30 November 2024). [CrossRef]
3. Kobelev, V. On a game approach to optimal structural design. *Struct. Multidisc. Optim.* **1993**, *6*, 194–199. [CrossRef]
4. von Neumann, J.; Morgenstern, O. *Theory of Games and Economic Behavior*; Princeton University Press: Princeton, NJ, USA, 2004. [CrossRef]
5. Werner, D. Normierte Räume. In *Funktionalanalysis*; Werner, D., Ed.; Springer Spektrum: Berlin/Heidelberg, Germany, 2018. [CrossRef]
6. Geering, H.P. *Optimal Control with Engineering Applications*; Springer: New, York, NY, USA, 2007; ISBN 978-3-540-69437-3.
7. Tadelis, S. *Game Theory. An Introduction*; Princeton University Press: Princeton, NJ, USA; Oxford, UK, 2013.
8. Szép, J.; Forgó, F. *Introduction to the Theory of Games*; Mathematics and Its Applications; Springer: Dordrecht, The Netherlands, 1985; Volume 17.
9. Zhang, F. *Matrix Theory, Basic Results and Techniques*; Springer: New York, NY, USA, 2011.
10. Banichuk, N.V. *Introduction to Optimization of Structures*; Springer: New York, NY, USA, 1990.
11. Lemke, C.E.; Howson, J.T. Equilibrium Points of Bi-Matrix Games. *SIAM J.* **1964**, *12*, 413–423.
12. Parlett, B.N. *The Symmetric Eigenvalue Problem*; Classics in Applied Mathematics, 20; Society for Industrial and Applied Mathematics: Philadelphia, PA, USA, 1998.
13. Biezeno, C.B.; Grammel, R. *Engineering Dynamics*; Blackie: London, UK, 1956.
14. Wilkinson, J.H. *The Algebraic Eigenvalue Problem*; Clarendon: Oxford, UK, 1965; Volume 662.
15. Kobelev, V. Optimization of Compressed Rods with Sturm Boundary Conditions. In *Fundamentals of Structural Optimization. Mathematical Engineering*; Kobelev, V., Ed.; Springer: Cham, Switzerland, 2023. [CrossRef]
16. Reddy, J.N. *Energy Principles and Variational Methods in Applied Mechanics*; Wiley: Hoboken, NJ, USA, 2002.
17. Courant, R.; Hilbert, D. *Methods of Mathematical Physics*; WILEY-VCH Verlag GmbH & Co. KGaA: Hoboken, NJ, USA, 2004.
18. Nash, J. Non-cooperative games. *Ann. Math. 2nd Ser.* **1951**, *54*, 286–295. [CrossRef]
19. Zettl, A. *Sturm–Liouville Theory*; American Mathematical Society: Providence, RI, USA, 2005.
20. Cox, S.J.; Overton, M.L. On the Optimal Design of Columns Against Buckling. *SIAM J. Math. Anal.* **1992**, *23*, 287–325. [CrossRef]
21. Polyanin, A.D.; Zaitsev, V.F. *Handbook of Exact Solutions for Ordinary Differential Equations*, 2nd ed.; CRC Press: Boca Raton, FL, USA; London, UK, 2003.
22. Bullen, P.S. *Handbook of Means and Their Inequalities*; Mathematics and Its Applications; Springer: Dordrecht, The Netherlands, 2003; Volume 560.
23. McIntosh, S.C.; Weisshaar, T.A.; Ashley, H. Progress in Aeroelastic Optimization—Analytical Versus Numerical Approaches. In *AIAA Structural Dynamics and Aeroelasticity Specialist Conference, SUDAAR No. 383, New Orleans, Lousiana, 16–17 April 1969*; AIAA: New Orleans, LA, USA, 1969.
24. Battoo, R. An introductory guide to literature in aeroelasticity. *Aeronaut. J.* **1999**, *103*, 511–518. [CrossRef]
25. Fernandez, F.R.; Monroy, L.; Puerto, J. Multicriteria Goal Games. *J. Optim. Theory Appl.* **1988**, *99*, 403–421. [CrossRef]
26. Monfared, M.S.; Monabbati, S.E.; Kafshgar, R.A. Pareto-optimal equilibrium points in non-cooperative multi-objective optimization problems. *Expert Syst. Appl.* **2021**, *178*, 114995. [CrossRef]
27. Ehrgott, M. *Multicriteria Optimization*; Springer: Berlin/Heidelberg, Germany, 2005. [CrossRef]
28. Kobelev, V. Comment to the Article "Several Examples of Application of Nash and Pareto Approaches to Multiobjective Structural Optimization with Uncertainties" of N. V. Banichuk, F. Ragnedda, M. Serra. *Mech. Based Des. Struct. Mach.* **2014**, *42*, 130–133. [CrossRef]
29. Borm, P.; Tijs, S.; van den Aarssen, J. *Pareto Equilibria in Multiobjective Games*; Tilburg University, School of Economics and Management: Tilburg, The Netherlands, 1988. Available online: https://research.tilburguniversity.edu/en/publications/pareto-equilibria-in-multiobjective-games (accessed on 4 November 2024).
30. Banichuk, N.V. On the game theory approach to problems of optimization of elastic bodies. *J. Appl. Math. Mech.* **1973**, *37*, 1042–1052. [CrossRef]
31. Greiner, D.; Periaux, J.; Emperador, J.M.; Galván, B.; Winter, G. Game Theory Based Evolutionary Algorithms: A Review with Nash Applications in Structural Engineering Optimization Problems. *Arch. Comput. Methods Eng.* **2017**, *24*, 703. [CrossRef]
32. Holmberg, E.; Thore, C.; Klarbring, A. Game theory approach to robust topology optimization with uncertain loading. *Struct. Multidisc. Optim.* **2017**, *55*, 1383–1397. [CrossRef]

eng

MDPI

Review

Optimization of Structures and Composite Materials: A Brief Review

André Ferreira Costa Vieira [1,*], Marcos Rogério Tavares Filho [2], João Paulo Eguea [2] and Marcelo Leite Ribeiro [2]

[1] Center for Mechanical and Aerospace Science and Technologies (C-MAST-UBI), University of Beira Interior, Marquês D'Ávila e Bolama St., 6201-001 Covilhã, Portugal

[2] Department of Aeronautical Engeneering, São Carlos School of Engineering, University of São Paulo, 1100 João Dagnone Ave., São Carlos 13563-120, SP, Brazil; marcosrtf@usp.br (M.R.T.F.); joao.eguea@usp.br (J.P.E.); malribei@usp.br (M.L.R.)

* Correspondence: andre.costa.vieira@ubi.pt

Abstract: Neural networks (NNs) have revolutionized various fields, including aeronautics where it is applied in computational fluid dynamics, finite element analysis, load prediction, and structural optimization. Particularly in optimization, neural networks and deep neural networks are extensively employed to enhance the efficiency of genetic algorithms because, with this tool, it is possible to speed up the finite element analysis process, which will also speed up the optimization process. The main objective of this paper is to present how neural networks can help speed up the process of optimizing the geometries and composition of composite structures (dimension, topology, volume fractions, reinforcement architecture, matrix/reinforcement composition, etc.) compared to the traditional optimization methods. This article stands out by showcasing not only studies related to aeronautics but also those in the field of mechanics, emphasizing that the underlying principles are shared and applicable to both domains. The use of NNs as a surrogate model has been demonstrated to be a great tool for the optimization process; some studies have shown that the NNs are accurate in their predictions, with an MSE of 1×10^{-5} and MAE of 0.007%. It has also been observed that its use helps to reduce optimization time, such as up to a speed 47.5 times faster than a full aeroelastic model.

Keywords: review; machine learning; aircraft structures; composite materials; genetic algorithm; topology optimization

Citation: Vieira, A.F.C.; Filho, M.R.T.; Eguea, J.P.; Ribeiro, M.L. Optimization of Structures and Composite Materials: A Brief Review. *Eng* **2024**, *5*, 3192–3211. https://doi.org/10.3390/eng5040168

Academic Editor: Antonio Gil Bravo

Received: 14 October 2024
Revised: 25 November 2024
Accepted: 26 November 2024
Published: 2 December 2024

1. Introduction

Over the years, the field of aeronautical engineering has experienced constant evolution in searching for solutions aimed at improving aircraft performance, efficiency, and safety. According to Bishop [1], in recent years, neural networks (NNs) have seen significant growth, leading to numerous practical applications across various fields and providing a robust set of tools for solving a wide range of problems. With this in mind, the use of NNs together with optimization methods can offer great potential for improving structural design and the speed of optimization processes.

In the aeronautical area, NNs have emerged as a powerful tool in several areas, one of which is the area of optimization. In this field, NNs can help improve the optimization process by integrating it with an optimization method, such as the genetic algorithm (GA). Thus, the following two areas have emerged with the potential for innovation, development, and advancement: structural and composite material optimization. This can be seen in Lagaros et al. [2], who used NNs to improve the performance of evolutionary strategies (ESs) in structural optimization, or Caixeta and Marques [3], who emphasized multi-objective optimization in the design of aircraft wings using NN metamodels, showing the integration of structural and material optimization in obtaining superior designs.

Structural optimization brings with it the potential to promote improvements in the performance of structures, that is, improving rigidity, strength, ductility, resilience, safety, and other points. Optimization can also help us find the best point at which the structure will deliver the best performance while reducing mass. Therefore, the search for better structural configurations is also aligned with the objective of increasing the performance, efficiency, and safety of aircrafts. In this context, the use of NNs integrated with an optimization method can open up frontiers such that it is possible to achieve new structural designs that were previously difficult to achieve through more conventional methods. Some articles have shown the use of NNs in the optimization of structures, such as those by Lagaros et al. [2] and Caixeta and Marques [3] (optimization using the evolutionary method), as well as Freitag et al. [4] and Jeong et al. [5] (topological optimization).

On the other hand, material selection is crucial for the development of a project; it is necessary to know the characteristics and properties of the material to choose the best one according to the objective. In aeronautical engineering, aluminum alloys are widely used but, currently, the use of composite materials (fiberglass and carbon fiber) is increasing. This is occurring due to the characteristics of composite materials that help them achieve specific performance requirements such as strength, stiffness, and weight reduction. Just as structural optimization aims to improve the performance of the structure, material optimization also seeks to obtain better performance by maximum strength, stiffness, and low weight. Similarly, NNs integrated with an optimization method can help in the optimization of composite materials, as can be seen in [6–8].

Therefore, it is noted that both material and structural optimization are necessary for improving structural properties. Structural and material optimization are processes that are connected, as one can influence the other. In structural optimization, it is possible to determine the best configurations and geometries that improve its performance, but material optimization can occur by optimizing the layer stacking or optimizing the fiber directions in each layer [9,10]. Furthermore, material optimization will bring benefits to structural design due to its influence on stiffness, strength, and weight. Therefore, optimizing both structure and material iteratively will help us to achieve the objective of improving the structural properties.

Over the years, technology has evolved and increasingly brings alternatives that contribute to enhancing processes, such as optimization. Computers are becoming more powerful, and this is helpful to increase the use of NNs, which is advantageous for engineering. This tool provides an alternative for structural optimization to accelerate processes. Because of this, research about the use of neural networks in optimization is increasing and there are many works about this topic. However, when searches for similar studies are carried out in aeronautics, there are fewer papers than in mechanics. The novelty of this article is to present not only aeronautical works but also mechanical works, as the ideas are similar and can be applied in both areas.

The sections of this article are as follows: A bibliometric analysis will be presented to provide an overview of the research landscape related to aircraft structural optimization and composite material optimization. This will be followed by a section on NNs, where the key concepts of NNs and deep neural networks (DNNs) will be explained. The third section will delve into training methods, covering both supervised and unsupervised learning techniques that are essential for the development and implementation of effective NN models. In the fourth section, a survey of the previous literature will be conducted, presenting a comprehensive review of the research on aircraft structural optimization and composite material optimization, emphasizing the integration of NNs in these areas. The fifth section will present a discussion on the findings, implications, and future directions for research and application. Finally, the last section will conclude the article, summarizing the key points and contributions of the study.

2. Bibliometric Survey

The significance of the union between NNs and optimization algorithms has been noted, so it is important to carry out a bibliometric survey to understand the use of NNs over the years and what the future projections are. The subsequent paragraphs present a survey on the optimization of aircraft structures using NNs, as well as more general research on the topic (also including articles without the use of NN), revealing interesting trends.

Although research into aircraft structure optimization using NN is more recent, as can be seen in Figure 1, it shows a notable growth trajectory. This growth is indicative of the growing interest and recognition of the potential benefits that NNs offer in optimizing aircraft structures. On the other hand, research into aircraft structural optimization in a general context has a longer history, dating back to 1918, as can be seen in Figure 2. This longer period reflects the traditional structural optimization approaches that have been employed over the years. Despite the recent emergence of NNs in this field, there continues to be a significant body of research conducted without the use of this tool. Furthermore, when analyzing the number of documents on optimization in both cases, two countries, China and the United States, emerge as the main contributors, as can be seen in Figures 3 and 4. This could highlight the global interest and involvement in the optimization of structures, regardless of the methodologies employed.

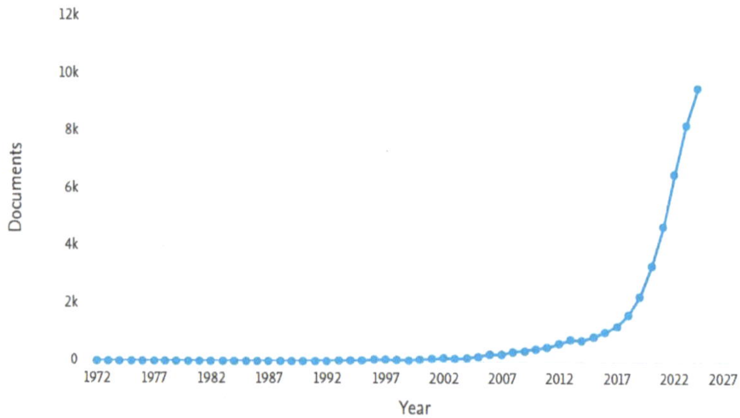

Figure 1. Structure Optimization Documents Using NNs by Year [11].

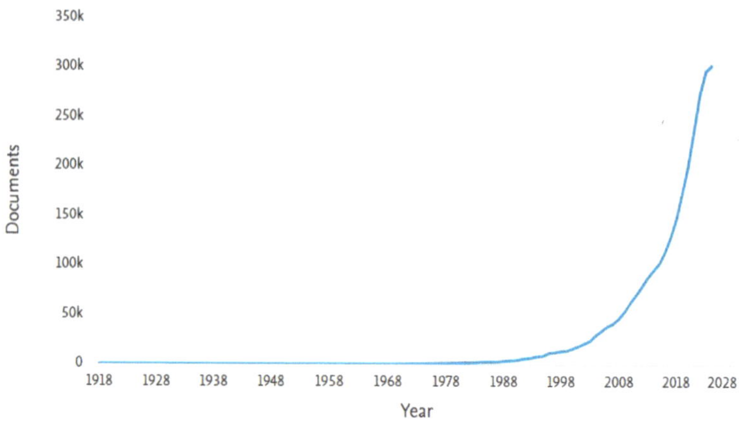

Figure 2. Structure Optimization Documents by Year [12].

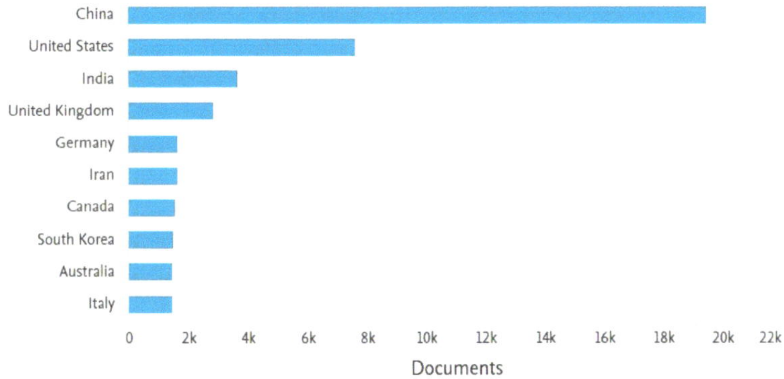

Figure 3. Structure Optimization Documents Using NNs by Country [11].

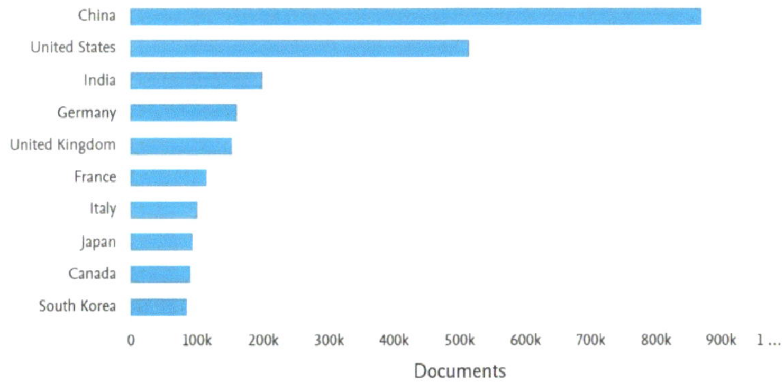

Figure 4. Structure Optimization Documents by Country [12].

On the other hand, it is also important to carry out a bibliometric analysis of the optimization of composite materials. The bibliometric analysis carried out includes research on the optimization of composite materials that used NNs in the process, as well as more comprehensive research on this topic that used or did not use NNs, in order to elucidate trends and patterns in the previous research activities within this domain.

Composite materials are one of the most used material on aircrafts. Due to that, a bibliometric analysis of the optimization of composite materials was carried out to provide insights into the evolution and trends of research in this field. Research focused on optimizing composite materials using NNs is more recent, beginning only in 1992, as can be seen in Figure 5. In contrast, optimizing composite materials without the use of NNs has a longer history, dating back to 1918, as can be seen in Figure 6. General bibliometric research on the optimization of composite materials reveals a growing trend in research activity over the years. When examining geographic distribution, research into the optimization of composite materials using NNs is prominent in China, India, and the United States, as can be seen in Figures 7 and 8.

Finally, delving deeper into NN applications and optimization methodologies is crucial to understanding the current research efforts. This paper offers invaluable insights into the methodologies, advancements, and applications in this field. Specifically, this article highlights the importance of understanding NNs along with optimization techniques such as GA and topology optimization. By exploring these technologies, we hope to present a more comprehensive understanding of their role in the design of aeronautical structures and in composite materials thus paving the way for future innovations and advances in this field.

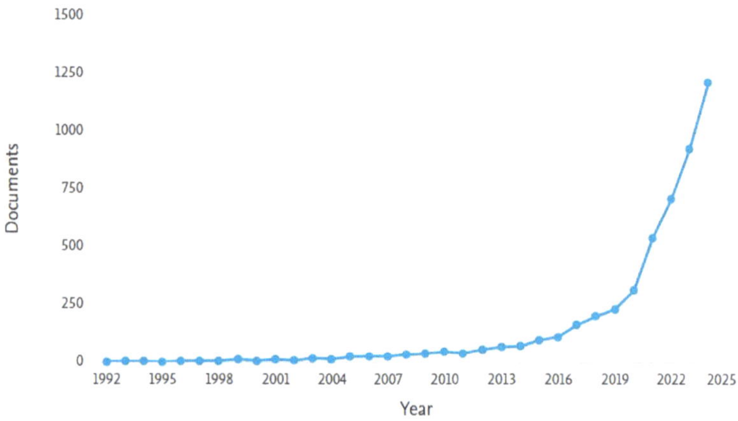

Figure 5. Composite Material Optimization Documents Using NNs by Year [11].

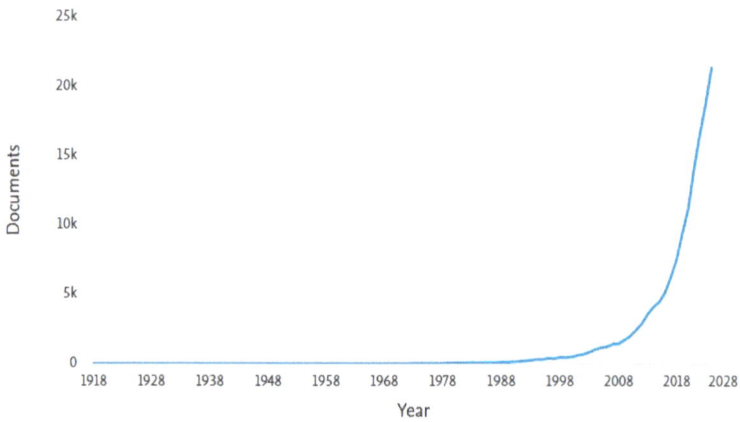

Figure 6. Composite Material Optimization Documents by Year [13].

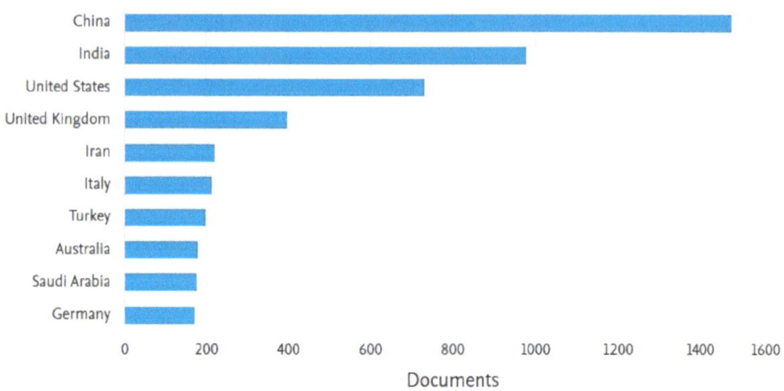

Figure 7. Composite Material Optimization Documents Using NNs by Country [11].

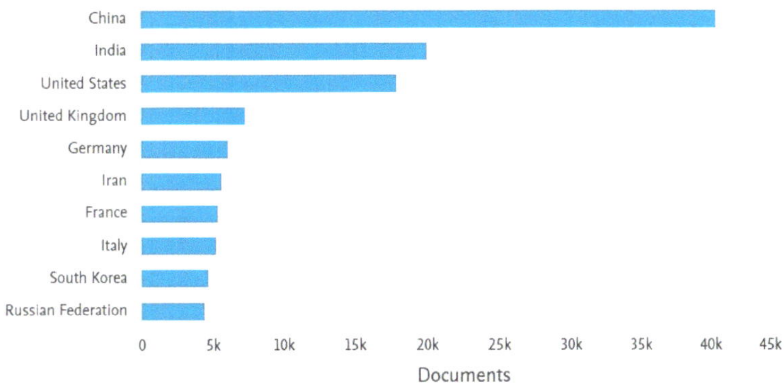

Figure 8. Composite Material Optimization Documents by Country [13].

3. Applications of Neural Networks in Aeronautical Engineering

This section aims to demonstrate the use of NNs in the aeronautical field. With this in mind, some of the articles that will be presented have used NNs in different fields of aeronautical engineering. Articles focused on structures or composite materials will be presented in Section 5.

NNs are a machine learning (ML) method inspired by the workings of the human brain, as can be seen in Figures 9 and 10. NNs are composed of interconnected nodes like neurons in the brain and are organized in layers, as can be seen in Figure 11. The training process is carried out with a set of data, which is composed of inputs and outputs; in this process, weights are assigned that are iteratively adjusted until the error is minimized. At the end of this process, NNs are capable of generalizing and predicting data or behavior with a certain accuracy. This can be seen in Reed [14], where a dataset was obtained and used to train the implemented NN. In aeronautical engineering, some of the applications of NNs include aerodynamic performance prediction, fault detection and system diagnosis, and structural optimization, among others. According to Brunton et al. [15], the impact of data science will be evident across multiple fields, including aerospace engineering, influencing areas such as inspection, design and performance, materials and composites, maintenance, and the development of future products.

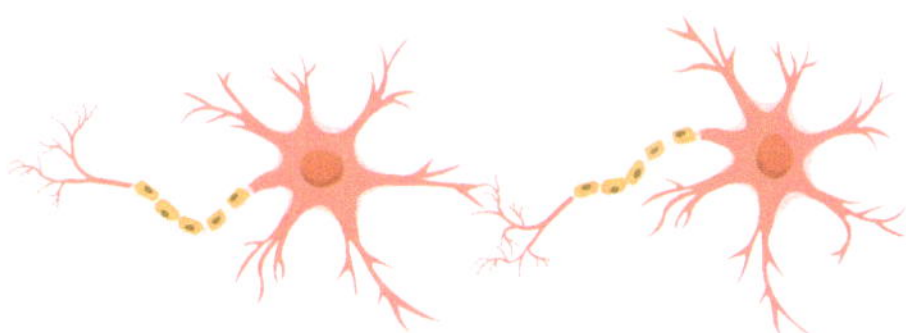

Figure 9. Representation of human neurons.

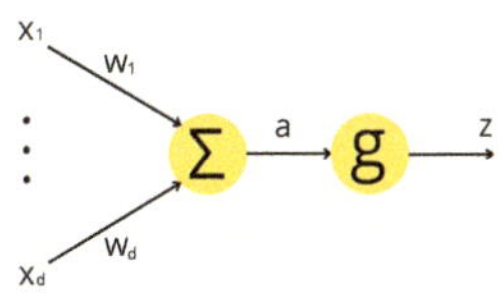

Figure 10. Representation of the model of a neuron in the NN.

The presented works showcase the innovative applications of NNs in diverse aerospace engineering domains. Wu et al. [16] presented a multi-fidelity neural network (MFNN) optimization framework for designing efficient propellers for an electric aircraft. Combining blade element momentum theory (BEMT) with high-fidelity computational fluid dynamics (CFD) simulations, the method improved propeller design accuracy while reducing computational costs. Results showed a notable increase in propeller cruise efficiency, from 82.3% to 87.1%, highlighting the effectiveness of the approach in achieving better optimization outcomes with enhanced efficiency.

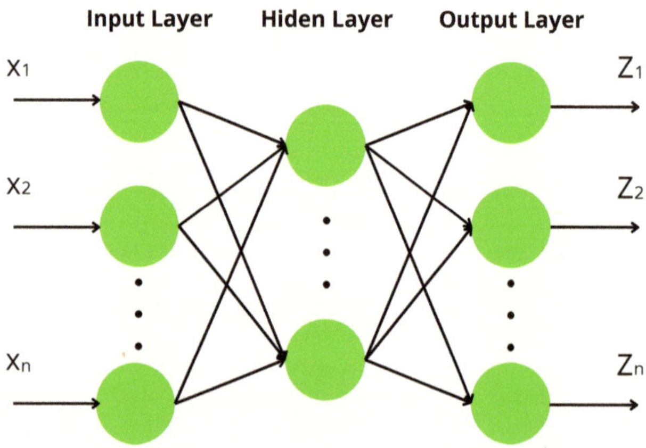

Figure 11. NN with 1 hidden layer.

Addressing challenges like limited communication and disturbance, Jia et al. [17] introduced a method for robust attitude synchronization in multi-spacecraft formations. It leveraged a dynamic event-triggered mechanism and a self-learning neural network control law (SLN2C), which utilized a radial basis function neural network (RBFNN) to compensate for disturbances. Numerical simulations validated the effectiveness of the method in achieving robust synchronization despite communication limitations and disturbances.

In spacecraft attitude stabilization, Zhang et al. [18] proposed a neural network-based fault-tolerant control scheme for spacecraft attitude stabilization in the presence of disturbances and actuator faults. A neuro-adaptive estimator approximated the disturbances, enabling an integrated event-based control scheme. This approach ensured system convergence and minimized actuator updates, conserving onboard resources. Numerical simulations validated the effectiveness of the proposed algorithms.

Lastly, Mazhar et al. [19] introduced a novel technique using an artificial neural network (ANN) to apply aerodynamic pressure loads on unmanned aerial vehicles (UAVs) for finite element (FE) analysis during structural design. By training the ANN models to approximate pressure functions based on aerodynamic pressure data, the method enabled the accurate application of pressure loads in FE analysis. Compared to conventional techniques, the ANN-based approach yielded a superior performance in matching the actual pressure profiles on the aircraft, offering a reliable and efficient solution for UAV structural design.

If more layers are added to the NN and the number of layers is greater than three, this NN could be named as a DNN. This happens because DNNs are a kind of NN; however, DNNs have more hidden layers compared to NNs, as can be seen in Figure 12. Due to their greater number of layers, DNNs are capable of learning more complex data structures than traditional NNs. According to Deng [20], deep learning (DL) is a subset of ML techniques that utilizes multiple layers of information processing in hierarchical architectures, which are leveraged for pattern classification and feature or representation learning. However, the training of DNNs requires a greater amount of data and computational capacity, such that the adjustment of the hyperparameters is carried out in the best way where the error is as small as possible. In aeronautical engineering, DNNs have the same applications as NNs; however, their greater capacity is highly desired.

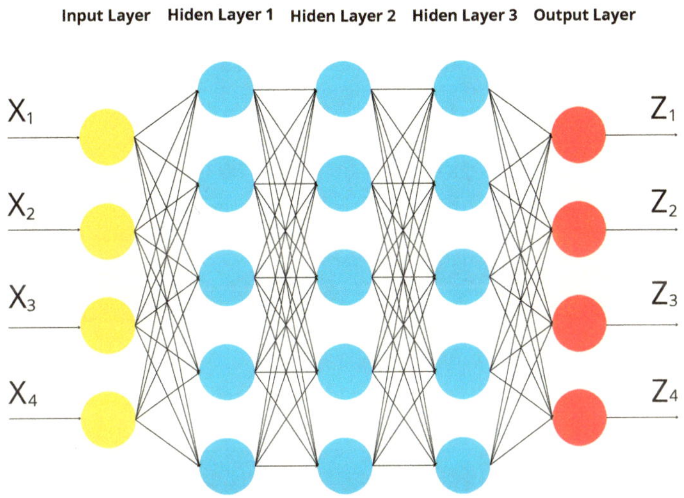

Figure 12. DNN with 3 hidden layers.

Recent advancements in aerospace engineering have showcased the transformative potential of integrating DNNs into various applications. Wang et al. [21] introduced DNN-MS, a method combining DNNs and multiagent synergism to evaluate high cycle fatigue (HCF) in aircraft engine compressor rotors. It improved accuracy and efficiency compared to other methods, offering a promising solution for reliability assessments.

In a similar vein, Zhou et al. [22] introduced a hybrid deep neural network (HDNN) for identifying hazards in aircraft auxiliary power units (APUs). By combining multiple CNN-BiLSTM models, the HDNN achieves high accuracy and stability in identifying potential risks. This method offers a promising approach to enhancing safety in civil aviation.

Investigating dynamic soaring, a technique inspired by albatross flight, Kim et al. [23] explored the extraction of energy from wind gradients. Their study introduced a DNN coupled with a feedback control law to execute dynamic soaring maneuvers based on mechanical energy extraction mechanisms. Results indicate that the trained network was proficient in maneuver execution across various wind profiles, highlighting the potential of NNs in replicating complex natural behaviors for aviation applications.

Meanwhile, Tao et al. [24] introduced the Abaqus-DNN mechanics system, coupling the Abaqus FE code with a DNNs. This system learned constitutive laws for fiber-reinforced composites without presuming specific functions, ensuring accuracy and adherence to physics laws. It accurately learned engineering constants and progressive damage laws for composites, offering a versatile approach for learning unknown physics within mechanical systems.

Furthermore, XIONG et al. [25] proposed a point cloud DL method to reduce costs in 3D aerodynamics simulations. Using PointNet architecture, it established an NN meta-model to map object surface positions to Computational Fluid Dynamics (CFD) quantities. The approach constructed point clouds from grid vertices, maintaining boundary smoothness and allowing for the detection of small geometric changes. Validation on the ONERA M6 wing demonstrated improved accuracy compared to the traditional methods, showcasing the effectiveness of the proposed approach for aerodynamic prediction and shape optimization.

4. Training Methods

As previously stated, both NNs and DNNs need to be trained so that they can perform the desired tasks. Shobha and Rangaswamy [26] highlighted the following:

> "ML is broadly classified as supervised, unsupervised, semisupervised, and reinforcement learning. A supervised learning model has two major tasks to be performed, classification and regression. Classification is about predicting a nominal class label, whereas regression is about predicting the numeric value for the class label. However, in the unsupervised learning, the model is trained without the use of labels in order for the model to identify and learn hidden patterns."

Also according to Shobha and Rangaswamy [26], in unsupervised learning, the objective is to identify regularities in the input data, recognizing patterns that occur more frequently than others, as well as to learn to distinguish common occurrences from anomalies. The choice of each method will depend on the nature of the data and the goal of the problem to be solved.

4.1. *Supervised Learning*

In supervised learning, model training is carried out in a guided way, as the relationships between the inputs and outputs are known. Therefore, training aims to reduce the error between the prediction and the real outcome.This type of learning may include linear regression and classification algorithms, such as a decision tree or logistic regression. Some of its applications include image and speech recognition, predictive analysis, etc.

A novel approach called aerodynamic strength prediction graph neural network (ASP-GNN), introduced by Li et al. [27], utilizes supervised graph learning to swiftly and accurately predict gas turbine characteristics. ASP-GNN efficiently forecasts the aerodynamic strength features and temperature fields of complex gas turbine blades, demonstrating superior performance and generalizability with limited training samples. It offers a fast analysis tool for turbomachinery design and analysis, potentially reducing the workload of complex engineering simulations.

For a shaker blower used in aeronautical systems, Cannarile et al. [28] developed a fault diagnostic system. It extracted features from condition monitoring signals using the ELastic NET (ELNET) algorithm and employed multinomial logistic regression (MLR) for classification. Experimental validation demonstrated a satisfactory diagnostic performance, highlighting the method's potential for developing robust classifiers with limited training data.

Using convolutional neural networks (CNNs), a form of supervised learning, [29] introduced a data-driven approach for predicting airfoil pressure distribution. Instead of relying on time-consuming CFD, this method trained the CNN to approximate the mapping between airfoil geometry and aerodynamic performance. By employing a flexible parametrization method called signed distance function, the CNN efficiently predicted pressure coefficients for unseen airfoils with less than 2% mean square error in seconds.

Incorporating a supervised ML model and fuzzy logic algorithm, the adaptive Kalman filter presented by Zhang and Hsu [30] demonstrated its ability to accurately classify global navigation satellite system (GNSS) accuracy levels and dynamically adjust noise covariance. This adaptive approach significantly improved the positioning accuracy of quadcopters operating in urban environments, ensuring reliable navigation despite challenging GNSS conditions.

4.2. Unsupervised Learning

Unlike supervised learning, unsupervised learning does not perform training with labeled data, as this approach aims to learn and identify patterns presented by the data. Some of the most used techniques are clustering and dimensionality reduction. According to Shobha and Rangaswamy [26], clustering is a mode of unsupervised learning used to infer patterns from datasets that lack class labels or target values, making it valuable for exploratory data analysis to uncover hidden structures. In dimensionality reduction, the focus is on reducing the number of features in the dataset without losing key information. Huang et al. [31] stated that although these dimensionality reduction techniques can eliminate redundant data and increase model accuracy, crucial features can lose their physical interpretation after reduction, making the model less explainable. Some of the uses of unsupervised learning are in anomaly detection, feature engineering tasks, and exploratory data analysis.

Introducing a method using unsupervised learning, particularly generative adversarial networks (GANs), to generate realistic inflow boundary conditions for turbulent channel flow simulations, Kim and Lee [32] presented an innovative approach. By learning from direct numerical simulation (DNS) data, the GAN can produce flow fields that match DNS statistics. The approach, called RNN-GAN, combines the GAN with a recurrent neural network (RNN) to generate time-varying fully developed flow, showcasing the effectiveness of unsupervised learning in synthesizing realistic turbulence fields.

Focusing on improving surrogate models for accurate aircraft fuel burn evaluation during missions, Liem et al. [33] utilized methods like adaptive sampling and a mixture of experts (MoE) approach.MoE proved superior, offering better approximation with fewer samples. Additionally, a cluster-based preprocessing step separated clustering in the y-space and the x-space using unsupervised and supervised learning algorithms like the Gaussian mixture model (GMM) and regularized Gaussian classifier, enhancing model efficiency and accuracy.

Introducing DeepESVDD-CNN, a system for the warning of a compressor rotating stall in aircraft power systems, Jin et al. [34] presented an innovative approach. By combining anomaly detection with DL, specifically unsupervised learning, it effectively identified stall precursors from dynamic pressure signals using a new method called Deep Ellipsoid Support Vector Data Description (DeepESVDD). This approach simplified model training and outperformed the traditional warning methods across various compressor operating modes, showing promise for real-time stall warning tasks.

For the super-resolution reconstruction of turbulent flows, Kim et al. [35] introduced an unsupervised learning model utilizing a cycle-consistent generative adversarial network (CycleGAN). Unlike supervised learning, which requires paired data, this model can be trained with unpaired turbulence data, making it more applicable in practical scenarios. Results demonstrate the model's effectiveness in reconstructing high-resolution flow fields from various turbulence datasets, showcasing its potential for wide application in the super-resolution reconstruction of turbulent fields.

5. Survey of Literature

An NN is a valuable tool that could predict the results of analysis via FEM or experimental loads. On the other hand, the optimization process inspired by Darwin's theory of evolution can be combined with an NN during the iterative optimization process. One of the proposals is to use the power of NNs to learn and predict FEM results. First, several simulations are carried out using FEM. Then, these results are used to train the NN. After training, the slow FEM calculations can be replaced by the NN predictions, with high accuracy and precision, and they are very fast compared to the FEM calculations. When the GA is executing the optimization, at each iteration, a set of parameters must be tested to evaluate its output in terms of optimization criteria. In this process, the mating pairs are selected using a selection technique such as elitism, in which the best individuals, those having the best outputs, are selected. In order to determine the best individuals, tests must be carried out to assess which are the best. Hence, each individual with a set of parameters can be tested using FEM analysis. Another costly alternative is actual mechanical testing, in which each sample, representing a set of parameters, is experimentally tested, and then these experimental results can also be used to train the NN. Although FEM analysis can speed up the learning process compared to the experimental results, it can be time-consuming to run a simulation if one has a trained NN capable of predicting the same results as the FEM. Hence, as mentioned above, NNs can replace FEM simulations during the iterative process of optimization. Therefore, the NN's only task is to evaluate the best individuals, and it does not take part in other processes such as crossover, so it has no interaction with the genes (strings) and chromosomes.

5.1. Structural Optimization

Optimizing an aircraft is extremely important, as this is a fundamental practice for improving it. When optimizing an aeronautical structure, the objective is to improve the aircraft's performance. To this end, there is a focus on improving stiffness, reducing weight, as well as aerodynamic efficiency. Such objectives can be achieved using the traditional optimization methods, but the integration between NNs and optimization methods can bring great benefits to the area.

The evolution of ML, such as NNs, is important for several industries. Its integration with GA has proven to be a great tool for structural design. According to Lagaros et al. [2], over the past decade, artificial intelligence and soft computing techniques have emerged as valuable tools for replacing time-consuming computational tasks in various scientific and engineering applications. NNs can be trained to predict the results of analyses via FE to speed up this process. Some works have studied this for structural design [36–39]. Therefore, the use of NNs will accelerate the FE analysis process of each optimized model obtained via GA.

This union has proven to be fundamental in structural design and this occurs due to the GA optimization process that is inspired by natural selection. According to Rao [40], genetic algorithms (GAs) are grounded in the principles of natural genetics and selection, incorporating the key elements of reproduction, crossover, and mutation in their search process. In this method, optimization occurs through the evolution of a population, evolving according to the potential solutions and thus iteratively converging to an optimal point. Combining NNs with GA allows for a better evaluation of each generation and the development of projects adapted to each predefined criteria throughout each generation.

Another example of an application is the prediction of load distribution along the aircraft. In this case, the NNs are trained with data from flight histories and simulations to be able to accurately predict loads, stress distribution, and strain in different flight conditions. Cao et al. [41] introduced a method using ANNs to identify loads on aircraft wings. Hoffman [42] used NNS to predict strain in critical parts of an aircraft's structure, bypassing the limitations of traditional equations.

In the context of structural optimization, many authors have used NNs together with some evolutionary method, as can be seen in [2,3,43–47]. Evolutionary methods are characterized by being inspired by principles of biological evolution and natural selection. According to Lagaros et al. [2], evolutionary algorithms (EAs), including GAs and evolutionary strategies (ESs), are now widely recognized as a powerful family of methods for addressing structural optimization problems.

Aiming to improve the computational efficiency of evolutionary algorithms in structural optimization, Lagaros et al. [2] presented an endeavor towards enhancing performance. They introduced an NN strategy to predict design feasibility, reducing the need for costly FE analysis. The adaptive NN was updated during the optimization process, offering computational advantages for optimizing skeletal structures in both sequential and parallel computing environments.

Exploring multidisciplinary design optimization (MDO) for flexible aircraft wings, with a focus on structural and aeroelastic characteristics, Caixeta and Marques [3] presented an in-depth investigation. They used ANNs to predict critical flutter speeds efficiently. Multi-objective optimization (MOO) with GA was aimed at maximizing flutter speed and minimizing structural mass. Results show the effectiveness of this approach in achieving optimal design trade-offs.

Discussing the optimization of aircraft components, such as composite panels of a vertical tail plane, using GAs and NNs, Ruijter et al. [45] presents an insightful examination. This approach adjusts the design variables to improve performance and reduce weight, with the NNs predicting constraint function values from FE analysis data. It offers flexibility for implementing new variables and is robust for high-dimensional design problems, with proposed modifications to enhance accuracy and reduce preparation time.

In addition to GAs, some works have also addressed topological optimization together with NNs [4,5,48–51]. In his optimization process, Hansen and Horst [51] also relied on an evolutionary algorithm to drive the advancements. Topology optimization is typically an iterative process that uses FE analysis to determine the optimal material distribution within a design domain by minimizing an objective function while satisfying one or more constraints [52].

Introducing topology optimization via the neural reparameterization framework (TONR), a method that combines DL with topology optimization, Zhang et al. [52] presented an innovative approach. TONR directly used NNs for optimization, transforming design variable updates into NN parameter updates. It addressed various optimization problems without needing a pre-constructed dataset and avoided structural disconnection issues. Numerical examples showed TONR's effectiveness compared to the conventional methods.

Introducing a methodology that considers uncertain load and material parameters, Freitag et al. [4] presented an innovative approach to topology optimization. It integrated compliance minimization and reliability-based design to handle uncertainties efficiently. To reduce computational costs, it used ANNs instead of FE simulations. Case studies demonstrated its effectiveness on different structural configurations.

Presenting a multilevel optimization method for implementing structural design changes in aircraft, Hansen and Horst [51] offered an alternative approach to the field. It optimized topology parameters and thicknesses/cross-sections using detailed FE models. Examples showcased its effectiveness in sizing and topology optimization of the framework structures and fuselage structures of a blended wing body aircraft.

As can be seen in the previous paragraphs, it is possible to understand the relevance of using NNs as a tool in structural optimization. The advanced computational capabilities of NNs can help us to obtain greater efficiency in the prediction and optimization of structural performance. Some of the above-mentioned articles are related to the mechanical field, while some are related to the aeronautical field.However, despite the few works in the aeronautical area, it is possible to see the importance of using NNs in the optimization of structures in this field. In addition to the works already presented, it is possible to see this application in other works such as [53,54]. These works are related to the structural optimization of unmanned aerial vehicles (UAVs), which is a type of aircraft that is growing in use. Below are some examples in which NNs have demonstrated an impact on the optimization of aeronautical or aerospace structures.

The paper by Park et al. [55] presented an approach for structural analysis using orthogonal decomposition and NNs to achieve the efficient and accurate optimization of wing structures using a reduced-order model. NN increased the ability to perceive and model the complex relationships between design variables and wing deformation due to fluid–structure interaction. This integration allowed for a more accurate prediction of aerodynamic and structural performance, ultimately leading to more efficient and effective wing designs. The impact of employing NNs in optimization lies in their ability to handle complex and nonlinear relationships and large datasets, resulting in more reliable and optimized design results compared to the conventional methods.

Furthermore, the combination of NNs with genetic algorithms can have an impact on the optimization of composite panels in aircraft structures, as seen in Ruijter et al. [45]. NNs provide a powerful tool for approximating complex functions and predicting performance metrics such as deformation and buckling multipliers. By training NNs with data from finite element analyses, researchers can create accurate models that predict the behavior of composite panels under various conditions. This capability enables the efficient exploration of the design space, enabling the identification of optimal configurations that balance performance and weight. Integration with genetic algorithms enhances the optimization process by enabling robust and efficient searches for optimal solutions in a high-dimensional design space. This combined approach reduces computational costs and setup time, making it feasible to optimize assemblies consisting of multiple components, thereby improving the overall design and performance of aircraft structures.

Finally, integrating NNs with modified backpropagation learning algorithms can significantly impact the optimization of aerospace components by increasing the efficiency and accuracy of the design processes, as can be seen in the study by Kodiyalam and Gurumoorthy [56]. By integrating a modified feedforward multilayer NN with an optimization algorithm, the approach provided a more sophisticated method for determining optimal design parameters. This is particularly crucial for aerospace components, which often require precise specifications to ensure performance, safety, and reliability. The modification to the standard backpropagation training allowed for faster convergence and required fewer hidden layer nodes, making the learning process more efficient and less resource-intensive. This led to faster prototyping and iteration cycles in the development of complex components such as aircraft engine guide vanes and satellite reflectors. Ultimately, this work could result in more robust and optimized aerospace designs, potentially reducing costs and time to market while improving the overall quality and performance of aerospace systems.

NNs are a very powerful tool that can assist in the design and optimization of structures due to its ability to work with complex data or problems, as can be seen in the previous paragraphs. Therefore, NNs can also be used to predict loads according to the mission of the optimized aircraft. This is important, as each type of aircraft will present different loading conditions, and the same aircraft can operate in different missions. Therefore, optimization should target a specific aircraft type and different missions so that the NN can predict the loads more accurately and thus the optimization will be more appropriate. Some of the articles presented here, such as those by Caixeta and Marques [3], Ruijter et al. [45], and Hansen and Horst [51], trained their NN models with different datasets so that the predictions obtained were appropriate.

The integration of NNs with GAs or other optimization methods represents a promising approach. However, it is essential to evaluate the effectiveness of such methodologies. Rujter demonstrated in his work that the trained NN achieved a mean squared error (MSE) of less than or equal to 1×10^{-5}, indicating high accuracy. Conversely, Mazhar highlighted challenges in his study, where the NN struggled to capture the exact pressure profile due to the non-linearity and significant variability of the data. Despite this, the final NN achieved a mean absolute error (MAE) of 0.007%, reflecting good performance. Meanwhile, Caixeta emphasized the computational efficiency of NN compared to a full aeroelastic model, reporting that the NN was approximately 47.5 times faster.

Therefore, the importance of integrating both techniques becomes evident, as there is the possibility of developing and improving configurations that were previously complex and, at the same time, speeding up design processes. This section presented the use of NNs in structural optimization; similarly, the next section will present articles on NNs in composite material optimization.

5.2. Composite Materials Optimization

The use of composite materials, such as carbon fiber-reinforced polymers (CFRPs), has emerged as a good alternative to the aluminum alloys that are commonly used in the aeronautical sector. According to Chen and Xu [6], composite materials, known for their exceptional design flexibility, have replaced traditional metallic materials in many applications. This phenomenon arises from the inherent characteristics of these materials, which are used in structural design, including high flexural modulus, high specific strength, low density, corrosion resistance, design flexibility, etc. [7].

However, there are other challenges to using these materials, such as more precise optimization to improve and maximize their structural performance. This is due to the fact that composite materials have anisotropic characteristics, which requires more sophisticated techniques for material design. Bisagni and Lanzi [57] introduced a post-buckling optimization method for designing composite stiffened panels under compression loads. Abouhamze and Shakeri [58] proposed a method to optimize the stacking sequence of laminated cylindrical panels for improved natural frequency and buckling load. It employed ANNs to simulate panel behavior and a GA for optimization. Marin et al. [59] introduced an optimization method for designing a composite material stiffened panel with a conventional stacking sequence, considering static analysis and hygrothermal effects. Pitton et al. [60] presented an optimization methodology for improving the buckling resistance of thin-walled cylindrical shells subjected to axial loads by using variable stiffness through curvilinear fibers. This approach aims to enhance the critical load associated with buckling.

In this context, NNs can be used as a tool capable of predicting the loads on a material under different working conditions, thereby determining a better design. To achieve this objective, it is necessary to train the model with an extensive amount of data that covers material properties and performance criteria. As a result, this predictive capability can facilitate better understanding and assist in selecting composite materials tailored to specific aircraft components.

In the field of optimizing composite materials, there is a predominant use of evolutionary methods, such as GA. Some works that have used this method are as follows: [6–8,57,61–65]. Some of these works also address the use of NNs [7,8,57,61–64,66,67] or DL [6,7] integrated with the use of GA.

Introducing an optimization framework for efficiently designing composite aircraft wings, Kilimtzidis et al. [64] offered a novel approach to the field. They employed a low-cost numerical approach and optimization techniques to minimize wing mass while meeting loading constraints. Results were validated against 3D finite element method (FEM) models, confirming accuracy. The framework provided the optimal lay-ups and dimensions for the wing components.

Introducing a new method that combines deep learning and GAs, Chen and Xu [6] offered an innovative approach to studying and optimizing the interfacial shear behavior in SiC fiber-reinforced SiC composites. By analyzing various microstructures and temperatures through molecular dynamics simulations, the approach accurately predicted shear properties and identified optimal microstructures, offering potential for advanced composite material development and intelligent manufacturing.

Combining ANNs with evolutionary algorithms (EAs), Tran-Ngoc et al. [7] introduced a new method for detecting damages in laminated composite structures. By integrating the global search capacity of EAs with the fast convergence of gradient descent, the approach efficiently found the optimal solutions and avoided local minima. Results showed improved accuracy and reduced computational time compared to traditional methods.

Introducing a novel optimization approach for composite sandwich structures, Mohammed Sahib and Kovács [65] utilized GAs and ANNs to advance the field. By minimizing weight and cost, the method targeted applications in aerospace and automotive industries. Through Monte Carlo simulation and ANN predictions, the study achieved multi-objective optimization, leading to Pareto optimal points. FEM validation confirmed the effectiveness of the approach, highlighting substantial weight reductions with CFRP or fiber metal laminate (FML) face sheets compared to all-aluminum structures.

For designing composite stiffened panels under compression loads, Bisagni and Lanzi [57] introduced a post-buckling optimization technique that has advanced the field. It employed NNs trained with FE analyses and GAs to reduce computational costs. The method achieved an 18% weight reduction by allowing local skin buckling between stiffeners.

In the pursuit of optimizing the weight of composite laminates, Liu et al. [63] introduced a streamlined methodology, offering a fresh perspective to the field. They utilized ANNs to predict buckling loads, reducing evaluation time. By incorporating lamination parameters and dimensional inputs, computational costs were minimized. The approach integrated ANN models with a GA for efficient laminate optimization, validated against other methods for buckling load prediction.

Describing an optimization process utilizing the ϵ-constraint method, Ehsani and Dalir [66] aimed to maximize the critical buckling load and minimize the structural weight of an angle grid plate. They employed an artificial ANN trained with data from the Mindlin plate theory and Ritz method to approximate the buckling load. Integrated with GAs, this approach identified optimized design variables for the angle grid structure, providing designers with options for maximizing buckling load while minimizing structural weight.

Exploring hierarchical origami-corrugation meta-sandwich (HOCM) structures, Yue et al. [67] illuminated their multifunctional potential, providing valuable insights for the field. They derived a compressive modulus, conducted experimental testing and simulations, and analyzed critical geometric parameters. The ANN-based surrogate model approximated the specific peak strength (SPS) and specific energy absorption (SEA), enabling a multi-objective optimization approach for designing HOCM structures with superior properties.

In addition to NNs, another point to be explored is GAs which, due to their approach, are capable of generating several sets of possible solutions that can evolve over generations until an optimal solution is found. However, for this to be possible, it is necessary to consider several design variables, such as the orientation of the fibers and thickness of the layers. This allows engineers to evaluate a set of design solutions that evolve into an ideal configuration that meets their requirements [9,10].

Another point to be explored here is the reliability of composite materials. As with other materials, composite materials are subject to failure. One of the techniques currently used to predict damage in composite materials and consequently to measure their reliability is continuous damage mechanics (CDM). CDM can measure the progressive load-bearing capacity when these materials are under stress as strain accumulates [68]. To predict the damage in the material, it is necessary to develop an FE model and implement the CDM as a user material subroutine (UMAT) [9,69]. Another option that is being explored is to train an NN with data from the FE model and replace the FE with the NN [70].

5.3. Future Research Directions and Emerging Trends

The articles presented in Section 5 show the different ways of using NNs in the optimization of structures or composite materials. These different ways indicate the future directions in research and trends in this field. One of the trends is the use of NNs to improve the optimization process, that is, using NNs to learn FE and perform analyses to accelerate the optimization process. Another perspective is the use of NNs to recognize patterns in topological optimization and thus determine the best material distribution that offers the best weight-to-strength ratio. Finally, there is also the possibility of using NNs to predict the distribution of loads in structures or composite materials in order to assist with the optimization process. It is noted that this field, optimization with the use of NNs, presents a growth trend over the years, as despite the fluctuation in the number of publications on the topic over the years, 2023 presented more publications in relation to 2022, as can be seen in Figures 1 and 5.

6. Conclusions

As seen previously, some works have presented the combination of NNs and GA for structural optimization. The articles presented show the different approaches to the use of NN during the optimization process. One of the approaches is the use of an NN to predict loads; in this case, training must be carried out with a large amount of data relating to different scenarios and in relation to the type of aircraft studied. This will help us to understand unobserved patterns and will also help to better predict the loads acting on the aircraft in that scenario in which the optimization will occur. Another approach is the use of NNs to predict FE analyses. In this case, the NNs must be trained with various data relating to the different configurations, types of structures, and load cases, so that the NNs understand the analysis process via FE and then carry out the process independently. Here, the objective is to speed up the FE process, because at the same time as the NNs provide the predictions, it will also optimize the process.

Still in the field of structure optimization, other authors have combined NNs or DL with topological optimization. Some works presented address the use of NNs or DL to assist in the optimization process, understanding patterns and determining the best material distribution in the structure. However, other authors have used NNs to analyze FE, as stated in the previous paragraph. It should be noted that, in both cases, the main objective is to minimize optimization time.

Just as previously seen, for structural optimization, the same techniques are observed in the optimization of composite materials. To optimize these materials, it is also necessary to predict the loads acting on the structures in which these materials are used in order to understand their capacity to withstand efforts and optimize them better. NNs are also used to carry out the FE analysis process, and the objective of accelerating the optimization process is also noted. Thus, the NNs are integrated with the GAs, such that the latter performs the optimization while the NNs analyze each optimized material at each stage of the optimization and thus help determine the best configuration.

Therefore, it is possible to notice that the use of NNs in the optimization of aeronautical structures and composite materials has shown remarkable potential and effectiveness. The main scientific contributions of this study include the following:

1. Enhanced Optimization Process: By employing NNs to learn finite element (FE) analysis and subsequently incorporating it into each stage of the GA optimization process, researchers have been able to effectively evaluate numerous structures and determine optimal configurations with greater speed. This integration allows for a more efficient and thorough exploration of the design space.
2. Pattern Recognition in Topology Optimization: NNs have been instrumental in identifying patterns during topology optimization, aiding in the selection of material distributions that offer ideal strength-to-weight ratios. This contributes to the development of lighter yet stronger aeronautical structures.
3. Prediction of Load Distributions: The use of NNs to predict load distributions on structures has facilitated the optimization of composite materials. This ensures that structures are tailored to different flight conditions and that materials used are optimally suited for their specific applications, improving both performance and safety.
4. Reduced Optimization Time: Leveraging NNs has significantly minimized optimization time, streamlining the iterative process and accelerating the design cycle. This reduction in time is crucial for meeting tight development schedules in the aerospace industry.

As previously stated, the use of NN shows good results. Even though, in certain circumstances, the NN may present difficulties, with cases where there is greater non-linearity, adjustments can be made to improve its performance. A well-trained NN can present good precision, such as an MSE of 1×10^{-5} and an MAE of 0.007%, as shown previously. In addition to presenting good precision, the NN shows a higher speed compared to other methods, being able to be 47.5 times faster, as mentioned, and thus showing that the NN can present good results.

Previously, the main scientific contributions and some practical results of this study were presented; however, some recommendations for designers and mechanical engineers are as follows:

1. Adoption of NN-GA Integration: Designers and engineers should consider integrating NNs with GAs in their optimization workflows to enhance the efficiency and effectiveness of their design processes. This approach allows for the rapid evaluation and optimization of complex structures.
2. Utilization in Topology and Material Selection: The pattern recognition capabilities of NNs can be leveraged to identify optimal material distributions, improving the strength-to-weight ratios of structures.
3. Predictive Load Analysis: Implementing NN-based predictive models for load distribution can lead to more precise and reliable designs, ensuring structural integrity under various flight conditions.
4. Streamlined Design Cycles: The time savings achieved through NN integration can be used to accelerate the design and development process, enabling faster iteration and refinement of designs.

Recapping the future research directions and emerging trends, the integration of neural networks (NNs) in optimizing structures and composite materials shows great promise. NNs can accelerate the optimization process by learning finite element (FE) analysis, enhance material distribution through pattern recognition in topological optimization, and predict load distributions to support the optimization process. The increase in publications from 2022 to 2023 highlights the growing interest and potential in this field, emphasizing its importance for future advancements in aeronautical engineering.

Overall, the use of NNs for optimization represents a powerful approach for the optimization of aircraft structures and composite materials. By leveraging the capabilities of NNs to learn and predict complex relationships within the optimization process, researchers can efficiently identify optimal solutions that meet performance requirements while minimizing weight and maximizing structural integrity. This approach holds significant promise for advancing the design and optimization of aircraft structures and composite materials in the future.

Funding: This research has been funded by Coordenação de Aperfeiçoamento de Pessoal de Nível Superior (88887.948686/2024-00) and supported by the Center for Mechanical and Aerospace Science and Technologies (C-MAST-UBI), through the Project reference UIDB/00151/2020, funded by the Fundação para a Ciência e a Tecnologia, IP/MCTES through national funds (PIDDAC), and DOI: 10.54499/UIDB/00151/2020.

Acknowledgments: We would like to express our sincere gratitude to all those who have contributed to this project. Special thanks to Matheus Urzedo Quirino, José Fernando Cárdenas Barbosa, Ênio Henrique Pires da Silva, and João Victor Barreto Netto, whose support and guidance were invaluable throughout the writing of this article. We also extend our thanks to all our other friends and family for their assistance and experience. This work would not have been possible without your dedication and collaboration.

Conflicts of Interest: The authors declare no conflicts of interest.

References

1. Bishop, C.M. Neural networks and their applications. *Rev. Sci. Instrum.* **1994**, *65*, 1803–1832. [CrossRef]
2. Lagaros, N.D.; Charmpis, D.C.; Papadrakakis, M. An adaptive neural network strategy for improving the computational performance of evolutionary structural optimization. *Comput. Methods Appl. Mech. Eng.* **2005**, *194*, 3374–3393. [CrossRef]
3. Caixeta, P.R.; Marques, F.D. Multiobjective optimization of an aircraft wing design with respect to structural and aeroelastic characteristics using neural network metamodel. *J. Braz. Soc. Mech. Sci. Eng.* **2018**, *40*, 17. [CrossRef]
4. Freitag, S.; Peters, S.; Edler, P.; Meschke, G. Reliability-based optimization of structural topologies using artificial neural networks. *Probabilistic Eng. Mech.* **2022**, *70*, 103356. [CrossRef]
5. Jeong, H.; Batuwatta-Gamage, C.; Bai, J.; Xie, Y.M.; Rathnayaka, C.; Zhou, Y.; Gu, Y.T. A complete Physics-Informed Neural Network-based framework for structural topology optimization. *Comput. Methods Appl. Mech. Eng.* **2023**, *417*, 116401. [CrossRef]
6. Chen, S.; Xu, N. The deep-learning-based evolutionary framework trained by high-throughput molecular dynamics simulations for composite microstructure design. *Compos. Struct.* **2023**, *318*, 117118. [CrossRef]

7. Tran-Ngoc, H.; Khatir, S.; Ho-Khac, H.; Roeck, G.D.; Bui-Tien, T.; Wahab, M.A. Efficient Artificial neural networks based on a hybrid metaheuristic optimization algorithm for damage detection in laminated composite structures. *Compos. Struct.* **2021**, *262*, 113339. [CrossRef]
8. Xu, Y.; Li, S.; Rong, X. Composite Structural Optimization by Genetic Algorithm and Neural Network Response Surface Modeling. *Chin. J. Aeronaut.* **1994**, *18*, 310–316. [CrossRef]
9. Almeida, J.H.S., Jr.; Ribeiro, M.L.; Tita, V.; Amico, S.C. Stacking sequence optimization in composite tubes under internal pressure based on genetic algorithm accounting for progressive damage. *Compos. Struct.* **2017**, *178*, 20–26. [CrossRef]
10. Almeida, J.H.S., Jr.; St-Pierre, L.; Wang, Z.; Ribeiro, M.L.; Tita, V.; Amico, S.C.; Castro, S.G. Design, modeling, optimization, manufacturing and testing of variable-angle filament-wound cylinders. *Compos. Part B Eng.* **2021**, *225*, 109224. [CrossRef]
11. Scopus. Available online: https://www.scopus.com/term/analyzer.uri?sort=plf-f&src=s&sid=4bf209762ffbb0c7aa2e6b98ffdf1113&sot=a&sdt=a&sl=115&s=%28+structure+OR+frame+%29+AND+optimization+AND+%28+aeronautic+OR+aerospace+OR+drone+OR+helicopter%29+AND+%22neural+networks%22&origin=resultslist&count=10&analyzeResults=Analyze+results (accessed on 13 November 2024).
12. Scopus. Available online: https://www.scopus.com/term/analyzer.uri?sort=plf-f&src=s&sid=a353339c3d1b243e982075b81427bb00&sot=a&sdt=a&sl=26&s=structure+AND+optimization&origin=resultslist&count=10&analyzeResults=Analyze+results (accessed on 13 November 2024).
13. Scopus. Available online: https://www.scopus.com/term/analyzer.uri?sort=plf-f&src=s&sid=6b84ae55377d0f10ed21c8f35a912af6&sot=a&sdt=a&sl=37&s=%22composite+material%22+AND+optimization&origin=resultslist&count=10&analyzeResults=Analyze+results (accessed on 13 November 2024).
14. Reed, S.C. Indirect aircraft structural monitoring using artificial neural networks. *Aeronaut. J.* **2008**, *112*, 251–265. [CrossRef]
15. Brunton, S.L.; Kutz, J.N.; Manohar, K.; Aravkin, A.Y.; Morgansen, K.; Klemisch, J.; Goebel, N.; Buttrick, J.; Poskin, J.; Blom-Schieber, A.W.; et al. Data-driven aerospace engineering: Reframing the industry with machine learning. *AIAA J.* **2021**, *59*, 2820–2847. [CrossRef]
16. Wu, X.; Zuo, Z.; Ma, L.; Zhang, W. Multi-fidelity neural network-based aerodynamic optimization framework for propeller design in electric aircraft. *Aerosp. Sci. Technol.* **2024**, *146*, 108963. [CrossRef]
17. Jia, Q.; Gao, J.; Zhang, C.; Ahn, C.K.; Yu, D. Dynamic event-triggered attitude synchronization of multi-spacecraft formation via a learning neural network control approach. *Aerosp. Sci. Technol.* **2023**, *142*, 108653. [CrossRef]
18. Zhang, C.; Dai, M.Z.; Wu, J.; Xiao, B.; Li, B.; Wang, M. Neural-networks and event-based fault-tolerant control for spacecraft attitude stabilization. *Aerosp. Sci. Technol.* **2021**, *114*, 106746. [CrossRef]
19. Mazhar, F.; Khan, A.M.; Chaudhry, I.A.; Ahsan, M. On using neural networks in UAV structural design for CFD data fitting and classification. *Aerosp. Sci. Technol.* **2013**, *30*, 210–225. [CrossRef]
20. Deng, L. Three Classes of Deep Learning Architectures and Their Applications: A Tutorial Survey. *Apsipa Trans. Signal Inf. Process.* **2012**, *57*, 58.
21. Wang, B.W.; Tang, W.Z.; Song, L.K.; Bai, G.C. Deep neural network-based multiagent synergism method of probabilistic HCF evaluation for aircraft compressor rotor. *Int. J. Fatigue* **2023**, *170*, 107510. [CrossRef]
22. Zhou, D.; Zhuang, X.; Zuo, H. A hybrid deep neural network based on multi-time window convolutional bidirectional LSTM for civil aircraft APU hazard identification. *Chin. J. Aeronaut.* **2022**, *35*, 344–361. [CrossRef]
23. Kim, S.; Lee, J.; Jung, S.; Lee, H.; Kim, Y. Deep neural network-based feedback control for dynamic soaring of unpowered aircraft. *IFAC-PapersOnLine* **2019**, *52*, 117–121. [CrossRef]
24. Tao, F.; Liu, X.; Du, H.; Yu, W. Learning composite constitutive laws via coupling Abaqus and deep neural network. *Compos. Struct.* **2021**, *272*, 114137. [CrossRef]
25. Xiong, F.; Zhang, L.; Hu, X.; Ren, C. A point cloud deep neural network metamodel method for aerodynamic prediction. *Chin. J. Aeronaut.* **2023**, *6*, 92–103. [CrossRef]
26. Shobha, G.; Rangaswamy, S. *Machine Learning*; Elsevier B.V.: Amsterdam, The Netherlands, 2018; Volume 38, pp. 197–228. [CrossRef]
27. Li, J.; Wang, Y.; Qiu, Z.; Zhang, D.; Xie, Y. Fast performance prediction and field reconstruction of gas turbine using supervised graph learning approaches. *Aerosp. Sci. Technol.* **2023**, *140*, 108425. [CrossRef]
28. Cannarile, F.; Compare, M.; Baraldi, P.; Diodati, G.; Quaranta, V.; Zio, E. Elastic net multinomial logistic regression for fault diagnostics of on-board aeronautical systems. *Aerosp. Sci. Technol.* **2019**, *94*, 105392. [CrossRef]
29. Hui, X.; Bai, J.; Wang, H.; Zhang, Y. Fast pressure distribution prediction of airfoils using deep learning. *Aerosp. Sci. Technol.* **2020**, *105*, 105949. [CrossRef]
30. Zhang, G.; Hsu, L.T. Intelligent GNSS/INS integrated navigation system for a commercial UAV flight control system. *Aerosp. Sci. Technol.* **2018**, *80*, 368–380. [CrossRef]
31. Huang, Y.; Liu, J.; Zhu, C.; Wang, X.; Zhou, Y.; Sun, X.; Li, J. An explainable machine learning model for superalloys creep life prediction coupling with physical metallurgy models and CALPHAD. *Comput. Mater. Sci.* **2023**, *227*, 112283. [CrossRef]
32. Kim, J.; Lee, C. Deep unsupervised learning of turbulence for inflow generation at various Reynolds numbers. *J. Comput. Phys.* **2020**, *406*, 109216. [CrossRef]
33. Liem, R.P.; Mader, C.A.; Martins, J.R. Surrogate models and mixtures of experts in aerodynamic performance prediction for aircraft mission analysis. *Aerosp. Sci. Technol.* **2015**, *43*, 126–151. [CrossRef]

34. Jin, H.J.; Zhao, Y.P.; Wang, Z.Q. A rotating stall warning method for aero-engine compressor based on DeepESVDD-CNN. *Aerosp. Sci. Technol.* **2023**, *139*, 108411. [CrossRef]
35. Kim, H.; Kim, J.; Won, S.; Lee, C. Unsupervised deep learning for super-resolution reconstruction of turbulence. *J. Fluid Mech.* **2021**, *910*, A29. [CrossRef]
36. Adeli, H.; Karim, A. Neural Network Model for Optimization of Cold-Formed Steel Beams. *J. Struct. Eng.* **1996**, *123*, 1535–1543. [CrossRef]
37. Arslan, M.A.; Hajela, P. Counterpropagation Neural Networks in Decomposition Based Optimal Design. *Comput. Struct.* **1997**, *65*, 641–650. [CrossRef]
38. Shieh, R.C. Massively Parallel Structural Design Using Stochastic Optimization and Mixed Neuralnet/Finite Element Analysis Methods. *Comput. Syst. Eng.* **1994**, *5*, 455–467. [CrossRef]
39. Papadrakakis, M.; Lagaros, N.D. Reliability-based structural optimization using neural networks and Monte Carlo simulation. *Comput. Methods Appl. Mech. Eng.* **2002**, *191*, 3491–3507. [CrossRef]
40. Rao, S.S. *Engineering Optimization: Theory and Practice*; John Wiley & Sons: Hoboken, NJ, USA, 2009.
41. Cao, X.; Sugiyama, Y.; Mitsui, Y. Application of artificial neural networks to load identification. *Comput. Struct.* **1998**, *69*, 63–78. [CrossRef]
42. Hoffman, M.E. *A Neural Network Prototype for Predicting F-14B Strains at the B.L. 10 Longeron*; Naval Air Systems Command, Department of the Navy: Washington, DC, USA, 1992.
43. Khorsand, A.R.; Akbarzadeh-T, M.R. Multi-objective meta level soft computing-based evolutionary structural design. *J. Frankl. Inst.* **2007**, *344*, 595–612. [CrossRef]
44. Ruijter, W.; Spallino, R.; Entzinger, J.; Hol, J. *Feedback-Based Neural Networks in Structural Optimisation of Aerospace Structures*; Civil-Comp Press: Stirlingshire, UK, 2003; Volume 78. [CrossRef]
45. Ruijter, W.; Spallino, R.; Wamet, L.; Boer, A.D. Optimization of composite panels using neural networks and genetic algorithms. In *Computational Fluid and Solid Mechanics 2003*; Elsevier: Amsterdam, The Netherlands, 2003.
46. Querin, O.M.; Steven, G.P.; Xie, Y.M. Evolutionary structural optimisation using an additive algorithm. *Finite Elem. Anal. Des.* **2000**, *34*, 291–308. [CrossRef]
47. Jingui, L.; Yunliang, D.; Bin, W.; Shide, X. An Improved Strategy for Gas in Structural Optimization. *Comput. Struct.* **1996**, *61*, 1185–1191.
48. Kallioras, N.A.; Lagaros, N.D. *DL-Scale: Deep Learning for Model Upgrading in Topology Optimization*; Elsevier B.V.: Amsterdam, The Netherlands, 2020; Volume 44, pp. 433–440. [CrossRef]
49. Zheng, S.; He, Z.; Liu, H. Generating three-dimensional structural topologies via a U-Net convolutional neural network. *Thin-Walled Struct.* **2021**, *159*, 107263. [CrossRef]
50. Liang, K.; Zhu, D.; Li, F. A Fourier neural operator-based lightweight machine learning framework for topology optimization. *Appl. Math. Model.* **2024**, *129*, 714–732. [CrossRef]
51. Hansen, L.U.; Horst, P. Multilevel optimization in aircraft structural design evaluation. *Comput. Struct.* **2008**, *86*, 104–118. [CrossRef]
52. Zhang, Z.; Li, Y.; Zhou, W.; Chen, X.; Yao, W.; Zhao, Y. TONR: An exploration for a novel way combining neural network with topology optimization. *Comput. Methods Appl. Mech. Eng.* **2021**, *386*, 114083. [CrossRef]
53. Karali, H.; Inalhan, G.; Tsourdos, A. Advanced UAV Design Optimization Through Deep Learning-Based Surrogate Models. *Aerospace* **2024**, *11*, 669. [CrossRef]
54. Sobota, M.; Skarka, W. *Multi-Objective Optimization of Composite Structure Using Rule-Based Parametrization*; IOS Press BV: Amsterdam, The Netherlands, 2022; Volume 28, pp. 453–462. [CrossRef]
55. Park, K.H.; Jun, S.O.; Baek, S.M.; Cho, M.H.; Yee, K.J.; Lee, D.H. Reduced-order model with an artificial neural network for aerostructural design optimization. *J. Aircr.* **2013**, *50*, 1106–1116. [CrossRef]
56. Kodiyalam, S.; Gurumoorthy, R. Neural networks with modified backpropagation learning applied to structural optimization. *AIAA J.* **1996**, *34*, 408–412. [CrossRef]
57. Bisagni, C.; Lanzi, L. Post-buckling optimisation of composite stiffened panels using neural networks. *Compos. Struct.* **2002**, *58*, 237–247. [CrossRef]
58. Abouhamze, M.; Shakeri, M. Multi-objective stacking sequence optimization of laminated cylindrical panels using a genetic algorithm and neural networks. *Compos. Struct.* **2007**, *81*, 253–263. [CrossRef]
59. Marin, L.; Trias, D.; Badallo, P.; Rus, G.; Mayugo, J.A. Optimization of composite stiffened panels under mechanical and hygrothermal loads using neural networks and genetic algorithms. *Compos. Struct.* **2012**, *94*, 3321–3326. [CrossRef]
60. Pitton, S.F.; Ricci, S.; Bisagni, C. Buckling optimization of variable stiffness cylindrical shells through artificial intelligence techniques. *Compos. Struct.* **2019**, *230*, 111513. [CrossRef]
61. Kicinger, R.; Arciszewski, T.; Jong, K.D. Evolutionary computation and structural design: A survey of the state-of-the-art. *Comput. Struct.* **2005**, *83*, 1943–1978. [CrossRef]
62. Fu, X.; Ricci, S.; Bisagni, C. Minimum-weight design for three dimensional woven composite stiffened panels using neural networks and genetic algorithms. *Compos. Struct.* **2015**, *134*, 708–715. [CrossRef]
63. Liu, X.; Qin, J.; Zhao, K.; Featherston, C.A.; Kennedy, D.; Jing, Y.; Yang, G. Design optimization of laminated composite structures using artificial neural network and genetic algorithm. *Compos. Struct.* **2023**, *305*, 116500. [CrossRef]

64. Kilimtzidis, S.; Kotzakolios, A.; Kostopoulos, V. Efficient structural optimisation of composite materials aircraft wings. *Compos. Struct.* **2023**, *303*, 116268. [CrossRef]
65. Mohammed Sahib, M.; Kovács, G. Multi-objective optimization of composite sandwich structures using Artificial Neural Networks and Genetic Algorithm. *Results Eng.* **2024**, *21*, 101937. [CrossRef]
66. Ehsani, A.; Dalir, H. Multi-objective optimization of composite angle grid plates for maximum buckling load and minimum weight using genetic algorithms and neural networks. *Compos. Struct.* **2019**, *229*, 111450. [CrossRef]
67. Yue, Z.; Han, B.; Wang, Z.; Yang, M.; Zhang, Q.; Lu, T.J. Data-driven multi-objective optimization of ultralight hierarchical origami-corrugation meta-sandwich structures. *Compos. Struct.* **2023**, *303*, 116334. [CrossRef]
68. Srinivasan, V.; Saxena, A. Creep Resistance in Nonferritic Metals. In *Comprehensive Structural Integrity*; Elsevier: Amsterdam, The Netherlands, 2007.
69. El Idrissi, H.; Seddouki, A. Modelling of progressive damage in a notched carbon/epoxy composite laminate subjected to tensile loading using different assessment methods coupled with FEM. *Fibers Polym.* **2022**, *23*, 3146–3162. [CrossRef]
70. Zhang, K.; Ma, L.H.; Song, Z.Z.; Gao, H.; Zhou, W.; Liu, J.; Tao, R. Strength prediction and progressive damage analysis of carbon fiber reinforced polymer-laminate with circular holes by an efficient Artificial Neural Network. *Compos. Struct.* **2022**, *296*, 115835. [CrossRef]

eng

MDPI

Article

Evaluation of Affordable Agricultural Drones for Small and Medium Farms

Adis Puška [1], Miroslav Nedeljković [2], Anđelka Štilić [3,*] and Darko Božanić [4]

[1] Department of Public Safety, Government of Brčko District of Bosnia and Herzegovina, 76100 Brčko, Bosnia and Herzegovina; adispuska@yahoo.com
[2] Institute of Agricultural Economics, 11060 Belgrade, Serbia; miroslavvnedeljkovic2015@gmail.com
[3] The College of Tourism, Academy of Applied Studies Belgrade, 11070 Belgrade, Serbia
[4] Military Academy, University of Defense in Belgrade, 11042 Belgrade, Serbia; dbozanic@yahoo.com
* Correspondence: andjelka.stilic@gmail.com

Abstract: Smart technologies are increasingly used in agriculture, with drones becoming one of the key tools in agricultural production. This study aims to evaluate affordable drones for agricultural use in the Posavina region, located in northern Bosnia and Herzegovina. To determine which drones deliver the best results for small and medium-sized farms, ten criteria were used to evaluate eight drones. Through expert evaluation, relevant criteria were first established and then used to assess the drones. The selected drones are designed for crop monitoring and are priced under EUR 2000. Using the fuzzy A-SWARA (Adapted Step-wise Weight Assessment Ratio Analysis) method, it was determined that the most important criteria for drone selection are control precision, flight autonomy, and ease of use, all of which are technical attributes. The fuzzy MARCOS method revealed that the best-performing drones are also the most affordable. The drones D5, D4, and D8 demonstrated the best results. These findings were confirmed through comparative analysis and sensitivity analysis. Their features are not significantly different from those of more expensive models and can, therefore, be effectively used for smart agriculture. This study demonstrates that drones can be a valuable tool for small farms, helping to enhance agricultural practices and productivity.

Keywords: drones; agriculture; Posavina; fuzzy methods

Citation: Puška, A.; Nedeljković, M.; Štilić, A.; Božanić, D. Evaluation of Affordable Agricultural Drones for Small and Medium Farms. *Eng* **2024**, 5, 3161–3173. https://doi.org/10.3390/eng5040166

Academic Editor: Antonio Gil Bravo

Received: 31 October 2024
Revised: 27 November 2024
Accepted: 28 November 2024
Published: 30 November 2024

1. Introduction

Recent changes in agriculture increasingly focus on using modern technology to improve production, especially through the development of smart systems. Drones in agriculture offer many benefits for small and medium-sized farms by helping to reduce costs and increase production efficiency. They can be used for tasks such as crop mapping, plant health monitoring, irrigation planning, and the careful use of pesticides and fertilizers [1]. Drones provide faster and more complete crop assessments than traditional methods while also reducing the need for human labor. The data they gather allow farmers to make better decisions, leading to improved agricultural outcomes.

A key advantage of drones is their ability to collect data in real time [2]. With this up-to-date information, farmers can react quickly and effectively to changes in crop conditions. Drones can also cover large areas in a short time, and by using multispectral cameras, they detect changes in crops before they are visible to the eye. This allows for precise actions, reducing the amount of pesticides and fertilizers needed, which helps lessen the environmental impact. By using drones, it is possible to identify whether certain crops lack water and to apply precise irrigation only to those crops. Protecting the environment should remain a central focus in agricultural production [3].

Drones in agriculture have attracted much attention from researchers who see their potential to boost production. Research by Javaid et al. [4], Panday et al. [5], Rejeb et al. [6], Nhamo et al. [7], and others shows that drones can help cut costs and increase crop yields.

Studies by Hafeez et al. [8] and Inoue [9] also suggest that drones make it faster and easier to assess crops, improving the use of resources like water, fertilizers, and pesticides. However, there are challenges when using drones on small farms. If precision farming methods are applied, the cost–benefit ratio may not always work in favor of smaller farms [10], raising questions about whether drones are worth the investment for them.

Although drones are widely used in large-scale farming [11], they are harder to apply on small and medium-sized farms. The main reasons are the high costs of buying and maintaining drones [12], which greatly influence farmers' decisions, as well as the technical knowledge required to operate them. These factors can be obstacles for smaller farms with limited budgets. However, as technology develops and market competition increases, drones are becoming more affordable and economically accessible [13], making them a more realistic option for smaller farms.

This study aims to identify affordable and practical drones for small and medium-sized farms in the Posavina region, located in northern Bosnia and Herzegovina. This region, the largest plain in the country, has the highest potential for agricultural growth [14]. The goal of this research is to evaluate affordable drones that meet the technical needs and budget limits of farmers in Posavina. Small and medium-sized farms are central to agriculture in this area, and the adoption of drones could greatly improve their productivity and sustainability [15]. However, to make drone use successful, it is necessary to identify models that best match the specific needs of these farms. This research provides an overview of affordable drones on the market, offering guidelines for their evaluation using decision-making methods. It will help answer the question of how to choose the best drone by balancing technical features and costs, especially when budgets are tight.

Based on this, the main goal of this paper is to evaluate drones suitable for small and medium-sized farms using a fuzzy approach with specialized methods. The fuzzy approach is used in this research because complete information is not available, and precise evaluations cannot be used; instead, assessments are expressed through linguistic values [16]. The application of multi-criteria decision-making (MCDM) methods is performed to evaluate drones based on the relevant criteria. The evaluation of both the criteria and drones is carried out subjectively, relying on expert opinions [17] while incorporating uncertainty into the decision-making process [18]. This approach aims to provide farmers with guidelines on choosing the drone that best fits their specific needs. The fuzzy approach relies on the A-SWARA (Adapted Step-wise Weight Assessment Ratio Analysis) and MARCOS (Measurement Alternatives and Ranking according to Compromise Solution) methods. Here, the fuzzy A-SWARA method helps determine the importance of various criteria based on subjective evaluations [19], while the fuzzy MARCOS method ranks drone options by finding a balance across different factors [15].

This research offers several practical and theoretical contributions, particularly in the application of drones in agriculture. The main contribution is to demonstrate that drones can be effectively used on small and medium-sized farms to enhance agricultural production and increase farmers' work efficiency. Additionally, the contributions of this research are reflected in the following:

- Guidance for farmers: The study provides clear, practical advice for small and medium-sized farmers on which drone models are best suited for agricultural tasks. By weighing technical and economic factors, this guidance highlights the strengths and weaknesses of specific models, making decision making easier.
- Innovative evaluation method: The use of the fuzzy A-SWARA and MARCOS methods offers a new way to evaluate drones in agriculture, focusing on subjective decision making that can be adapted for similar studies in the future.
- Sustainable agriculture: This drone evaluation contributes directly to sustainable farming by helping optimize resource use in agriculture, which can reduce costs, improve efficiency, and promote ecologically and economically sustainable production practices—vital for the long-term health of agriculture.

- Supporting small farm modernization: This research encourages a broader conversation about applying new technologies to small farms and positioning drones as an effective modernization tool. This can increase the competitiveness of small and medium farms.
- Economic development in rural areas: By lowering costs and boosting productivity, this research can support economic growth in rural regions. Drones help farmers use resources more efficiently, increase income, and improve living standards in rural communities.

Following the introduction, this paper is organized into four main sections that build upon one another. First, the materials and research methods will be explained, detailing the steps and tools used in the study. The results section will apply these methods to a real-world example, guiding the selection of the best-suited drone for small and medium-sized farms in the Posavina region. The discussion will examine the findings in detail, comparing them with similar research. Finally, the last section will summarize the key results, discuss limitations, and suggest directions for future research.

2. Materials and Methods

This research was conducted in several phases:

- Selection of experts, drones, and criteria;
- Evaluation of criteria and alternatives;
- Application of fuzzy SWARA and MARCOS methods;
- Analysis of results using comparative and sensitivity analysis.

In the first phase, drones were assessed by a panel of experts based on key characteristics and selected criteria. Ten experts were chosen for their knowledge of agricultural production in Posavina and extensive practical experience with drones. Most of these experts are professors from agricultural faculties. Drones for observation were selected based on the price range and availability in Bosnia and Herzegovina, with a maximum price limit set at around EUR 2000. Therefore, this study focuses on drones used primarily for crop monitoring.

To ensure objectivity, drones were labeled rather than named directly, so none were rated "best" in an absolute sense. It is important to note that this research relies on the subjective evaluations of experts familiar with the Posavina region, and different regions or farming conditions could lead to different rankings of the drones. The basic characteristics of the drones examined are summarized in Table 1. However, these characteristics do not encompass all of the criteria used in this study. Some criteria can only be evaluated after these drones are tested in practice. For example, image quality is not determined solely by the number of megapixels but is also influenced by factors such as sensors, hardware and software used by the cameras, and various other parameters. As a result, certain criteria can only be assessed when these drones are applied in real-world conditions.

The features of these drones indicate that they are designed for small and medium-sized farms, as they can cover small areas on a single battery charge. They are intended for crop monitoring on these farms, while advanced functions such as spraying or seeding are not available. With a single battery charge, they can typically cover areas of a few hectares, although this coverage can be increased by replacing the batteries. Additionally, these drones are lightweight and cannot carry extra loads. In most cases, it is possible to attach certain sensors to them, provided they do not exceed the drone's weight limit. Due to their specific features and affordability, these drones are a good choice for monitoring and recording crop conditions on farms. This allows farmers to gather useful information and take appropriate measures to improve agricultural production.

Table 1. Technical characteristics of selected drones.

Designation	Camera	Range	Flight Autonomy	Weight	Obstacle Sensors	Max Speed	Price Range	Payload
D1	20 MP	7 km	30 min	1375 g	Yes	72 km/h	EUR 1500–1800	500 g
D2	21 MP	4 km	25 min	320 g	No	55 km/h	EUR 700–900	200 g
D3	48 MP	10 km	34 min	570 g	Yes	68 km/h	EUR 900–1000	200 g
D4	8 MP	8 km	43 min	700 g	No	60 km/h	EUR 400–600	100 g
D5	12 MP	8 km	35 min	790 g	No	65 km/h	EUR 500–600	200 g
D6	48 MP	12 km	34 min	249 g	Yes	57 km/h	EUR 1000–1100	100 g
D7	50 MP	10 km	28 min	249 g	Yes	54 km/h	EUR 800–900	300 g
D8	4 K	1 km	26 min	495 g	No	45 km/h	EUR 300–500	100 g

The observed drones are evaluated based on ten criteria, focusing on technical and economic characteristics. These criteria are as follows:

- Criterion (C1) camera quality [20,21]: Assesses camera resolution, stabilization, and image quality, rated from very low to excellent quality.
- Criterion (C2) flight autonomy [21–23]: Evaluates flight duration on a single charge, rated from very short to long-term autonomy.
- Criterion (C3) steering precision [22,24]: Measures accuracy of controls and GPS, rated from unreliable to highly precise steering.
- Criterion (C4) stability in air [22,25]: Evaluates drone stability in poor weather, rated from very poor to excellent stability.
- Criterion (C5) obstacle avoidance sensors [24,26]: Assesses the effectiveness of sensors, rated from no sensors to superior crash protection.
- Criterion (C6) control range [21,23,24]: Measures maximum control distance, rated from very short to extremely large range.
- Criterion (C7) ease of use [25,27]: Evaluates control simplicity, rated from very complex to very easy to use.
- Criterion (C8) portability and weight [22,24]: Assesses weight and ease of transport, rated from heavy and hard to carry to light and portable.
- Criterion (C9) battery charging time [21,23]: Measures battery charging speed, rated from slow charging to very fast charging.
- Criterion (C10) value for money [20,27]: Rates the price–performance ratio from very poor to excellent value.

Each criterion and its importance were rated on a nine-level scale, ranging from least to most important. Due to differences across criteria, the scale was adjusted as needed for each. To make these ratings consistent, linguistic evaluations were converted to numerical values from 1 to 9. Likewise, the alternatives were scored using this same scale. Because these ratings are given in linguistic terms, they were converted into fuzzy numbers to apply fuzzy evaluation methods. For example, the lowest rating was converted into a fuzzy number (1, 1, 2) and the highest into (8, 9, 9). Other values were converted in a similar manner.

To practically apply these ratings, fuzzy methods were used to determine the importance of each criterion and to rank the alternatives. Since many methods are available, this research focused on two that have proven effective in similar studies: the fuzzy SWARA [28] and fuzzy MARCOS [29] methods. The fuzzy SWARA method established the importance of each criterion based on expert evaluations, while the fuzzy MARCOS method ranked the selected drones according to these criteria. Through this approach, we aimed to recommend which drones are best suited for agricultural production in the Posavina region.

The fuzzy SWARA method, initially developed by Keršulienė et al. [30], provides a way to determine the importance of various criteria through a step-by-step approach. However, this study uses a simplified variant known as A-SWARA, which facilitates group evaluation of criteria importance. The A-SWARA method includes the following steps:

Step 1. Experts assess the importance of criteria using linguistic values.

Step 2. Linguistic values are converted into fuzzy numbers.

Step 3. An aggregate rating for each criterion is formed.

Step 4. Criteria are ranked based on their overall scores.

Step 5. Normalized values are generated by dividing the aggregate scores by the highest fuzzy number of the top-ranked criterion.

Step 6. Final weights for each criterion are determined by dividing individual values by the sum of all normalized values.

Once the importance of each criterion is set, the ranking of alternatives is performed using the fuzzy MARCOS method developed by Stević et al. [31]. This approach assesses the relationship between each alternative and reference value representing ideal and anti-ideal points, following these steps [32]:

Step 1. Experts evaluate the alternatives based on the chosen criteria using linguistic values.

Step 2. Linguistic values are transformed into fuzzy numbers, creating the initial fuzzy decision matrix.

Step 3. The initial decision matrix is expanded.

Step 4. The matrix is normalized using the following:

$$\tilde{n} = \left(n_{ij}^{l}, n_{ij}^{m}, n_{ij}^{u}\right) = \left(\frac{x_{id}^{l}}{x_{ij}^{u}}, \frac{x_{id}^{l}}{x_{ij}^{m}}, \frac{x_{id}^{l}}{x_{ij}^{l}}\right)\ if\ j \in C \tag{1}$$

$$\tilde{n} = \left(n_{ij}^{l}, n_{ij}^{m}, n_{ij}^{u}\right) = \left(\frac{x_{ij}^{l}}{x_{id}^{u}}, \frac{x_{ij}^{m}}{x_{id}^{u}}, \frac{x_{ij}^{u}}{x_{id}^{u}}\right)\ if\ j \in B \tag{2}$$

where *l*, *m*, and *u* are the lower, middle, and upper bounds of the fuzzy numbers, respectively.

Step 5. Weighted values of the normalized matrix are calculated as follows:

$$\tilde{v}_{ij} = \left(v_{ij}^{l}, v_{ij}^{m}, v_{ij}^{u}\right) = \tilde{n}_j \times \tilde{w}_j \tag{3}$$

Step 6. Calculation of *Si* matrix is performed as follows:

$$S_i = \sum\nolimits_{i=1}^{n} v_{ij} \tag{4}$$

Step 7. Calculation of the utility degree K_i is carried out using the following:

$$\tilde{K}_i^{-} = \left(\frac{\tilde{S}_i}{\tilde{S}_{ai}}\right) = \left(\frac{s_i^l}{s_{ai}^u}, \frac{s_i^m}{s_{ai}^u}, \frac{s_i^u}{s_{ai}^u}\right) \tag{5}$$

$$\tilde{K}_i^{+} = \left(\frac{\tilde{S}_i}{\tilde{S}_{id}}\right) = \left(\frac{s_i^l}{s_{id}^u}, \frac{s_i^m}{s_{id}^u}, \frac{s_i^u}{s_{id}^u}\right) \tag{6}$$

Step 8. Calculation of the fuzzy matrix $\tilde{T}_i$ is carried out using the following:

$$\tilde{T}_i = \tilde{t}_i = \left(t_i^l, t_i^m, t_i^u\right) = \tilde{K}_i^{-} + \tilde{K}_i^{+} = \left(\tilde{k}_i^{-l} + \tilde{k}_i^{+l}, \tilde{k}_i^{-m} + \tilde{k}_i^{+m}, \tilde{k}_i^{-u} + \tilde{k}_i^{+u}\right), \tag{7}$$

and by setting the fuzzy number $\tilde{D}$ as follows:

$$\tilde{D} = \left(d^l, d^m, d^u\right) = \max_i \tilde{t}_{ij} \tag{8}$$

Step 9. Defuzzification of fuzzy numbers is performed with the following:

$$df_{def} = \frac{l + 4m + u}{6} \tag{9}$$

Step 10. Determination of the utility function $f(K_i)$ is carried out as follows:

$$f\left(\overset{\sim+}{K_i}\right) = \frac{\overset{\sim-}{K_i}}{df_{def}} \tag{10}$$

$$f\left(\overset{\sim-}{K_i}\right) = \frac{\overset{\sim+}{K_i}}{df_{def}} \tag{11}$$

Step 11. Calculation of the final utility function is performed as follows:

$$f(K_i) = \frac{K_i^+ + K_i^-}{1 + \frac{1-f(K_i^+)}{f(K_i^+)} + \frac{1-f(K_i^-)}{f(K_i^-)}} \tag{12}$$

3. Results

The selection process for a suitable drone for agricultural production in Posavina begins with evaluating the importance of specific criteria. This assessment relies on expert ratings, where each criterion is scored on a scale of 1 to 9, with 1 representing the lowest importance and 9 indicating the highest, or extreme, importance (Table 2).

Table 2. Criteria importance evaluation.

Experts	C1	C2	C3	C4	C5	C6	C7	C8	C9	C10
Expert 1	8	8	9	8	7	7	9	6	7	8
Expert 2	7	8	9	7	7	6	9	6	7	9
Expert 3	7	9	9	7	7	6	9	6	8	9
Expert 4	6	9	9	7	7	8	9	6	9	9
Expert 5	6	8	9	6	7	5	9	5	8	9
Expert 6	5	9	9	7	6	6	7	6	8	6
Expert 7	6	9	9	6	5	6	7	6	8	9
Expert 8	7	8	8	6	6	6	8	7	8	8
Expert 9	6	9	8	7	6	6	8	6	8	6
Expert 10	6	8	8	6	6	5	8	5	7	8

After assigning ratings, linguistic values are converted to fuzzy numbers, and a total score across all criteria is calculated. The next step involves calculating the weights for each criterion. First, criteria are ranked by the total score. Then, each total score is divided by the highest fuzzy value, and the weight for each criterion is determined by dividing individual normalized values by the sum of these normalized values. According to this approach, the results show that experts consider criterion C3, steering precision, as the most important factor in drone ranking (Table 3), followed by C2, flight autonomy, and C7, ease of use. The least important, according to experts, is C8, portability and weight.

The next phase evaluates the selected drones based on the defined criteria (Table 4). Experts first rate each drone, and these linguistic ratings are then converted to fuzzy numbers. This allows for the formation of a summary decision matrix, which is the basis for ranking drones. In forming this matrix, each expert is given equal importance, averaging all matrices created from expert evaluations. With the aggregate decision matrix complete, the fuzzy MARCOS method is applied, including steps for expanding the aggregate matrix, normalizing values, and weighting these normalized values according to criteria weights. This paper does not detail the calculations of the fuzzy MARCOS method, as it has already been validated in over 1000 studies.

Table 3. Criteria importance results.

Id	Sum (s_j)	Normalization (n_j)	Weight (w_j)
C3	(77, 87, 90)	(0.86, 1.00, 1.00)	(0.10, 0.11, 0.11)
C2	(75, 85, 90)	(0.83, 0.94, 1.00)	(0.09, 0.11, 0.11)
C7	(73, 83, 88)	(0.81, 0.92, 0.98)	(0.09, 0.10, 0.11)
C10	(71, 81, 86)	(0.79, 0.90, 0.96)	(0.09, 0.10, 0.11)
C9	(68, 78, 87)	(0.76, 0.87, 0.97)	(0.08, 0.10, 0.11)
C4	(57, 67, 77)	(0.63, 0.74, 0.86)	(0.07, 0.08, 0.10)
C5	(54, 64, 74)	(0.60, 0.71, 0.82)	(0.07, 0.08, 0.09)
C1	(54, 64, 74)	(0.60, 0.71, 0.82)	(0.07, 0.08, 0.09)
C6	(51, 61, 71)	(0.57, 0.68, 0.79)	(0.06, 0.08, 0.09)
C8	(49, 59, 69)	(0.54, 0.66, 0.77)	(0.06, 0.07, 0.09)

Table 4. Drone evaluation by selected criteria.

Expert 1	C1	C2	C3	C4	C5	C6	C7	C8	C9	C10
Drone 1	5	6	5	4	6	6	6	7	8	6
Drone 2	6	5	5	4	6	5	4	5	7	5
Drone 3	5	5	5	5	7	8	5	8	5	6
Drone 4	5	6	7	3	5	8	6	8	6	5
Drone 5	6	6	7	8	5	5	4	5	6	8
Drone 6	5	6	7	5	6	8	6	8	5	6
Drone 7	6	6	6	6	6	5	8	7	7	5
Drone 8	5	8	6	4	8	7	5	6	4	5
⋮	⋮	⋮	⋮	⋮	⋮	⋮	⋮	⋮	⋮	⋮
Expert 10	C1	C2	C3	C4	C5	C6	C7	C8	C9	C10
Drone 1	4	5	5	3	5	5	6	6	7	6
Drone 2	5	5	4	3	5	4	5	4	6	6
Drone 3	4	4	4	4	4	7	4	7	4	5
Drone 4	5	5	6	2	5	7	5	7	5	4
Drone 5	5	6	6	7	4	6	4	4	5	7
Drone 6	4	4	6	4	4	7	5	7	4	5
Drone 7	6	5	6	5	5	4	7	6	6	6
Drone 8	5	7	5	3	7	6	4	5	3	5

Once values for each alternative are calculated, the utility degree and utility function are determined, followed by the defuzzification process to obtain crisp numbers. Finally, a utility function is calculated, which indicates that, according to expert evaluations, Drone 5 performs best, followed by Drones 4 and 8, with Drone 2 ranked the lowest (Table 5).

Table 5. Fuzzy MARCOS method results.

	$d\tilde{K}_i^{-}$	$d\tilde{K}_i^{+}$	$df(\tilde{K}_i^{-})$	$df(\tilde{K}_i^{+})$	K_i	Rank
Drone 1	0.819	1.245	0.562	0.370	0.592	6
Drone 2	0.777	1.181	0.533	0.351	0.525	8
Drone 3	0.817	1.242	0.560	0.369	0.589	7
Drone 4	0.871	1.324	0.597	0.393	0.682	2
Drone 5	0.879	1.337	0.603	0.397	0.697	1
Drone 6	0.841	1.278	0.577	0.379	0.629	5
Drone 7	0.854	1.299	0.586	0.386	0.653	4
Drone 8	0.868	1.319	0.595	0.392	0.676	3

This ranking reflects a balance among the criteria considered. To confirm this ranking, a comparative analysis is conducted using the same decision matrix and criteria weights but employing different fuzzy methods [33,34]. Five additional methods are used in this analysis: fuzzy RAWEC (Ranking of Alternatives with Weights of Criteria), fuzzy WASPAS (Weighted Aggregated Sum Product Assessment), fuzzy SAW (Simple Additive Weighting), fuzzy MABAC (Multi-Attributive Border Approximation Area Comparison), and fuzzy ARAS (Additive Ratio Assessment). Each method has unique steps: for instance, fuzzy RAWEC applies two types of normalization, fuzzy WASPAS balances two methods, fuzzy SAW ranks based on individual values, fuzzy MABAC ranks relative to the average, and fuzzy ARAS uses a new utility function relative to the maximum value. As these methods also differ in normalization approaches, results may vary.

This analysis reveals some differences in ranking, especially for Drone 4 (Figure 1). According to one method, this drone ranks first, while another method places it fourth. However, most methods agree that Drone 5 is the best performer. Additionally, the ranking obtained by the fuzzy MARCOS method aligns closely with the other methods, thereby supporting the results produced by this approach.

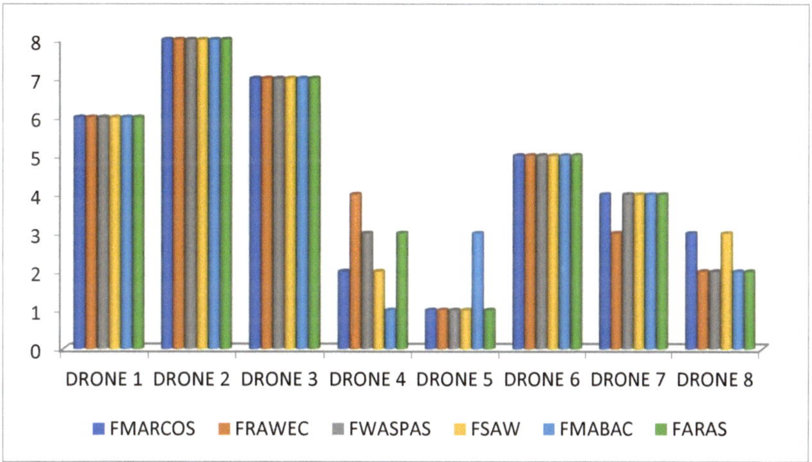

Figure 1. Comparative analysis results.

Following the comparison with other fuzzy methods, a sensitivity analysis is conducted. This analysis helps determine how shifts in criteria weights affect the final ranking of the drone alternatives. Sensitivity analysis can be performed in various ways, and for this study, specific criteria weights are reduced by 30%, 60%, and 90%, with the weights of the remaining criteria adjusted to balance these reductions. This approach allows us to see how a reduction in the weight of one criterion influences the overall ranking. Since there are 10 criteria with three levels of reduction each, this results in 30 distinct scenarios for analysis.

The sensitivity analysis results indicate that even slight changes in criteria weightings can impact the ranking order (Figure 2). For instance, Drone 5, which originally ranked highest, maintained the top position in 13 scenarios but ranked lower in the others. In Scenario 12, Drone 5 dropped to fifth place when criterion C4, stability in air, was reduced by 90%. This outcome suggests that Drone 5 excels in stability; however, when the importance of this feature is minimized, its ranking declines. A similar examination of Drone 8 reveals that it performs well in flight autonomy (C2) and obstacle avoidance (C5) but could improve in steering precision (C3) and battery charging time (C9) to become the top-ranked option. Based on these findings, Drones 4, 5, and 8 emerge as the top choices, with experts highlighting their strengths for small to medium farms.

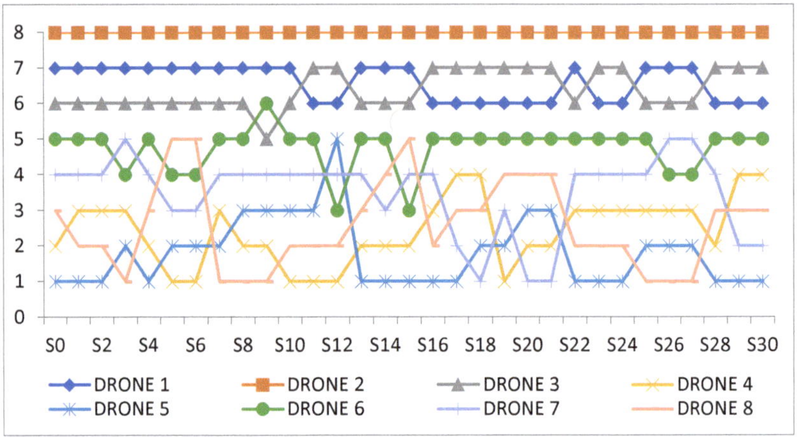

Figure 2. Sensitivity analysis results.

4. Discussion

With the rise of technology, agriculture is also undergoing substantial transformation [35]. Agriculture in many regions, including Posavina [36], is moving toward adopting smarter technologies. These advancements were initially limited to large farms but are now reaching medium and smaller farms due to the availability of more affordable technologies [37]. However, the adoption rate of smart technologies is between 15 and 20 percent, as indicated by research conducted by Gabriel and Gandorfer [38]. Drones, in particular, are gaining traction in the agricultural sector [6]. They serve multiple functions and are becoming integral to farming operations. Research by McCarthy et al. [39] has shown that drones should be used on small and medium-sized farms to increase productivity, reduce costs, and improve food security. This research focused on selecting suitable drones for medium and small farms in Posavina for several reasons:

- Posavina, the largest lowland area in Bosnia and Herzegovina, holds significant potential for expanding agricultural production.
- Bosnia and Herzegovina, as a developing country, relies heavily on traditional farming practices, with limited adoption of smart technologies.
- Drone prices have decreased, making this technology more accessible to farmers.
- Drone use in Bosnia and Herzegovina's agriculture sector remains limited, highlighting the need for studies like this to guide adoption in the region.

The motivations behind this research underscore its potential contribution to advancing agricultural practices in Bosnia and Herzegovina. Given the diverse range of drones on the market, each designed with specific applications in mind, this study focused on eight drones priced under EUR 2000, each intended for crop monitoring. However, crop monitoring is not achievable without the use of specific software, and the future of this field lies in the integration of artificial intelligence, as noted by Guebsi et al. [40]. Therefore, future research should pay particular attention to software solutions for crop monitoring analysis. It is also essential to emphasize that artificial intelligence is becoming increasingly common in practice [41], including in agriculture.

To determine which of these drones would best meet the needs of agricultural producers in Posavina, fuzzy methods were used, as expert evaluations were expressed through linguistic values. Research by Więckowski et al. [42] demonstrated that linguistic values and expert evaluations are most commonly applied when qualitative criteria are used. Additionally, MCDM methods are often used to obtain such results [43]. This research leveraged the SWARA method for collective decision making, based on its initial principles [30], with certain steps removed and others added for a more streamlined approach. The modified A-SWARA method used collective expert evaluations to determine the weight of the criteria.

According to the evaluations and the results obtained with A-SWARA, drones needed to have strong control, high flight autonomy, and ease of use. Consequently, the focus on drone selection centered on technical characteristics that make them easy to operate, even for agricultural producers who may have limited technical knowledge. To implement these features effectively in agriculture, Merz et al. [21] suggested the development of appropriate software interfaces that allow for both online drone management and the advancement of automatic systems. Such systems would enable drones to continuously monitor crop conditions on farms independently. The less demanding the drones are in terms of required knowledge, the simpler they are to use. Results from this method showed that portability and weight were the least important criteria; all drones in this selection were lightweight, with a maximum weight of 1375 g, making them highly portable. Research by Delavarpour et al. [25] has proven that lighter drones are more commonly used in agriculture because they ascend more quickly and operate at relatively lower altitudes. Additionally, such drones are easier for a single person to handle.

For drone selection, the fuzzy MARCOS method was also applied, and it was used to rank the drones. As a relatively new multi-criteria analysis method with proven practical effectiveness, MARCOS was an appropriate choice for this study. The results revealed that avoidance sensors were not a significant factor in drone selection, as the two top-ranked drones did not have this feature. However, in complex agricultural production, where diverse farming activities are present, it is necessary for drones to have these sensors, as noted by Ahmed et al. [24]. This may be due to Posavina's vast lowland terrain, where drones are likely used for cereal cultivation with few obstacles to navigate. Additionally, the most favorable drones were among the least expensive of those reviewed, an important factor in Bosnia and Herzegovina as a developing country where cost plays a major role in technology adoption. Farmers are likely to prefer affordable drones over costlier models. Another interesting finding was that camera megapixels did not significantly impact selection; even a 2 MP camera provides adequate quality for monitoring crop conditions [44].

Further analysis confirmed that more drones could be effectively used in Posavina's agricultural production. In particular, the sensitivity analysis showed that three drones performed best, suggesting that major investments in high-end models are not necessary. These findings indicate that drones are no longer a luxury restricted to large farms but are increasingly accessible for all farm sizes. However, alongside these benefits, certain challenges remain. Effective use of drone photography requires specialized software and a capable computer. Additionally, farmers must have the necessary knowledge to maximize the benefits drones offer in agriculture. Moreover, drones enable the implementation of sustainable agricultural production and enhance the competitiveness of small and medium-sized farms [45].

5. Conclusions

This research applied fuzzy methods to evaluate cost-effective drones suitable for agricultural production, specifically tailored to the needs of the Posavina region, the largest lowland area in Bosnia and Herzegovina. By incorporating expert evaluations, this study determined the relative importance of various criteria and ranked drones accordingly. Using the A-SWARA method, it was established that the most critical criteria in drone selection are C3—steering precision, C2—flight autonomy, and C7—ease of use. The fuzzy MARCOS method further identified that drones D5, D4, and D8 exhibit the most favorable characteristics and are the top choices for agricultural use in Posavina. It should be noted that this result was obtained solely based on expert opinions and the use of qualitative criteria. Therefore, future research should incorporate a combination of both qualitative and quantitative criteria to make better-informed decisions. This study aimed to demonstrate how drones can be used in agricultural production in Posavina and how more affordable drones can also be effectively utilized. For this reason, the primary goal of this research was not to select the best drone but to show that drones are essential for improving agricultural production.

Like any study, this research has certain limitations. Specifically, the criteria and selection of drones included may affect the results; alternative criteria or additional drones might lead to different rankings. Future research should aim to identify the most relevant criteria for drone selection and apply those in evaluations. Furthermore, it is necessary to use quantitative criteria in future studies, with a focus on selecting specific types of drones. The choice of drones was influenced by the models available on the Bosnia and Herzegovina market and limited by budget considerations, meaning that other drones could potentially be included in future studies. The primary focus here was to demonstrate the feasibility of drones for medium and small farms. Sensitivity analysis showed that each drone has distinct characteristics that may appeal differently based on individual farm requirements, suggesting that future research should develop adaptable decision-making models based on specific research objectives and farm needs.

Author Contributions: Conceptualization, A.P. and M.N.; methodology, A.P.; software, D.B.; validation, A.P., M.N. and A.Š.; formal analysis, A.P.; investigation, A.Š.; resources, M.N.; data curation, M.N.; writing—original draft preparation, A.P.; writing—review and editing, A.Š.; visualization, A.Š.; supervision, D.B.; project administration, M.N.; funding acquisition, A.Š. All authors have read and agreed to the published version of the manuscript.

Funding: This paper is a part of research financed by the MSTDI RS, agreed in decision no. 451-03-66/2024-03/200009 from 5.2.2024.

Institutional Review Board Statement: Not applicable.

Informed Consent Statement: Not applicable.

Data Availability Statement: The data and methods used in the research are presented in sufficient detail in the manuscript.

Conflicts of Interest: The authors declare no conflicts of interest. Drones in this article are intended for research purposes only and are not associated with any commercial advertising or promotions.

References

1. Li, W.Q.; Han, X.X.; Lin, Z.B.; Rahman, A. Enhanced Pest and Disease Detection in Agriculture Using Deep Learning-Enabled Drones. *Acadlore Trans. AI Mach. Learn.* **2024**, *3*, 1–10. [CrossRef]
2. Maghazei, O.; Netland, T. Drones in manufacturing: Exploring opportunities for research and practice. *J. Manuf. Technol. Manag.* **2020**, *31*, 1237–1259. [CrossRef]
3. Wang, C.; Fu, X. Environmental Cost Accounting in the Sugar Industry: An MFCA Perspective on "Sweet" Environmental Burdens. *J. Green Econ. Low-Carbon Dev.* **2024**, *3*, 45–55. [CrossRef]
4. Javaid, M.; Khan, I.H.; Singh, R.P.; Rab, S.; Suman, R. Exploring Contributions of Drones towards Industry 4.0. *Ind. Robot Int. J. Robot. Res. Appl.* **2022**, *49*, 476–490. [CrossRef]
5. Panday, U.S.; Shrestha, N.; Maharjan, S.; Pratihast, A.K.; Shahnawaz; Shrestha, K.L.; Aryal, J. Correlating the Plant Height of Wheat with Above-Ground Biomass and Crop Yield Using Drone Imagery and Crop Surface Model, A Case Study from Nepal. *Drones* **2020**, *4*, 28. [CrossRef]
6. Rejeb, A.; Abdollahi, A.; Rejeb, K.; Treiblmaier, H. Drones in Agriculture: A Review and Bibliometric Analysis. *Comput. Electron. Agric.* **2022**, *198*, 107017. [CrossRef]
7. Nhamo, L.; Magidi, J.; Nyamugama, A.; Clulow, A.D.; Sibanda, M.; Chimonyo, V.G.P.; Mabhaudhi, T. Prospects of Improving Agricultural and Water Productivity through Unmanned Aerial Vehicles. *Agriculture* **2020**, *10*, 256. [CrossRef]
8. Hafeez, A.; Husain, M.A.; Singh, S.P.; Chauhan, A.; Khan, M.T.; Kumar, N.; Chauhan, A.; Soni, S.K. Implementation of drone technology for farm monitoring & pesticide spraying: A review. *Inf. Process. Agric.* **2023**, *10*, 192–203. [CrossRef]
9. Inoue, Y. Satellite- and drone-based remote sensing of crops and soils for smart farming—A review. *Soil Sci. Plant Nutr.* **2020**, *66*, 798–810. [CrossRef]
10. Loures, L.; Chamizo, A.; Ferreira, P.; Loures, A.; Castanho, R.; Panagopoulos, T. Assessing the Effectiveness of Precision Agriculture Management Systems in Mediterranean Small Farms. *Sustainability* **2020**, *12*, 3765. [CrossRef]
11. Hassler, S.C.; Baysal-Gurel, F. Unmanned Aircraft System (UAS) Technology and Applications in Agriculture. *Agronomy* **2019**, *9*, 618. [CrossRef]
12. Gül, A.; Cakmak, E.; Karakas, A. Drone Selection for Forest Surveillance and Fire Detection Using Interval Valued Neutrosophic EDAS method. *Facta Univ. Ser. Mech. Eng.* 2024, online first. [CrossRef]

13. Radovanović, M.; Božanić, D.; Tešić, D.; Puška, A.; Hezam, I.; Jana, C. Application of Hybrid DIBR-FUCOM-LMAW-Bonferroni-GREY-EDAS Model in Multicriteria Decision-Making. *Facta Univ. Ser. Mech. Eng.* **2023**, *21*, 387–403. [CrossRef]
14. Roy, S.; Ciobotaru, A.-M. Does integrated transport topology act as a stimulus for inclusive growth and regional development in Bosnia and Herzegovina? *Reg. Sci. Policy Pract.* **2024**, *16*, 12732. [CrossRef]
15. Nedeljković, M.; Puška, A.; Jeločnik, M.; Božaniž, D.; Subić, J.; Stilić, A.; Maksimović, A. Enhancing fruit orchard establishment: A multicriteria approach for plum variety selection. *Yugosl. J. Oper. Res.* **2024**, *34*, 355–380. [CrossRef]
16. Tešić, D.; Marinković, D. Application of fermatean fuzzy weight operators and MCDM model DIBR-DIBR II-NWBM-BM for efficiency-based selection of a complex combat system. *J. Decis. Anal. Intell. Comput.* **2023**, *3*, 243–256. [CrossRef]
17. Kizielewicz, B.; Sałabun, W. SITW Method: A New Approach to Re-identifying Multi-criteria Weights in Complex Decision Analysis. *Spectr. Mech. Eng. Oper. Res.* **2024**, *1*, 215–226. [CrossRef]
18. Sarfraz, M. Application of Interval-valued T-spherical Fuzzy Dombi Hamy Mean Operators in the antiviral mask selection against COVID-19. *J. Decis. Anal. Intell. Comput.* **2024**, *4*, 67–98. [CrossRef]
19. Kamran, M.; Ashraf, S.; Kalim Khan, S.; Hussain Khan, A.; Zardi, H.; Mehmood, S. Integrated Decision-Making Framework for Hospital Development: A Single-Valued Neutrosophic Probabilistic Hesitant Fuzzy Approach with Innovative Aggregation Operators. *Yugosl. J. Oper. Res.* **2024**, *34*, 515–550. [CrossRef]
20. Ecer, F.; Ögel, İ.Y.; Krishankumar, R.; Tirkolaee, E.B. The q-rung fuzzy LOPCOW-VIKOR model to assess the role of unmanned aerial vehicles for precision agriculture realization in the Agri-Food 4.0 era. *Artif. Intell. Rev.* **2023**, *56*, 13373–13406. [CrossRef]
21. Merz, M.; Pedro, D.; Skliros, V.; Bergenhem, C.; Himanka, M.; Houge, T.; Matos-Carvalho, J.P.; Lundkvist, H.; Cürüklü, B.; Hamrén, R.; et al. Autonomous UAS-Based Agriculture Applications: General Overview and Relevant European Case Studies. *Drones* **2022**, *6*, 128. [CrossRef]
22. Ukaegbu, U.F.; Tartibu, L.K.; Okwu, M.O.; Olayode, I.O. Development of a Light-Weight Unmanned Aerial Vehicle for Precision Agriculture. *Sensors* **2021**, *21*, 4417. [CrossRef] [PubMed]
23. Kim, S.-B.; Lee, S.-H. Battery Balancing Algorithm for an Agricultural Drone Using a State-of-Charge-Based Fuzzy Controller. *Appl. Sci.* **2020**, *10*, 5277. [CrossRef]
24. Ahmed, S.; Qiu, B.; Ahmad, F.; Kong, C.-W.; Xin, H. A State-of-the-Art Analysis of Obstacle Avoidance Methods from the Perspective of an Agricultural Sprayer UAV's Operation Scenario. *Agronomy* **2021**, *11*, 1069. [CrossRef]
25. Delavarpour, N.; Koparan, C.; Nowatzki, J.; Bajwa, S.; Sun, X. A Technical Study on UAV Characteristics for Precision Agriculture Applications and Associated Practical Challenges. *Remote Sens.* **2021**, *13*, 1204. [CrossRef]
26. Yu, Y.; Liu, Y.; Wang, J.; Noguchi, N.; He, Y. Obstacle avoidance method based on double DQN for agricultural robots. *Comput. Electron. Agric.* **2023**, *204*, 107546. [CrossRef]
27. Parmaksiz, O.; Cinar, G. Technology Acceptance among Farmers: Examples of Agricultural Unmanned Aerial Vehicles. *Agronomy* **2023**, *13*, 2077. [CrossRef]
28. Aytekin, A.; Korucuk, S. Evaluating Performance-Based Logistics in Manufacturing Through Polytopic Fuzzy SWARA: A Criterion Assessment Approach. *J. Eng. Manag. Syst. Eng.* **2024**, *3*, 65–71. [CrossRef]
29. Shahid, T.; Ashraf, S.; Mashat, D.S. Enhancing Urban Development with Picture Fuzzy Sets: A Strategic Decision Support Framework. *J. Urban Dev. Manag.* **2023**, *2*, 172–180. [CrossRef]
30. Keršulienė, V.; Zavadskas, E.K.; Turskis, Z. Selection of rational dispute resolution method by applying new step-wise weight assessment ratio analysis (SWARA). *J. Bus. Econ. Manag.* **2010**, *11*, 243–258. [CrossRef]
31. Stević, Ž.; Pamučar, D.; Puška, A.; Chatterjee, P. Sustainable supplier selection in healthcare industries using a new MCDM method: Measurement Alternatives and Ranking according to COmpromise Solution (MARCOS). *Comput. Ind. Eng.* **2020**, *140*, 106231. [CrossRef]
32. Huskanović, E.; Bjelić, D.; Novarlić, B. Optimising the Efficiency of Municipal Utility Vehicle Fleets Using DEA-CRITIC-MARCOS: A Sustainable Waste Management Approach. *J. Intell. Manag. Decis.* **2024**, *3*, 139–149. [CrossRef]
33. Więckowski, J.; Kizielewicz, B.; Shekhovtsov, A.; Sałabun, W. How Do the Criteria Affect Sustainable Supplier Evaluation?—A Case Study Using Multi-Criteria Decision Analysis Methods in a Fuzzy Environment. *J. Eng. Manag. Syst. Eng.* **2023**, *2*, 37–52. [CrossRef]
34. Więckowski, J.; Sałabun, W. Comparative Sensitivity Analysis in Composite Material Selection: Evaluating OAT and COMSAM Methods in Multi-criteria Decision-making. *Spectr. Mech. Eng. Oper. Res.* **2025**, *2*, 1–12. [CrossRef]
35. Khan, N.; Ray, R.L.; Sargani, G.R.; Ihtisham, M.; Khayyam, M.; Ismail, S. Current Progress and Future Prospects of Agriculture Technology: Gateway to Sustainable Agriculture. *Sustainability* **2021**, *13*, 4883. [CrossRef]
36. Čadro, S.; Marković, M.; Hadžić, A.; Hadžić, A.; Žurovec, O. Assessing the impact of climate change on extreme hydrological events in Bosnia and Herzegovina using SPEI. *J. Cent. Eur. Agric.* **2024**, *25*, 531–541. [CrossRef]
37. Sharma, V.; Tripathi, A.K.; Mittal, H. Technological revolutions in smart farming: Current trends, challenges & future directions. *Comput. Electron. Agric.* **2022**, *201*, 107217. [CrossRef]
38. Gabriel, A.; Gandorfer, M. Adoption of digital technologies in agriculture—An inventory in a european small-scale farming region. *Precis. Agric.* **2023**, *24*, 68–91. [CrossRef]
39. McCarthy, C.; Nyoni, Y.; Kachamba, D.J.; Banda, L.B.; Moyo, B.; Chisambi, C.; Banfill, J.; Hoshino, B. Can Drones Help Smallholder Farmers Improve Agriculture Efficiencies and Reduce Food Insecurity in Sub-Saharan Africa? Local Perceptions from Malawi. *Agriculture* **2023**, *13*, 1075. [CrossRef]

40. Guebsi, R.; Mami, S.; Chokmani, K. Drones in Precision Agriculture: A Comprehensive Review of Applications, Technologies, and Challenges. *Drones* **2024**, *8*, 686. [CrossRef]
41. Borissov, B.; Hristozov, Y. Potential for using artifical intelligence in public administration. *Econ. Innov. Econ. Res. J.* **2024**, *12*, 409–423. [CrossRef]
42. Więckowski, J.; Kizielewicz, B.; Shekhovtsov, A.; Sałabun, W. RANCOM: A novel approach to identifying criteria relevance based on inaccuracy expert judgments. *Eng. Appl. Artif. Intell.* **2023**, *122*, 106114. [CrossRef]
43. Trung, D.D.; Dudić, B.; Dung, H.T.; Truong, N.X. Innovation in financial health assessment: Applying MCDM techniques to banks in Vietnam. *Econ. Innov. Econ. Res. J.* **2024**, *12*, 21–33. [CrossRef]
44. Kondo, S.; Yoshimoto, N.; Nakayama, Y. Farm Monitoring System with Drones and Optical Camera Communication. *Sensors* **2024**, *24*, 6146. [CrossRef] [PubMed]
45. Škuflić, L.; Šokčević, S.; Bašić, M. Sustainable development and competitiveness: Is there a need for GCI reconstruction? *Econ. Innov. Econ. Res. J.* **2024**, *12*, 153–173. [CrossRef]

eng

MDPI

Article

Sustainable Charging Stations for Electric Vehicles

Carlos Armenta-Déu * and Luis Sancho

Facultad de Ciencias Físicas, Universidad Complutense de Madrid, 28040 Madrid, Spain; luisan04@ucm.es
* Correspondence: cardeu@fis.ucm.es

Abstract: In this work, we develop a detailed analysis of the current outlook for electric vehicle charging technology, focusing on the various levels and types of charging protocols and connectors used. We propose a charging station for electric cars powered by solar photovoltaic energy, performing the analysis of the solar resource in the selected location, sizing the photovoltaic power plant to cover the demand completely, and exploring different configurations such as grid connection or physical and virtual electric energy storage. Despite the current development applying for specific working conditions, operating voltage, charging rate, power demand, etc., the proposed configuration is modular, adaptable, and resilient. The simulated system operates within the 360 V to 800 V range of direct current for charging the electric vehicles, with a selectable power range between 20 and 180 kW. The basic layout includes four charging poles, each servicing all working voltages. An oversized PV plant powers the charging station at any time of the year, saving money compared to the alternative of the electric storage unit. In addition, we build simulation tools and algorithms that optimize the design of future projects, providing a solid basis for sustainable energy infrastructure planning and design.

Keywords: electric vehicle; charging station; technology; modular design; PV power plant; simulation; sizing

Citation: Armenta-Déu, C.; Sancho, L. Sustainable Charging Stations for Electric Vehicles. *Eng* **2024**, *5*, 3115–3136. https://doi.org/10.3390/eng5040163

Academic Editor: Antonio Gil Bravo

Received: 2 September 2024
Revised: 18 November 2024
Accepted: 25 November 2024
Published: 27 November 2024

1. Introduction

Today, the continuous growth of carbon emissions, which are responsible for climatic change, requires a transitional process to power sources more respectful of the environment. Among the feasible solutions to implement this process, renewable energy sources are a priority for politicians, technicians, energy developers, and citizens.

The global GHG emissions into the atmosphere come from the industrial, commercial, residential, and transportation sectors; this last source contributes 30% of the total emissions, representing a significant fraction [1,2]. The implementation of electric vehicles represents a positive action in the process of reducing greenhouse gas emissions [3–5].

People opposing the transformation of the vehicle fleet from fossil fuel engines to electric motors argue that the required electricity to charge the electric vehicle battery comes from fossil fuel power plants, generating carbon emissions in the energy generation process [6–8]. Therefore, it is necessary to power the electric vehicle charging stations from energy plants powered by renewable energies like photovoltaic plants, wind farms, geothermal and hydroelectric power plants, etc. [9–12].

The atmosphere's increased pollution level is a relevant problem in cities, where the concentration of particles harmful to health is higher due to the high vehicle density. Local authorities apply measurements to reduce these particle emissions by promoting and implementing laws, norms, and regulations to favor electric vehicle use and penalizing conventional fossil fuel engine cars [13–15].

The European Union leads the transportation sector electrification process, developing a protocol about the infrastructure of alternative fuels, which specifically includes the following action (sic) [16]:

"Starting in 2025, fast electric vehicle charging stations, with a minimum power of 150 kW for light vehicles and small vans, should be implemented every 60 km in the main European Union traffic corridors; the so-called Trans-European Transportation Network (RTE-T)".

"The charging stations for heavy vehicles, with a minimum charging power of 350 kW, should also be implemented every 60 km in the basic RTE-T network and every 100 km in the enlarged global RTE-T network, should be completed by 2030".

The European directive aims to solve the problem of the low number of charging stations in some countries that are a long way from completing the directions included in the protocol; one of the most relevant cases is Spain, the fourth economy of the European Union with only thirty thousand charging stations in the national transportation network, far below the European Union average. To make matters worse, more than 25% of the charging stations are out of order, aggravating the problem of vehicle fleet electrification [17]. The Spanish case is not the only one in the European Union, as we show in Figure 1. We notice that some other countries suffer from the same low density of electric vehicle charging stations [18].

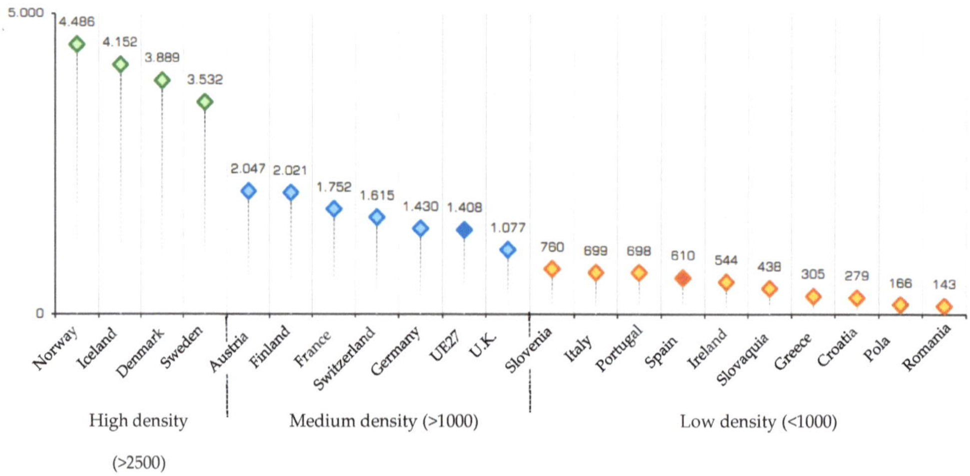

Figure 1. Electric vehicle charging point density in the European Union (public access) (number of points per million inhabitants) [19].

Readers can notice the difference between Norway (highest level) and Romania (lowest level), with a factor of 30 in the charging point density. On the other hand, countries in the high-density range—Norway, Iceland, Denmark, and Sweden—only represent 0.65% of the total number of charging points in the European Union, since the population density in these countries is very low, reinforcing the need to implement a dense electric vehicle charging station network to compete with fossil fuel engine cars.

Another issue for consideration is the high number of low-charging-rate points, 83%, with a maximum output power of 22 kW, representing a one-hour charging time for a driving range of 100 km (60 miles) in a standard electric vehicle, which discourages drivers from using electric cars in long-distance trips [19].

Despite the many obstacles to implementing electric vehicles in modern society, the number of cars powered by electricity is continuously growing, as shown in Figure 2. This situation generates a conflict because the number of charging stations is not growing at the same rate as electric vehicle sales.

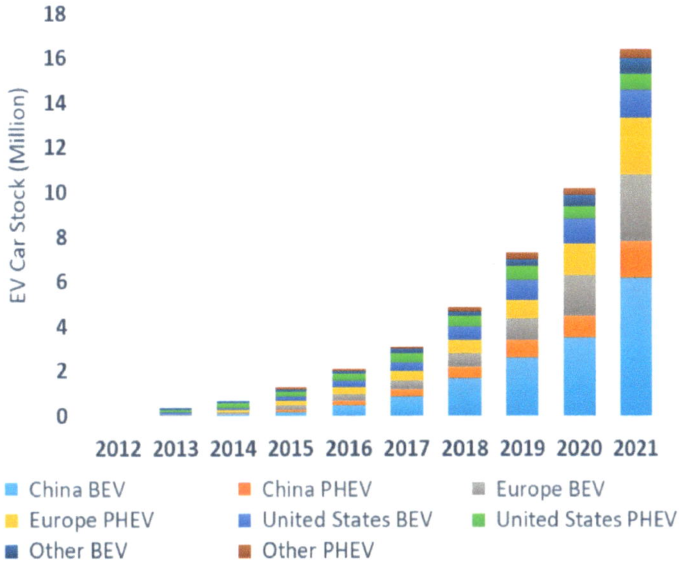

Figure 2. Global electric car stock (2012–2021) [20].

2. Charging Point

An electric vehicle charging point is characterized by the charging rate, the charging mode, and the charging connector. The charging rate is regulated by the norms SAE J1772 [21,22] and IEC 61851-1 [23]; the former includes three types of rate and the latter four types. Table 1 shows the specifications of the different charging rates depending on the power supply.

Table 1. Charging rate specifications for variable power supply [21–23].

Specification	Level 1	Level 2	Level 3	Extreme Fast Charging (XFC)
Charging power	1.44 kW–1.9 kW	3.1 kW–19.2 kW	20 kW–350 kW	>350 kW
Charger type	Onboard—Slow charging	Onboard—Semi-fast charging	Offboard—Fast cahrging	Offboard—Ultrafast charging
Charge location	Residential	Private and commercial	Commercial	Commercial
Charging time	200 km: +/−20 h	200 km: +/−5 h	80% of 200 km: +/−30 min	Approximately 5 min with high energy density
Power supply	120/230 Vac, 12 A–16 A Single phase	208/240 Vac, 12 A–80 A Single phase/Split phase	120/230 Vac & 300–800 Vdc, 250 A–300 A Three phase	1000 Vdc and above, 400 A and higher Polyphase
Supply interface and protection type	Convenience outlet (Breaker in cable)	Dedicated EV supply equipment (Breaker in the cable and pilot function)	Dedicated EV supply equipment (communication & event monitoring between EV and charging station)	Dedicated EV supply equipment (communication & event monitoring between EV and charging station)
Standards	SAE J1772 [24], IEC 62196-2 [25], IEC 61851-22/23 [26], GB/T 20234-2 [27]		IEC 61581-23/24 [26], IEC 62196-3 [28]	IEC 62196 [29], SAE J2839/2 & J2847/2 [30]

2.1. *Charging Rate*

Level 1 corresponds to a domestic power socket with a maximum power supply of 2 kW. Level 2 requires a specific power socket for electric vehicle charging with a maximum power supply of 19.2 kW. These two levels operate in alternating monophasic current. Level 3 corresponds to a fast charging rate and operates only in triphasic alternating current up to 350 kW power supply. The level 3 charging points are in road or street public charging stations. Extreme fast charging, or ultra-fast charging, is an extension of the fast charging rate for a power supply above 350 kW; this type of charging point is scarce.

2.2. *Charging Mode*

The IEC defines four operational modes for electric vehicle charging depending on the specifications and configuration: alternating or direct current, maximum charging current, voltage, power, and connection type.

In levels 1 to 3, the charging mode uses alternating current (AC), monophasic for levels 1 and 2, and triphasic for level 3, as mentioned previously. The operating voltage ranges from 220 V for level 1 to 480 V for level 3. Levels 2 and 3 also operate with intermediate voltages of 360 and 400 V. The power supply is 7.6, 15.3, and 120 kW for levels 1, 2, and 3, respectively. In the XFC mode, the current ranges up to 400 ADC (direct current), and the voltage operates between 600 and 1000 VDC.

Figure 3 shows the various types of charging mode.

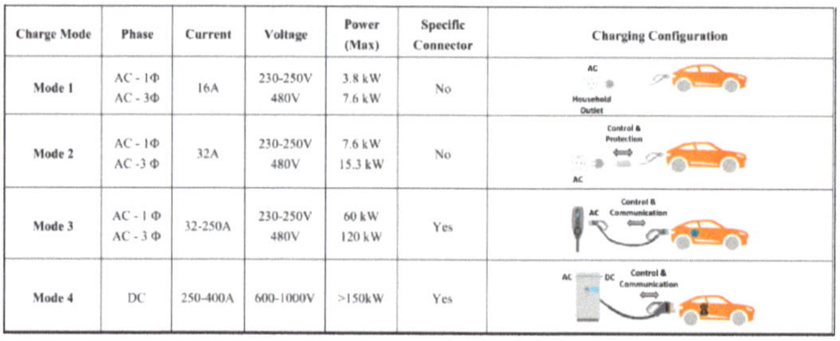

Charge Mode	Phase	Current	Voltage	Power (Max)	Specific Connector	Charging Configuration
Mode 1	AC - 1Φ AC - 3Φ	16A	230-250V 480V	3.8 kW 7.6 kW	No	AC Household Outlet
Mode 2	AC - 1Φ AC -3 Φ	32A	230-250V 480V	7.6 kW 15.3 kW	No	Control & Protection AC
Mode 3	AC - 1 Φ AC - 3 Φ	32-250A	230-250V 480V	60 kW 120 kW	Yes	AC Control & Communication
Mode 4	DC	250-400A	600-1000V	>150kW	Yes	AC DC Control & Communication

Figure 3. Electric vehicle charging modes [31].

2.3. *Charging Socket*

Level 1 does not require a specific socket for charging the electric vehicle; the charging point can use a domestic one; nevertheless, levels 2 and 3, and the XFC mode, require a specific socket of the type shown in Figures 4 and 5.

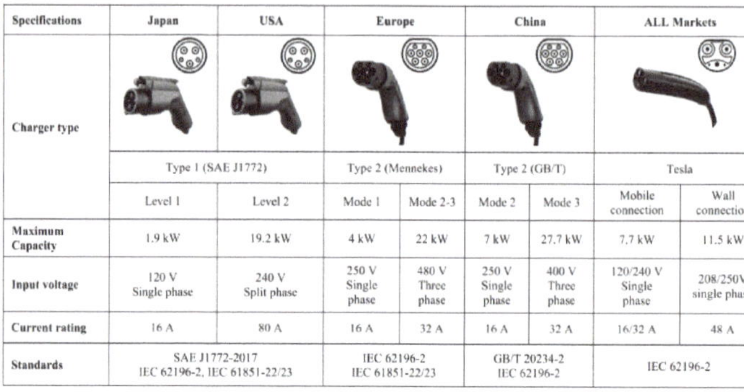

Specifications	Japan	USA	Europe		China		ALL Markets	
Charger type	Type 1 (SAE J1772)		Type 2 (Mennekes)		Type 2 (GB/T)		Tesla	
	Level 1	Level 2	Mode 1	Mode 2-3	Mode 2	Mode 3	Mobile connection	Wall connection
Maximum Capacity	1.9 kW	19.2 kW	4 kW	22 kW	7 kW	27.7 kW	7.7 kW	11.5 kW
Input voltage	120 V Single phase	240 V Split phase	250 V Single phase	480 V Three phase	250 V Single phase	400 V Three phase	120/240 V Single phase	208/250V single phase
Current rating	16 A	80 A	16 A	32 A	16 A	32 A	16/32 A	48 A
Standards	SAE J1772-2017 IEC 62196-2, IEC 61851-22/23		IEC 62196-2 IEC 61851-22/23		GB/T 20234-2 IEC 62196-2		IEC 62196-2	

Figure 4. Socket types for electric vehicle charging in alternating current (AC) [24–30,32].

Specifications	Japan	USA	Europe	China	ALL Markets	
Charger type						
	CHAdeMO	CCS - Combo 1	CCS - Combo 2	GB/T	Tesla Supercharger	CHAdeMO
Capacity	50 - 400 kW	150 - 350 kW	350 kW	60 - 237 kW	250 - 350 kW	50 - 400 kW
Input voltage	50 - 1000 V	200 - 1000 V	200 - 1000 V	250 – 950 V	300 - 480 V	50 - 1000 V
Maximum Current	400 A	500 A	500 A	250 – 400 A	800 A	400 A
Standards	IEC 61851-23/4 IEC 62196-3 JEVS G105	SAE J1772 IEC 61851-23/24 IEC 62196-3	IEC 61851-23/24 IEC 62196-3 DIN EN 62196-3	GB/T 20234-3 IEC 62196-3	IEC 62196-3	IEC 61851-23/4 IEC 62196-3 JEVS G105

Figure 5. Socket types for electric vehicle charging in direct current (DC) [24–30,32].

The DC socket type allows higher charging voltage and current, operating at a power supply of up to 350–400 kW. This configuration reduces the charging time for a 100 km driving range in a standard electric vehicle from 1 h in AC mode to 15 min in DC mode.

3. Solar Power Supply

A sustainable charging station for electric vehicles should collect energy from renewable power sources like photovoltaic, wind, geothermal, hydroelectric, and others. Since analyzing every type of renewable power plant is beyond the scope of this work, we prioritize photovoltaic power plants as the primary energy source for powering the charging station prototype.

Photovoltaic power plant location is selected from the solar radiation database provided by the Geographical Information System (GIS) from the European Commission Service for Photovoltaic Systems (PVGIS) [33]. The selection criteria are based on maximizing the monthly average solar radiation on site; following these criteria, the photovoltaic array is oriented to the equator and tilted 35°. Figure 6 shows the monthly solar radiation distribution in the selected location (Madrid, Spain).

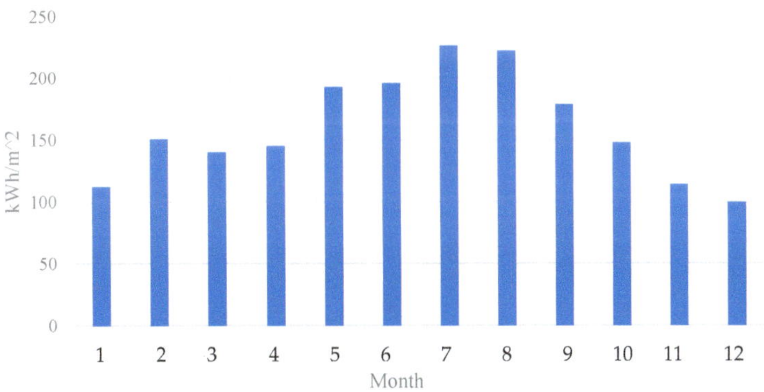

Figure 6. Monthly distribution of solar radiation in Madrid, Spain [34].

Considering a uniform behavior over time, we calculate the daily average solar radiation, obtaining the following (Table 2):

Table 2. Estimated average monthly and daily solar radiation in the selected location (Madrid, Spain).

Month	Monthly Average (kWh/m^2)	Daily Average (kWh/m^2)
January	112.13	3.74
February	150.45	5.02
March	140.02	4.67
April	145.05	4.84
May	192.73	6.42
June	195.55	6.52
July	226.09	7.54
August	222.10	7.40
September	178.51	5.95
October	147.50	4.92
November	113.85	3.80
December	99.48	3.32

Because solar radiation changes every day, we apply the estimated daily average value for the calculations.

4. Energy Demand

The energy demand depends on the number of charging points, the maximum power supply at every charging point, the daily operating time, and the daily occupancy factor. Considering a charging pole with four charging points, one per power supply, a 14 h daily operating time, and a 75% occupancy factor, the daily energy demand is as follows (Table 3):

Table 3. Daily energy demand (charging station).

Charging Point Number	Maximum Power Supply (kW)	Daily Energy Demand (kWh)
1	22	308
2	48	672
3	96	1344
4	180	2520
Total	346	4844

5. Photovoltaic Array Design

The photovoltaic array consists of a set of PV modules grouped in series and parallel to generate the required voltage and current. We select a panel of 460 W with 21% efficiency. Figure 7 shows the PV module's electric response.

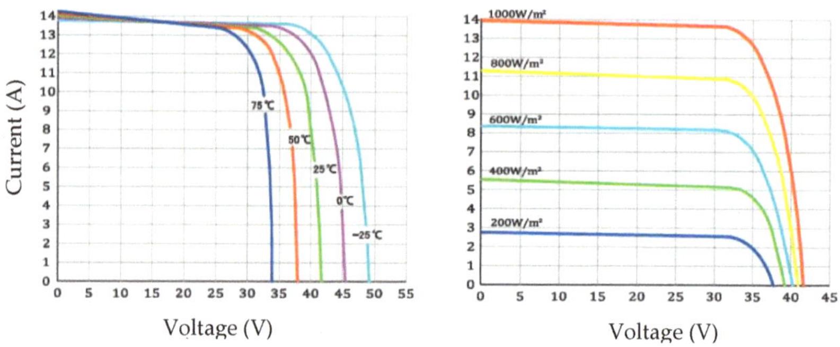

Figure 7. PV module electric response (I–V curve). (**Left**) Temperature dependence. (**Right**) Solar radiation level dependence.

Technical characteristics for the PV module are provided by the manufacturer (Table 4).

Table 4. PV module technical characteristics [35].

Parameter	Value
Maximum power	460 Wp
Voltage at maximum power	35.10 V
Current at maximum power	13.11 A
Open-circuit voltage	41.65 V
Short-circuit current	13.99 A
Efficiency	21.25
Power tolerance	0/+5
PV array maximum voltage	DC 1.500 V

Data have been retrieved for the selected location of Madrid, Spain [36]; therefore, we can estimate the temperature correction for the operational PV module voltage and current. The maximum and minimum temperatures over the year are listed below:

- Minimum average monthly temperature: −1.0 °C (January).
- Maximum average monthly temperature: 33.5 °C (July).
- Maximum yearly absolute temperature: 43.5 °C (10 August).
- Minimum yearly absolute temperature: −12.5 °C (12 January).

Applying the following equation, we determine the PV module operating temperature [37]:

$$T_{cell} = T_{amb} + \frac{NOCT - 20}{800} G \tag{1}$$

T is the operating temperature, with subscripts *cell* and *amb* indicating PV cell temperature and ambient temperature; $NOCT$ is the normal operating cell temperature, being the standard temperature of 45 °C; and G is the solar radiation level.

According to Equation (1), and applying values for maximum and minimum temperatures, the operating PV module temperature is as follows (Table 5):

Table 5. PV module working temperature.

Condition	Value
Maximum	69.8 °C
Minimum	8.9 °C

Since the working temperature modifies the output voltage and current, we should apply the correction factor for these two parameters to obtain the operational values for the working temperature. The new values for the voltage and current derive from the following expressions [30]:

$$V_M = V_M^{STD} + \beta\left(T_{cell}^{\min} - 25\right) \tag{2}$$

$$I_M = I_M^{STD} + \alpha(T_{cell}^{\max} - 25) \tag{3}$$

V and I are the PV module voltage and current, and coefficients α and β are the temperature correction factors for current and voltage. Subscripts M and *cell* account for maximum power point and PV cell, and superscripts STD, min, and max correspond to standard working conditions and minimum and maximum cell temperature.

Applying values from Tables 4 and 5 to Equations (2) and (3), we obtain the following (Table 6):

Table 6. PV module electric operating parameters.

Parameter	Value
α (A/°C)	0.05%/°C
β (V/°C)	−0.30%/°C
Voltage (V)	46.00
Current (A)	9.65
Power (W)	444

Using data from Table 6, we can determine the PV array energy generation by applying the following expression:

$$\xi_{AC} = P_{PV}(sph)\eta_T \tag{4}$$

P_{PV} is the PV module's peak power, *sph* is the sun's peak hours, and η_T is the global efficiency including electric and thermal losses.

The global efficiency in a standard PV array is about 80% (η_T = 0.8) [38].

The PV array configuration depends on the charging station operating voltage. Because the electric vehicle manufacturers disagree on a common voltage, we decided to operate with a standard 400 V for the PV array and line transmission and convert this voltage into the required one at the charging point, depending on the pole configuration.

Combining the output voltage for a single panel with the global 400 V voltage for the power supply to the charging station, the number of serial PV modules is as follows:

$$n_s = \frac{400}{46} = 8.69 \tag{5}$$

If we round the result from Equation (5), the effective number is nine.

Repeating the process for the current, considering the maximum power supply required by the charging station (Table 3), the number of parallel strings is as follows:

$$n_s = \frac{346000}{9 \times 444} = 86.58 \tag{6}$$

If we round the result from Equation (5), the effective number is 87.

Therefore, the PV array configuration results in a set of 87 parallel strings of nine panels in series for a total of 783 panels, which generate a global output power of 347.65 kW, 0.5% above the required value.

The above calculation is valid for a single pole with four charging points of power supply, as shown in Table 3; if the charging station includes multiple charging poles, we should resize the system according to the number of poles. In the next section, we develop the case of numerous charging poles in a charging station.

6. Charging Station Layout Engineering

The charging station layout engineering is based on a modular functioning that can be replicated and enlarged depending on current requirements. As we have mentioned before, the basic unit is a charging pole with four sockets, one per power supply; this configuration permits the charging of four electric vehicles simultaneously, provided the charging power requirements are different for each one, with a slow (22 kW), medium (48 kW), fast (96 kW), and extra fast (180 kW) charging rate. The supply voltage range is 360 to 800 V. Table 7 summarizes the main characteristics of the charging station layout.

The charging station operates under direct current (DC) since it allows a higher charging rate than alternating current.

The basic design of the charging station includes four charging poles with four sockets for each pole. Figure 8 shows the voltage supply for each pole and socket.

Table 7. Electric characteristics of the charging station.

Power (kW)	22			48			96			180		
Voltage (V)	360	400	480	360	400	480	400	480	800	400	480	800
Current (A)	61	55	45	133	120	100	266	200	60	450	375	225

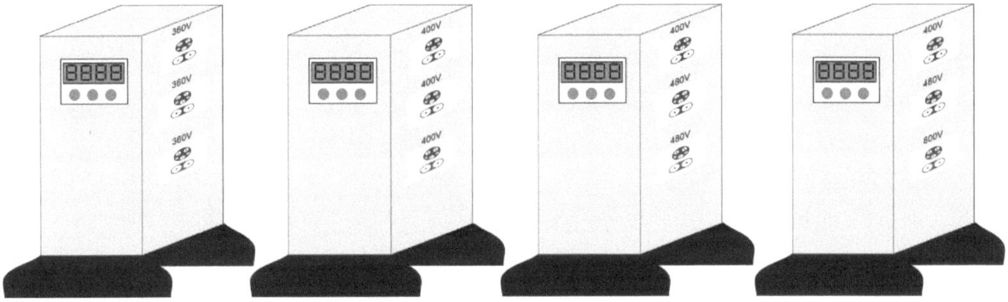

Figure 8. Voltage supply distribution at every pole and socket for the charging station prototype.

We observe that one charging pole is exclusively dedicated to a voltage supply of 360 V (first from the left) and another to 400 V (second from the left) because these are the most common electric vehicle motor voltages in the vehicle fleet. The other two charging poles combine a voltage supply from 400 V to 800 V due to the lower number of electric vehicles with 480 and 800 V electric motors.

Since the electric transmission line operates at 135 kV alternating current, the engineering requires a multiple voltage conversion; to do so, we use a first voltage conversion from 135 kV to 10 kV and a second conversion from 10 kV to 400 V. These two conversions are in alternating current. The reason for converting from high to low voltage in two steps is to reduce energy losses, since a drastic voltage reduction increases energy losses in the voltage converter. A higher number of voltage conversions in more than two steps would require multiple voltage converters, meaning higher investment and maintenance costs. Therefore, the two-step voltage conversion optimizes the energy losses to investment ratio. Figure 9 shows the voltage conversion layouts' engineering.

According to the previous statement of reducing the transmission line voltage conversion in two steps, the electric engineering operates with a reference voltage. We decided to operate at 360, 400, 480, or 800 V as the reference voltage, analyzing the energy losses derived from the initial conversion from 10 kV to the selected voltage, and from this value to the required ones at the pole sockets. The analysis concludes that the optimum reference value is 400 V.

Once the voltage has been reduced to 400 VAC, we convert it into direct current using specific rectifiers depending on the voltage supply; therefore, the electric circuit engineering requires four types of voltage rectifiers: 400 VAC to 360 VDC, 400 VAC to 400 VDC, 400 VAC to 480 VDC, and 400 VAC to 800 VDC. We decided to duplicate each type of rectifier to avoid unexpected service stops in case of failure.

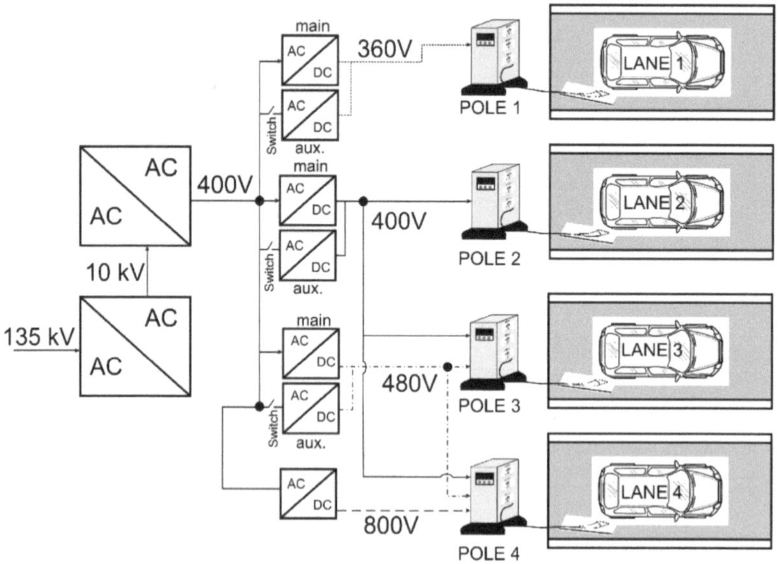

Figure 9. Voltage conversion layouts' engineering for the charging station prototype.

7. PV Power Plant Layout Engineering

Based on data for a single charging pole, shown in the Energy Demand section, and considering the charging station layout from Figure 9, the prototype requires a global power of 1776 kW, corresponding to four charging poles of 444 kW each. On the other hand, if we transport this power at the 400 VDC corresponding to the PV array output voltage, as described in a previous section, the energy losses are enormous; indeed, for every kilometer of transmission line from the PV power plant to the charging station, operating under these values, we have the following power losses:

$$\xi_L = I^2R = \left(\frac{P}{V}\right)^2 R = \left(\frac{1776 \times 10^3}{400}\right)^2 1 = 19.7 \times 10^6 W = 19.7MW \tag{7}$$

Nevertheless, if we operate at a high voltage, 135 kV, the power losses are reduced to the following:

$$\xi_L = \left(\frac{1776 \times 10^3}{135 \times 10^3}\right)^2 1 = 137W \tag{8}$$

The new value represents a negligible power loss amount.

Since the basic PV array layout produces 347.65 kW, we need to replicate this basic configuration to reach the global power; therefore, the number of PV array blocks in the PV power plant matrix is as follows:

$$m = \frac{1776}{347.65} = 5.11 \tag{9}$$

Because of the fractional number, two options arise for the PV power plant configuration; the first is to install five blocks with the same characteristics as for the PV array and a smaller PV array to complete the fractional number, while the second is to design five identical PV array blocks incrementing the number of PV modules in each block with regard to the original configuration.

Since the voltage conversion from 400 VDC to 135 kV alternating current is too high, we decided to group every two serial strings in pairs, doubling the string output voltage, reducing the voltage conversion factor, and lowering the energy losses at the DC/AC converter.

Table 8 summarizes the PV power plant characteristics of the two options.

Table 8. PV power plant characteristics (per block).

Configuration	PV Array Blocks	PV Modules Number	Serial String Number	Parallel Row Number	Output Voltage (V)	Output Current (A)	Global Output Power (kW)
A	5	783	18	44	828	424.6	351.6
	1	90	18	3	828	28.95	24
B	5	792	18	46	828	443.9	367.5

Configuration B uses more PV modules, 4140 versus 3969 for configuration A, but its design is less complex since it uses the same structure for every block. On the other hand, configuration B generates 1837.5 kW versus 1782.0 kW for configuration A, which is 3% higher, representing a safety factor in case of increasing demand at the charging station. Therefore, we decided to adopt configuration B as the reference for our prototype.

The PV power plant uses an area of 3054 m^2, corresponding to a rectangular space of 61.08 m $\times$ 50 m, as shown in Figure 10.

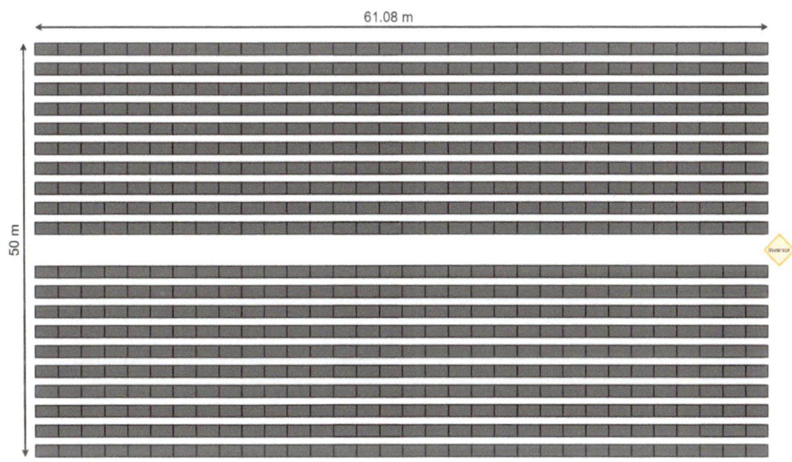

Figure 10. Schematic representation of the PV power plant layout.

Figure 10 shows a simulated view of the PV power plant prototype layout; therefore, the number of parallel rows and serial strings do not correspond to current values.

The PV power plant uses four DC/AC converters, model INGECON® SUN 330TL M12 from the INGETEAM company (Milwaukee, WI, USA) [39], for a total of 24 inputs and 12 MPPT units; however, due to design specifications, the converters only use 20 inputs and 10 MPPT units.

Every converter output connects to the voltage transformation station that elevates the voltage to 135 kV for the transmission line. The process develops in two steps; the first elevates the voltage from 800 VDC to 10 kV alternating current and the second from 10 to 135 kV AC. Figure 11 shows the schematic representation of the voltage transformation process.

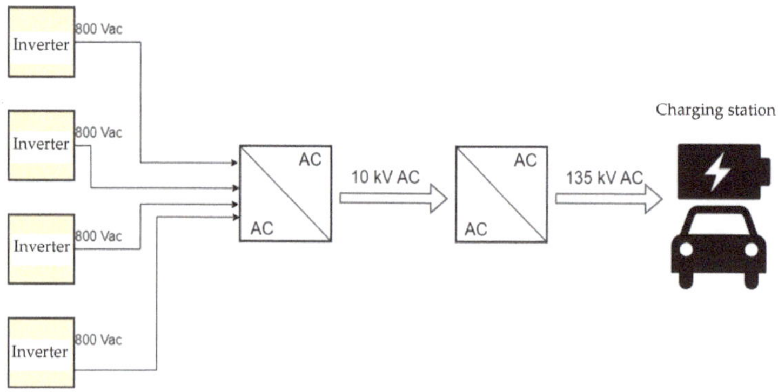

Figure 11. Schematic view of the voltage transformation process at the PV power plant.

8. Transmission Line to Charging Station Electric Engineering

The electric connection between the transmission line and the charging station depends on the power source, renewable energy plant, grid, or storage unit. Therefore, there may be an alternating-current bus configuration, a direct-current bus configuration, or a combination of the two. We can see the AC and DC bus configurations in Figures 12 and 13.

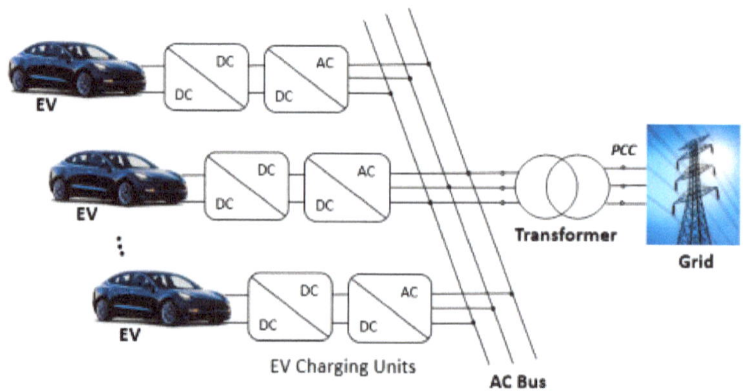

Figure 12. AC bus configuration for the transmission line to charging station connection.

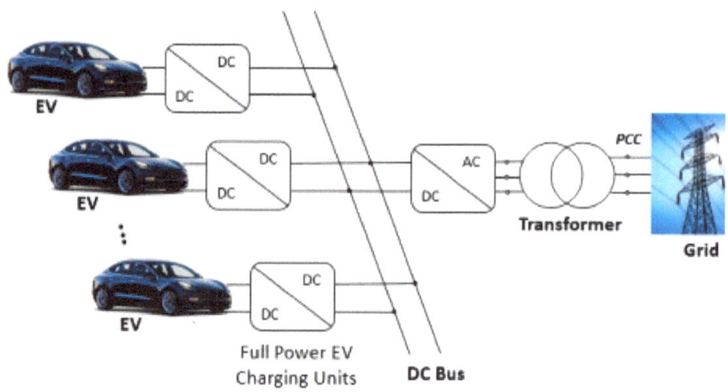

Figure 13. DC bus configuration for the transmission line to charging station connection.

The AC bus configuration is characterized by an AC/AC transformer that reduces the 135 kV from the line transmission to 10 kV in the AC bus. The 10 kV alternating current is then converted into direct current at the reference voltage, 400 V, using an AC/DC converter; the process is replicated in three lines to service the final DC voltage supply sockets at 360, 480, and 800 volts.

The DC bus configuration is characterized by an AC/AC transformer that reduces the 135 kV from the line transmission to 10 kV and an AC/DC converter, which reduces the 10 kV alternating current to 400 volts direct current at the DC bus. The 400 V is reduced to 360 V or elevated to 480 or 800 V using the corresponding DC/DC converter to service the final DC voltage supply sockets at 360, 480, and 800 volts.

9. Charging Station Wiring Design and Engineering

The voltage drop in the transmission lines is due to the Joule effect. We need to size the wires correctly to minimize the energy losses during transmission. The wire section derives from the following expression [40–42]:

$$S = \frac{1}{28}\frac{I^2}{\Delta V} \tag{10}$$

S is the wire section, I is the transported current, and ΔV is the voltage drop percentage.

Because the national legislation differs for every country, we selected a 1.5% voltage drop percentage, a current value admitted in many Western countries [43,44]. Since we operate with different power and voltage supply values, we size the wire section according to every configuration. Considering a 5 m length for every power supply hose, the wire sizing is as follows (Table 9):

Table 9. Wire sizing for the charging points.

Power (kW)		22			48			96			180	
Voltage (V)	360	400	480	360	400	480	400	480	800	400	480	800
Current (A)	61	55	45	133	120	100	266	200	60	450	375	225
Section (mm^2)	25	25	25	60	50	35	150	120	25	300	240	120
Percentage losses in 5 m	0.8	0.7	0.5	1.38	1.34	1.33	2.21	2.24	1.34	3.15	2.73	1.97

10. Simulation

Applying data from the previous development, we run a simulation process based on the engineering design to validate the proposed prototype. The simulation uses data from previous development and focuses on determining the feasibility of the system and its autonomy.

The first simulation step evaluates the solar energy coverage factor, defined as the ratio between the photovoltaic power generation and energy demand. Considering the data from the designed PV power plant and the charging station, calculated in previous sections, we obtain the following results (Figure 14):

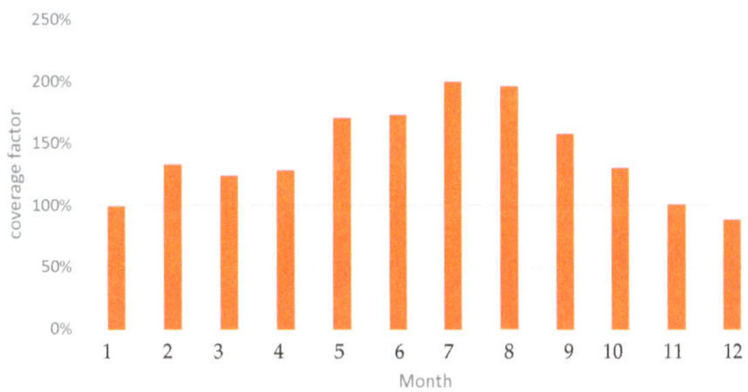

Figure 14. Monthly solar photovoltaic energy coverage factor.

The proposed PV power plant produces a solar photovoltaic energy coverage factor above 100% for the entire year except in December, where the coverage factor barely reaches 88%. The coverage factor in January was 99%, near the required 100% to ensure the functioning of the charging station. On the other hand, the energy excess generated from February to November can be stored in electric batteries and used in January and December to guarantee complete energy coverage from the PV power plant. The power supply is guaranteed because the coverage factor exceeds the 100% coverage factor in various months of the year.

We also run a daily energy coverage to verify the system feasibility; to do so, we evaluate the daily hourly evolution for the characteristic day of every month. We only show the results for the lowest and highest energy coverage months to validate the simulation process (Figure 15).

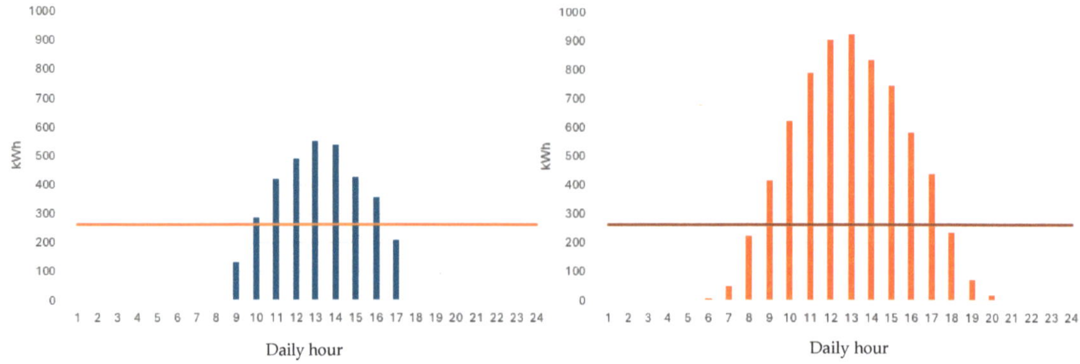

Figure 15. Hourly evolution of power generation and energy demand. Left side: lowest PV power generation; right side: highest PV power generation.

The horizontal line in Figure 15 represents the charging station energy demand, which is considered to be constant. The bars represent the photovoltaic power supply. It is clear that solar photovoltaic power does not cover the energy demand from sunset to sunrise (night hours); nevertheless, this is the period of lowest activity at the charging station, matching the time when the charging station requires low or no energy because of the small number of charging sockets being used. On the other hand, the excess power generation during the central hours of the day, from 10:00 to 16:00 in the poorest month (left side) and from 09:00 to 17:00 in the best month (right side), compensates for the lack or absence of photovoltaic power supply.

Energy Compensation

Based on statistical data from many countries, the occupancy of charging poles in an electric vehicle charging station follows human activity, which reduces drastically during night hours (Figure 16) [45–47]. Therefore, the lack or absence of solar photovoltaic power is minimized and compensated for by the excess during the central hours of the day, considering the balance between the grid and PV facility and the grid as a virtual battery.

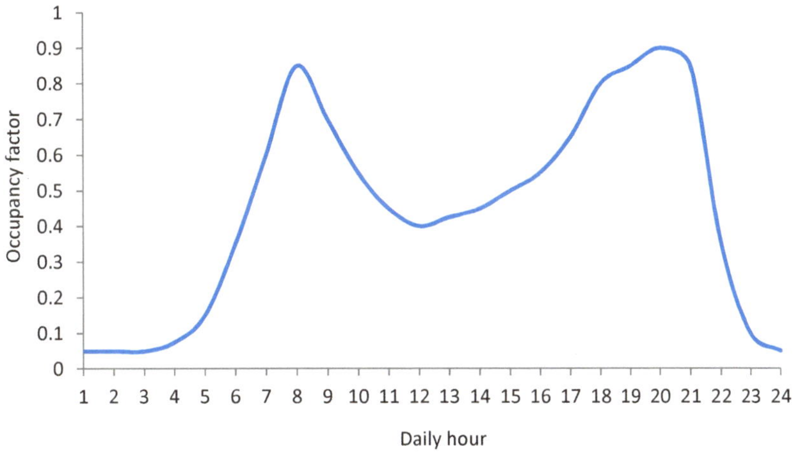

Figure 16. Charging station daily hourly occupancy factor.

Combining the maximum power demand and the occupancy factor, the daily energy demand distribution in the charging station yields the following (Figure 17):

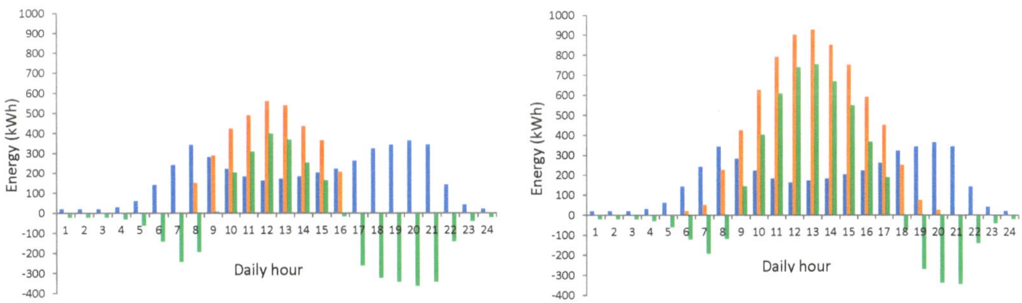

Figure 17. Daily hourly energy evolution for the system prototype. (**Left**) Lowest PV power supply. (**Right**) Highest PV power supply.

The bars in Figure 17 represent the PV power supply (orange), the charging station energy demand (blue), and the energy balance (green). We observe that some green bars are negative, indicating a lack of power supply from the PV power plant during these hours; nevertheless, the cumulative negative energy balance is partially or completely compensated for by the hours when the PV power supply exceeds the energy demand from the charging station.

We analyzed the daily energy balance compensation for each month, and we found that only in January and December were the monthly energy balance results negative; however, if we extend the energy balance compensation throughout the year, the result becomes positive, indicating the system is energetically feasible provided we implement a storage unit to compensate for the negative and positive monthly energy balance.

11. Energy Storage Engineering Design and Sizing

We have only two months with a negative energy balance, December and January; therefore, we determine the cumulative energy balance for these two months. Because the two months are consecutive, the storage unit energy capacity should match the cumulative energy balance.

Using data from the simulation, the cumulative energy balance for the two months is −952.7 kWh. The battery capacity is given by the following formula [48]:

$$C_n = \frac{\xi}{V_{bat}} \tag{11}$$

C_n, V_{bat}, and ξ are the battery nominal capacity, operating voltage, and energy capacity, respectively.

Considering that the battery operates at the reference voltage for the system, 400 V, the nominal capacity is as follows:

$$C_n = \frac{952.7 \times 10^3}{400} = 2381.75\ Ah \tag{12}$$

This capacity corresponds to a discharge current of 119 A, matching a discharge rate at 20 h.

The battery capacity, however, depends on the discharge rate according to the following expression [49]:

$$C_r = f_C C_n \tag{13}$$

C_r is the current battery capacity for a given discharge rate, and f_C is the capacity correction factor, which is expressed as follows:

$$f_C = \left(\frac{C_n V_{bat}}{20 P_t}\right)^b \tag{14}$$

P_t is the power demand, and b is a coefficient that depends on the type of battery. For lithium-ion batteries, b = 0.0148, and for lead–acid ones, b = 0.1701 [50].

Retrieving data from the simulation for the maximum negative power demand, we obtain the following:

$$f_C = \begin{cases} \left[\frac{(2381.75)(400)}{20(363.3\times 10^3)}\right]^{0.0148} = 0.970 \rightarrow Li-ion \\ \left[\frac{(2381.75)(400)}{20(363.3\times 10^3)}\right]^{0.1701} = 0.708 \rightarrow Pb \end{cases} \tag{15}$$

Therefore, the corrected battery capacity for the storage unit is as follows:

$$C_n = \begin{cases} \frac{2381.75}{0.970} = 2455.4\ Ah\ (Li-ion) \\ \frac{2381.75}{0.708} = 3364.1\ Ah\ (Pb) \end{cases} \tag{16}$$

The lead–acid battery capacity is 37% higher; however, the lithium-ion battery costs 40% more, compensating for it. We therefore base the selection on maintenance and replacement costs, where lithium-ion batteries are cheaper; our selection is a lithium-ion battery block of 2500 Ah at 400 V.

No single cell fulfills the mentioned capacity and voltage; therefore, we design the battery block by grouping lithium-ion cells in series and parallel to fit the required values. The commercial units may reach up to 22.1 kWh per battery block, grouping cells of 3 Ah and 4.2 volts per cell. The battery block configuration consists of 19 parallel strings of 95 cells for a total energy capacity of 22.7 kWh. The final battery layout consists of 16 blocks of 22.7 kWh coupled in parallel to reach the required energy capacity of 363.3 kWh for negative energy balance compensation.

Figures 18 and 19 show the engineering design of the battery block and storage unit.

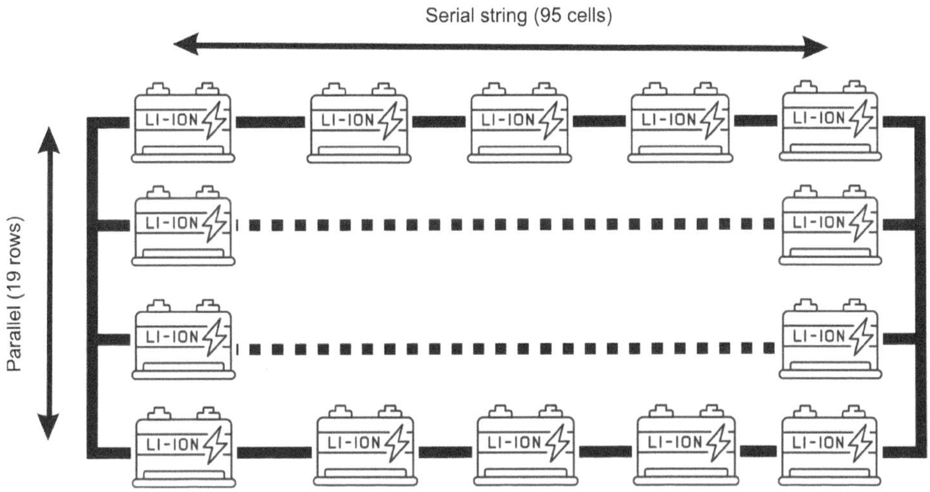

Figure 18. Schematic view of the battery block.

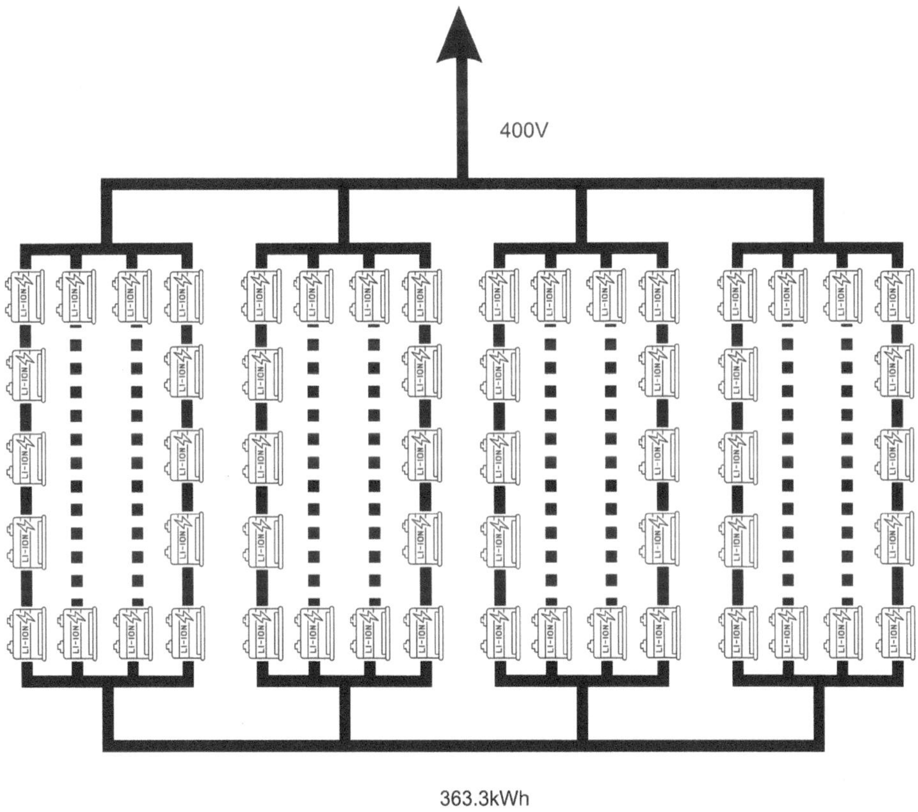

Figure 19. Engineering design of the storage unit.

12. Control Unit

The charging station operates automatically without on-site operator supervision. Users manipulate the charging socket following instructions specifically addressed to guide the users to recharge the electric vehicle according to their needs. The charging protocol appears on the charging pole screen, where the users may select the charging mode and the charging parameters like the driving range, charging time, charging rate, and operational voltage.

The charging process develops according to one of the following parameters to be selected by the user: charging time, specific driving range, or battery capacity percentage. If the parameter does not correspond to electric vehicle battery characteristics, the protocol shows an error and asks the user again to introduce a valid value for the selected parameter.

Once the user selects the parameter and introduces a valid value, the protocol asks the user to choose the charging power depending on the electric vehicle battery voltage (see Table 7). The protocol calculates the charging parameters depending on the charging rate and shows the relevant charging process parameter on the charging pole screen for the user's information.

Figure 20 shows the protocol that controls the charging process.

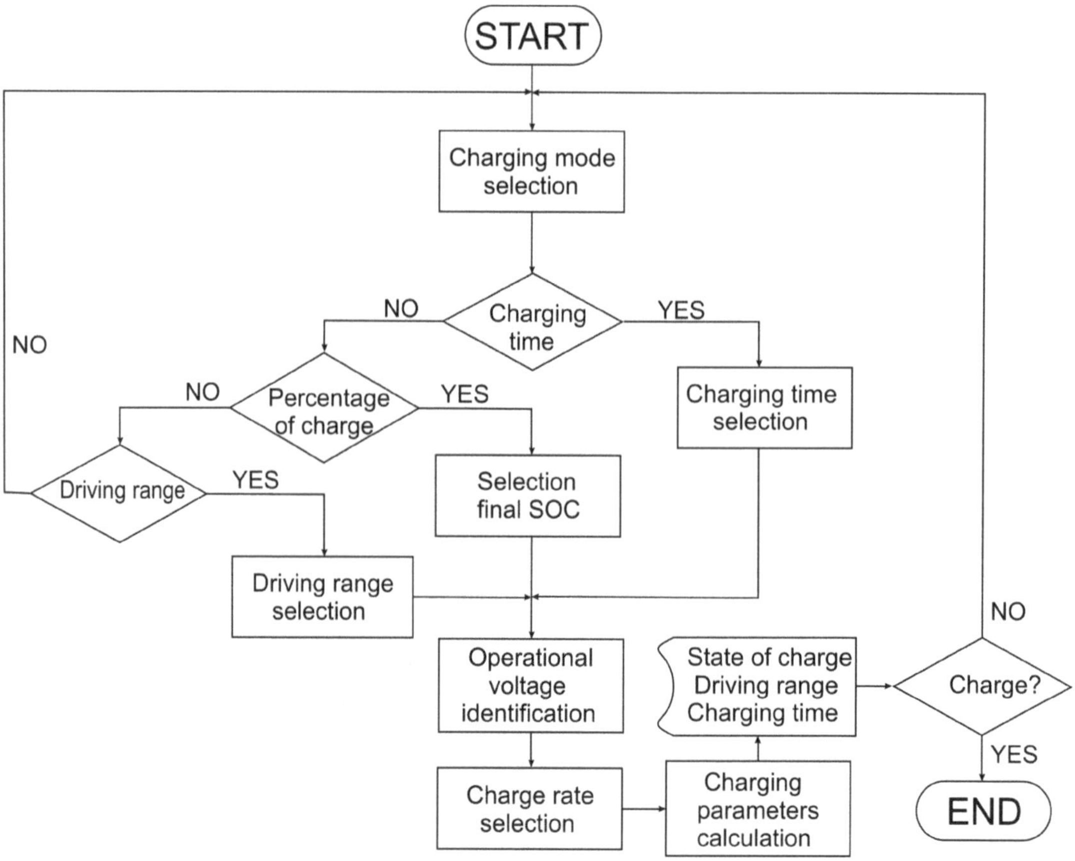

Figure 20. Charging process protocol flow rate.

13. Life Cycle Cost

The feasibility of the proposed installation depends on the economic reliability; with this goal in mind, we develop a Life Cycle Cost (LCC) analysis to prove the project's viability. The LCC process is based in the software developed by the Alaska Department of Education & Early Development [51]. Table 10 shows the LCC development.

Table 10. Life Cycle Cost analysis of the proposed system.

Concept	Cost (EUR)
Initial investment	298,455
Operations	44,223
Maintenance and repair	92,130
Replacement	31,576
Residual value	0
Total LCC	466,384

Considering a 25-year exploitation, the yearly revenue represents EUR 18,655, or EUR 1555 per month, an amount that is perfectly feasible.

Details of the initial investment costs are as follows:

Concept	Cost (EUR)
Construction management	4875
Land acquisition	94,674
Site investigation	1020
Design services	5100
Construction	29,250
Equipment	110,088
Technology	17,049
Indirect/administration	6250
Art	17,649
Contingency	12,500
TOTAL	298,455

The internal rate derives from the following expression:

$$F_{RC} = \frac{r}{\left[1 - (1+r)^{-N}\right]} \quad (17)$$

r is the yearly interest rate, and N the investment payback time.

Considering an average interest rate of 3%, the internal rate as a function of the payback time is as follows (Table 11):

Table 11. Yearly internal rate as a function of investment payback time.

N	1	2	3	4	5	6	7	8	9	10	11	12	13	14	15
F_{RC}	1.03	0.52	0.35	0.27	0.22	0.18	0.16	0.14	0.13	0.12	0.11	0.10	0.09	0.09	0.08

Using the F_{RC} value, the required electricity cost for the payback time is as follows:

$$\zeta_{el} = \frac{C_c F_{RC} + C_{OM}}{\xi} \quad (18)$$

C_c is the investment cost, C_{OM} is the operation and maintenance costs, and ξ is the yearly energy demand.

Applying data for our system, we obtain the following (Table 12):

Table 12. Electricity cost vs. payback time.

N	2	3	4	5	6	7	8
ζ_{el}	0.1764	0.1511	0.1481	0.1540	0.1644	0.1773	0.1918
N	9	10	11	12	13	14	15
ζ_{el}	0.2074	0.2237	0.2406	0.2579	0.2754	0.2933	0.3028

On the other hand, considering the average price of electric energy, 0.1781 EUR/kWh [52], the investment payback time is around seven years, which we consider a suitable value for an investment since the system lifespan is 25 years.

14. Conclusions

Using renewable energy sources to supply electric vehicles makes this transport solution sustainable, resilient, and future-proof. The design proposed in this work covers current and future standards for electric cars, also making it possible to replicate the design in different locations, increasing or decreasing the number of stations and the generator to meet demand based on the expected influx and the solar resource present at the location.

In places where solar resources are high and with great potential for use in electric vehicle charging infrastructure, this work demonstrates the feasibility of their use as an alternative to traditional energy sources, with CO_2 emissions, which are increasingly falling into disuse.

The modular design allows for expansions, either in the number of stations or in the supply power, at a relatively low cost, making this installation especially attractive.

The independence from external sources, battery packs, or grid connection allows this type of installation to adapt to areas where access to the grid is complex, expensive, or impossible, as well as in areas where the power supply from the grid is subject to periodic failures or interruptions.

This independence also makes expansions of the installation economically feasible because they do not require the expansion of the supply network, which in many cases limits the installation or expansion of renewable energies. The greater complexity of systems connected to the grid with or without a storage unit makes them more prone to failure and requires higher maintenance.

For all the above reasons, electric vehicle charging stations powered by photovoltaic solar energy are a solution for the present and the future, not requiring a grid connection to build an installation suitable for present and future electric mobility standards.

The LCC analysis shows the economic viability of the proposed project.

Author Contributions: Conceptualization, C.A.-D.; Methodology, C.A.-D. and L.S.; Software, L.S.; Validation, C.A.-D.; Formal analysis, C.A.-D. and L.S.; Investigation, L.S.; Data curation, L.S.; Writing—original draft, C.A.-D.; Visualization, L.S.; Supervision, C.A.-D. All authors have read and agreed to the published version of the manuscript.

Funding: This research received no external funding.

Informed Consent Statement: Not applicable.

Data Availability Statement: The original contributions presented in the study are included in the article, further inquiries can be directed to the corresponding author.

Conflicts of Interest: The authors declare no conflict of interest.

References

1. Albuquerque, F.D.; Maraqa, M.A.; Chowdhury, R.; Mauga, T.; Alzard, M. Greenhouse gas emissions associated with road transport projects: Current status, benchmarking, and assessment tools. *Transp. Res. Procedia* **2020**, *48*, 2018–2030. [CrossRef]
2. Andrés, L.; Padilla, E. Driving factors of GHG emissions in the EU transport activity. *Transp. Policy* **2018**, *61*, 60–74. [CrossRef]
3. Casals, L.C.; Martinez-Laserna, E.; García, B.A.; Nieto, N. Sustainability analysis of the electric vehicle use in Europe for CO2 emissions reduction. *J. Clean. Prod.* **2016**, *127*, 425–437. [CrossRef]
4. Taljegard, M.; Thorson, L.; Odenberger, M.; Johnsson, F. Large-scale implementation of electric road systems: Associated costs and the impact on CO2 emissions. *Int. J. Sustain. Transp.* **2020**, *14*, 606–619. [CrossRef]
5. Sudjoko, C.; Sasongko, N.A.; Utami, I.; Maghfuri, A. Utilization of electric vehicles as an energy alternative to reduce carbon emissions. *IOP Conf. Ser. Earth Environ. Sci.* **2021**, *926*, 012094. [CrossRef]
6. Li, P.; Xia, X.; Guo, J. A review of the life cycle carbon footprint of electric vehicle batteries. *Sep. Purif. Technol.* **2022**, *296*, 121389. [CrossRef]
7. Xia, X.; Li, P.; Xia, Z.; Wu, R.; Cheng, Y. Life cycle carbon footprint of electric vehicles in different countries: A review. *Sep. Purif. Technol.* **2022**, *301*, 122063. [CrossRef]
8. Muha, R.; Perosa, A. Energy consumption and carbon footprint of an electric vehicle and a vehicle with an internal combustion engine. *Transp. Probl.* **2018**, *13*, 49–58. [CrossRef]
9. Reed, G.F.; Grainger, B.M.; Sparacino, A.R.; Kerestes, R.J.; Korytowski, M.J. Advancements in medium voltage DC architecture development with applications for powering electric vehicle charging stations. In Proceedings of the 2012 IEEE Energytech, Cleveland, OH, USA, 29–31 May 2012; pp. 1–8.
10. Danese, A.; Torsæter, B.N.; Sumper, A.; Garau, M. Planning of high-power charging stations for electric vehicles: A review. *Appl. Sci.* **2022**, *12*, 3214. [CrossRef]
11. Singh, A.; Letha, S.S. Emerging energy sources for electric vehicle charging station. *Environ. Dev. Sustain.* **2019**, *21*, 2043–2082. [CrossRef]
12. Eze, V.H.U.; Eze, M.C.; Ogbonna, C.C.; Ugwu, S.A.; Emeka, K.; Onyeke, C.A. Comprehensive Review of Recent Electric Vehicle Charging Stations. *Glob. J. Sci. Res. Publ.* **2021**, *1*, 16–23.
13. Reche, C.; Querol, X.; Alastuey, A.; Viana, M.; Pey, J.; Moreno, T.; Quincey, P. New considerations for PM, Black Carbon and particle number concentration for air quality monitoring across different European cities. In *Air Quality*; Apple Academic Press: Palm Bay, FL, USA, 2016; pp. 203–244.
14. Molina, M.J.; Molina, L.T. Megacities and atmospheric pollution. *J. Air Waste Manag. Assoc.* **2004**, *54*, 644–680. [CrossRef] [PubMed]
15. Colls, J.; Tiwary, A. *Air Pollution: Measurement, Modelling and Mitigation*; CRC Press: Boca Raton, FL, USA, 2017.
16. Mammonas, D. Infrastructure for the Alternative Fuels: The Council Adopts New Regulations to Increase Charging Stations and Refueling in Europe. European Union Council, 25 July 2023. Available online: https://www.consilium.europa.eu/es/press/press-releases/2023/07/25/alternative-fuels-infrastructure-council-adopts-new-law-for-more-recharging-and-refuelling-stations-across-europe/ (accessed on 2 September 2024).
17. Inactive Electric Vehicle Charging Points in Spain as of the Third Quarter of 2023, by Autonomous Community. Transportation & Logistics. Vehicles & Road Traffic. Statista. Available online: https://www.statista.com/statistics/1446152/spain-inactive-ev-chargers-by-autonomous-community/ (accessed on 2 September 2024).
18. Foley, A.M.; Winning, I.J.; Gallachóir, B.Ó.Ó. State-of-the-art in electric vehicle charging infrastructure. In Proceedings of the 2010 IEEE Vehicle Power and Propulsion Conference, Lille, France, 1–3 September 2010; pp. 1–6.
19. ANFAC. Available online: https://anfac.com/categorias_publicaciones/informe-anual/ (accessed on 2 September 2024).
20. IEA. *Global EV Outlook 2022*; IEA: Paris, France, 2022; Available online: https://www.iea.org/reports/global-ev-outlook-2022 (accessed on 2 September 2024).
21. *SAE J1772*; SAE Electric Vehicle and Plug in Hybrid Electric Vehicle Conductive Charge Coupler J1772_201710 (DOC). SAE International: Warrendale, PA, USA, 13 October 2017.
22. *Basics of SAE J1772*; Open EVSE: Monroe, NC, USA. Available online: https://openev.freshdesk.com/support/solutions/articles/6000052074-basics-of-sae-j1772 (accessed on 1 September 2024).
23. *IEC 61851-1*; IEC 61851-1: Electric Vehicle Conductive Charging System-Part 1: General Requirements. IEC: London, UK, 2010; Chapter 2. pp. 1–99.
24. SAE J1772. Available online: https://www.sae.org/standards/content/j1772_201710/preview/ (accessed on 1 September 2024).
25. IEC 62196-2. Available online: https://webstore.iec.ch/en/publication/64364 (accessed on 1 September 2024).
26. IEC 61851-22/23. Available online: https://webstore.iec.ch/en/publication/32973 (accessed on 1 September 2024).
27. GB/T 20234-2. Available online: https://www.chinesestandard.net/PDF.aspx/GBT20234.2-2015 (accessed on 1 September 2024).
28. IEC 62196-3. Available online: https://webstore.iec.ch/en/publication/59923 (accessed on 1 September 2024).

29. IEC 62196. Available online: https://webstore.iec.ch/en/publication/59922 (accessed on 1 September 2024).
30. SAE J2839/2 & SAE J2847/2. Available online: https://www.sae.org/standards/content/j2839_202001/ (accessed on 1 September 2024).
31. Acharige, S.S.G.; Haque, M.E.; Arif, M.T.; Hosseinzadeh, N.; Hasan, K.N.; Oo, A.M.T. Review of Electric Vehicle Charging Technologies, Standards, Architectures, and Converter Configurations. *IEEE Access* **2023**, *11*, 41218–41255. [CrossRef]
32. EV Charging Connector Types: A Complete Guide. Electric Vehicle Energy Storage Company (EVESCO). Available online: https://www.power-sonic.com/blog/ev-charging-connector-types/ (accessed on 2 September 2024).
33. PHOTOVOLTAICGEOGRAPHICALINFORMATIONSYSTEMPVGIS European Commission (01/03/2022). Available online: https://re.jrc.ec.europa.eu/pvg_tools/es/ (accessed on 25 February 2024).
34. García-Badell, J.J. Cálculo de la Energía Solar. (IGME, Ed.). 1982. Available online: https://books.google.es/books?hl=es&lr=&id=P_bLE_m-ik4C&oi=fnd&pg=PA7&dq=calculo+temperatura+operacion+celula+fotovoltaica&ots=d1bOUWXGU9&sig=sNXlLXsQY5jcWw4Lu4IAi2XsgmE#v=onepage&q=calculo%20temperatura%20operacion%20celula%20fotovoltaica&f=false (accessed on 26 April 2024).
35. ATERSA. Available online: https://atersa.shop/app/uploads/2023/12/MU-M6M-6x20-M7-10BB-GS-ES-A-440-460W_460W-1.pdf (accessed on 18 August 2024).
36. Meteorological State Agency. AEMET. Available online: https://www.aemet.es/es/serviciosclimaticos/datosclimatologicos/valoresclimatologicos (accessed on 29 August 2024).
37. Luque, A.; Hegedus, S. (Eds.) *Handbook of Photovoltaic Science and Engineering*; John Wiley & Sons: Hoboken, NJ, USA, 2011.
38. Luis Vilariño García, J.V. Rendimiento Global de Sistemas Fotovoltaicos Conectados a la Red Electrica. *Tecnologi@ y Desarrollo* **2014**, *XII*, 20–24.
39. INGECON® SUN 330TL M12. Ingeteam. Electrifying a Sustainable Future. Available online: https://www.ingeteam.com/es-es/sectores/energia-fotovoltaica/p15_24_727/330-350tl-m12.aspx (accessed on 1 September 2024).
40. Blinov, I.; Zaitsev, I.O.; Kuchanskyy, V.V. Problems, methods and means of monitoring power losses in overhead transmission lines. In *Systems, Decision and Control in Energy I*; Springer International Publishing: Cham, Switzerland, 2020; pp. 123–136.
41. Harting, C. Ac Transmission Line Losses. 2010. Available online: http://large.stanford.edu/courses/2010/ph240/harting1/ (accessed on 25 January 2024).
42. Wang, H.; Wang, L.; Wang, Y.; Xue, H.; Yang, C.; Yan, T. The electric energy loss in overhead ground wires of 110kV six-circuit transmission line on the same tower. In Proceedings of the IEEE PES Innovative Smart Grid Technologies, Tianjin, China, 21–24 May 2012; pp. 1–5.
43. How Much Power Loss in Transmission Lines, C.H.N.T. Empower the World. Available online: https://chintglobal.com/blog/how-much-power-loss-in-transmission-lines/ (accessed on 2 September 2024).
44. Wirfs-Brock, J. Lost In Transmission: How Much Electricity Disappears Between A Power Plant And Your Plug? Inside Energy. IE. Available online: https://insideenergy.org/2015/11/06/lost-in-transmission-how-much-electricity-disappears-between-a-power-plant-and-your-plug/ (accessed on 2 September 2024).
45. Trends in Electric Vehicle Charging. International Energy Agency (IEA). Global EV Outlook 2024. Available online: https://www.iea.org/reports/global-ev-outlook-2024/trends-in-electric-vehicle-charging (accessed on 2 September 2024).
46. Ma, T.Y.; Faye, S. Multistep electric vehicle charging station occupancy prediction using hybrid LSTM neural networks. *Energy* **2022**, *244*, 123217. [CrossRef]
47. Douaidi, L.; Senouci, S.-M.; El Korbi, I.; Harrou, F. Predicting Electric Vehicle Charging Stations Occupancy: A Federated Deep Learning Framework. In Proceedings of the VTC2023-Spring, Florence, Italy, 20–23 June 2023; pp. 1–5. [CrossRef]
48. Linden, D. Handbook of batteries. *Fuel Energy Abstr.* **1995**, *4*, 265.
49. Armenta-Deu, C. Prediction of battery behaviour in SAPV applications. *Renew. Energy* **2003**, *28*, 1671–1684. [CrossRef]
50. Armenta-Deu, C.; Carriquiry, J.P.; Guzman, S. Capacity correction factor for Li-ion batteries: Influence of the discharge rate. *J. Energy Storage* **2019**, *25*, 100839. [CrossRef]
51. Life Cycle Cost Software. LCC Analysis Worksheet Tool. Alaska Department of Education & Early Development, 2018. Available online: https://www.google.com/url?sa=t&rct=j&q=&esrc=s&source=web&cd=&ved=2ahUKEwix9JOfw7OJAxV0fKQEHeHyJB0QFnoECBYQAQ&url=https://education.alaska.gov/facilities/publications/LCCWorksheet.xlsx&usg=AOvVaw2McMLqBFYU4kg6xDt9PmYZ&opi=89978449 (accessed on 15 December 2023).
52. Price of Electricity by y 18/11 Per Hour and Yearly Trend. Precio de la Luz (kWh) Hoy Por Horas 18/11/24 | Comparadorluz. Available online: https://comparadorluz.com/tarifas/precio-kwh#:~:text=El%20precio%20medio%20de%20la%20luz%20hoy%20en,20-21%20horas%20con%20un%20precio%20de%200,2482%20%E2%82%AC/kWh (accessed on 18 November 2024).

eng

Article

Understanding the Musculoskeletal Demand of Ride-On Mowing Using Wearable Technology

Kevin Netto [1,*], Garry Francis-Pester [2], Peter Benazic [3] and Peter Edwards [1]

1 Curtin School of Allied Health, Curtin University, Perth, WA 6102, Australia; peter.edwards1@curtin.edu.au
2 Physiotherapy and Human Movement Clinic, Melbourne, VIC 3056, Australia; phmc.gfp@gmail.com
3 Cardinia Shire Council, Cardinia, VIC 3810, Australia; p.benazic@cardinia.vic.gov.au
* Correspondence: kevin.netto@curtin.edu.au

Abstract: This study aimed to quantify the postures and muscle activity while parks and gardens workers operated ride-on mowers during a typical shift. Eight participants operated ride-on mowers in the same park but on different terrains (flat and undulating). Body postures and muscle activity were collected wirelessly and unobtrusively. Participants adopted a forward-flexed seated posture with the predominant movement being head rotation. Oscillatory movements (20–40° from neutral) of the thorax in all three planes of movement were noted in all participants. Low levels (<30% MVIC) of muscle activity were recorded in all muscles tested. These levels were elicited for most (>90%) of the recording time. Higher (>50% MVIC) activation levels were interspersed through the data, but these were not sustained. There was no difference in posture or muscle activity between the flat and undulating terrain. The forward-flexed posture combined with vibration can increase the risk of discomfort and injury in the low back while ride-on mowing. The low levels of muscle activity suggest participants did not actively brace for the occupational situation and task. The large inter-participant difference in posture attests to subjective variation to accommodate muscular stress, and this may not be optimal for injury mitigation.

Keywords: musculoskeletal injury risk; occupational safety; biomechanical assessment

Citation: Netto, K.; Francis-Pester, G.; Benazic, P.; Edwards, P. Understanding the Musculoskeletal Demand of Ride-On Mowing Using Wearable Technology. *Eng* **2024**, *5*, 3108–3114. https://doi.org/10.3390/eng5040162

Academic Editors: Antonio Gil Bravo and Huanyu Cheng

Received: 30 September 2024
Revised: 13 November 2024
Accepted: 25 November 2024
Published: 27 November 2024

1. Introduction

Ride-on mowers are often used in agricultural, landscaping and ground maintenance work, while the public also use such machines in domestic settings [1]. Research into the safe use of ride-on mowers has focused on roll over and amputation injuries as well as noise and vibration exposure [1,2]. A US epidemiological investigation showed nationally, more than 80,000 presentations to the emergency department were recorded yearly for the period between 2005 and 2015 [3]. These data do not clearly differentiate between push or ride-on mowers; however, the most common injuries reported were laceration (23.1%) followed by musculoskeletal strain and sprain injuries (18.8%) [3]. The risk of musculoskeletal injury in mowing has been examined via vibration analysis [4] but the kinematics and muscle activity while operating ride-on mowers has yet to be quantified. This leaves a gap in our knowledge of the total risk of using these machines.

The paucity of information regarding the musculoskeletal demands of ride-on mower use may, in part, be due to the challenges of performing ergonomic and biomechanical analyses. Traditionally, these forms of analyses would be performed in a laboratory setting or with outdoor, optical motion capture cameras. This setup limits our ability to perform the analyses with environmental fidelity (when in a laboratory) or with only limited capture volumes (when using outdoor cameras).

Wireless inertial measurement units (IMUs) and surface electromyography (sEMG) technology provides the opportunity to assess kinematics and muscle activity unobtrusively, and in a work-specific context. IMUs contain a combination of three-dimensional accelerometers, rate gyroscopes, inclinometers and a magnetometer. They can be used singularly to assess the orientation of an object they are affixed to or as an array of interconnected sensors where the measurement of human segment kinematics is enabled. In terms of their validity in biomechanical studies, spinal kinematics from IMUs have been shown to be accurate compared to traditional optoelectronic motion capture [5,6]. sEMG has been commonly used to assess muscle activity in human movement. The wireless advancement of these sensors has allowed the application of this technology to a variety of more previously challenging environments such as in firefighting, during rapid sporting movements, and in long-duration analysis such as office desk use. Recently, in combination, the suite of IMUs and sEMG has been used to quantify the posture and muscle activity of waste drivers [7], giving invaluable insight into a challenging job task while also not interfering with the natural process of the work.

Considering the gaps in our knowledge with ride-on mowing, especially in the area of musculoskeletal demand, this project aimed to quantify the specific spinal and upper body postures as well as selected muscle activity when participants operated ride-on mowers during a typical mowing session (30 min). This information has the potential to increase the understanding of the work demand of ride-on mowing, allowing interventions to mitigate accident and injury to be formulated.

2. Materials and Methods

2.1. Participants

Eight participants (2 female; 6 male; mean ± SD; age: 35 ± 8 years; mass: 82 ± 12 kg; height: 1.73 ± 0.1 m; experience: 3 ± 2.3 years) volunteered for this study. Participants were recruited from a pool of parks and gardens workers within a local government organisation and were included if they presented with no current or recent (within six months) musculoskeletal injury and were trained in the operation of a ride-on mower. Participants in the study gave informed consent prior to the commencement of any data collection. Ethical approval (RDHS-62-15) was obtained from the Institutional Human Research Ethics Committee prior to all participant recruitment and testing procedures.

2.2. Data Collection

An initial observation of the task revealed marked differences in the types of terrains mowed. Thus, it was decided to conduct the analysis in a single location (Remembrance Park, Banksia & Studley Streets, Heidelberg, Victoria, Australia) where both flat and hilly terrains are available to be mowed. A cross-sectional study design was used to examine three-dimensional spinal and shoulder postures and muscle activity from three muscles while participants operated the ride-on mower. Participants performed two types of mows, a flat terrain mow and an undulating terrain mow in the same park. All participants wore their typical work wear which comprising self-supplied overalls, safety boots, and eye and ear protection. They also used their council-supplied ride-on mower (Kubota F3690SN-72, Osaka, Japan). No familiarisation trial was conducted as participants were well versed with their equipment and the task.

Wireless IMUs (Noraxon Myomotion, Noraxon, Scottsdale, AZ, USA) was used to analyse the postures and joint angles associated with operating the mowers. An array of six wireless and lightweight sensors were attached to the participants' head (on the occiput in line with the midline of the body), over C7, over T12, on the left and right upper arms (lateral aspect) and on the pelvis (over L5/S1) according to manufacturer specifications and using manufacturer-supplied straps with pouches or double-sided tape. All kinematic data were collected at 100 Hz.

From the array of IMUs, a selection of joint angles was extracted using manufacturer-supplied software (MR 3.10, Noraxon, Scottsdale, AZ, USA). This included flexion/extension, lateral flexion, and axial rotation angles of the lumbar, thoracic and cervical (head and neck) spine. These data were then processed to obtain an average posture adopted during the data collection period. Further, a range of movement (ROM) value for all the joint angles was obtained by calculating the absolute range of the joint from the peak excursion adopted during the capture period.

Muscle activation from three specific muscles was measured using wireless surface electromyography (sEMG) (Noraxon DTS, Noraxon, Scottsdale, AZ, USA) at 1500 Hz. The muscles of interest were *m. flexor carpi radialis, m. upper trapezius and m. lumbar erector spinae*. All sEMG readings were measured unilaterally from the right side of the body as all participants reported they were right-handed. The locations were found using industry-standard guidelines to ensure accurate and consistent placement [8] (Figure 1).

Figure 1. Screen capture from data collection depicting participant on a ride-on mower with IMUs and sEMG attached.

Each muscle site was prepared to ensure a clean and accurate sEMG signal could be recorded. This involved the area being shaved, gently abraded and wiped with a sterile swab. Once all sEMG equipment was affixed and tested, each participant completed a maximal voluntary isometric contraction (MVIC) of each of the muscles being recorded. This process was completed by pulling on an anchored, inextensible exercise strap (TRX, Fitness Anywhere LLC, San Francisco, CA, USA) in specific reference movements according to published guidelines [8].

All sEMG signals were full-wave rectified and root-mean squared smoothed using a 200 msec window [7]. A peak value was obtained from the MVIC trials, and these were used to normalise the trials for each participant. To further understand the muscle activity, average and peak normalised sEMG were calculated using manufacturer-supplied software (MR 3.10, Noraxon, Scottsdale, AZ, USA). Also, muscle activity data were segregated into the time spent in a specific zone. For this study, we performed a secondary time-in-zone analysis using the amplitude distribution profile of <30% MVIC, 30–50% MVIC and >50% MVIC. These data showed how much time participants were spending in each activation zone. sEMG processing procedures were adopted from previous work investigating muscle activity while driving waste vehicles [7].

2.3. Data Analysis and Statistics

All data were segregated into the terrain mowed where group averages and standard deviations were obtained and reported. As the sEMG data appeared quite varied, we also calculated interquartile ranges for these. To answer the question of whether terrain caused a change in posture or muscle activity, we used a one-way ANOVA to contrast the average postures, ROM, average and peak normalised sEMG and time in specific sEMG zones (Microsoft Excel 2013, Microsoft, Redmond, WA, USA). Statistical significance was set at $p \leq 0.05$.

3. Results

The average posture analysis revealed participants adopted forward flexion posture originating at the lumbar spine (45° lumbar flexion; Figure 2). The range of motion analysis showed the largest range of movement was cervical axial rotation with participants twisting their head over 70° to both the left and right during the data collection period (Figure 3). Much smaller (22–42°) ranges were recorded in the other spinal segments and planes of motion (Table 1). The ANOVA revealed no significant differences in average postures or ROM when these comparisons were made between the flat and hilly terrains.

Figure 2. Screen capture from data collection depicting forward-flexed posture.

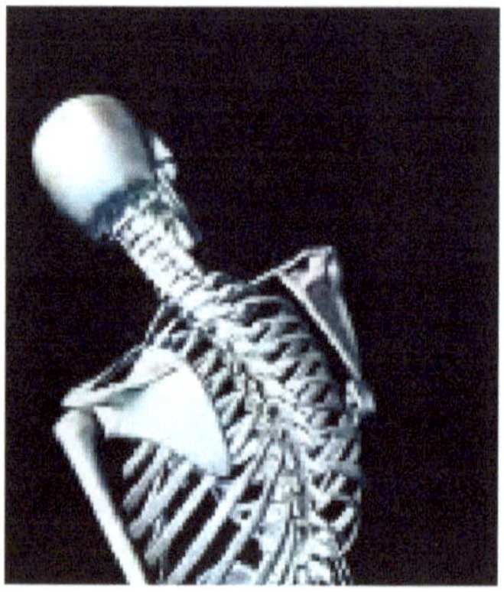

Figure 3. Screen capture from data collection depicting head rotation.

Table 1. Average (±SD) and range (±SD) of spinal postures adopted while ride-on mowing.

	Cervical			Thoracic			Lumbar		
	Flexion	**Lateral Flexion**	**Axial Rotation**	**Flexion**	**Lateral Flexion**	**Axial Rotation**	**Flexion**	**Lateral Flexion**	**Axial Rotation**
Average Posture (°)	−5.6 (±8.2)	2.5 (±13.9)	−1.3 (±44.3)	−6.7 (±16.5)	3.1 (±14)	10.8 (±32.9)	45.6 (±6.2)	5.5 (±6.7)	−2.2 (±19.5)
Range (°)	41.7 (±9.3)	41.1 (±8.1)	141.8 (±13.3)	22.3 (±9.3)	18.8 (±11.3)	35.8 (±9.2)	31.3 (±13.2)	38.1 (±6.2)	34.6 (±11.3)

In our trials, *m. flexor carpi radialis* displayed the highest average (14%MVIC in hilly) and peak muscle activity (98%MVIC in hilly), while m. upper trapezius displayed the lowest average (5% in hilly) and peak (33% in flat) muscle activity (Table 2). There were instances of higher (>50% MVIC) activation recorded in all muscles, but these were not sustained for any length of time. Specifically, most of the muscle activity recorded was <30% MVIC. The time-in-zone analysis revealed that this low level of muscle activity accounted for the majority (>90%) of the assessment time (Table 3). The ANOVA showed no significant differences in mean and peak muscle activation levels between flat and hilly terrains.

Table 2. Normalised sEMG for each of the muscles investigated.

	Average ± SD (%MVIC)		Interquartile Range (%MVIC)		Peak ± SD (%MVIC)		Interquartile Range (%MVIC)	
	Flat	Hilly	Flat	Hilly	Flat	Hilly	Flat	Hilly
Flexor carpi radialis	12 ± 7	14 ± 7	8	4	81 ± 38	98 ± 47	64	40
Upper trapezius	6 ± 3	5 ± 3	4	4	33 ± 22	41 ± 19	11	20
Lumbar erector spinae	12 ± 7	14 ± 4	9	4	65 ± 52	66 ± 15	28	15

Table 3. Percentage of trial time spent in the different muscle activation zones.

	Flexor Carpi Radialis	Upper Trapezius	Lumbar Erector Spinae
<30% MVIC	92 ± 4.5%	99 ± 1.3%	98 ± 3.8%
30–50% MVIC	5 ± 3.6%	0.5 ± 1.1%	1 ± 2.1%
>50% MVIC	3 ± 1.8%	0.5 ± 0.1%	1 ± 1.8%

4. Discussion

The main findings of our study showed that while ride-on mowing, our participants adopted a forward-flexed posture originating at the lumbar spine. They also used large ranges of movement to rotate their heads while mowing. The sEMG readings showed that low-level muscle activity interspersed with short bursts of higher activation were recorded from the trunk and upper limb muscles.

The forward-flexed posture of the trunk recorded in this study is similar to other reports of driving heavy vehicles [7,9]. These postures in isolation are not associated with the risk of developing musculoskeletal injury. However, when factors such as whole-body vibrations, shocks and jolts that come with mowing uneven, bumpy terrain, the risk of musculoskeletal injury may increase [10]. A study of grass harvesting using a tractor showed significant increases in vibration especially where working characteristics were severe [11]. The risk of developing musculoskeletal injury in these situations is further increased if the seat posture and exposure to vibration/undulations are prolonged (>2 h) [12]. Johnson and colleagues [13] showed active suspension seats significantly decreased vibration in heavy trucks compared to air-suspended seats.

The predominant movement recorded in our study was axial rotation of the head about the trunk. Similar kinematics have been recorded in drivers of waste collection vehicles [7] and also in bus drivers [14]. These studies have attributed this movement to the attention demand of the work, and this is also similar in operating ride-on mowers. The mowers used in our study did not have any means of rearward vision. Thus, operators had to rotate their head, sometimes to near end-of-range, to facilitate visual information.

Data from the ride-on mowers suggest participants may have encountered a combination of vibration of the mower as well as jolts from the bumps in the terrain being mowed. The ranges of movement (>30°) in all three planes in the lumbar spine (Table 1) along with the low levels of muscle activity (Tables 2 and 3) recorded suggest participants did not brace for these perturbations [15]. Bracing for sudden perturbations has been shown to reduce intersegmental movement in the lumbar spine [15]. Education and specific practice about bracing may be valuable when operating ride-on mowers, and managers of this workforce may wish to schedule targeted training in this area.

The vibration from the mower and the sudden perturbations from the terrain combined with the pedestal-like nature of the ride-on mower seat may have compromised participants' ability to stabilise their posture. Slota and colleagues [16] showed impaired trunk postural control after whole-body vibration in unstable sitting. Also, given the open nature of the ride-on mower seat, participants would have no opportunity to brace themselves. This can be evidenced from the low levels of muscle activity exhibited while mowing. Manufacturers of ride-on mowers may wish to lower the seat of the mower and better design seats to allow participants to brace themselves.

There was a large inter-participant difference in seated postures, and this can also be attributed to the vibration and undulation of the terrain and mower [17]. More specifically, the oscillatory movement in the spinal kinematics measured during mowing may be attributed to these factors.

This study used wearable sensors that allowed field-based kinematic and electromyographic analysis in a task that would have been impossible to simulate in a laboratory. Our study was limited by the small and homogenous nature of our sample. We also did not consider individual participant characteristics such as age, experience, sex and body mass index in our analysis. Further, the design of our study does not allow causative factors of injury to be assessed. Future investigations may wish to consider these limitations and study this important workforce in a longitudinal manner such that factors associated with increase injury risk can be ascertained.

Further research in ride-on mowers could examine whole-body vibration or attempt to manipulate standard fixtures on the mower and test the resulting changes in musculoskeletal load. Also, future designs of ride-on mowers may wish to incorporate mirrors or rearward facing cameras to enhance operator visual information and minimise the frequency of cervical rotations.

5. Conclusions

Our study shows operators of ride-on mowers are faced with several stressors that may impact their musculoskeletal system. Solutions such as active damping seats, rear vision, targeted training strategies and appropriately scheduled breaks can have a positive impact on the health and well-being of this vital work group. Our findings were limited by a small and homogenous sample, and the design of the study did not allow injury causation to be examined. Future research may wish to focus on these as well as examine the role whole-body vibration has on injury in this vital cohort of workers.

Author Contributions: Conceptualization, K.N. and G.F.-P.; methodology, K.N., G.F.-P., P.B. and P.E.; formal analysis, K.N., G.F.-P., P.B. and P.E.; investigation, K.N.; resources, P.B.; writing—original draft preparation, K.N.; writing—review and editing, G.F.-P., P.B. and P.E.; project administration, K.N.; funding acquisition, K.N. All authors have read and agreed to the published version of the manuscript.

Funding: This research was funded by Banyule City Council, Banyule, Australia. Grant number RES-56499.

Institutional Review Board Statement: The study was conducted in accordance with the Declaration of Helsinki, and approved by the Human Research Ethics Committee of Curtin University (protocol code RDHS-62-15 on 21 April 2015).

Informed Consent Statement: Informed consent was obtained from all subjects involved in the study.

Data Availability Statement: The datasets presented in this article are not readily available because technical limitations. Requests to access the datasets should be directed to kevin.netto@curtin.edu.au.

Acknowledgments: The authors would like to acknowledge Caleb Lewis and Patrina Dunnill for their assistance in data collection and processing.

Conflicts of Interest: The authors declare no conflicts of interest.

References

1. Tint, P.; Tarmas, G.; Koppel, T.; Reinhold, K.; Kalle, S. Vibration and noise caused by lawn maintenance machines in association with risk to health. *Agron. Res.* **2012**, *10*, 251–260.
2. Myers, M.L. Ride-On Lawnmowers: The Hazards of Overturning. *Prof. Saf.* **2009**, *54*, ASSE-09-05-52.
3. Harris, C.; Madonick, J.; Hartka, T.R. Lawn mower injuries presenting to the emergency department: 2005 to 2015. *Am. J. Emerg. Med.* **2018**, *36*, 1565–1569. [CrossRef] [PubMed]
4. Park, M.-S.; Fukuda, T.; Kim, T.G.; Maeda, S. Health Risk Evaluation of Whole-body Vibration by ISO 2631-5 and ISO 2631-1 for Operators of Agricultural Tractors and Recreational Vehicles. *Ind. Health* **2013**, *51*, 364–370. [CrossRef] [PubMed]
5. Li, S.S.W.; Chow, D.H.K. Multi-objective analysis for assessing simultaneous changes in regional spinal curvatures under backpack carriage in young adults. *Ergonomics* **2016**, *59*, 1494–1504. [CrossRef] [PubMed]
6. Hölzel, C.; Bengler, K.; Dressel, T. Ergonomic Evaluation of upper limb movements in the automobile production measured by means of motion capturing. In Proceedings of the 3rd International Digital Human Modeling Symposium DHM, Tokyo, Japan, 20–22 May 2014.
7. Netto, K.; Francis-Pester, G.; Lewis, C.; Dunnill, P.; Darling, R. Posture and Muscle Activity during Waste Collection Work. *Ann. Work Expo. Health* **2021**, *66*, 119–123. [CrossRef] [PubMed]
8. Perotto, A.O. *Anatomical Guide for the Electromyographer: The Limbs and Trunk*; Charles C Thomas Publisher: Springfield, IL, USA, 2011.
9. Lis, A.M.; Black, K.M.; Korn, H.; Nordin, M. Association between sitting and occupational LBP. *Eur. Spine J.* **2007**, *16*, 283–298. [CrossRef] [PubMed]
10. Bovenzi, M. Metrics of whole-body vibration and exposure–response relationship for low back pain in professional drivers: A prospective cohort study. *Int. Arch. Occup. Environ. Health* **2009**, *82*, 893–917. [CrossRef] [PubMed]
11. Kang, T.-H.; Kaizu, Y. Vibration analysis during grass harvesting according to ISO vibration standards. *Comput. Electron. Agric.* **2011**, *79*, 226–235. [CrossRef]
12. Carvalho, D.E.D.; Callaghan, J.P. Spine Posture and Discomfort During Prolonged Simulated Driving with Self-Selected Lumbar Support Prominence. *Hum. Factors* **2015**, *57*, 976–987. [CrossRef] [PubMed]
13. Johnson, P.W.; Zigman, M.; Ibbotson, J.; Dennerlein, J.T.; Kim, J.H. A Randomized Controlled Trial of a Truck Seat Intervention: Part 1—Assessment of Whole Body Vibration Exposures. *Ann. Work Expo. Health* **2018**, *62*, 990–999. [CrossRef] [PubMed]
14. Albert, W.J.; Everson, D.; Rae, M.; Callaghan, J.P.; Croll, J.; Kuruganti, U. Biomechanical and ergonomic assessment of urban transit operators. *Work* **2014**, *47*, 33–44. [CrossRef] [PubMed]
15. Norrie, J.P.; Brown, S.H.M. Brace yourself: How abdominal bracing affects intersegmental lumbar spine kinematics in response to sudden loading. *J. Electromyogr. Kinesiol.* **2020**, *54*, 102451. [CrossRef] [PubMed]
16. Slota, G.P.; Granata, K.P.; Madigan, M.L. Effects of seated whole-body vibration on postural control of the trunk during unstable seated balance. *Clin. Biomech.* **2008**, *23*, 381–386. [CrossRef] [PubMed]
17. Amari, M.; Caruel, E.; Donati, P. Inter-individual postural variability in seated drivers exposed to whole-body vibration. *Ergonomics* **2015**, *58*, 1162–1174. [CrossRef]

eng

MDPI

Article

The Role of Productization in End-To-End Traceability

Janne Harkonen [1,2,*], Javier Mauricio Guerrero Rodriguez [1] and Erno Mustonen [1]

1 Industrial Engineering and Management, Faculty of Technology, University of Oulu, FI-90014 Oulu, Finland
2 International School for Social and Business Studies, Mariborska cesta 7, 3000 Celje, Slovenia
* Correspondence: janne.harkonen@oulu.fi

Abstract: End-to-end traceability offers significant opportunities for product lifecycle visibility, sustainability enhancement, and regulatory compliance in product management. However, it faces challenges in data integration and management, supplier collaboration, cost and complexity, and the sharing of information across the supply chain. Productization refers to the representation of a product and connects commercial and technical aspects to the systemic perspective of product management. This includes a focus on the engineering lifecycle with inherent linkages to product data. The product management perspective, specifically in relation to the connection between end-to-end traceability and the productization concept, has not been extensively studied. This study explores the role of both productization and traceability in the context of end-to-end traceability. It combines an extensive literature review and an empirical example of applying productization logic across company borders to support end-to-end traceability. The key findings indicate that productization logic with a product structure focus can support end-to-end traceability in product management by providing consistency and a foundation for tracking both technical and operational data across the engineering lifecycle of a product. By focusing on productization, companies can overcome traceability challenges and unlock the benefits of end-to-end traceability.

Keywords: traceability; end-to-end traceability; technical traceability; operational traceability; supply chain traceability; PDM/PLM traceability; productization; productisation; product structure; traceability data; product data; product management

Citation: Harkonen, J.; Guerrero Rodriguez, J.M.; Mustonen, E. The Role of Productization in End-to-End Traceability. *Eng* **2024**, *5*, 2943–2965. https://doi.org/10.3390/eng5040153

Academic Editor: Antonio Gil Bravo

Received: 3 October 2024
Revised: 7 November 2024
Accepted: 8 November 2024
Published: 12 November 2024

1. Introduction

End-to-end traceability is essential in product management to ensure quality [1,2], regulatory compliance [3,4], and accountability [5,6] across the entire product lifecycle. This enables companies to manage risks more effectively [3,7], respond quickly to problems [8,9], and meet customer expectations for safety and sustainability [10–12]. Although various studies have elaborated on end-to-end traceability, there is a notable lack of focus on the integration of technical and operational traceability.

Technical traceability, or product data management (PDM)/product lifecycle management (PLM) traceability, relates to the ability to accurately capture and recover technical product design and manufacturing information and how it may have changed over the lifecycle of a product. Specifically, the ability to recall and review information when key decisions are made is crucial [13]. In contrast, *operational traceability*, or supply chain traceability, is the recording of operational data of a specific product and its components' journeys across the value-adding stages of a supply chain network. This type of traceability requires addressing fundamental questions, such as what information needs to be collected/recorded and, crucially, by whom [14]. Companies within a supply network are independent economic actors and either need to voluntarily agree to participate in a supply chain traceability system or are forced to participate by regulatory or other mandatory requirements [15]. Examples of mandatory requirements include legal compliance in the healthcare and pharmaceutical industries [16] and stakeholder power [17] by a major automotive original equipment manufacturer (OEM) [14]. Nevertheless, organisations could benefit from combining technical and operational traceability to address traceability challenges and from working towards end-to-end traceability.

These two traceability systems, technical traceability and operational traceability, are related yet distinct. Both contribute to ensuring traceability over a product's engineering lifecycle. However, they deal with different sets of data and information. The lifecycle spans multiple processes carried out by various actors, posing challenges to general-purpose traceability efforts [18]. Technical traceability focuses on a product's design and development history through PDM/PLM systems [19], whereas supply chain traceability ensures visibility over its manufacturing, logistics, and delivery processes, even if the concepts of traceability and visibility may not be fully interchangeable [20]. Technical traceability data include product design, development, and change data [21,22], such as design documents and computer-aided design (CAD) files, bills of materials (BOMs) and part lists, change orders and revision histories, material specifications, testing and validation data, manufacturing instructions, compliance and certification records, and supplier and component data. On the other hand, operational traceability data involve records of raw material sourcing, manufacturing processes, transportation, storage, and handling activities [14,23,24]. These data may include batch numbers, lot codes, supplier information, production dates, and compliance certifications [25–27]. The combination of technical and operational traceability can be considered a key component of end-to-end traceability, with end-to-end traceability emphasising the integrated flow of data over a product's engineering lifecycle. End-to-end traceability offers potential benefits in areas such as sustainability, product quality, regulatory compliance, supply chain transparency, and informed decision-making; however, its realisation requires addressing significant challenges. A product management-focused approach could serve as a starting point for achieving end-to-end traceability.

Challenges related to product management considerations within companies and traceability include inconsistent product understanding [28], inconsistent product structure [29], unclear or non-defined processes or issues [28], and issues with data and product data ownership [30]. Additionally, batch operating logic, which is often used in manufacturing, does not support linking data to individual units and is a challenge for item-level traceability [31]. The unique identification of components is necessary for item-level traceability [32], which is necessary for identifying the source of any individual component [33]. Scoping traceability requirements and identifying opportunities for end-to-end traceability have also been recognised as significant challenges [34]. A structural approach that enables consistency between products and data can support end-to-end traceability.

Productization can be defined as a concept that focuses on representing a product by aligning and integrating its commercial and technical aspects with systemic product management practices. This approach manages the scope of the offering throughout the engineering lifecycle, ensuring consistency, efficiency, and support for product data.

The significance of product structure and productization has been studied in the context of product management, including product portfolio management [35], product data, business processes and information systems [21], data and fact-based product profitability analysis [29,36], product and supply chain-related data, processes and information systems [22], the engineering lifecycle of products [37], BOM configurations [38], business model orientations [39], and productization strategies [40]. Productization is a concept that helps ensure that products are well articulated and documented with visualised features and a defined structure [41–43]. The primary goal of productization is to ensure that products can be efficiently produced, delivered, sold, purchased, and used. It provides consistency in a product's structural focus, aligning both its commercial and technical representations [44,45]. Productization acknowledges the varying focus along the stages of a product's engineering lifecycle [37] and can be applied to gain control over products to enable processes to perform [46]. Decisions on products and data management affect vital company processes [47]. However, the significance of productization and product structure logic for traceability has been deficiently studied, particularly in the context of combining technical and operational traceability, leading to end-to-end traceability.

This study aims to clarify the role of productization in the end-to-end traceability context by focusing on combining technical and operational traceability. This study specifically focuses on product structure and data as key elements for achieving end-to-end traceability. This study combines an extensive literature review and an empirical example of applying productization logic across company borders to support end-to-end traceability. This aim is described by the following research question (RQ):

RQ. What are the roles and potential of productization and product lifecycle management (PLM) in enabling end-to-end traceability?

2. Methodology

This study combines an extensive literature review and company analyses to clarify end-to-end traceability in conjunction with the application of productization logic. This involves addressing the current reality and related future potential. The literature review covers traceability in general, end-to-end traceability, technical and operational traceability, the traceability of products, productization, product structure logic, and data for traceability. The literature was reviewed by conducting keyword searches using Google Scholar. The keywords included but were not limited to "Productization", "Traceability", "End-to-end traceability", "Product Traceability", "Technical traceability", "Product structure", "Product management", "Operational traceability", "BOM", "Supply chain", "Value chain", "Data management", " Product data", and "Master data". The use of different keywords and their combinations was guided by the identified content. The specific interest was in the link between traceability, productization, product structure, and data. The inclusion criterion was the suitability of the context discussed with the focus of this study. The inclusion of journal articles was favoured over other publication types when possible. Studies without a suitable fit for the chosen focus were excluded. A systematic literature review was not conducted. The content of the included articles was analysed, first through in-article searches and then by reading the articles, first around the relevant content and then the entire article when necessary. Notes were made by linking them to the reference information. The content was organised by themes, and the literature findings were synthesised and distilled. The empirical part of this study contained a real-life exploration of the role of productization in the context of end-to-end traceability. The empirical example involved applying productization logic across company borders to support end-to-end traceability. This included studying the applicability of productization and product structure logic, traceability, product data, and product data management.

The analysis focuses on five companies in a business-to-business environment. The context is a partnered supply chain and projects to develop productization practices and traceability, with a focus on new commercial products and business generation. The focus was on a specific smart windscreen product to which different companies contributed. The product was related to smart transparent surfaces that support digital services with two-directional data and opportunities for infotainment and the display of information for purposes in the automotive sector and elsewhere. The main company is a well-known OEM that provides products to the automotive sector and operates in a highly regulated environment. The third company provides lamination technology, and the fourth and fifth companies provide complementary systems that either provide additional use value or enable functionalities. Cooperation involves motivation for new business generation.

Qualitative research was conducted to focus on productization, traceability management, and related data management. The existing productization focus and traceability management were assessed to construct a combined productization logic for companies to support a combined business perspective and traceability. Relevant data management was also addressed. Semi-structured interviews [48] were used to gain the necessary understanding while allowing interviewees to explain issues as entities. The interviewees were guided to remain within the focus but were allowed to provide responses as entities.

Seven interviews were conducted, with one to three interviewees representing each company. The interviewees included new business development, IT, product, and production managers. Snowball sampling was applied to benefit the interviewees by proposing participants [48], simultaneously ensuring that different groups and key people in a variety of product management and supply chain roles were included. New interviewees were introduced until information saturation was achieved. The utilised data also included internal documentation and relevant information from the data systems. Publicly available materials were also used to understand the offerings. The interviews were recorded and transcribed to enable a thorough analysis and focus on the interview and discussion during the interview sessions. The interview data were analysed manually by generating codes while reading the transcripts. The primary target of the initial coding was to support the finding of complementarities in productizing a combined offering. Traceability practices and relevant data management were also considered. As a result, offerings by five different interlinked companies were productised and a common product structure logic was considered. The constructed productization logic and findings were presented to a focus group for validation purposes. Traceability was then further considered in the context of validated productised offerings and product structure logic. Individual component suppliers were not included in this study.

3. Literature Review

3.1. Benefits and Challenges of End-To-End Traceability

The benefits and challenges of end-to-end traceability vary. For example, *sustainability reporting* is becoming an increasingly timely issue [49] because companies are called on to adopt sustainable behaviour owing to global concerns about the environment and social responsibility [50]. End-to-end traceability enables the tracking of environmental impacts across the product engineering lifecycle, from raw material sourcing to end-of-life disposal [51–53]. In addition, other key areas of ESG criteria (Environmental, Social, and Governance) can be tracked [54,55]. However, the current discussion is extremely limited in terms of addressing the data perspective on sustainability. Sustainability-related data enable companies to produce accurate sustainability reports [56] and meet the demands of consumers, regulators, and investors for transparency and accountability [57]. It is necessary to operate based on facts to avoid greenwashing [58]. Companies could benefit from support for complying with environmental regulations, demonstrating commitment to sustainability, and potentially gaining competitive advantage [59,60]. Companies would also gain support in pursuing certifications, such as ISO 14001 [61] or B Corp status, a certification awarded to businesses that meet high standards of social and environmental performance, accountability, and transparency [62].

Improved supply chain transparency enables visibility in supply chain stages [34,63]. This would benefit companies by aiding in identifying inefficiencies and areas for improvement [64]. Better visibility over supplier, manufacturer, and logistics partner activities [65] would allow sustainability and resilience benefits [66] but also support regulatory alignment [3,4,67]. Improved transparency can enhance trust between customers and business partners [68], strengthen supplier relationships [69], and reduce the risk of supply chain disruptions due to non-compliance or unethical practices [15,70].

Another potential benefit is the *improvement in product quality* [1,2] and safety [10–12] through the traceability of components and materials throughout the engineering lifecycle [51–53]. End-to-end traceability helps companies identify defects or quality issues early and trace them back to their source [71]. End-to-end traceability capability is highlighted in regulated industries such as the pharmaceutical, automotive, and food industries [72]. It reduces the likelihood of costly recalls [1], improves product quality [1,2], and enhances customer trust [72].

Support for risk management [3,63] *and regulatory compliance* [6,73] revolve around end-to-end traceability, which provides a full record of a product's journey over its lifecycle [2]. This is highlighted in industries with strict compliance standards [74], such as the automotive and pharmaceutical industries. Traceability can ensure compliance with safety, environmental, or other relevant laws [3,75]. This will reduce the risk of non-compliance penalties [76], improve the ability to respond to audits [77], and increase confidence in the integrity of the product [78], among other benefits.

End-to-end traceability provides *support for circular economy initiatives* [79,80]. Businesses can implement circular economy practices by tracking products and materials throughout their lifecycles [81]. This can involve reuse, refurbishment, and recycling [82,83]. The design for disassembly is a related example that supports tracking the flow of materials to ensure that they are properly recycled or reused [84]. End-to-end traceability can help reduce waste, conserve resources, and support sustainability goals by closing the product lifecycle loop [85].

The contribution to *improved decision-making* is also among the benefits [86,87]. Integrated data can enable informed decision-making [88]. Companies can analyse the product lifecycle and optimise factors such as costs, efficiency, and sustainability [89]. This can enhance operational efficiency, reduce costs, and provide analytical opportunities to support decision-making, potentially enhancing competitiveness [90].

The challenges in end-to-end traceability include *data integration and interoperability*. This is because end-to-end traceability requires the integration of multiple data sources across different systems and deals with different sets of data [18]. This can include technical traceability data on a product's design and development history through PDM/PLM systems [19,21,22] and operational traceability data involving records of raw material sourcing, manufacturing processes, transportation, storage, and handling activities [14,23,24] over the supply chain. This can be challenging because of legacy systems [2,91], inconsistent data formats [92,93], and technological gaps between suppliers and manufacturers [94]. In addition, the complexity of managing and unifying data from different stages and actors in the product lifecycle and supply chain can result in data integration and interoperability challenges [95].

Reluctance to share information is a significant challenge because companies within a supply network are independent economic actors [15]. Achieving end-to-end traceability necessitates organisations to collaborate with multiple stakeholders in the supply chain, including suppliers, manufacturers, logistics providers, and distributors [63]. Companies are reluctant to share detailed information because of competitive concerns and confidentiality [96]. Data availability and information-sharing among actors may pose significant challenges [97]. Thus, deficient transparency and data-sharing across the supply chain can hinder traceability and challenges in achieving sustainability and compliance goals [12].

End-to-end traceability can be *costly* because its implementation may involve significant investments in technology, infrastructure, and personnel [12]. Traceability is costly and time-consuming because it requires long-term investments in technologies and information systems, company engagement and commitment, and coordination between supply chain actors [98].

Complex regulatory landscapes can also be viewed as a challenge [99]. Different industries and regions have varying regulatory requirements for traceability [100]. Companies operating in global supply chains must comply with diverse standards [101]. Managing compliance with complex regulations can be resource-intensive and time-consuming. Non-compliance can lead to fines, legal consequences, and reputational damage [102].

Finding 1. *End-to-end traceability combines technical and operational traceability to cover the entire product lifecycle and can offer benefits in terms of sustainability, supply chain transparency, product quality, risk management, and regulatory compliance. In addition, support for circular economy initiatives and improved decision-making are among the potential benefits. The challenges include but are not limited to reluctance to share information, cost–value balance, data integration and interoperability, and complex regulatory landscapes.*

3.2. Product Perspective on Traceability

The product perspective on traceability is vital, as the products and services drive revenue, are the source of competitive advantage, and are fundamental to companies' success. The *technical traceability* of a product's design and development history through PDM/PLM systems [19,21,22] and the *operational traceability* of products involving records of raw material sourcing, manufacturing processes, transportation, storage, and handling activities [14,23,24] over the supply chain are focal to end-to-end traceability, focusing on the stages of the engineering lifecycle of products [37].

Traceability is important for smooth product flow through the supply chain and product safety [103]. It enables the verification of product origins, integrity of components, and sustainability of production practices, while also ensuring high product quality [104]. Traceability relates to the availability of product information throughout the product lifecycle from raw material sourcing to disposal [105,106]. The rise in counterfeit products further underscores the need for robust traceability systems [107]. The importance of traceability is emphasised, particularly in safety-critical industries such as food production, medical industries, and aviation, while traceability is often mandated by industry standards [108–111] and enforced by government regulations.

However, for comprehensive traceability, technical and operational traceability must be combined to ensure end-to-end coverage of the supply chain [112]. Definitions vary, but traceability generally refers to the ability to document and track a product and its components as they move along the production chain, including stages such as transport, storage, processing, distribution, and sales [108–113]. Excessive traceability data can also be counterproductive, highlighting the need for a balanced approach that ensures an acceptable cost while maintaining the right level of traceability [114]. Moreover, without proper management, traceability systems can backfire because of misconduct, such as the sale of product labels to unregulated parties. Therefore, it is essential to carefully evaluate the suitability of any traceability system [115].

3.2.1. Information and Architecture for Traceability

Product lifecycle management information is not always integrated or automatically shared among cooperating parties in the supply chain, which poses challenges for achieving true traceability [116]. Additionally, the structural differences between supply chains across industries and the varying legal and customer requirements for traceability must be understood and considered [100]. In some cases, traceability is facilitated by extracting bill-of-material (BOM) data from enterprise resource planning (ERP) systems. BOM data representing the components that constitute a product can be viewed along with the product's information requirements [91]. Effective traceability systems are essential for tracking products, verifying their authenticity, and monitoring components or locations during manufacturing and distribution [117]. Furthermore, linking traceability data to the as-built BOM configuration has been considered for lifecycle traceability [118]. For effective and scalable traceability solutions, a framework based on blockchain has been presented that specifies the type and granularity of information to be collected at each value-adding stage of the supply chain [14].

It is product structure that supports efficient product configuration management and helps one understand the evolution when managed in a PDM/PLM system [119]. Product structure is also necessary to consider data models that provide a clear structure for organising, managing, and understanding data [120]. Product structure can help improve IT integration and play a role in supporting traceability [121]. However, product structures are typically not addressed beyond company borders. Some efforts have been made to link traceability data and product structure in the context of an extended enterprise [122], without demonstrating the role of product structure in a generic manner, while providing links to IT systems. In addition, bills of materials have been linked to production batch numbers in the manufacturing context [123]; however, product structures have not been discussed in detail. Nevertheless, some studies have realised the significance of product structures in the context of IT systems and their relevance to traceability [124–126]. However, establishing capabilities for traceability and structure management across engineering disciplines has been suggested [127]. Traceability and its role in evolving from a company's internal paradigm of technical traceability to a collaborative paradigm in the extended enterprise are seen as important [128]. However, inconsistent identification and limited traceability remain as challenges [129].

A well-designed data flow architecture is essential for visualising end-to-end traceability, considering the product structure, and expanding traceability across the entire product lifecycle. Various traceability architectures with different abstraction levels have been proposed. For example, traceability has been discussed across four layers: the application layer (data usage and processing), network layer (data exchange), integration layer (data collection, compression, analysis, and system integration), and infrastructure layer (data identification and processing infrastructure) [130]. In addition, data readers and sensors have been linked to IoT contexts, with layers focusing on interconnection (IoT integration), data (storage, conversion, and intelligence), and services (managing processed data) [131].

Another proposed architecture features a data generation layer (physical world), data gathering and preprocessing, data management and application, and service layers [132]. Data-driven approaches have also highlighted cyber–physical visibility and traceability using smart technologies [133]. However, one of the most straightforward models is the four-layered data flow architecture proposed by [134], which includes a data carrier layer (using suitable technologies to carry product-related data), data capture layer (involving data readers and middleware), data-sharing layer (manipulation and exchange of product-identifying information among stakeholders), and application layer (communication protocols and interfaces for applications).

Traceability relates to many considerations relevant to product management, business processes, IT systems, and data and how their combinations are addressed. This can be a challenging development target from a technical traceability perspective alone, but more so if operational traceability is considered. Consequently, a clear methodology for achieving traceability would be valuable to many actors, particularly if possible to automate [135]. However, the field is evolving as digital technologies are being applied for traceability in modern supply chains to enable data sharing and the availability of relevant data with end-to-end visibility over products and components [134]. In addition, recent technological developments and the potential of AI to support traceability deserve attention [136,137]. Therefore, system perspectives and systematics are necessary to support end-to-end traceability, effectively address challenges, and ensure compatibility with product management. For example, solid data management benefits from technical and operational traceability.

3.2.2. Technologies for Traceability

Various technologies have been used for traceability in industry, such as barcodes, Radio Frequency Identification (RFID), and Quick Response (QR) codes, with the primary purpose of identifying products or groups of products (e.g., production lots) [138]. However, it is challenging to implement these technologies to achieve effective traceability. Barcodes, for example, assign the same Universal Product Code (UPC) to all stock-keeping units (SKU), making it difficult to distinguish between individual items. In contrast, RFID provides each item with a unique identification number, allowing for differentiation between similar items and avoiding double counting [139]. In addition, the potential for automatic information collection contributes to more timely traceability [140]. Despite its potential, RFID has not fully lived up to its promise and remains costly, with various challenges [141]. Nonetheless, it has been applied to tracking trade items [28]. Similarly, other barcode systems such as the European Article Number (EAN) are inadequate for situations in which the product structure or component interdependence varies [142]. RFID can provide information on an item's origin, packaging location, packaging date, best-before date, current location, and time spent at specific points in the supply chain [139]. Recent discussions in the traceability field have highlighted the role of RFID in smart digital networks [143]. QR codes are useful for internal processes and customer interaction but have limitations, often requiring other methods such as barcodes for external use. In recent years, cloud technologies that rely on object identification have begun to supersede older technologies in the traceability context [144,145]. As a result, understanding the strengths and weaknesses of different technologies is essential for effective traceability, and it may be that no single technology suits end-to-end traceability needs [138]. Therefore, the suitability of specific technologies must be carefully assessed.

Traceable objects, trade units, and items are often referred to as traceable resource units (TRUs) [146]. However, differences exist between technical and operational traceability because traceable objects outside specific companies must be identifiable by multiple parties along the supply chain. In technical traceability, units or items are typically assigned internal codes that are meaningful to the company but may not be easily understood by external actors [147]. The identification of traceable objects across the supply chain can be supported by technologies such as barcodes, RFID, QR codes, bulk numbers, or raw material markers [26]. However, traceability is not always transparent because products may carry the same identification, whereas their raw material origins or manufacturing locations differ [123]. However, this study does not provide a definitive position regarding the technologies that should be used.

End-to-end traceability offers significant benefits to the supply chain, where the traceability of individual components can be ensured through a chain of trust among manufacturers, distributors, and users [148]. Blockchain, as an ordered list of blocks containing transactions, holds the potential for traceability because transactions on a blockchain cannot be deleted or altered, providing strong defences against data tampering [76,149,150]. However, blockchain-based traceability also faces challenges owing to the reluctance of independent actors to share data [96]. Many studies have explored the deployment of blockchain for various purposes [151], and it is often presented as a solution to traceability issues in supply chain management (SCM) [117].

A key issue in using blockchain for traceability is that private data cannot be publicly stored on the blockchain, raising concerns about the information that should remain on or off the chain [116]. This challenge has been addressed by proposing a framework that integrates product lifecycle management (PLM) with blockchain, IT systems, and other traceability methods while considering off-chain activities [116]. Such frameworks may provide opportunities for application-independent data usage [152,153]. Additionally, the Internet of Things (IoT) and blockchain can potentially work together by utilising sensor data, although the integration faces challenges related to scalability, security, and data privacy [154]. Despite these hurdles, IT systems, processes, and traceability solutions can be integrated to achieve comprehensive data integration [155].

Artificial intelligence (AI) is also gaining attention because of its potential role in traceability, particularly in handling large datasets and enabling advanced analytics [156,157]. Smart technologies can further enhance architecture design and cross-industry collaboration [158], although low-cost, reliable, and efficient interconnectivity is seen as crucial from the Industry 4.0 perspective [159], making them particularly relevant for developing traceability frameworks. Ultimately, traceability holds business value and benefits product management and other stakeholders in the supply chain.

3.3. Productization and Potential Support for Traceability

Productization is an offering-centric concept that can be linked to a holistic product management perspective. The necessary whole can be seen as involving products with commercial and technical representation [37,44,45]; product data, where master data play a key role [21,30,160–162]; business processes [21,46,47,163]; IT systems [21,22,29,36,47], such as PDM/PLM, CAD, and ERP; and customer relationship management (CRM). Productization can be viewed as having the objective of transforming an ad hoc offering into a well-defined one [42,44,45,164], meeting key requirements cost-effectively [37], managing the scope of the offering [46], increasing efficiency, reducing inefficiencies [43,165,166], and increasing understanding of the offering [167]. Productization has been seen to benefit active product portfolio management over the engineering lifecycle of products [35,37], which is a wider focus than traditional new product development [168]. Productization benefits data analytics by providing consistency [22,29,152]. It ensures that products that are produced, delivered, sold, purchased, and used remain the same throughout the lifecycle [37,44,45]. This is the structural focus along with the commercial and technical representation of the product [38,44,45], which is necessary to gain control over the product. The products must be controlled before the processes are able to perform [46]. Productization is also necessary to gain control over a product's master data [21]. The revenue model is a potential starting point for considering a commercial offering to clarify the core of the product in terms of what a customer is willing to pay [169]. The revenue model supports a structural approach to a product [170,171]. Understanding product architecture at the company level is also beneficial before focusing on product structure [172,173]. The technical product structure is related to modules and components [172–177] or processes and resources [43,45,46,178], depending on the product type. From an engineering lifecycle perspective, productization acknowledges a varying focus along the stages of the lifecycle [37]. Traceability has been referred to in relevant productization discussions [45,173,175,179] but has not been elaborated upon further. The significance of productization and product structure logic on traceability has not been studied in detail or considered when combining technical and operational traceability for end-to-end traceability. The level of detail in describing product structure as correlating with the delivery structure and logistics model [180] is a potential point of connection between productization and traceability, but it has not been studied from this perspective.

It can be assumed that focusing on productization and product structure logic helps bridge the gap between technical traceability and operational traceability by creating a unified framework for tracing components, parts, and materials from the design phase through production and delivery until disposal. Figure 1 illustrates the potential connection between productization and end-to-end traceability. Productization and logistics models are interlinked through the product structure. Product design and the applied productization model impact the delivery process and model, and vice versa. The selected delivery model affects product design. Productization logic is important because it affects the entire engineering lifecycle of a product. Therefore, it can be assumed that productization logic affects end-to-end traceability, as well as both technical and operational traceability.

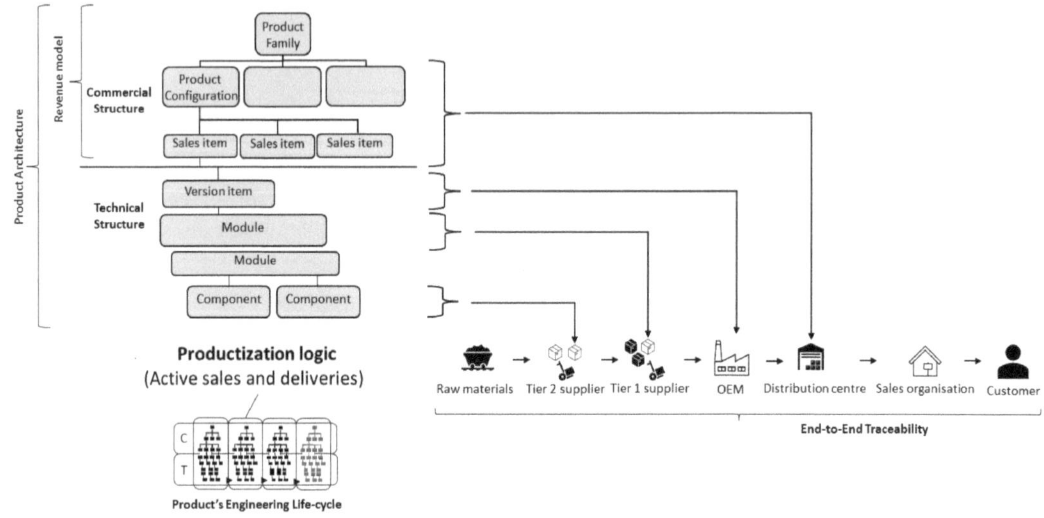

Figure 1. Potential connection between productization and end-to-end traceability.

Finding 2. *The interface between the product and delivery structures creates a key reference point for operational traceability and productization.*

3.4. Literature Synthesis

End-to-end traceability requires the integration of multiple data sources across different traceability systems, involving the *technical traceability* of a product's design and development history through PDM/PLM systems and the *operational traceability* of products involving records of raw material sourcing, manufacturing processes, transportation, storage, and handling activities over the supply chain. Productization supports end-to-end traceability throughout the product structure by aligning commercial and technical product representations with systemic product management practices and linking them to product data. This approach enhances the data perspective and supports traceability over the product lifecycle.

Productization can be linked to the technical traceability relevant to individual actors in the value chain through product structure and systemic product management (Figure 2). Systemic product management combines products, business processes, IT, and data in the engineering lifecycle. The value chain links the product structures of multiple actors, meaning that productization logic is relevant for connecting the actors within the value chain. Operational traceability relies on the data framework and lifecycle support provided by productization.

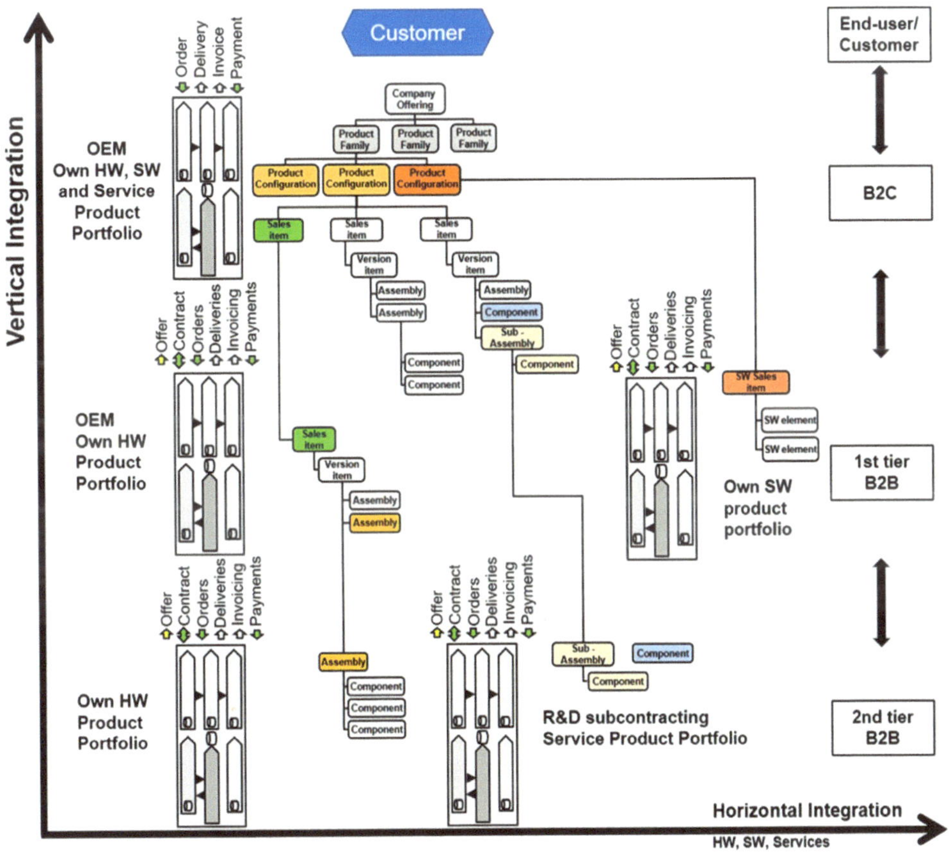

Figure 2. Product structure and value chain.

The productization of each individual company along the value chain, whose participation is necessary for end-to-end traceability, is significant because consistency in productization and product structure logic affect traceability. The product structure provides a frame for the data and enables data consistency. End-to-end traceability leverages data from multiple companies, and each participating company is part of the value chain in which each actor plays a role. Productization logic and product structure connect the actors (Figure 2).

Value chain actor-specific business processes define how products are developed, sold, supplied, manufactured, ordered, delivered, invoiced, installed, and serviced. Additionally, a specific actor's role in the value chain may influence business processes. Productization logic connects actors within a value chain. Product definition is necessary so that processes can be defined and developed. Product data are linked to product structure and enriched in the value chain. If parts of the product are supplied by an actor lower in the value chain, the company higher in the chain must manage that item, and the company lower in the chain must manage the content. This should influence traceability. What is considered a product varies according to the company's position in the value chain.

The product structure provides a frame for the master data within each actor. Master data, being critical business information and relating to products, also make its quality significant for traceability. Nevertheless, other data assets, including transactional and interactional assets, can be beneficial for traceability. Operational traceability necessitates the voluntary sharing of data between actors, motivated by the benefits, regulations, or requirements of a powerful stakeholder. Hence, end-to-end traceability entails considering the data to be shared and addressing any data availability-related challenges.

Finding 3. *Technical traceability, relevant to individual actors in the value chain, can be supported by a systemic product management perspective that combines products, business processes, IT, and data in the context of the engineering lifecycle. Productization is linked to the whole through commercial and technical product representation. Productization logic and traceability can be interlinked in the technical traceability context. Operational traceability relates to a collaborative paradigm among companies that can provide benefits in lifecycle visibility and prominent matters such as sustainability but is challenged by issues such as reluctance to share information.*

Finding 4. *Understanding the value chain, nature of products, and product flow is necessary for effective end-to-end traceability. Traceability also necessitates considering how to combine different traceability systems, what data need to be shared, the benefits of end-to-end traceability, and the motivations for sharing data among independent economic actors. Consistent productization logic may support item-level traceability, whereas traceability data, data flow, data integration, and suitable technologies are necessary for end-to-end traceability.*

4. Empirical Study

Figure 3 illustrates the interlinked product structure of the combined offerings of five different companies constructed based on interviews and product information. The dotted line separates the commercial and technical product structures. The main product sold by Company 1 to car manufacturers was a type of smart windscreen that could contain a display and the windscreen itself, including a heating system and camera system. Product configurations were car manufacturer-specific, owing to differences in design and desired functionalities. The LED displays were built into the windshield with options regarding the display colour. The sub-parts of the products of Company 1 were provided by Companies 2–5 (main suppliers), with each of their offerings being a version item from the perspective of Company 1. Version item (VI) relates to the specific product versions that may change due to, for example, quality improvements or cost reductions and, hence, affect product structure. The required LED technology was provided by Company 2. Company 3 provided the technology related to the required glass, with options in terms of glass colour and shading. (The red dotted line divides the commercial and technical structure). The windscreen colour was based on the colour of the glass, and shading could be integrated into lamination. Company 4 provided a heating system to protect the windscreen from icing. The built-in camera system provided by Company 5 provided various innovative possibilities for utilising glass surfaces. The products sold were not limited to windshields; the offering was generally possible on glass surfaces.

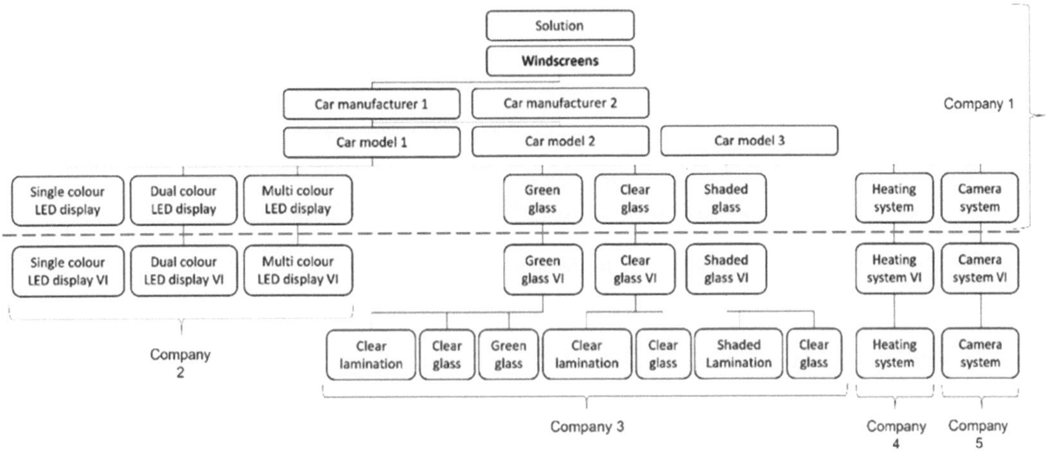

Figure 3. Interlinked product structure by five companies.

Figure 4 illustrates a more detailed product structure for Company 2. The company provided a built-in screen. One such product is a screen built within glass. The LED product consisted of LED foil, wired foil, and LEDs, depending on the colour options needed. A solder mask was also necessary and had colour-based options. Delivery options were based on the type of adhesive used. LED foil was customised in terms of shape. The wired foil, LED colour, and solder mask had delivery links to module variants. Several suppliers provided the necessary materials and components. (The red dotted line divides the commercial and technical structure).

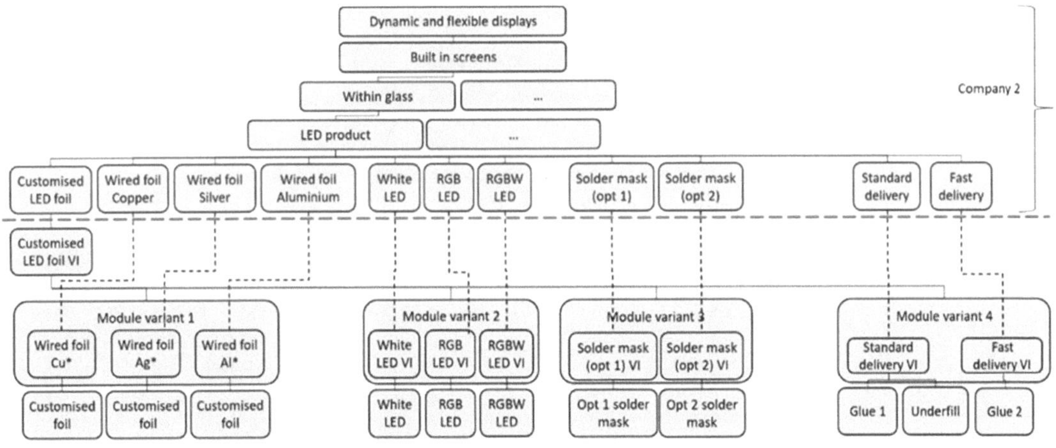

Figure 4. The product structure by Company 2.

The companies involved had shortcomings in their traceability, many of which were related to the integrity of the related information and data management. Some of these shortcomings were related to IT tools and responsibilities regarding traceability and data. The regulatory side did not affect all companies equally. Some were obliged to comply with stringent industry standards and legal implications. Those who were more regulated placed heavier emphasis on technical traceability with corresponding activities. Those who were more regulated had better-defined responsibilities and processes. Not all companies originally considered product structure and its role from a traceability perspective. Quality and related standards are drivers of traceability. Only some utilised unique numbers for shipments. The numberings enabled traceability to a certain extent to link the materials used. Customer requirements were mentioned as the driving force behind these practices. The traceability information relating to the order–delivery process was recorded and integrally linked to the product data. Technical traceability was linked to PDM/PLM for those that had adopted it. Traceability codes related to shipments were recorded in the ERP. Barcodes were applied to shipments, enabling them to link data internally and backtrace internally. Some other technologies, such as RFID, were considered by companies, but, generally, no bulletproof technologies or their combinations were applied to ensure the integrity of the supply chain.

In a supply chain sense, the original situation was challenged by a variety of practices and levels of maturity in terms of productization in companies. Some lacked an official product structure and systematic logic for products and their variations. Certain types of informalities prevailed. Some utilised a product structure motivated by the PDM/PLM, and certain rules and practices were deliberate. Consistency in logic could not be confirmed. The supply chain was challenged by weaker links (companies) lacking product definitions and logic of variations and revisions. This further contributed to deficiencies in instruction and product management. Suppliers and customers were seen to be in their own information silos, and the visibility of product data was mostly limited to an internal perspective.

The extent of the companies' data models and a lack of such thinking could not be confirmed. However, product data management received considerable attention. Product data were understood by the companies, but the structure of product data was lacking in some cases. There were also deficiencies in data collection for some companies. The same variety was visible in the applied processes, with clear deficiencies in definitions in some and people's centricity in others. IT infrastructure was also considered better by more advanced companies. Some companies used PDM/PLM but focused on product data in an ERP-centric manner. Other companies that used PDM/PLM focused on traceability and had foundations for product structure but were driven towards unnecessary manual work in traceability. The versatility of the IT used and the product data being distributed to multiple sources hindered traceability, indicating a lack of a data model for companies. Training personnel for PDM/PLM use were considered necessary to ensure the technical traceability and timeliness of the product data.

The constructed productization logic that formed a combined structure for the developed offering was seen to have potential by the focus group, validating the construction. Some adjustments were made to the initial proposal based on these comments. The focus group seemed to understand the benefits of combining structural logic with productization systematics. They seemed to understand the benefits of consistency, specifically those of using more systematic technical traceability. The scalability potential and modularity were particularly appreciated. Company 1 realised the role of industrial standards in traceability and the possibility of introducing systematics for compliance. They also understood the significance of deliberately considering structures to support traceability. The others did not follow similar systematics to the same extent but applied traceability more from the perspective of "customer requirements". Nevertheless, they seemed to be enlightened by their potential to improve their systematics. Company 1 had traceability-related processes and roles, defined procedures, and an embedded traceability information model for technical traceability. Operational traceability was highlighted as a means of leveraging complementarity and synergies between relevant actors. However, other companies seemed to have operational traceability in a better order than technical traceability. The potential reason for this seemed to involve customers, who played a strong role in their traceability. There had been ongoing efforts to improve the formalisation of technical traceability. Data management processes in general, and specifically traceability, were seen to benefit from the partnership and combined productization logic by focusing on the data that would be created, distributed, and shared.

The interviewees viewed productization and product data as having further potential in terms of traceability in all analysed companies individually, but also across the supply chain. This entails careful consideration of company processes, IT use, and product data. These factors influence traceability-related processes and responsibilities. Nevertheless, the "synchronised" productization was seen as beneficial by aligning the product focus. Synchronised productization among key companies was beneficial for providing consistency in logic. However, it was not seen to be enough, but data from the entire supply chain were seen as needing to be "synchronised" to allow the necessary integrity in traceability. This was understood to necessitate the use of additional technologies aside from the currently used ones that enable technical traceability to further address the processes, IT, and data and cover the whole supply chain. The availability of data was not a major issue for the five studied companies, as the example involved a partnered supply chain with a common interest in new business generation. Figure 5 illustrates the need to synchronise productization and data for supply chain traceability. (The arrows (↑) indicate suppliers contributing to the technical offering, which necessitates data availability to enable synchronised data).

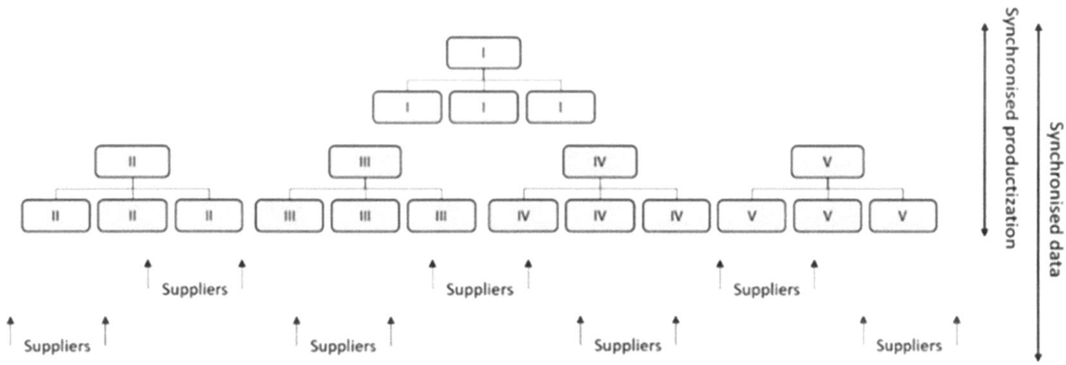

Figure 5. Synchronised productization and data.

Finding 5. *Consistent productization and product structure may support traceability. Traceability is supported by consistency in the product structure, which provides consistency in the data. This consistency can extend beyond company borders and support supply chain traceability.*

5. Discussion

The key contribution of this study is that productization with a product structure focus can benefit end-to-end traceability by creating a structured and scalable approach to commercial and technical product definitions to manage products from concept to end-of-life. This can be achieved by addressing the varying focus at different life-cycle stages. Productization involves the formalisation of the product structure, which can help formalise business processes and supply chain workflows to ensure scalability and traceability. Through this formalised product structure logic—including components, modules, and processes—productization ensures that products are consistently traceable from design to end-of-life. A clear structure is provided to manage product definitions across different lifecycle stages, reduce uncontrolled variability, and allow better control over products. This formalisation enables PDM/PLM systems to capture and manage data related to products, their development, and changes throughout the lifecycle more effectively, benefitting technical traceability. Simultaneously, operational traceability is simplified by standardising the processes, modules, and components. The linkage between productization logic and the delivery process is established through a product's structural level, providing a touchpoint for approaching both operational traceability and productization. Overall, it appears that the focus on product structure through productization has the potential to support the integration of technical and operational traceability by promoting data consistency and facilitating seamless traceability across different traceability systems. This is a well-considered product structure logic that relates to the representation of modules, assemblies, components, and raw materials within the systems used and acts as a foundation for managing data. The structured approach focuses on product structure levels and related data across the engineering lifecycle and supports both technical traceability and operational traceability integration.

Productization can help reduce inefficiencies and support a focus on productivity by promoting the standardisation of components, materials, and processes. This involves defining and using a fixed set of parts, modules, and processes to create scalable products and further support traceability by promoting unified data in terms of technical and operational traceability. Productization logic involves considering what is allowed to change in different lifecycle stages. Process standardisation can promote the use of approved suppliers and validated methods to enhance operational traceability by simplifying processes. Operational traceability is also simplified by scalability through consistent productization logic, product structure, and data. Nevertheless, it is productization that promotes the perspective of the engineering lifecycle and its stages, consisting of commercial and technical aspects that potentially mediate the view of both technical and operational data. Hence, productization may encourage the integration of technical and operational traceability systems, PDM/PLM managing product lifecycle data, and supply chain traceability systems that manage operational data. This creates a digital thread for end-to-end traceability from raw materials to end-of-life disposal.

The scientific implications include providing original contributions to the context of end-to-end traceability by exploring the role of productization in a context that combines technical and operational traceability. This should provide new perspectives on end-to-end traceability [1]. This study contributes by directly combining technical and operational traceability to achieve an end-to-end perspective. Nevertheless, it can be argued that the coverage of combined technical and operational traceability is equal to that of end-to-end traceability. Studies focusing on technical traceability [13,19,21,22] provide ideas on the role and potential of productization to support traceability in terms of technical traceability and potential linkages to operational traceability. Similarly, studies focusing on operational traceability [14,23,24,90] provide ideas on the role and potential of productization for operational traceability and potential linkages to technical traceability. Naturally, it is possible that the discussion on technical or operational traceability contains elements that may have purposes similar to those of productization, but the perspective of productization inherently involves the product management perspective, which has been insufficiently discussed. This study contributes to previous product management research [21,22,28–30,35–40] by providing a direct discussion on traceability through a productization lens. Previous discussions of productization [21,22,29,35,37,41–45,152,164,173,175,179] gain a new contribution with this study, which explored the role and potential of productization in supporting traceability. Previous product structure discussions in the productization context [21,22,29,35–37,39,40,45–47,152,170,172–176] are expanded to a discussion of the traceability context. Nevertheless, the role of product structure is also further understood, especially for technical traceability [91,118,120–122]. However, the discussion has not been as extensive and has not covered an end-to-end perspective. Specifically, the inclusion of commercial product representation is lacking. Additionally, a productization logic focus has not yet been applied. Earlier studies have focused on the IT system context and its relevance to traceability by applying a structural approach [124–126]. The applied productization perspective also contributes to the earlier related focus on the engineering lifecycle [21,35–37] by including the traceability perspective. The lifecycle perspective is not new to traceability [2,18,19,51–53,81,85,89,95]; however, this study provides new perspectives through commercial and technical focus and logic and emphasises varying focus over lifecycle stages. This study contributes to previous discussions on traceability data in different forms [12,14,18,19,23,27,56,88,92,93,95], pointing out how productization can aid in standardising data structures and provide consistency to facilitate traceability, ultimately facilitating end-to-end traceability.

5.1. Managerial Implications

Managers can benefit from understanding the potential benefits of productization logic over the engineering lifecycle in general and specifics relating to end-to-end traceability. Productization supports addressing the challenges of end-to-end traceability. For example, the reluctance to share information might be alleviated by productization logic promoting the formalisation of product structure, the standardisation of components and materials, and the establishment of standardised data structures, potentially enabling selective information sharing. Productization may also promote further clarity on data ownership and control over product data, enabling the building of the necessary trust for sharing data. The modularity of a product structure may also enable separate data-sharing structures to maintain confidential details. The following implications 1–4 can be particularly valuable:

(1) Addressing the cost–value balance could be supported by productization, which promotes economies of scale in product designs that reduce complexity, thereby lowering the cost of implementing traceability solutions. Implementing traceability within the product architecture during productization is also a possibility. When implemented consistently, this can lead to cost reductions in traceability, while providing value. Data integration from technical and operational systems may be supported by productization by potentially standardising product data models so that all stakeholders can align. This supports the unification of data from different systems.

(2) Productization focusing on the engineering lifecycle may support cross-functional collaboration and the alignment of technical and operational traceability data. Additionally, navigating the variety in regional and industry-specific regulations might be facilitated by productization logic by supporting embedding compliance into product designs and regulatory reporting aided by a formalised product structure. This approach would provide consistency in data collection, thereby aiding compliance reporting.

(3) Managers can benefit from understanding how motivations for end-to-end traceability can be promoted through considerations such as sustainability. Sustainability reporting is becoming an increasingly timely issue for companies in many parts of the world and could act as a driving force for improving traceability, should the business perspective provide sufficient support. A productization focus may play a crucial role in supporting sustainability reporting and compliance through a product structure-centric approach, which may support the ability to collect, analyse, and report relevant sustainability information. This approach promotes the consistent connection of sustainability data to products and may support related traceability over the engineering lifecycle.

(4) In summary, productization can enhance the integration of technical and operational traceability, leading to improved product-related insights that can inform decision-making.

5.2. Limitations and Future Studies

The limitations include the conceptual nature of the findings and the limited empirical focus on a specific sector, analysing a limited number of companies with a limited number of interviewees. The methodological approach of combining an extensive literature review with a limited number of examples sets the boundaries for generalisability. These findings can be considered conceptual, as the full application of this concept has not been studied for end-to-end traceability. In addition, this study did not consider the suitability of different technologies for traceability or their dynamics when combined while considering the productization approach and end-to-end traceability. However, the chosen approach appears to be sufficient to indicate the role and potential of productization for end-to-end traceability. This study did not delve into the intricate details of buyer–supplier relationships or consider the contractual arrangements necessary for traceability between separate companies.

Some industry-specific factors are expected to require further research. Future empirical studies are needed to confirm the benefits and applicability of productization logic for traceability. In addition, the potential and limitations of different technologies for combining technical and operational traceability should be clarified. Future studies can clarify potential application-independent and data-centric approaches to traceability by focusing on the potential of cloud technology and productization support for traceability. Furthermore, the intersection of sustainability reporting, productization, and end-to-end traceability deserves further investigation. The suitability of product structures in different contexts should be studied further from a traceability perspective.

Author Contributions: Conceptualization, J.H., J.M.G.R. and E.M.; Methodology, J.H., J.M.G.R. and E.M.; Validation, J.H., J.M.G.R. and E.M.; Formal analysis, J.H., J.M.G.R. and E.M.; Investigation, J.H., J.M.G.R. and E.M.; Writing—original draft, J.H., J.M.G.R. and E.M.; Writing—review & editing, J.H.; Visualization, J.H. and E.M.; Supervision, J.H. All authors have read and agreed to the published version of the manuscript.

Funding: This research received no external funding.

Institutional Review Board Statement: Not applicable.

Informed Consent Statement: Not applicable.

Data Availability Statement: The original contributions presented in the study are included in the article, further inquiries can be directed to the corresponding author.

Acknowledgments: The authors would like to thank the companies that participated in the interviews.

Conflicts of Interest: The authors declare no conflict of interest.

References

1. Leal, F.; Chis, A.E.; Caton, S.; González–Vélez, H.; García–Gómez, J.M.; Durá, M.; Sánchez–García, A.; Sáez, C.; Karageorgos, A.; Gerogiannis, V.C.; et al. Smart Pharmaceutical Manufacturing: Ensuring End-to-End Traceability and Data Integrity in Medicine Production. *Big Data Res.* **2021**, *24*, 100172. [CrossRef]
2. Alsadi, M.; Arshad, J.; Ali, J.; Prince, A.; Shishank, S. TruCert: Blockchain-Based Trustworthy Product Certification within Autonomous Automotive Supply Chains. *Comput. Electr. Eng.* **2023**, *109*, 108738. [CrossRef]
3. Bründl, P.; Stoidner, M.; Nguyen, H.; Abrass, A.; Franke, J. Traceability in Engineer-to-Order Manufacturing SMEs. In Proceedings of the 32nd Mediterranean Conf. Control Autom. (MED), Chania-Crete, Greece, 11–14 June 2024; pp. 57–62.
4. Munasinghe, U.J.; Halgamuge, M.N. Supply Chain Traceability and Counterfeit Detection of COVID-19 Vaccines Using Novel Blockchain-Based VacLedger System. *Expert Syst. Appl.* **2023**, *228*, 120293. [CrossRef] [PubMed]
5. Kavasidis, I.; Lallas, E.; Mountzouris, G.; Gerogiannis, V.C.; Karageorgos, A. A Federated Learning Framework for Enforcing Traceability in Manufacturing Processes. *IEEE Access* **2023**, *11*, 57585–57597. [CrossRef]
6. Malik, M.; Gahlawat, V.K.; Mor, R.S.; Singh, M.K. Unlocking Dairy Traceability: Current Trends, Applications, and Future Opportunities. *Future Foods* **2024**, *10*, 100426. [CrossRef]
7. Zhou, X.; Pullman, M.; Xu, Z. The Impact of Food Supply Chain Traceability on Sustainability Performance. *Oper. Manag. Res.* **2022**, *15*, 93–115. [CrossRef]
8. Pytel, N.; Ziegler, C.; Winkelmann, A. From Dissonance to Dialogue: A Token-Based Approach to Bridge the Gap Between Manufacturers and Customers. *ACM Trans. Manag. Inf. Syst.* **2024**, *15*, 3. [CrossRef]
9. Wu, C.K.; Tsang, K.F.; Liu, Y.; Zhu, H.; Wei, Y.; Wang, H.; Yu, T.T. Supply Chain of Things: A Connected Solution to Enhance Supply Chain Productivity. *IEEE Commun. Mag.* **2019**, *57*, 78–83. [CrossRef]
10. Kulshrestha, N.; Agrawal, S.; Shree, D. Spare Parts Management in Industry 4.0 Era: A Literature Review. *J. Qual. Maint. Eng.* **2024**, *30*, 248–283. [CrossRef]
11. Winkelmann, S.; Guennoun, R.; Möller, F.; Schoormann, T.; van der Valk, H. Back to a Resilient Future: Digital Technologies for a Sustainable Supply Chain. *Inf. Syst. E-Bus. Manag.* **2024**, *22*, 315–350. [CrossRef]
12. Zhou, X.; Lu, H.; Mangla, S.K. The Impact of Digital Traceability on Sustainability Performance: Investigating the Roles of Sustainability-Oriented Innovation and Supply Chain Learning. *Supply Chain Manag. Int. J.* **2024**, *29*, 497–522. [CrossRef]
13. MacCarthy, B.L.; Pasley, R.C. Group Decision Support for Product Lifecycle Management. *Int. J. Prod. Res.* **2021**, *59*, 5050–5067. [CrossRef]

14. Ahmed, W.A.; MacCarthy, B.L. Blockchain-Enabled Supply Chain Traceability–How Wide? How Deep? *Int. J. Prod. Econ.* **2023**, *263*, 108963. [CrossRef]
15. Sodhi, M.S.; Tang, C.S. Research Opportunities in Supply Chain Transparency. *Prod. Oper. Manag.* **2019**, *28*, 2946–2959. [CrossRef]
16. Dudczyk, P.; Dunston, J.K.; Crosby, G.V. Blockchain Technology for Global Supply Chain Management: A Survey of Applications, Challenges, Opportunities and Implications. *IEEE Access* **2024**, *12*, 70065–70088. [CrossRef]
17. Timmer, S.; Kaufmann, L. Conflict Minerals Traceability—A Fuzzy Set Analysis. *Int. J. Phys. Distrib. Logist. Manag.* **2017**, *47*, 344–367. [CrossRef]
18. Corallo, A.; Latino, M.E.; Menegoli, M.; Pontrandolfo, P. A Systematic Literature Re-view to Explore Traceability and Lifecycle Relationship. *Int. J. Prod. Res.* **2020**, *58*, 4789–4807. [CrossRef]
19. Stark, J. The Importance of Product Data in PLM. In *Product Lifecycle Management*; Decision Engineering; Springer: Cham, Switzerland, 2024; Volume 2, pp. 155–191.
20. Roy, V. Contrasting Supply Chain Traceability and Supply Chain Visibility: Are They Interchangeable? *Int. J. Logist. Manag.* **2021**, *32*, 942–972. [CrossRef]
21. Hannila, H.; Tolonen, A.; Harkonen, J.; Haapasalo, H. Product and supply chain related data, processes and information systems for product portfolio management. *Int. J. Prod. Lifecycle Manag.* **2019**, *12*, 1–19. [CrossRef]
22. Hannila, H.; Silvola, R.; Harkonen, J.; Haapasalo, H. Data-driven begins with DATA; potential of data assets. *J. Comput. Inf. Syst.* **2020**, *16*, 29–38. [CrossRef]
23. Rejeb, A.; Rejeb, K.; Simske, S.; Keogh, J.G. Exploring Blockchain Research in Supply Chain Management: A Latent Dirichlet Allocation-Driven Systematic Review. *Information* **2023**, *14*, 557. [CrossRef]
24. Lv, G.; Song, C.; Xu, P.; Qi, Z.; Song, H.; Liu, Y. Blockchain-Based Traceability for Agricultural Products: A Systematic Literature Review. *Agriculture* **2023**, *13*, 1757. [CrossRef]
25. Moysiadis, T.; Spanaki, K.; Kassahun, A.; Kläser, S.; Becker, N.; Alexiou, G.; Zotos, N.; Karali, I. AgriFood Supply Chain Traceability: Data Sharing in a Farm-to-Fork Case. *Benchmarking Int. J.* **2023**, *30*, 3090–3123. [CrossRef]
26. Kumar, V.; Hallqvist, C.; Ekwall, D. Developing a Framework for Traceability Implementation in the Textile Supply Chain. *Systems* **2017**, *5*, 33. [CrossRef]
27. Liao, Y.; Kwaramba, C.S.; Kros, J.F. Supply Chain Traceability: An Institutional Theory Perspective. *Int. J. Logist. Econ. Global.* **2020**, *8*, 193–223. [CrossRef]
28. Stark, J. PLM and Products. In *Product Lifecycle Management*; Decision Engineering; Springer: Cham, Switzerland, 2020; Volume 1.
29. Hannila, H.; Koskinen, J.; Harkonen, J.; Haapasalo, H. Product-level profitability—Current challenges and preconditions for data-driven, fact-based Product Portfolio Management. *J. Enterp. Inf. Manag.* **2022**, *33*, 214–237. [CrossRef]
30. Silvola, R.; Harkonen, J.; Vilppola, O.; Kropsu-Vehkapera, H.; Haapasalo, H. Data quality assessment and improvement. *Int. J. Bus. Inf. Syst.* **2016**, *22*, 62–81. [CrossRef]
31. Musa, A.; Gunasekaran, A.; Yusufa, Y. Supply chain product visibility: Methods, systems and impacts. *Expert Syst. Appl.* **2014**, *41*, 176–194. [CrossRef]
32. Karadgi, S.; Kulkarni, V.; Doddamani, S. Traceable and Intelligent Supply Chain based on Blockchain and Artificial Intelligence. *J. Phys. Conf. Ser.* **2021**, *2070*, 012158. [CrossRef]
33. Khabbazi, M.R.; Ismail, M.D.Y.; Ismail, N.; Mousavi, S.A. Modeling of Traceability Information System for Material Flow Control Data. *Aust. J. Basic Appl. Sci.* **2010**, *4*, 208–216.
34. Ahmed, W.A.H.; Maccarthy, B.L. Blockchain-Enabled Supply Chain Traceability in the Textile and Apparel Supply Chain: A Case Study of the Fiber Producer, Lenzing. *Sustainability* **2021**, *13*, 10496. [CrossRef]
35. Tolonen, A.; Shahmarichatghieh, M.; Harkonen, J.; Haapasalo, H. Product Portfolio Management—Targets and Key Performance Indicators for Product Portfolio Renewal over Life Cycle. *Int. J. Prod. Econ.* **2015**, *170*, 468–477. [CrossRef]
36. Harkonen, J.; Mustonen, E.; Hannila, H. Productization and Product Structure as the Backbone for Product Data and Fact-Based Analysis of Company Products. In Proceedings of the 2019 IEEE International Conference on Industrial Engineering and Engineering Management (IEEM2019), Macau, China, 15–18 December 2019.
37. Lahtinen, N.; Mustonen, E.; Harkonen, J. Commercial and technical productization for fact-based product portfolio management over life-cycle. *IEEE Trans. Eng. Manag.* **2021**, *68*, 1826–1838. [CrossRef]
38. Harkonen, J.; Mustonen, E.; Haapasalo, H. Construction Related Data Management—Classification and Description of Data from Different Perspectives. *Int. J. Manag. Knowl. Learn.* **2019**, *8*, 195–220.
39. Mustonen, E.; Harkonen, J.; Haapasalo, H. From Product to Service Business: Productization of Product-oriented, Use-oriented, and Result-oriented Business. In Proceedings of the 2019 IEEE International Conference on Industrial Engineering and Engineering Management (IEEM2019), Macau, China, 15–18 December 2019; pp. 985–989.
40. Harkonen, J.; Tolonen, A.; Haapasalo, H. Modelling of Construction Products and Services for Effective Productisation. *Management* **2018**, *13*, 335–353. [CrossRef]
41. Dubois, L.-E.; Pine, B.J., II; Harkonen, J. Beyond the ephemeral: Scaling experiences through productization. *Bus. Horiz.* 2024, *in press*. [CrossRef]
42. Wirtz, J. Viewpoint: Service products, development of service knowledge and our community's target audience. *J. Serv. Mark.* **2021**, *35*, 265–270. [CrossRef]

43. Wirtz, J.; Fritz, M.P.; Jaakkola, E.; Gelbrich, K.; Hartley, N. Service products and productization. *J. Bus. Res.* **2021**, *137*, 411–421. [CrossRef]
44. Harkonen, J.; Haapasalo, H.; Hanninen, K. Productisation: A Review and Research Agenda. *Int. J. Prod. Econ.* **2015**, *164*, 65–82. [CrossRef]
45. Harkonen, J.; Tolonen, A.; Haapasalo, H. Service productisation: Systematising and defining offering. *J. Serv. Manag.* **2017**, *28*, 936–971. [CrossRef]
46. Harkonen, J. Exploring the benefits of service productisation: Support for business processes. *Bus. Process Manag. J.* **2021**, *27*, 85–105. [CrossRef]
47. Wings, S.; Harkonen, J. Decentralised or centralised management of data and products: Influence on revenue-generating processes. *Int. J. Manag. Decis. Making* **2023**, *22*, 74–105. [CrossRef]
48. Harrell, M.C.; Bradley, M.A. *Data Collection Methods: Semi-Structured Interviews and Focus Groups*; RAND Corporation: Santa Monica, CA, USA, 2009; pp. 1–140.
49. Farkas, M.; Matolay, R. Designing the CSRD System: Insights from Management Systems to Advance a Strategic Approach. *J. Decis. Syst.* **2024**. [CrossRef]
50. Sparacino, A.; Merlino, V.M.; Brun, F.; Blanc, S.; Massaglia, S. Corporate social responsibility communication from multinational chocolate companies. *Sustain. Futures* **2024**, *7*, 100–151. [CrossRef]
51. Lin, K.; Chavalarias, D.; Panahi, M.; Yeh, T.; Takimoto, K.; Mizoguchi, M. Mobile-Based Traceability System for Sustainable Food Supply Networks. *Nat. Food* **2020**, *1*, 673–679. [CrossRef] [PubMed]
52. Sarkar, B.D.; Sharma, I.; Shardeo, V. A Multi-Method Examination of Barriers to Traceability in Industry 5.0-Enabled Digital Food Supply Chains. *Int. J. Logist. Manag.* **2024**. [CrossRef]
53. Basia, A.; Simeu-Abazi, Z.; Gascard, E.; Zwolinski, P. A Conceptual Framework Based on Current Directives to Design Lithium-Ion Battery Industrial Repurposing Models. *Machines* **2024**, *12*, 440. [CrossRef]
54. Almadadha, R. Blockchain Technology in Financial Accounting: Enhancing Transparency, Security, and ESG Reporting. *Blockchains* **2024**, *2*, 312–333. [CrossRef]
55. Chauhan, S.; Singh, R.; Gehlot, A.; Akram, S.V.; Twala, B.; Priyadarshi, N. Digitalization of Supply Chain Management with Industry 4.0 Enabling Technologies: A Sustainable Perspective. *Processes* **2023**, *11*, 96. [CrossRef]
56. Safari, M.; Areeb, A. A Qualitative Analysis of GRI Principles for Defining Sustainability Report Quality: An Australian Case from the Preparers' Perspective. *Account. Forum* **2020**, *44*, 344–375. [CrossRef]
57. Laszlo, C.; Zhexembayeva, N. *Embedded Sustainability: The Next Big Competitive Advantage*; Routledge: New York, NY, USA, 2017.
58. Nemes, N.; Scanlan, S.J.; Smith, P.; Smith, T.; Aronczyk, M.; Hill, S.; Lewis, S.L.; Montgomery, A.W.; Tubiello, F.N.; Stabinsky, D. An Integrated Framework to Assess Greenwashing. *Sustainability* **2022**, *14*, 4431. [CrossRef]
59. Solaimani, S. From Compliance to Capability: On the Role of Data and Technology in Environment, Social, and Governance. *Sustainability* **2024**, *16*, 6061. [CrossRef]
60. Ahmadi-Gh, Z.; Bello-Pintado, A. Why Is Manufacturing Not More Sustainable? The Effects of Different Sustainability Practices on Sustainability Outcomes and Competitive Advantage. *J. Clean. Prod.* **2022**, *337*, 130392. [CrossRef]
61. Margaret, I.; Schoubben, F.; Verwaal, E. When Do Investors See Value in International Environmental Management Certification of Multinational Corporations? A Study of ISO 14001 Certification after the Paris Agreement. *Int. J. Environ. Sustain.* **2024**, *14*, 25–55. [CrossRef]
62. Paelman, V.; Van Cauwenberge, P.; Vander Bauwhede, H. Mission Alignment with Employees and Financiers: Probing into the Workings of B Corp Certification. *J. Bus. Ethics* **2023**, *30*, 1632–1644. [CrossRef]
63. Garcia-Torres, S.; Rey-Garcia, M.; Sáenz, J. Enhancing Sustainable Supply Chains through Traceability, Transparency and Stakeholder Collaboration: A Quantitative Analysis. *Bus. Strat. Environ.* **2024**. [CrossRef]
64. Ahmed, W.; Omar, M. Drivers of Supply Chain Transparency and Its Effects on Performance Measures in the Automotive Industry. *Int. J. Serv. Oper. Manag.* **2019**, *33*, 159–186.
65. Wood, L.C.; Reiners, T.; Srivastava, H.S. Think Exogenous to Excel: Alternative Supply Chain Data to Improve Transparency and Decisions. *Int. J. Logist. Res. Appl.* **2016**, *20*, 426–443. [CrossRef]
66. Jia, F.; Li, K.; Chen, L.; Nazrul, A.; Yan, F. Supply Chain Transparency: A Roadmap for Future Research. *Ind. Manag. Data Syst.* **2024**, *124*, 2665–2688. [CrossRef]
67. Alexander, A.; Kumar, M.; Walker, H.; Gosling, J. Innovation for Zero-Deforestation Sustainable Supply Chain Management Services: A Performance Measurement and Management Approach. *Supply Chain Manag. Int. J.* **2024**, *29*, 620–641. [CrossRef]
68. Iranmanesh, M.; Maroufkhani, P.; Asadi, S.; Ghobakhloo, M.; Dwivedi, P.K.; Tseng, M.-L. Effects of Supply Chain Transparency, Alignment, Adaptability, and Agility on Blockchain Adoption in Supply Chain among SMEs. *Comput. Ind. Eng.* **2022**, *176*, 108931. [CrossRef]
69. Feng, B.; Zheng, M.; Shen, Y. The Effect of Relational Embeddedness on Transparency in Supply Chain Networks: The Moderating Role of Digitalization. *Int. J. Oper. Prod. Manag.* **2024**, *44*, 1621–1648. [CrossRef]
70. Yang, L.; Lu, L. Improving Supply Chain Transparency: From the Perspective of Suppliers. *Ann. Oper. Res.* **2024**. [CrossRef]
71. Wohlrab, R.; Knauss, E.; Steghöfer, J.P.; Maro, S.; Anjorin, A.; Pelliccione, P. Collaborative Traceability Management: A Multiple Case Study from the Perspectives of Organization, Process, and Culture. *Requir. Eng.* **2020**, *25*, 21–45. [CrossRef]

72. Romano, M.; Cavaleiro Reis, B.M.; Santos, L.F.F.M.; Carvalho, P. 3D Printing and Blockchain: Aeronautical Manufacturing in the Digital Era. *Prod. Manuf. Res.* **2024**, *12*, 2368731. [CrossRef]
73. Srivastava, A.; Dashora, K. Application of Blockchain Technology for Agrifood Supply Chain Management: A Systematic Literature Review on Benefits and Challenges. *Benchmarking Int. J.* **2022**, *29*, 3426–3442. [CrossRef]
74. Ullagaddi, P. Leveraging Digital Transformation for Enhanced Risk Mitigation and Compliance in Pharma Manufacturing. J. *Adv. Med. Pharm. Sci.* **2024**, *26*, 75–86. [CrossRef]
75. Vashishth, T.K.; Sharma, V.; Sharma, K.K.; Kumar, B.; Chaudhary, S.; Panwar, R. Intelligent Resource Allocation and Optimization for Industrial Robotics Using AI and Blockchain. In *AI and Blockchain Applications in Industrial Robotics*; IGI Global: Hershey, PA, USA, 2024; pp. 82–110.
76. Ullagaddi, P. Safeguarding Data Integrity in Pharmaceutical Manufacturing. *J. Adv. Med. Pharm. Sci.* **2024**, *26*, 64–75. [CrossRef]
77. Aggarwal, M.; Rani, P.; Rani, P.; Sharma, P. Revolutionizing agri-food supply chain management with blockchain-based traceability and navigation integration. *Cluster Comput.* **2024**, *27*, 12919–12942. [CrossRef]
78. Vazquez Melendez, E.I.; Bergey, P.; Smith, B. Blockchain Technology for Supply Chain Provenance: Increasing Supply Chain Efficiency and Consumer Trust. *Supply Chain Manag. Int. J.* **2024**, *29*, 706–730. [CrossRef]
79. Onyeaka, H.; Tamasiga, P.; Nwauzoma, U.M.; Miri, T.; Juliet, U.C.; Nwaiwu, O.; Akinsemolu, A.A. Using Artificial Intelligence to Tackle Food Waste and Enhance the Circular Economy: Maximising Resource Efficiency and Minimising Environmental Impact: A Review. *Sustainability* **2023**, *15*, 10482. [CrossRef]
80. Rumetshofer, T.; Fischer, J. Information-Based Plastic Material Tracking for Circular Economy—A Review. *Polymers* **2023**, *15*, 1623. [CrossRef]
81. Alves, L.; Ferreira Cruz, E.; Lopes, S.I.; Faria, P.M.; Rosado da Cruz, A.M. Towards Circular Economy in the Textiles and Clothing Value Chain through Blockchain Technology and IoT: A Review. *Waste Manag. Res.* **2022**, *40*, 3–23. [CrossRef] [PubMed]
82. Ellsworth-Krebs, K.; Rampen, C.; Rogers, E.; Dudley, L.; Wishart, L. Circular Economy Infrastructure: Why We Need Track and Trace for Reusable Packaging. *Sustain. Prod. Consum.* **2022**, *29*, 249–258. [CrossRef]
83. Nag, U.; Sharma, S.K.; Kumar, V. Multiple Life-Cycle Products: A Review of Antecedents, Outcomes, Challenges, and Benefits in a Circular Economy. *J. Eng. Des.* **2021**, *33*, 173–206. [CrossRef]
84. El Jalbout, S.; Keivanpour, S. Development of a Body of Knowledge for Design for Disassembly and Recycling of High-Tech Products: A Case Study on Lithium-Ion Batteries. *J. Ind. Prod. Eng.* **2023**, *41*, 19–39. [CrossRef]
85. Gonçalves da Silva, S.B.; Barros, M.V.; Radicchi, J.Â.Z.; Puglieri, F.N.; Piekarski, C.M. Opportunities and Challenges to Increase Circularity in the Product's Use Phase. *Sustain. Futures* **2024**, *8*, 100297. [CrossRef]
86. Santana, S.; Ribeiro, A. Traceability Models and Traceability Systems to Accelerate the Transition to a Circular Economy: A Systematic Review. *Sustainability* **2022**, *14*, 5469. [CrossRef]
87. Sira, M. Potential of Advanced Technologies for Environmental Management Systems. *Manag. Syst. Prod. Eng.* **2024**, *32*, 33–44. [CrossRef]
88. Meier, S.; Klarmann, S.; Thielen, N.; Pfefferer, C.; Kuhn, M.; Franke, J. A Process Model for Systematically Setting Up the Data Basis for Data-Driven Projects in Manufacturing. *J. Manuf. Syst.* **2023**, *71*, 1–19. [CrossRef]
89. Zhang, Y.; Ren, S.; Liu, Y.; Sakao, T.; Huisingh, D. A Framework for Big Data Driven Product Life Cycle Management. *J. Clean. Prod.* **2017**, *159*, 229–240. [CrossRef]
90. Kache, F.; Seuring, S. Challenges and Opportunities of Digital Information at the Intersection of Big Data Analytics and Supply Chain Management. *Int. J. Oper. Prod. Manag.* **2017**, *37*, 10–36. [CrossRef]
91. Lee, D.; Park, J. RFID-based traceability in the supply chain. *Ind. Manag. Data Syst.* **2008**, *108*, 713–725. [CrossRef]
92. Dietrich, F.; Louw, L.; Palm, D. Blockchain-Based Traceability Architecture for Mapping Object-Related Supply Chain Events. *Sensors* **2023**, *23*, 1410. [CrossRef] [PubMed]
93. Sugandh, U.; Nigam, S.; Khari, M.; Misra, S. An Approach for Risk Traceability Using Blockchain Technology for Tracking, Tracing, and Authenticating Food Products. *Information* **2023**, *14*, 613. [CrossRef]
94. Stute, M.; Sardesai, S.; Parlings, M.; Senna, P.P.; Fornasiero, R.; Balech, S. Technology Scouting to Accelerate Innovation in Supply Chain. In *Next Generation Supply Chains, Lecture Notes in Management and Industrial Engineering*; Fornasiero, R., Sardesai, S., Barros, A.C., Matopoulos, A., Eds.; Springer International Publishing: Cham, Switzerland, 2021; pp. 129–145.
95. Bianchini, D.; Fapanni, T.; Garda, M.; Leotta, F.; Mecella, M.; Rula, A.; Sardini, E. Digital Thread for Smart Products: A Survey on Technologies, Challenges, and Opportunities in Service-Oriented Supply Chains. *IEEE Access* **2024**, *12*, 125284–125305. [CrossRef]
96. Essien, A.; Chukwukelu, G.O.; Kazantsev, N.; Subramanian, N. Unveiling the Factors Influencing Transparency and Traceability in Agri-Food Supply Chains: An Interconnected Framework. *Supply Chain Manag.* **2024**, *29*, 602–619. [CrossRef]
97. Berger, K.; Baumgartner, R.J.; Weinzerl, M.; Bachler, J.; Preston, K.; Schöggl, J.P. Data Requirements and Availabilities for a Digital Battery Passport—A Value Chain Actor Perspective. *Cleaner Prod. Lett.* **2023**, *4*, 100032. [CrossRef]
98. Schöneich, S.; Saulich, C.; Müller, M. Traceability and Foreign Corporate Accountability in Mineral Supply Chains. *Regul. Gov.* **2023**, *17*, 954–969. [CrossRef]
99. Garfield, S. The Impact of Regulations on the Sustainability of the Food Industry: From Safety to Traceability. *Curr. Res. Law. Pract.* **2023**, *1*, 44–50.
100. Gartner, P.; Benfer, M.; Kuhnle, A.; Lanza, G. Potentials of Traceability Systems—A Cross-Industry Perspective. *Procedia CIRP* **2021**, *104*, 987–992. [CrossRef]

101. Nadvi, K.; Waltring, F. Making Sense of Global Standards. In *Local Enterprises in the Global Economy*; Schmitz, H., Ed.; Edward Elgar: Cheltenham, UK, 2004; pp. 53–94.
102. Muirhead, J.; Porter, T. Traceability in Global Governance. *Glob. Netw.* **2019**, *19*, 423–443. [CrossRef]
103. Frendi, M.; Nachet, B.; Adla, A. Physical Internet Based Ontology for Supporting Traceability in Logistic IoT. *Int. J. Comput. Digit. Syst.* **2024**, *15*, 427–440. [CrossRef]
104. Myae, A.C.; Goddard, E. Importance of traceability for sustainable production: A cross-country comparison. *Int. J. Consum. Stud.* **2012**, *36*, 192–202. [CrossRef]
105. Samarasinghe, Y.M.P.; Kumara, B.A.M.S.; Kulatunga, A.K. Traceability of Fruits and Vegetables Supply Chain towards Efficient Management: A Case Study from Sri Lanka. *Int. J. Ind. Eng. Oper. Manag.* **2021**, *3*, 89–106. [CrossRef]
106. Bahrami, S.; Zeinali, D. The sustainability challenge of product information quality in the design and construction of facades: Lessons from the Grenfell Tower fire. *Smart Sustain. Built Environ.* **2023**, *12*, 488–506. [CrossRef]
107. Agrawal, T.K.; Koehl, L.; Campagne, C. A secured tag for implementation of traceability in textile and clothing supply chain. *Int. J. Adv. Manuf. Technol.* **2018**, *99*, 2563–2577. [CrossRef]
108. *ISO 8402:1994*; Quality Management and Quality Assurance—Vocabulary. International Organization for Standardization: Geneva, Switzerland, 1994.
109. *ISO 9000:2015*; Quality Management and Quality Assurance—Vocabulary. International Organization for Standardization: Geneva, Switzerland, 2015.
110. *BS EN ISO 9000:2005*; Quality Management Systems—Fundamentals and Vocabulary. British Standards Institution: London, UK, 2005.
111. *Global Standards One*; GS1 Standards Document GS1 Global Traceability Standard: Brussels, Belgium, 2012.
112. Schuitemaker, R.; Xu, X. Product traceability in manufacturing: A technical review. *Procedia CIRP* **2020**, *93*, 700–705. [CrossRef]
113. van der Vorst, J.G. Product traceability in food-supply chains. *Accredit. Qual. Assur.* **2006**, *11*, 33–37. [CrossRef]
114. Cheng, M.; Simmons, J. Traceability in Manufacturing Systems. *Int. J. Oper. Prod. Manag.* **1994**, *14*, 4–16. [CrossRef]
115. Yao, S.; Zhu, K. Combating product label misconduct: The role of traceability and market inspection. *Eur. J. Oper. Res.* **2020**, *282*, 559–568. [CrossRef]
116. Liu, X.L.; Wang, W.M.; Guo, H.Y.; Barenji, A.V.; Li, Z.; Huang, G.Q. Industrial blockchain based framework for product lifecycle management in industry 4.0. *Robot. Comput. Integr. Manuf.* **2020**, *63*, 101897. [CrossRef]
117. Lu, Q.; Xu, X. Adaptable Blockchain-Based Systems: A Case Study for Product Traceability. *IEEE Softw.* **2017**, *34*, 21–27. [CrossRef]
118. Tekin, E. A Method for Traceability and "As-Built Product Structure" In Aerospace Industry. *Procedia CIRP* **2014**, *17*, 351–355. [CrossRef]
119. Eynard, B.; Gallet, T.; Nowak, P.; Roucoules, L. UML based specifications of PDM product structure and workflow. *Comput. Ind.* **2004**, *55*, 301–316. [CrossRef]
120. Li, Y.; Wan, L.; Xiong, T. Product data model for PLM system. *Int. J. Adv. Manuf. Technol.* **2011**, *55*, 1149–1158. [CrossRef]
121. Kropsu-Vehkapera, H.; Haapasalo, H.; Harkonen, J.; Silvola, R. Product data management practices in high-tech companies. *Ind. Manag. Data Syst.* **2009**, *109*, 758–774. [CrossRef]
122. Campos, J.G.; Míguez, L.R. Digital Traceability from Design to Manufacturing in Extended Enterprises. *IFAC Proc. Vol.* **2006**, *39*, 529–534. [CrossRef]
123. Ospital, P.; Masson, D.H.; Beler, C.; Legardeur, J. Toward total traceability and full transparency communication in textile industry supply chain. *INCOSE Int. Symp.* **2022**, *32*, 1–7. [CrossRef]
124. Heber, D.; Groll, M. Towards a digital twin: How the blockchain can foster e/e-traceability in consideration of model-based systems engineering. In Proceedings of the 21st International Conference on Engineering Design: Product, Services and Systems Design, Vancouver, BC, Canada, 21–25 August 2017; Volume 3, pp. 321–330.
125. Giddaluru, M.P.; Gao, J.X.; Bhatti, R. A Modular Product Structure Based Methodology for Seamless Information Flow in PLM System Implementation. *Comput. Aided Des. Appl.* **2015**, *12*, 742–752. [CrossRef]
126. Giddaluru, M.P.; Gao, J.X.; Bhatti, R. Integrating product knowledge with modular product structures in PLM. In Proceedings of the International Conference on Innovative Design and Manufacturing, Milan, Italy, 17–19 July 2017; pp. 1–6.
127. Bitzer, M.; Vielhaber, M.; Kaspar, J. Product Lifecycle Management—How to adapt PLM to support changing product development processes in industry? In Proceedings of the NordDesign 2016, Trondheim, Norway, 10–12 August 2016; Volume 1, pp. 360–369.
128. Campos, J.G.; Martin, R.M.; Lopez, J.S.; Ignacio, J. e-Traceability: Traceability for collaborative spread CAD-CAM-CNC manufacturing chains. In Proceedings of the 5th WSEAS International Conference on E-activities, Venice, Italy, 20–22 November 2006; pp. 425–431.
129. Zsifkovits, H.; Kapeller, J.; Reiter, H.; Weichbold, C.; Woschank, M. Consistent Identification and Traceability of Objects as an Enabler for Automation in the Steel Processing Industry. In *Industry 4.0 for SMEs*; Matt, D., Modrák, V., Zsifkovits, H., Eds.; Palgrave Macmillan: Cham, Switzerland, 2020.
130. Appelhanz, S.; Osburg, V.S.; Toporowski, W.; Schumann, M. Traceability system for capturing, processing and providing consumer-relevant information about wood products: System solution and its economic feasibility. *J. Clean. Prod.* **2016**, *110*, 132–148. [CrossRef]
131. Campos, L.B.; Cugnasca, C.E. Towards an IoT-based architecture for wine traceability. In Proceedings of the International Conference on Distributed Computing in Sensor Systems, Fortaleza, Brazil, 10–12 June 2015; pp. 212–213.

132. Benatia, M.A.; Baudry, D.; Louis, A. Detecting counterfeit products by means of frequent pattern mining. *J. Ambient. Intell. Humaniz. Comput.* **2022**, *13*, 3683–3692. [CrossRef]
133. Guo, D.; Li, M.; Lyu, Z.; Kang, K.; Wu, W.; Zhong, R.Y.; Huang, G.Q. Synchroperation in industry 4.0 manufacturing. *Int. J. Prod. Econ.* **2021**, *238*, 108171. [CrossRef]
134. Syed, N.M.; Shah, S.W.; Trujillo-Rasua, R.; Doss, R. Traceability in supply chains: A Cyber security analysis. *Comput. Secur.* **2022**, *112*, 102536. [CrossRef]
135. Aizenbud-Reshef, N.; Nolan, B.T.; Rubin, J.; Shaham-Gafni, Y. Model traceability. *IBM Syst. J.* **2006**, *45*, 515–526. [CrossRef]
136. Brintrup, A.; Kosasih, E.; Schaffer, P.; Zheng, G.; Demirel, G.; MacCarthy, B.L. Digital supply chain surveillance using artificial intelligence: Definitions, opportunities and risks. *Int. J. Prod. Res.* **2023**, *62*, 4674–4695. [CrossRef]
137. Charles, V.; Emrouznejad, A.; Gherman, T. A critical analysis of the integration of blockchain and artificial intelligence for supply chain. *Ann. Oper. Res.* **2023**, *327*, 7–47. [CrossRef] [PubMed]
138. Barata, J.; da Cunha, P.R.; Gonnagar, A.S.; Mendes, M. Product Traceability in Ceramic Industry 4.0: A Design Approach and Cloud-Based MES Prototype. In *Advances in Information Systems Development*; Paspallis, N., Raspopoulos, M., Barry, C., Lang, M., Linger, H., Schneider, C., Eds.; Lecture Notes in Information Systems and Organisation; Springer: Cham, Switzerland, 2018; Volume 26.
139. Sahin, E.; Dallery, Y.; Gershwin, S. Performance evaluation of a traceability system: An application to radio frequency identification technology. In Proceedings of the IEEE International Conference on Systems, Man and Cybernetics, Yasmine Hammamet, Tunisia, 6–9 October 2002.
140. Green, K.W.; Zelbst, P.J.; Sower, V.E.; Bellah, J.C. Impact of Radio Frequency Identification Technology on Environmental Sustainability. *J. Comput. Inf. Syst.* **2017**, *57*, 269–277. [CrossRef]
141. Zuo, J.; Feng, J.; Gameiro, M.G.; Tian, Y.; Liang, J.; Wang, Y.; Ding, J.; He, Q. RFID-Based Sensing in Smart Packaging for Food Applications: A Review. *Future Foods* **2022**, *6*, 100198. [CrossRef]
142. Dietrich, A.J.; Kirn, S.; Timm, I.J. Implications of mass customisation on business information systems. *Int. J. Mass. Cust.* **2006**, *1*, 218–236. [CrossRef]
143. Valdivia, C.A.S.; Mamédio, D.F.; Loures, E.d.F.R.; Tortato, U. Dimensions of Digital Transformation for Digital Supply Chains—Evidence from an Automotive OEM Group. *Res.-Technol. Manag.* **2024**, *67*, 57–68.
144. Khan, S.I.; Kaur, C.; Al Ansari, M.S.; Muda, I.; Borda, R.F.C.; Bala, B.K. Implementation of Cloud-Based IoT Technology in Manufacturing Industry for Smart Control of Manufacturing Process. *Int. J. Interact. Des. Manuf.* **2023**. [CrossRef]
145. Hrustek, L. Sustainability Driven by Agriculture through Digital Transformation. *Sustainability* **2020**, *12*, 8596. [CrossRef]
146. Olsen, P.; Borit, M. The components of a food traceability system. *Trends Food Sci. Technol.* **2018**, *77*, 143–149. [CrossRef]
147. Gunawan, I.; Vanany, I.; Widodo, E.; Mulyana, I.J. Improving Traceability System in Indonesian Coconut Oil Company. In Proceedings of the IEEE International Conference on Industrial Engineering and Engineering Management (IEEM), Bangkok, Thailand, 16–19 December 2018; pp. 51–55.
148. Zhang, Y.; Guin, U. End-to-End Traceability of ICs in Component Supply Chain for Fighting Against Recycling. *IEEE Trans. Inf. Forensics Secur.* **2020**, *15*, 767–775. [CrossRef]
149. Li, J.; Wang, Z.; Guan, S.; Cao, Y. ProChain: A Privacy-Preserving Blockchain-Based Supply Chain Traceability System Model. *Comput. Ind. Eng.* **2024**, *187*, 109831. [CrossRef]
150. Alnafrah, I.; Mouselli, S. Revitalizing Blockchain Technology Potentials for Smooth Academic Records Management and Verification in Low-Income Countries. *Int. J. Educ. Dev.* **2021**, *85*, 102460. [CrossRef]
151. Alnafrah, I.; Bogdanova, E.; Maximova, T. Text Mining as a Facilitating Tool for Deploying Blockchain Technology in the Intellectual Property Rights System. *Int. J. Intellect. Prop. Manag.* **2019**, *9*, 120–135. [CrossRef]
152. Harkonen, J.; Mustonen, E.; Koskinen, J.; Hannila, H. Digitizing Company Analytics—Digitalization Concept for Valuable Insights. In Proceedings of the 2020 IEEE International Conference on Industrial Engineering and Engineering Management (IEEM2020), Singapore, 14–17 December 2020; pp. 1012–1016.
153. Koskinen, J.; Mustonen, E.; Harkonen, J.; Haapasalo, H. Product-level profitability analysis and decision-making: Opportunities of IT application-based approach. *Int. J. Prod. Lifecycle Manag.* **2020**, *12*, 210–225. [CrossRef]
154. Reyna, A.; Martín, C.; Chen, J.; Soler, E.; Díaz, M. On blockchain and its integration with IoT: Challenges and opportunities. *Future Gener. Comput. Syst.* **2018**, *88*, 173–190. [CrossRef]
155. Sopadang, A.; Chonsawat, N.; Ramingwong, S. Smart SME 4.0 Implementation Toolkit. In *Industry 4.0 for SMEs*; Matt, D., Modrák, V., Zsifkovits, H., Eds.; Palgrave Macmillan: Cham, Switzerland, 2020.
156. Khan, M.; Parvaiz, G.S.; Dedahanov, A.T.; Abdurazzakov, O.S.; Rakhmonov, D.A. The Impact of Technologies of Traceability and Transparency in Supply Chains. *Sustainability* **2022**, *14*, 16336. [CrossRef]
157. Dedeoglu, V.; Malik, S.; Ramachandran, G.; Pal, S.; Jurdak, R. Blockchain meets edge-AI for food supply chain traceability and provenance. *Compr. Anal. Chem.* **2023**, *101*, 251–275.
158. Chatterjee, R.; Gamota, D. The Convergence of Technologies and Standards Across the Electronic Products Manufacturing Industry (SEMI, OSAT, and PCBA) to Realize Smart Manufacturing. In Proceedings of the 2020 Pan Pacific Microelectronics Symposium (Pan Pacific), Big Island, HI, USA, 10–13 February 2020; pp. 1–6. [CrossRef]
159. Lee, J.; Singh, J.; Azamfar, M.; Pandhare, V. Industrial AI and predictive analytics for smart manufacturing systems. In *Smart Manufacturing*, 1st ed.; Soroush, M., Baldea, M., Edgar, T.F., Eds.; Elsevier: Amsterdam, The Netherlands, 2020; pp. 213–244.

160. Ofner, M.H.; Straub, K.; Otto, B.; Oesterle, H. Management of the Master Data Lifecycle: A Framework for Analysis. *J. Enterp. Inf. Manag.* **2013**, *26*, 472–491. [CrossRef]
161. Silvola, R.; Tolonen, A.; Harkonen, J.; Haapasalo, H.; Mannisto, T. Defining One Product Data for a Product. *Int. J. Bus. Inf. Syst.* **2019**, *30*, 489–520. [CrossRef]
162. Silvola, R.; Jaaskelainen, O.; Kropsu-Vehkapera, H.; Haapasalo, H. Managing One Master Data—Challenges and Preconditions. *Ind. Manag. Data Syst.* **2011**, *111*, 146–162. [CrossRef]
163. Wuttke, T.; Haskamp, T.; Perscheid, M.; Uebernickel, F. Building the Processes Behind the Product: How Digital Ventures Create Business Processes That Support Their Growth. *Bus. Inf. Syst. Eng.* **2024**. [CrossRef]
164. Jaakkola, E. Unraveling the Practices of "Productization" in Professional Service Firms. *Scand. J. Manag.* **2011**, *27*, 221–230. [CrossRef]
165. Wirtz, J.; Kowalkowski, C. Putting the "Service" into B2B Marketing: Key Developments in Service Research and Their Relevance for B2B. *J. Bus. Ind. Mark.* **2023**, *38*, 272–289. [CrossRef]
166. Elia, V.; Gnoni, M.G.; Tornese, F. Exploring the Benefits of Productization in the Utilities Sector. *Sustainability* **2019**, *11*, 5864. [CrossRef]
167. Valminen, K.; Toivonen, M. Seeking Efficiency through Productisation: A Case Study of Small KIBS Participating in a Productisation Project. *Serv. Ind. J.* **2012**, *32*, 273–289. [CrossRef]
168. Cooper, R.G.; Edgett, S.J.; Kleinschmidt, E.J. New Product Portfolio Management: Practices and Performance. *J. Prod. Innov. Manag.* **1999**, *16*, 333–351. [CrossRef]
169. Remeňová, K.; Kintler, J.; Jankelová, N. The General Concept of the Revenue Model for Sustainability Growth. *Sustainability* **2020**, *12*, 6635. [CrossRef]
170. Kunwar, P.J.; Luukkonen, T.; Haapasalo, H.; Majava, J. Addressing adsorbent materials commercialization challenges for water treatment in European markets through productization. *Cogent Eng.* **2024**, *11*, 2320952. [CrossRef]
171. Riesener, M.; Keuper, A.; Boßmann, C.; Schuh, G. Derivation of targets for the portfolio planning of a solution provider in machinery engineering. In Proceedings of the 2024 Portland International Conference on Management of Engineering and Technology (PICMET), Portland, OR, USA, 4–8 August 2024; pp. 1–10.
172. Mustonen, E.; Harkonen, J. Commercial and technical productization for design reuse in engineer-to-order business. *IEEE Trans. Eng. Manag.* **2024**, *71*, 1271–1284. [CrossRef]
173. Shamsuzzoha, A.; Blomqvist, H.; Takala, J. Service productisation through standardisation and modularisation: An exploratory case study. *Int. J. Sustain. Eng.* **2023**, 1–19. [CrossRef]
174. Mammela, J.; Mustonen, E.; Harkonen, J.; Pakkanen, J.; Juuti, T. Productization as a link to combining product portfolio management and product family development. *Proc. CIRP* **2022**, *109*, 25–30. [CrossRef]
175. Mansoori, S.; Harkonen, J.; Haapasalo, H. Productization and product structure enabling BIM implementation in construction. *Eng. Constr. Archit. Manag.* **2023**, *30*, 2155–2184. [CrossRef]
176. Leppänen, T.; Mustonen, E.; Saarela, H.; Kuokkanen, M.; Tervonen, P. Productization of industrial side streams into by-products—Case: Fiber sludge from pulp and paper industry. *J. Open Innov. Technol. Mark. Complex.* **2020**, *6*, 185. [CrossRef]
177. Chauhan, K.; Peltokorpi, A.; Lavikka, R.; Seppänen, O. The monetary and non-monetary impacts of prefabrication on construction: The effects of product modularity. *Buildings* **2022**, *12*, 459. [CrossRef]
178. Rodrigues, R.N.; Pinto da Silva Júnior, A.R.; Barroso, M.B.C.; Bagno, R.B. Productizing methods and tools for innovation management and entrepreneurship: A process proposal. *Prod. Manag. Dev.* **2023**, *21*, e20240001. [CrossRef]
179. Kangas, N.; Kropsu-Vehkapera, H.; Haapasalo, H.; Kinnunen, K. Empirical aspects on defining product data for rapid productisation. *Int. J. Synerg. Res.* **2013**, *2*, 107–128.
180. Kaski, T.; Heikkila, J. Measuring product structures to improve demand-supply chain efficiency. *Int. J. Technol. Manag.* **2002**, *23*, 578–598. [CrossRef]

eng

MDPI

Article

Examining Sentiment Analysis for Low-Resource Languages with Data Augmentation Techniques

Gaurish Thakkar *, Nives Mikelić Preradović * and Marko Tadić

Faculty of Humanities and Social Sciences, University of Zagreb, Ivana Lučića 3, 10000 Zagreb, Croatia; marko.tadic@ffzg.hr
* Correspondence: gthakkar@ffzg.hr (G.T.); nmikelic@ffzg.hr (N.M.P.)

Abstract: This investigation investigates the influence of a variety of data augmentation techniques on sentiment analysis in low-resource languages, with a particular emphasis on Bulgarian, Croatian, Slovak, and Slovene. The following primary research topic is addressed: is it possible to improve sentiment analysis efficacy in low-resource languages through data augmentation? Our sub-questions look at how different augmentation methods affect performance, how effective WordNet-based augmentation is compared to other methods, and whether lemma-based augmentation techniques can be used, especially for Croatian sentiment tasks. The sentiment-labelled evaluations in the selected languages are included in our data sources, which were curated with additional annotations to standardise labels and mitigate ambiguities. Our findings show that techniques like replacing words with synonyms, masked language model (MLM)-based generation, and permuting and combining sentences can only make training datasets slightly bigger. However, they provide limited improvements in model accuracy for low-resource language sentiment classification. WordNet-based techniques, in particular, exhibit a marginally superior performance compared to other methods; however, they fail to substantially improve classification scores. From a practical perspective, this study emphasises that conventional augmentation techniques may require refinement to address the complex linguistic features that are inherent to low-resource languages, particularly in mixed-sentiment and context-rich instances. Theoretically, our results indicate that future research should concentrate on the development of augmentation strategies that introduce novel syntactic structures rather than solely relying on lexical variations, as current models may not effectively leverage synonymic or lemmatised data. These insights emphasise the nuanced requirements for meaningful data augmentation in low-resource linguistic settings and contribute to the advancement of sentiment analysis approaches.

Keywords: sentiment analysis; language models; data augmentation

Citation: Thakkar, G.; Preradović, N.M.; Tadić, M. Examining Sentiment Analysis for Low-Resource Languages with Data Augmentation Techniques. *Eng* **2024**, *5*, 2920–2942. https://doi.org/10.3390/eng5040152

Academic Editor: Antonio Gil Bravo

Received: 13 September 2024
Revised: 1 November 2024
Accepted: 4 November 2024
Published: 7 November 2024

1. Introduction

"A neural network is a computational model inspired by the way biological neural networks in the human brain function. It consists of layers of interconnected nodes (called neurons), where each node performs a simple computation, and information is passed from one layer to another". In the context of a neural network, parameters refer to the internal variables that the model learns from the training data. These include the weights and biases associated with the neurons in each layer [1]. "Hyperparameters are configuration settings that are used to control the learning process of a machine learning model but are not learnt from the data itself. They differ from model parameters in that they are set before training begins and remain constant during the training process" [1,2]. In contrast to learnt parameters (such as weights and biases), which are modified during training, hyperparameters govern the learning process and must be defined before model training commences. Examples encompass the learning rate, batch size, layer count, number of neurons per layer, and dropout rate. In neural networks, parameters refer to the values learned by the model during the training process.

The performance of a neural network is completely dependent on its hyperparameters and the training set-learned parameters [3]. It is commonly believed that having more data points is the default method for improving performance [4]. A direct approach requires running an annotation campaign, which is expensive, time-consuming, and labour-intensive in terms of annotation and training [5]. Because these models rely on large parameters that necessitate many training instances to perform the intended task, this requirement cannot be eliminated.

In the reverse direction, new data points are generated from existing supervised or unsupervised text bodies [6]. To date, numerous techniques for data generation have been identified. Ref. [7] reported using contextual language under the assumption that sentences are invariant when original words are replaced by words with paradigmatic relations [8]. When compared to original texts, in-context predicted words were deemed to be better options for creating data samples that vary in terms of pattern. Attempts [9,10] have also been made at using data augmentation for different text classifications in large English-language datasets. The augmentations were derived from an English thesaurus and then trained using various machine learning and deep learning algorithms. Ref. [6] described simple augmentation operations (such as insertion, deletion, swap, and replacement) that produced comparable results when only half of the original dataset was used.

In data-driven research, these techniques focus primarily on resolving low-data scenarios, mitigating the phenomenon of class imbalance, or serving as regularising terms to make systems more resistant to adversarial attacks. The purpose of enhancing a neural network model's resistance against adversarial attacks is to guarantee its robustness and reliability in practical applications. Adversarial attacks entail the creation of minor, frequently undetectable, modifications to input data that can lead the model to produce erroneous predictions. This presents considerable security threats in applications like autonomous driving, medical diagnosis, and facial recognition [11].

Existing data augmentation strategies for other tasks in languages with abundant resources (especially English) have also been investigated. To detect event causality, Ref. [12] employed a remote annotator, followed by filtering, relabelling, and annealing on instances with noisy labels. For the common-sense reasoning task, Ref. [13] used a pre-trained task model (XLM-R) and a generative language model (GPT-2) to generate synthetic data instances. Data selection was conducted using filtering functions that considered the quality and diversity of synthetic instances. The approaches that have been published for high-resource languages, such as English, are constructed using other linguistic resources as primary building blocks. To produce facts from an existing knowledge repository or knowledge graph, such a resource must be available in the target language. Therefore, a language with limited resources may lack these dependent resources, thereby rendering the method inapplicable. Empirical evidence regarding the effectiveness of these interventions in low-resource settings is still lacking. Even though data augmentation techniques such as EDA (Easy Data Augmentation) [6] are simple to implement, it is essential to conduct additional research on their applicability in low-resource settings.

This paper aims to investigate the efficacy of various data augmentation (DA) strategies in enhancing sentiment analysis for low-resource languages, particularly South Slavic languages. The article presents a novel strategy termed "expand-permute-combine" and assesses its efficacy in comparison to other methods to evaluate their influence on classification accuracy for under-resourced languages. We hypothesise that DA strategies are equivalent to cross-lingual and cross-family configurations. For the task of sentiment classification, we experiment with various data augmentation techniques on a set of low-resource languages from the same language family (i.e., South Slavic languages). To analyse each of these facets, we employ three distinct data augmentation techniques that rely on synonymy [14] and pre-trained large language models [15,16]. In addition, we propose a straightforward method of augmentation that requires no additional resources. To determine the effectiveness of these techniques, evaluation was performed on the task of sentiment classification. Experiments were conducted on South-Slavic languages (i.e., Bulgarian, Croatian, Slovak, and Slovene). To enable a three-class classification of the dataset for the Croatian language, we also conducted an annotation campaign to label instances that were claimed to be noisy by the original authors of the dataset.

2. Research Question

In this study, we explore DA methods as a means to artificially increase the instance space and compare the performance with that when using resources from the same language family. This study has the following main research question: can data augmentation be utilised effectively for sentiment analysis in low-resource languages? Additionally, 3–4 more specific questions are used, as follows:

(1) Can the data augmentation technique improve the performance metric?
(2) What is the effect of using augmented data generated from different techniques? We explore three different data augmentation techniques and compare their performances with each other.
(3) Can WordNet-based augmentation techniques work better with sentiment classification tasks?
    - Does training with Lemma-based instances work for Croatian?

We hypothesise that the accuracy of the data augmentation techniques is comparable to that of supervised methods when applied to typologically related languages.

3. Literature Review

The section commences by examining the key methodologies and advancements in the field that are relevant to this investigation. This includes an analysis of different approaches, including data augmentation, adversarial attacks, and distant supervision, that have been used to enhance the performance of NLP tasks. In the subsequent subsections, we will explore specific techniques and models, emphasising their application in a variety of domains, with a particular emphasis on their relevance to sentiment analysis and low-resource languages. This investigation establishes the groundwork for understanding the landscape of previous research and the context for the methodologies proposed in this work.

Data Augmentation

Distant supervision is a method for curating labelled data instances by utilising an existing knowledge base [17]. Ref. [18] reported the first instance of using distant supervision in NLP. The work entailed curating datasets for the task of relation extraction. The authors used Freebase, a large database that stores the relationships between two entities. The assumption was that any sentence containing two freebase entities could express the relationship. As a result, Freebase was used as an unsupervised lookup table. Various features were designed, ranging from POS tag, NER, and n-words within the context window. Ref. [17] introduced a similar approach in the BioNLP domain, in which knowledge from a database is used to label sentences containing two entities to generate a dataset based on remote supervision. In the same work, heuristics (trigger words and high confidence patterns) were proposed to reduce noise in the sentence augmentation process. A CNN trained with an automatically created dataset and then trained on a manually annotated dataset achieved the highest score. The authors hypothesised that the direct union of two datasets (distant supervision-based and manually annotated) is not advantageous because noisy datasets lead to a decline in the final performance.

Two types of augmentation methods for NLP can be broadly distinguished: (1) text-based augmentation and (2) feature-based augmentation. The text-based enhancements operate at the text level. The process of augmentation can be implemented at various linguistic levels (morphological, syntactic, and semantic). Another branch of research focuses on adversarial attacks against the trained model. This is accomplished by generating text instances X' similar to the training data X, such that the model attempting to perform the intended task fails. Instances X and X' should have identical human predictions, with X' containing minimal textual changes relative to the original instance. All adversarial attack techniques [19–21] on classification tasks rely on text-augmenters as their primary component when supplying augmented instances for adversarial attacks.

Ref. [22] experimented with various synonym replacement methods to generate adversarial samples. The synonyms were obtained from WordNet. The method for choosing a synonym for a word ranged from random selection to a more sophisticated method based on Word Saliency [23] score. Another way of finding a replacement for a given word is to use a pre-trained language model that uses context to predict the replacement word. Ref. [7] altered the language model so that it integrates the label in the model along with the context during the word prediction stage. The language mode was trained on the WikiText-103 corpus of English Wikipedia articles. Ref. [19] used contextual perturbations from a BERT masked language model to replace and insert tokens at masked locations. Ref. [24] extended the work using RoBERTa and three contextualised perturbations, i.e., replace, insert, and merge. All of these studies were published in English datasets.

In the field of Neural Machine Translation (NMT), the technique of translating a target language into a source language is known as back-translation [25]. The ultimate goal of this procedure is to increase the number of samples by paraphrasing using the translation module. The final system is trained using both the parallel synthetic corpus and the original training data. Although back-translation is an easy-to-use technique, it necessitates the training of a machine-translation model for low-resource languages, which may not be a viable option given the required volume of data. Ref. [26] showed through experiments that sampling and noisy beam outputs (delete, swap, and replace words) are better for making fake data than pure beam and greedy search. Ref. [6] introduced EDA (Easy Data Augmentation), a set of augmentation techniques consisting of multiple processes including synonym replacement, random replacement, random swap, and random deletion. On five distinct datasets, the processes were executed and benchmarked. The authors conducted experiments with an augmentation parameter named α whose values were in the range [0.05, 0.1, 0.2, 0.3, 0.4, 0.5] and discovered that small α values provided greater gain than large values. The same work was expanded by [27] to include two additional datasets for examining the impact of data augmentations using pre-trained language models (BERT, XL-NET, and ROBERTA). EDA and back-translation are two task-independent data augmentation techniques. According to reports, data-augmentation methods do not provide any consistent improvement for pre-trained transformers. The authors attributed this phenomenon to large-scale, unsupervised, domain-spanning pre-training, although all datasets utilised in the study were English-based.

Consistency training is based on the premise that small changes or noise in the input should not impact model predictions. Ref. [28] used data augmentation in place of noise signals to enforce consistency constraints during training. The overall loss consisted of classification losses and consistency losses between the original input and the enhanced version of the same. The consistency loss is only computed for instances in which the model has high confidence. The author used back-translation, RandAugment (for image classification), and TF-IDF word replacement for augmentations. A data filter was implemented within the domain to prevent domain mismatch.

Ref. [29] proposed the first method for classifying the sentiment of tweets using emoticons as remote supervisors. The technique was based on the premise that the emoticons ":)" and ":(" (and their variants) are poor indicators of positive and negative emotions. Therefore, each tweet containing these emoticons was tagged with their respective classes. There was an assumption that the statements in Wikipedia and newspaper headlines were neutral. The neutral class was not classified because it had no emoticons associated with it. The dataset was used to train the machine learning algorithms Naive Bayes, Maximum Entropy (MaxEnt), and Support Vector Machines (SVM). The entire setup was studied using English as the study language. Ref. [30] compared multiple data augmentation strategies (such as WordNet and Bert-based) for the generation of news headlines in Croatian, Finnish, and English. In addition to ROGUE, the authors employed two additional methods to assess the performance score. One technique was the computation of semantic similarity using a sentence transformer trained in the task of paraphrasing. The second method employed a metric based on natural language inference to quantify the similarity between the original and generated headlines. The authors did note that there was no NLI model covering Croatian and Estonian. The other branch of data augmentation directly focuses on the latent space. Training as a whole aims to add new latent information without altering the original class representation. This enables difficult-to-input semantic cases with limited training data to be induced. Ref. [31] proposed that difficult-to-classify samples are the best candidates for data augmentation because they contain more information. Latent space augmentations were created using interpolation, extrapolation, noise addition, and the difference transform. Table 1 presents a summary of all the aforementioned approaches.

Table 1. Literature review.

Author	Purpose	Method	Sample Size	Key Findings
[18]	relation extraction in NLP	Distant supervision (DS) using Freebase as a lookup table	800 K	Multi-instance learning framework.
[29]	Classifying sentiment in tweets	Remote supervision using emoticons as labels	1600 K	Emoticons were used as labels for the SA of tweets.
[25]	Enhancing NMT with synthetic data	Back-translation	100 K	Used machine translation as paraphraser.
[7]	Improving adversarial attack performance	Altered language model trained on WikiText-103 corpus	7 K–540 K	Contextual DA method outperforms traditional DA methods
[26]	Improving NMT sample quality	Sampling and noisy beam outputs for back-translation	29 M	Noisy beam outputs, create better synthetic data than beam or greedy search.
[17]	Curating datasets for BioNLP tasks	Distant supervision with heuristics to reduce noise	25 K–77 K	Proposed heuristics to reduce noise.
[22]	Generating adversarial samples for NLP	Synonym replacement using WordNet	25 K –1.4 M	Saliency-based methods for detecting important words.
[6]	Simplifying data augmentation	EDA: synonym replacement, random replacement, swap, deletion	500–5 K	Found small augmentation values (α) produced better performance gains than large values.
[19]	Improving adversarial sample generation	Contextual perturbations using BERT masked language model	10 K–598 K datasets	Used BERT for replacing and inserting tokens at masked locations.
[27]	Examining the impact of pre-trained language models on data augmentation	Augmentation with BERT, XL-NET, and RoBERTa	500–10 K	DA did not provide consistent improvements for pre-trained transformers.
[28]	Enforcing consistency in model predictions with augmented data	Consistency training with back-translation and TF-IDF	25 K	Used consistency loss to improve model predictions.
[24]	Extending adversarial attack methods	Contextualized perturbations with RoBERTa	105 K–560 K	Introduced replace, insert, and merge operations for adversarial attacks.
[31]	Proposing data augmentation using latent space for difficult-to-classify samples	Latent space augmentation using interpolation and noise addition	50 K–120 K	Difficult-to-classify samples contain more information, making them ideal for DA in low-data settings.

Table 1. *Cont.*

Author	Purpose	Method	Sample Size	Key Findings
[30]	Comparing augmentation strategies for headline generation in various languages	WordNet and Bert-based augmentation	10 K–260 K	Domain-specific data benefit more from data augmentation and pretraining schemes
Ours	Comparing multiple DA strategies for SA in various low-resourced languages	Expansion and permutation-based techniques	10 K–40 K	Transformer-based models do not benefit from DA based on synonymy.

Techniques dependent on external knowledge bases [18] encounter challenges in disambiguating and resolving contexts for a singular matched item. This introduces noisy labels, which impact the system's accuracy. Ref. [29] faced challenges with noisy text and informality, as well as the effect of emoticons as labels. Methods employing NMT presume the existence of an NMT system and a substantial monolingual corpus within the domain. The NMT system generates noisy back-translations mostly characterised by lexical inaccuracies [25]. Previous research indicates that sentiment analysis using augmented data for low-resource languages has received little attention.

Morphology is the examination of the structure of words and the process by which they are assembled from lesser elements, known as morphemes [32]. In relation to their grammatical function within a sentence, the morphological features of words, such as their tense, case, and number, are substantially modified in a number of low-resource languages. These transformations have the potential to substantially alter the form of words, which poses a challenge for models that were trained on smaller datasets. Inflection systems are a component of morphological structure and the process by which words alter their form to convey various grammatical categories, such as tense, mood, or number. For instance, in highly inflected languages, a single word can take on numerous forms based on its function in a sentence, which complicates the process of generalising machine learning models across various forms of the same word. In the absence of sufficient data to account for these variations, models may encounter difficulty in generalising, which may result in inaccurate classifications. The situtuation is further complicated as these languages' grammars are not simple and their morphology and inflexion systems are complex.

4. Data

This study employed a mixed-methods research approach, combining both qualitative and quantitative methods to provide a comprehensive understanding of the research phenomenon. The quantitative component of the study entailed the collection and modelling of a dataset, which provided a comprehensive understanding of performance. In contrast, the qualitative component involved an in-depth analysis of the predictions from the trained classification systems, which offered contextualised and nuanced perspectives on the research phenomenon. To address the research questions, this mixed-methods approach was considered necessary, as it enabled the triangulation of data modelling and the validation of findings through error analysis, thereby enhancing the reliability and validity of the results.

We used sentiment classification datasets to answer our research questions, employing existing datasets from the previous studies. However, we targeted only low-resource languages in our experiment: Bulgarian, Croatian, Slovak, and Slovene. A single dataset was selected for each language in the study. In Table 2, the sizes of the original training, development, and test dataset splits are displayed.

Table 2. The original distribution of sentiment analysis datasets.

Language	Dataset	Train	Val	Test
Bulgarian	Cinexio	5520	614	682
Croatian	Pauza	2050	227	1033
Slovak	Reviews3	3834	661	1235
Slovene	KKS	3977	200	600

4.1. Croatian Re-Annotation

The authors of the Pauza dataset [33] eliminated reviews with a rating between 2.5 and 4.0 because these reviews were noisy. Therefore, ratings below 2.5 are considered negative, whereas ratings above 4.0 are considered positive. The reviews with ratings ranging from 2.4 to 4.0 have instances where the text is positive but has ratings that might tag it as a positive instance, and vice versa. We hypothesise that this might lead to semantic drift, meaning that the model might learn to classify instances incorrectly. Our methodology involves artificially augmenting data using multiple techniques; however, a text with contradictory labels, when excessively enhanced, may hinder the model's learning process. Hence, we take up the activity of re-annotating our Croatian dataset. We re-evaluated the ratings between 2.5 and 4.0 and asked three native speakers to annotate particular instances. Annotators were asked to classify the given text as positive, negative, or neutral/mixed. Only two annotators manage to complete the annotation of all the provided instances. The instances devoid of consensus were eliminated through filtering. Nine instances of the text were not included in the final set, as collective agreement about these instances was not reached by the annotators.

4.2. Sentiment Analysis Datasets

This section provides a detailed overview of the dataset's characteristics, including size, source, and distribution across different sentiment classes, which form the foundation for training the sentiment classification models.

- Bulgarian The Cinexio [34] dataset is composed of film reviews with 11-point star ratings: 0 (negative), 0.5, 1, . . . , 4.5, 5 (positive). Other meta-features included in the dataset were film length, director, actors, genre, country, and various scores.
- Croatian Pauza [33] contains restaurant reviews from Pauza.hr4, the largest food-ordering website in Croatia. Each review is assigned an opinion rating ranging from 0.5 (worst) to 6 (best). User-assigned ratings are the benchmark for the labels. The dataset also contains opinionated aspects.
- Slovak The Review3 [35] is composed of customer evaluations of a variety of services. The dataset is categorised using the 1–3 and 1–5 scales.
- Slovene The Opinion corpus of Slovene web commentaries KKS 1.001 [36] includes web commentaries on various topics (business, politics, sports, etc.) from four Slovene web portals (RtvSlo, 24ur, Finance, Reporter). Each instance within the dataset is tagged with one of three labels (negative, neutral, or positive).

The following two sections explains the overall methodology: data generation and model training. First, we used tools for natural language processing and data augmentation to create samples of the data. Then, we used the samples to train a transformer-based classification model on the data.

4.3. Data Generation and Augmentation

To answer the questions posed in earlier sections, we utilised four simple language processing techniques and three existing data augmentation methods. The aforementioned existing data augmentation strategies are used in adversarial attacks against trained classification models and can be utilised to obtain samples that are more semantically similar to the original dataset. Next, we describe the individual techniques for augmenting data and the overall procedure for augmenting and training the classifier.

- $Data_{lemma}$ based on lemmatisation.
- $Data_{expanded}$ based on sentence tokenisation [ours].
- $Data_{expanded-combined}$ based on sentence tokenisation [ours].
- $Data_{expanded-permuted}$ based on sentence tokenisation [ours].
- WordNet [22].
- Masked Language Model (MLM) based Clare [24].
- Causal Language Model (CLM)-based Generative Pre-trained Transformer (GPT)-2 [37].

4.4. Lemmatisation

After performing a morphological analysis, the lemmatisation process returned the word's morphological base. The output was the canonical form of the original word. Since South Slavic languages are rich in morphology, we decided to create a lemma-form variant of the original dataset. Previous studies [38,39] fed lemmas into machine learning classification algorithms as input features (such as Support Vector Machines and Random Forests). Transformers-based models use byte-pair encoding to reduce the vocabulary size, which is required to avoid sparse vector representations of the input text.

For instance, the word running is converted to run + ##ing and the neural network learns to weight individual byte-pairs based on the dataset and the requirements of the task. Therefore, the affixes may be useful for tasks that take the additional information into account. However, this requirement has not been looked at in pre-trained models with languages that are rich in morphology, or for sentiment analysis in particular. We made a lemmatised version of the original dataset to see how lemmatisation affects the final performance of a language model that has already been trained.

- Original HR: super, odlicni cevapi.
- Lemmatised: super, odličan ćevap.

4.5. Expansion [Ours]

Every labelled instance D^i from the train-set, i.e., the document or text, consists of one or more sentences $D^i_{1..n}$ and a single instance $D^i \in L$, where L can be positive, negative, or neutral/mixed.

$$D^i = D^i_{1..n} \tag{1}$$

$$D^i \in L \tag{2}$$

$$D^i_{1..n} \in L \tag{3}$$

$$D^1 D^2 D^n \in L \implies D^i \in L \tag{4}$$

From (4), it follows that each of the sentences $(D^1, D^2, .., D^n)$ of a single training instance can be weakly assumed to be labelled with the same class. Therefore, every sentence from a review can be individually treated as a new labelled instance. For example:

1. (Original HR): "Pizze Capriciosa i tuna, dobre. Inače uvijek dostava na vrijeme i toplo jelo".
2. (Translated EN): "Pizza Capricios and tuna, good. Otherwise always delivery on time and hot food".

This example belongs to the positive class, and individual sentences may be treated as reviews of the positive class. Theoretically, this assumption may hold true for extremely polar classes, such as positive and negative, but may fail for classes that are mixed or neutral. The mixed and neutral instances are indistinguishable. A mixed review consists of both positive and negative elements that are either connected by a conjunction or presented as two distinct phrases. There is no clear mechanism to differentiate between the positive and negative components. As the polar components are indiscernible without further processing, employing a positive statement from a mixed review and exponentially augmenting it would immediately lead to the inclusion of positive instances for the case of mixed classes in varying proportions. This would eventually lead to the misrepresentation of mixed classes during the training of neural networks. In practice, we are also presented with instances in which the service was poor, but the reviewer still awarded a high rating due to previous positive experiences.

4.5.1. Expansion-Combination [Ours]

Based on the previous technique for expansion, we propose a straightforward extension. Assuming that all individual sentences from all reviews for a given class also belong to the same parent class, we can now create a brand-new dataset by randomly sampling from this set of individual sentences. Here, we consider the entire $D^i_{1..n}$ range to be the universal set. We obtained the new dataset by sorting the instances using combinations denoted by mathematical (5). For a more intuitive explanation, assume ABCD to be four positive sentences from various positive reviews. Combination ordering produces a new sampled dataset represented by the combinations (ABCD′, 2) > AB AC AD BC BD CD." Elements are treated as unique based on their position, not on their value. So, if the input elements are unique, there will be no repeat values in each combination" [40]. This indicates that AB and BA will not be present in the final sampled dataset.

$$ {}^{n}C_k = \frac{n!}{k!(n-k)!} - combination \tag{5} $$

4.5.2. Expansion-Permutation [Ours]

We also propose a second simple method that replaces the previous combination sampling with a permutational process. Mathematically, this is denoted by Equation (6), in which the universal set of individual sentences belonging to a single class can be combined, as depicted by permutations (ABCD′, 2)—> AB AC AD BA BC BD CA CB CD DA DB DC. According to the order of the input iterable, the permutation tuples are returned in lexicographic order. Therefore, if the input iterable is sorted, the output combination tuples will also be sorted. "Elements are treated as unique based on their position, not on their value. So if the input elements are unique, there will be no repeat values in each permutation" [41]. In other words, AB and BA will represent two distinct instances of the generated dataset.

$$ {}^{n}P_k = \frac{n!}{(n-k)!} - permutation \tag{6} $$

4.5.3. WordNet Augmentations

WordNet [14,42–44] provides a straightforward formal synonym model for locating replacement words in context. This method replaces each word in a given text with its synonym. The assumption that a word's synonym will not affect the polarity of the given instance makes this one of the most straightforward data enhancement techniques. Synonyms are derived from synsets by querying WordNet with candidate keywords. The synset includes words with equivalent meanings. Notably, the word being searched may belong to multiple synsets, necessitating additional processing, such as word-sense disambiguation, to prevent incorrect synset selection (Due to the limited resources available, we did not pursue a more sophisticated synset selection).

(1) Lemma HR: Jako dobar pizza. (Translation: very good pizza.)
(2) Augmented HR: jako divan pizza.
(3) Augmented HR: jako krasan pizza.

Here, the word dobra ("good") has been replaced with its synonyms, 'divan' and 'krasan'. WordNet's entries are in lemmatised form, which is an important detail to note.

Therefore, in order to obtain more results for the words in context, they must be lemmatised. The lemma can then be used to retrieve the synonym set. The retrieved results are also in lemma form. Although this is not a necessary condition, we can still obtain a significant number of terms to replace the words in the dataset. This is illustrated by the following examples:

(1) HR: Jako dobra pizza i brza dostava. (Translation: Very good pizza and fast delivery.)
(2) Augmented HR: Jako dobra pizza i brza dostavljanje.
(3) Augmented HR: Jako dobra pizza i brza doprema.

To prevent semantic drift, no additional relations were employed. To reimplement a custom WordNet augmentor for each of the languages (Bulgarian, Croatian, Slovak, and Slovene), we used the textattack (https://github.com/QData/TextAttack, accessed on 21 July 2022) library, and derived a new class from the Augmentor (https://tinyurl.com/wz85rf43, accessed on 21 July 2022) base class. In the augmentor, we introduced constraints to prevent modifications to stopwords and words that were already modified. Based on the recommendation reported by [6], the pct-words-swap parameter (i.e., percentage of words to swap) was set to 0.05, limiting the number of words that were to be replaced with synonyms. The number of augmentations per instance was set at 16. We used Open Multilingual WordNet (http://compling.hss.ntu.edu.sg/omw, accessed on 24 July 2022) to find replacements for synonyms.

4.6. Language Tools

Each dataset for each of the four languages was required to undergo tokenisation, part of the speech extraction and lemmatisation. The Classla (https://github.com/clarinsi/classla, accessed on 26 July 2022) library was used for processing Bulgarian, Croatian, and Slovene, while the Stanza (https://stanfordnlp.github.io/stanza/, accessed on 26 July 2022) library was utilised for Slovak (https://huggingface.co/stanfordnlp/stanza-sk, accessed on 26 July 2022). We used the tokenised and lemmatised data to generate the lemmatised ($Data_{lemma}$) and expanded ($Data_{expanded}$) versions of the dataset. The expanded version was converted into $Data_{expanded\text{-}combined}$ and $Data_{expanded\text{-}permuted}$ by combining two individual sentences into a single training instance via sampling.

4.7. MLM Augmentations

CLARE (ContextuaLized AdversaRial Example) [24] is an adversarial attack text generation technique. In this method, each word in the given sentence is greedily masked, followed by an infill procedure that is used to obtain a replacement word for the masked word. The method permits data enhancement through the replace, insert, and merge operations. This method makes locally optimal choices, which may not always lead to globally optimal solutions, as it replaces all the words in a sentence with substitutes. This typically results in augmentations with a different semantic meaning than the original, so it relies on multiple constraints to generate meaningful data. These constraints eliminate enhancements that do not meet the given criteria. Checking the semantic similarity of the augmented sentence with the original input using an existing process is one of these constraints. Using a neural network already trained on sentence similarity, cosine distance (i.e., 1—Cosine Similarity) can be used to compute the semantic similarity in its most basic form. This distance ranges from 0 to 2, where a value of 0 indicates that the vectors are identical (i.e., the angle between them is 0°). A value of 1 indicates that the vectors are orthogonal (i.e., the angle between them is 90°). A value of 2 indicates that the vectors are diametrically opposed (i.e., the angle between them is 180°) [45]. To compute the similarity between the encoding of original sentences and augmentations, the authors utilised the Universal Sentence Encoder, a text encoder model that maps variable-length English input to a fixed-size 512-dimensional vector. In addition to the encoding model, there are dataset-dependent parameters such as minimum confidence, window size, and maximum candidates. To prevent semantic drift due to arbitrary deletions and insertions, we only used the Replace method.

(1) HR: Ne narucivat chilly. (Translation: Do not order chilly.)
(2) Augmented HR: Ne narucivat meso. (Translation: Do not order meat.)

Initially, we compared each augmentation to the original sentence using a second pre-trained language model. The authors suggested using the Universal Sentence Encoder, a pre-trained language model, to compute the similarity between the encoding of original sentences and augmentations. The Universal Sentence Encoder (https://tfhub.dev/google/universal-sentence-encoder-multilingual/3, accessed on 28 July 2022) has been trained in 16 languages, but none of them is South Slavic; as a result, it was not a good candidate for encoding our data. Consequently, we utilised LaBSE (https://tfhub.dev/google/LaBSE/2, accessed on 28 July 2022), which has been trained in 109 languages. We used cosine scores as a similarity measure and eliminated all sentences that had a cosine similarity of less than 0.80. This was to obtain augmentations with the same class label as the original sentence due to their similar meaning. We implemented a custom MLM-CLARE augmentor with the constraints using the CLARE (https://tinyurl.com/wz85rf43, accessed on 28 July 2022) base class from the textattack library. The percentage of exchanged words was set at 0.5 percent. For Croatian, MLM augmentations were performed using a variety of pre-trained language models, including EMBEDDIA/crosloengual-bert, Andrija/SRoBERTa-F, macedonizer/hr-roberta-base, and classla/bcms-bertic. In terms of perplexity score, EMBEDDIA/crosloengual-bert, xlm-roberta-base, and Andrija/SRoBERTa-F performed the best. Ultimately, EMBEDDIA/crosloengual-bert was selected after examining its enhanced output. Similar procedures were repeated for additional languages.

4.8. CLM Augmentations

Language generation tasks are competitively performed by causal language models such as GPT-2. During training, the model is tasked with predicting the next word in a text sequence. This causes the model to generate the next suitable word based on the previous words or context. During the inference stage, a model is fed an initial prompt and instructed to predict the next word. The entire procedure can be easily used to generate training resources for a model. This method was reported by [37] using a small supervised English dataset. Typically, a single model is trained with data from multiple classes in such a way that the generated text depends on the label. For instance, to generate a positive review, we instructed the model, during training, with the start token, class label, and text (i.e., '< | startoftext | > | review pos | > WHOLE TEXT | endoftext | >'). During the inference, only a few initial words (such as ' | startoftext | > | review pos | > PROMPT-TEXT') are needed to produce the entire text. Using a single model to generate data for all classes with a large amount of data is possible. After training in this environment, we noticed that the model began to generate negative reviews for the mixed/neutral class. Consequently, we trained three distinct models for each of the individual classes. Due to the fact that each class has its own model, the model can only generate text for the class in question. Since they are discussed in the reviews, we decided to use nouns as prompts to capture the overall context during the generation process. Typically, the context is food, such as pizza or risotto, or a service, such as delivery. Using morphosyntactic (MSD) tags, we extracted all nouns from the dataset. The nouns were manually inspected for pipeline-annotated false-positive artefacts. The obtained nouns were then used as inputs for the three fine-tuned GPT-2 models to generate the datasets.

(1) HR: naručili salatu, dostava je bila na vrijeme, dostavljac simpatican.
(2) translation: pizza arrived, no complaints just ordered a salad in advance, delivery was on time, the delivery man was nice.

Using the original and WordNet-augmented datasets, we optimised three distinct GPT-2 models for each of the three classes. The model was independently optimised for each dataset label to generate positive, negative, and mixed reviews. For the purpose of training the language generator, we eliminated all reviews longer than five words. We utilised GPT-2 models trained in the respective languages as the initial backbone encoder. We optimised the model for the language generation task using a learning rate of 0.001,

1 epoch, a batch size of 4, and 1000 warm-up steps. We employed a decoding strategy with a penalty for bi-gram repetition and a beam search with five beams for text generation. Using this method, we created three different datasets that grew larger so we could study the size of the corpus as a dependent feature.

4.9. Experiments

Using a transformer-based classifier, we compared the efficacy of various data generation methods. Two distinct dataset versions were created: two-class, which is the binary version (positive and negative), and three-class, which is the ternary version (positive, negative, or neutral—We refer to the class as neutral despite the fact that it consists of both positive and negative elements). Using the various training sets, the parameters of entire networks were optimised. We trained a separate model for each language in the study and for each dataset generated using the previously described methods (including the original dataset) while maintaining the same network parameters. When the dataset was not balanced, labels from the training set were used to determine the class weight, which was then used as a rescaling weight parameter in the cross-entropy loss. This allowed for a greater penalty if a class with few instances made an incorrect prediction. We trained the model with a learning rate of 1×10^{-5}, a weight decay of 0.01, early stopping on validation loss, and a patience of four to five epochs. Utilising the softmax classifier, the class probabilities were calculated. The final scores for the original set of manually administered tests associated with the dataset are reported. Table 3 presents various transformer-based models used for MLM and CLM augmentations. We utilised the "unsloth/gemma-7b-bnb-4bit" model to perform instruction fine-tuning on all datasets under examination. This is a large-language multilingual model and is a quantised version of Gemma-7b [46].

Table 3. Transformer models used in the training as base encoders for CLM and MLM.

Language	Method	Model Name
Croatian	CLM	macedonizer/hr-gpt2
	MLM	EMBEDDIA/crosloengual-bert
Bulgarian	CLM	rmihaylov/gpt2-medium-bg
	MLM	rmihaylov/bert-base-bg
Slovak	CLM	Milos/slovak-gpt-j-405M
	MLM	gerulata/slovakbert
Slovene	CLM	macedonizer/sl-gpt2
	MLM	EMBEDDIA/sloberta

4.10. Training Set Size

Table 4 displays the final distribution of the original, expanded–combined, and expanded–permuted datasets. For the expanded–combined and expanded–permuted datasets, we varied the training set by sampling 10k, 20k, and 40k instances for each class. In the cases of WN, MLM, and CLM, the augmentation methods affected the final size of the training set, as the process of augmentation is influenced by several factors, including the nature of the original text, the matching of the words, WordNet, and semantic constraints. We obtained 10,000 and 20,000 (and, in some cases, 25,000 and 40,000) samples to be trained and tested for all languages, except for Bulgarian, where the number of instances remained low.

Table 4. Train–development–test distribution of the original and expanded datasets: pos—positive; neg—negative; neu—neutral.

Language	Version	Train			Dev			Test		
		neg	**pos**	**neu**	**neg**	**pos**	**neu**	**neg**	**pos**	**neu**
Croatian	Original	467	1586	145	47	159	14	236	719	78
	lemma	467	1586	145	47	159	14	236	719	78
	expanded	1523	3979	436	44	398	152	742	1787	254
Bulgarian	Original	864	3898	710	96	436	80	107	486	88
	lemma	864	3898	710	96	436	80	107	486	88
	expanded	1435	6321	1060	154	686	116	185	803	133
		neg	**pos**	**neu**	**neg**	**pos**	**neu**	**neg**	**pos**	**neu**
Slovak	Original	297	1337	1926	46	211	265	80	416	545
	lemma	297	1337	1926	46	211	265	80	416	545
	expanded	879	2493	2397	136	352	326	279	841	627
Slovene	Original	2722	749	506	138	37	25	431	112	57
	lemma	2722	749	506	138	37	25	431	112	57
	expanded	13,676	2165	2073	559	170	141	2183	400	229

5. Results and Discussion

Our findings indicate that augmentation methods do not contribute directly to sentiment classification. We found that the performance of augmentations based on pre-trained contextualised language models is inferior to that of methods constructed by combining multiple datasets from the same and different languages. Factors that indirectly affect the final classification score include noisy text and code-mixing. In addition, we found that WordNet-based augmentations are more effective than those based on the Masked Language Model or Causal Language. In seven instances, the expansion–permutation–combination technique resulted in an improvement. The results of the experiments are shown in Tables 5–7. The F1-score and accuracy values for the original, lemma, and expanded versions are shown in Table 5. The results of all the experiments for all the languages are shown in Figures 1–4. The performance of the original version of the dataset was superior to that of two other datasets.

The performance of the binary-lemmatised version was 1% worse than that of the original dataset. This performance decline is greater in a three-class setting. This demonstrates that the pre-trained models, in this case, XLM-R, which were trained on unprocessed text, prefer a grammatically correct form over a lemma form for the given text. We conclude that non-lemmatised data should be used when using pre-trained models like XLM-R. In contrast, separating reviews into individual sentences and using them for training did not lead to a better performance than the other two settings. In conclusion, treating opinionated text as a sum of parts does not make any contribution to training classification models. In addition, we compared the scores obtained with augmentation techniques with scores trained on a large-language model, i.e., Gemma. The Gemma model provided higher overall scores than other models without any additional data.

In all languages except Croatian, the *nary-original *nary-lemmatised settings outperformed the simple expansion technique. The results of using permuted and combined versions of the datasets are presented in Table 6. Using the 20k/class version of the dataset yielded a slight improvement in the F1 score for Croatian compared to the original training dataset, based on the data presented in the table. There were no significant changes to the Bulgarian language. For Slovak, the expanded–permuted 10k-class version produced a four-point improvement in binary classification, but no improvement was observed for ternary classification. The performance of Slovene decreased when permuted and combined versions of the dataset were utilised. Except for Slovak, all other languages scored higher on the expanded combined train set.

According to the data in Table 7, training on the three augmented datasets did not improve the final classification scores. Some cells in the table were left blank because the augmentation technique did not generate the required number of training instances. In the final column, we present the scores for the data points for each class that were either less than 10,000 or greater than 40,000. We performed random approximation tests [47] using the sigf package with 10,000 iterations to determine the statistical significance of differences between the models. For all the languages, none of the models showed a statistically significant improvement ($p < 0.05$) in score compared to the model trained with the original data. Our findings related to the MLM-based DA techniques are very similar to the ones for Norwegian reported by [48]. The authors indicate that augmentation strategies frequently yield gains; nevertheless, the impacts are moderate, and the significant volatility complicates the ability to draw definitive conclusions.

Table 5. Results of original, lemmatised, and expanded (ours) versions of the dataset.

Language	Version	Binary		Ternary	
		F1	ACC	F1	ACC
Croatian	Original	94.11	95.86	75.04	88.18
	lemma	93.61	95.53	60.95	77.77
	expanded	73.99	78.76	73.31	86.93
	gemma	98.05	98.03	90.84	90.99
Bulgarian	Original	90.00	94.43	72.90	83.55
	lemma	88.82	93.76	68.31	81.20
	expanded	84.44	91.09	65.89	80.55
	gemma	96.41	96.45	80.39	84.43
Slovak	Original	94.83	97.17	79.50	81.07
	lemma	94.65	96.97	79.43	81.84
	expanded	88.07	90.98	71.60	72.46
	gemma	98.99	98.99	76.07	76.65
Slovene	Original	80.92	87.84	68.70	79.33
	lemma	79.25	87.29	66.38	77.16
	expanded	68.05	85.63	49.96	67.03
	gemma	93.57	93.73	85.8	85.83

Table 6. Results of expanded–combined (ours) and expanded–permuted (ours) datasets for all languages.

Lang	Ver	Binary_10k		Ternary_10k		Binary_20k		Ternary_20k		Binary_40k		Ternary_40k	
		F1	ACC	F1	ACC	F1	ACC	F1	ACC	F1	ACC	F1	ACC
Hr	expanded-combined	95.37	96.84	73.17	87.41	95.84	97.16	72.96	85.96	94.26	96.07	71.84	87.6
	expanded-permuted	95.53	96.84	73.87	87.99	94.79	96.4	68.72	84.99	93.06	95.31	71.63	86.93
Bg	expanded-combined	90.16	94.26	66.18	76.35	89.88	93.92	72.23	81.93	89.41	93.76	72.27	82.96
	expanded-permuted	89.85	94.26	71.7	80.91	89.17	93.76	71.69	81.64	89.08	93.76	70.5	79.29
Sk	expanded-combined	97.76	98.79	76.58	77.52	96.92	98.38	77.55	78.09	96.72	98.18	79.34	80
	expanded-permuted	98.12	98.99	76.4	76.94	97.37	98.58	78.31	79.05	97.8	98.79	77.86	79.05
Sv	expanded-combined	75.89	81.76	59.73	70.16	77.9	84.16	62.89	74.88	77.67	83.6	58.8	67
	expanded-permuted	75.57	81.21	53.66	60.16	74.07	79.92	54.62	59.33	77.84	83.24	61.5	73.5

Table 7. Results when using augmented datasets using WordNet, MLM, and CLM. Bold values represent best performing system.

		10k				20k				25k				40k				All			
Lang	**Version**	**Binary**		**Ternary**		**Binary**		**Ternary**		**Binary**		**Ternary**		**Binary**		**Ternary**		**Binary**		**Ternary**	
		F1	ACC	F1	ACC	F1	ACC	F1	ACC	F1	ACC	F1	ACC	F1	ACC	F1	ACC	F1	ACC	F1	ACC
Hr	WN	94.18	95.96	**71.90**	87.12	93.09	95.31	68.73	84.80					**94.20**	95.96	61.78	84.31	93.94	95.86	69.43	86.73
	MLM	92.30	94.55	67.74	81.31	90.26	93.35	70.63	83.93	90.76	93.68	69.36	83.15								
	CLM	92.06	94.44	64.96	81.89	90.74	93.89	6235	81.80									89.73	93.02	67.11	83.83
Bg	WN																	**91.56**	94.94	**70.64**	84.43
	MLM																	88.73	93.76	70.07	81.49
	CLM	87.07	92.58	61.87	79.73	84.15	90.55	59.05	77.09	82.76	88.87	58.43	80.02					84.10	91.23	58.35	76.65
Sk	WN	96.00	97.78	74.86	79.82	95.61	97.58	**79.35**	82.32					95.22	97.37	77.67	80.97	**97.37**	98.58	76.50	78.96
	MLM	96.19	97.98	77.24	78.67	94.93	97.17	76.49	76.75	96.27	97.98	73.44	74.25								
	CLM	92.31	95.96	70.01	72.14	90.54	94.55	69.8	71.85					91.63	95.56	68.79	71.66	91.40	95.16	68.66	70.50
Sv	WN	73.47	79.18	59.39	68.83	78.25	84.71	53.33	65.00					78.25	84.71	58.53	69.5	**77.83**	86.37	59.87	73.5
	MLM	63.02	66.11	**62.00**	72.16	73.99	79.37	60.827	72.33	76.152	82.13	56.11	64.16								
	CLM	74.29	81.03	55.16	65.33	67.19	72.19	54.46	69.83	90.26	73.66	56.38	65.83					65.89	69.98	47.68	57.83

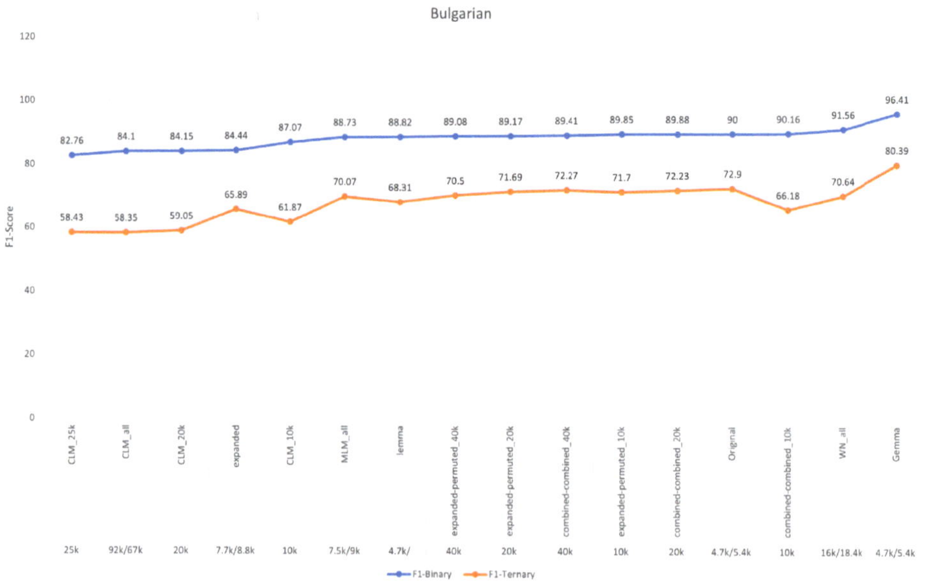

Figure 1. Comparison of F1 scores for Bulgarian datasets. Our proposed methods are labelled with prefix "expanded".

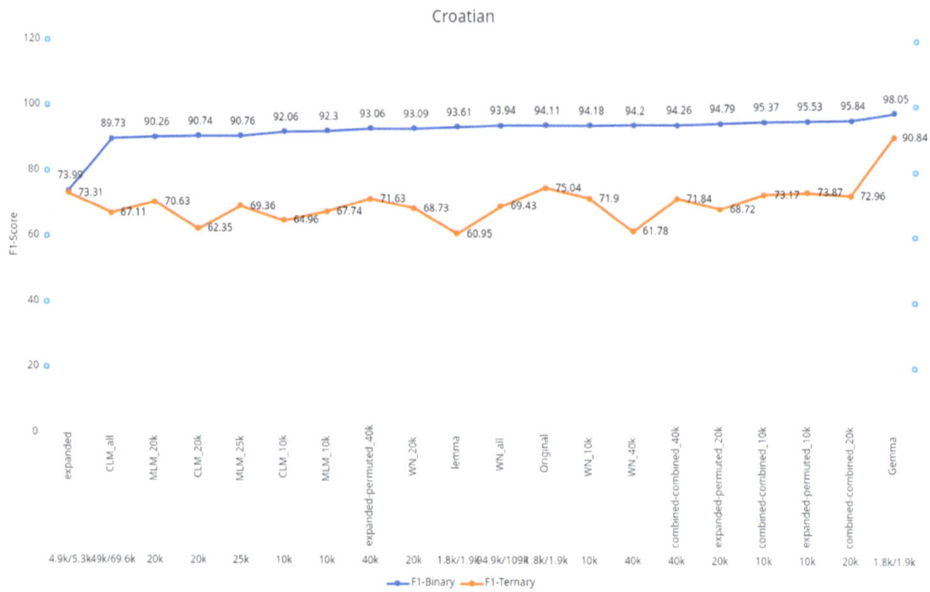

Figure 2. Comparison of F1 scores for Croatian datasets. Our proposed methods are labelled with prefix "expanded".

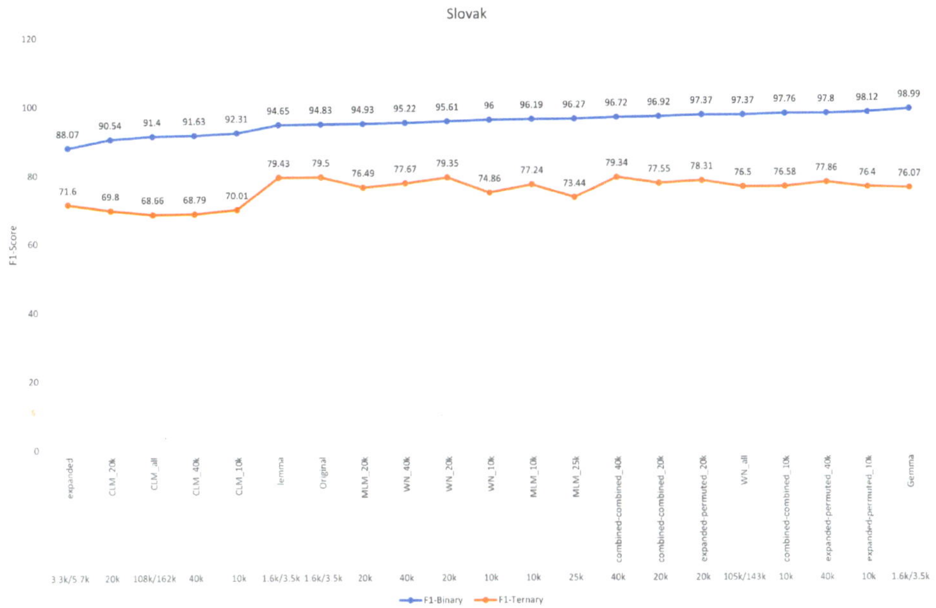

Figure 3. Comparison of F1 scores for Slovak datasets. Our proposed methods are labelled with prefix "expanded".

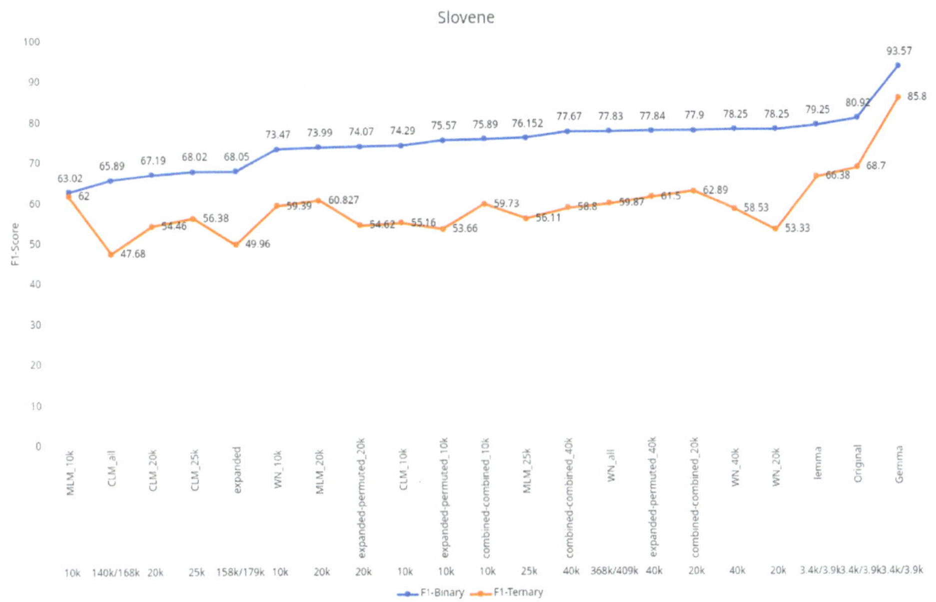

Figure 4. Comparison of F1 scores for Slovene datasets. Our proposed methods are labelled with prefix "expanded".

5.1. *Error Analysis*

For the best scoring models, we randomly sampled incorrectly classified instances from the test set for each language. We manually examined the cases and present a summary of the results. A majority of the issues encountered throughout the evaluations were previously reported in other studies [49].

5.1.1. Text Accompanied by Additional Context

In this category of incorrectly classified instances, the statement begins with a premise or speculation (I believe it will be good) and ends with the user's opinion (But I did not like it). Alternatively, the text might start with an opinion and then move on to speculation. The additional information may or may not justify the users' feelings. The user discusses audience members leaving the theatre in the following example, then he provides his own review. The original label of the review is positive, but the predicted label is negative.

- (Original BG) Половината салон си тръгна на 30тата минута. Аз следя сериала от както го има и филма ми хареса.
- (Transliteration BG) Polovinata salon si trgna na 30tata minuta. Az sledya seriala ot kakto go ima i filma mikharesa.
- (Translation EN) Half the salon left at the 30 min mark. I've been following the series since it started and I liked the movie.
- Original label: positive; predicted: negative.

The sentence "I liked the movie" points to the final user sentiment, while the first sentence causes the model to predict the review to be negative.

5.1.2. Reviews with aspect ratings

In this type of text, each aspect is evaluated separately by the user. The current classifier fails to classify these formats, and a specialised process may be required to classify them.

- (Original BG) 1 за декорите ... Начосът заслужава 5.
- (Transliteration BG) 1 za dekorite ... Nachost zasluzhava 5
- (Translation EN) 1 for the decorations ... The nachos deserve a 5.
- original label: negative; predicted: positive.

5.1.3. Mixed Aspects

The majority of cases fall into this category. The text comprises a compound or a complex sentence with multiple targets.

- (Original BG) Твърде много ненужно пеене,но всичко останало е супер!:)
- (Transliteration BG) Tvrde mnogo ne nuzhno peene,no vsichko ostanalo e super!:)
- (Translation EN) Too much unnecessary singing, but everything else is great!:)
- original rating: negative; predicted: positive.

5.1.4. Contradictory Expressions

The conflicting sub-parts of a sentence are presented as a single unit rather than a compound sentence, as in the previous error type.

- (Original BG) Красив филм с безкрайно несъстоятелен сценарий.
- (Transliteration BG) Krasiv film s bezkraino nesstoyatelen stsenarii.
- (Translation EN) A beautiful film with an endlessly unworkable script.
- Original rating: negative; predicted: positive.

The neutral/mixed-class instances in the Croatian test set have the highest number of misclassifications. We used the SHAP (https://github.com/shap/shap, accessed on 1 June 2022) (SHapley Additive exPlanations) tool to observe and study the model predictions. The text of binary-classified reviews consists of only positive or negative words. When used with the Transformer encoder, these polar words receive aheightened focus, which ultimately determines whether the final classification is positive or negative. In the case of the mixed-class, the text is composed of both positive and negative polar words, with one group receiving a disproportionate amount of attention, resulting in an incorrect classification. We discovered that 'ali'-containing sentences were misclassified because the model could not identify compound sentences. As specified by [50], dealing with mixed-class sentences is difficult because the assumption that the document or sentence has a single target is false. Further examination of the test-set predictions and ground-truth labels yielded the following findings:

(1) Some reviews contain sentences that are lengthy. The XLM-R accepts 512 (-2) tokens that have been processed by a tokeniser [16]. Due to the omission of these text tokens, the model performs poorly when the text is exceedingly long. This phenomenon is notable in the Slovene and Croatian datasets.
(2) Cases in which the author gave the review a positive rating, but the text contains many unrelated negative statements. This occurs when the author rants about many other stores and writes one positive line about the target entity [50].
(3) We also found that the greater the distance between the negation cue and the scope of the negation, the less likely the model is to capture the negation. For example, "Pizza dola mlaka, i ne ukusna", vs. "Pizza dola mlaka, i ne ba ukusna", and "Pizza dola mlaka, i ne ba previe ukusna". The first sample was correctly classified, but the second and third samples were not [51].
(4) People write negative reviews but rate the restaurant highly because they had a pleasant experience there [52].
(5) Code-mixing and English text in Croatian and Slovene [53].

Additionally, we observe that customers may rate the overall review positively even if something was missing from the delivery.

(1) Brza dostava, ok hrana. Jedino kaj su zaboravili coca colu :(. (Translation EN) Fast delivery, ok food. Only what they forgot about Coca Cola :(.
(2) Nisam vidjela prut na pizzi special, al nema veze, vratina je bila sasvim dovoljna! (Translation EN) I did not see the prosciutto on the pizza special, but it does not matter, the door was enough!
(3) Malo gumasto tijesto, inace OK pizza. (Translation EN) A little rubber dough, otherwise ok pizza.

The MLM model augmentor generated "Treba narucivat chilly" as the correct augmentation for "Ne narucivat chilly", despite paraphrasing the constraints. This may be due to the LaBSe model misclassifying texts as paraphrases of one another. Therefore, improved constraints are recommended. For Slovak, we identified cases that contained positive phrases but were labelled neutral by the authors.

(1) Bol som vemi spokojný. (Translation EN) I have been very satisfied.
(2) super super super. (Translation EN) Super Super Super.
(3) Bola vemi príjemná a milá. (Translation EN) She was very pleasant and nice.
(4) Vemi ústretová a ochotná. (Translation EN) Very helpful and willing.
(5) Bagety, ktoré som kúpila boli perfektné ... akujem. (Translation EN) Baguettes I bought were perfect ... Thank you.

In addition to classification errors, the following text-processing errors were observed: Using the Classla package, errors are introduced at three stages (sentence tokenisation, lemmatisation, and POS). For instance, garbled tokens are identified as nouns in the text, and improper sentence boundary detection is also detected. Typically, the user-text lacks diacritics (narucívati -> naruívati). Therefore, processing is required to correct the spelling in order to reduce the number of failed WordNet lookups. The Bulgarian dataset consists of movie reviews with emoticons included in the text. This calls for an emoticon-aware tokenizer. Classla did not support the processing of non-standard text types for Bulgarian, so standard mode was used for sentence splitting, lemma, and POS. This is a potential entry point for errors.

5.2. Revisiting Research Questions

We can answer our research questions after conducting the experiments and analysing the data.

Can the data augmentation techniques improve the performance metric? According to our findings, using a pre-trained contextualised language encoder reduces the impact of an augmented dataset. As previously reported by [27], these transformer-based models are invariant to certain transformations, such as synonym substitution. This is attributable to the proximity of synonyms in the representation space of these encoders. Therefore, using synonyms obtained from WordNet or other sources and encoding them in these spaces does not result in a significant gain. The only way to improve performance is to generate novel linguistic structures that were not encountered during the Transformer model's pre-training.

What is the effect of having augmented data generated from different techniques? We investigated three distinct data augmentation techniques in addition to three text expansion techniques. Comparing their performance reveals that training with augmented data does not lead to a performance improvement compared with training with the original dataset alone. Although binary class performance improved by a few points, this improvement was not consistent. In addition, increasing the size of the augmented data has little effect on the performance of the techniques.

Can WordNet-based augmentation techniques work better with sentiment classification tasks? Although WordNet-based augmentation techniques appear to be more effective than MLM and CLM-based techniques, they provided no significant improvement for the downstream task. Training with lemma-based instances decreased system performance by one point for binary classification but drastically decreased system performance for ternary classification. Also, as [28] pointed out, it is easy to improve the performance of binary sentiment classification by adding more data, but fine-grained classification faces the same problem as training on the whole dataset.

6. Conclusions

In summary, this investigation assessed the efficacy of data augmentation methodologies in enhancing sentiment analysis in low-resource languages, with a particular emphasis on Slovene, Slovak, Croatian, and Bulgarian. Our results suggest that traditional augmentation methods, such as WordNet-based synonym replacement, MLM-based augmentations, and sentence permutation and combination, provide limited benefits to model performance, particularly when transformer-based encoders are used. Although the results of the WordNet-based augmentation were marginally superior to those of other methods, none of the techniques achieved significant improvements over the original datasets. In practical terms, this implies that existing augmentation strategies may require modification to accommodate the distinctive complexities and linguistic variability in low-resource languages. In theory, these results suggest that more innovative methods, such as the development of syntactic diversity rather than lexical diversity, may be necessary to more accurately simulate real-world language use in order to effectively augment sentiment analysis in these languages. Therefore, future research should investigate innovative augmentation methods that integrate syntactic transformations and intricate language structures, as these have the potential to provide more significant enhancements in sentiment analysis in low-resource language contexts.

Author Contributions: Conceptualization, G.T.; Methodology, G.T.; Software, G.T.; Validation, G.T.; Formal analysis, G.T.; Investigation, N.M.P.; Resources, M.T.; Writing—original draft, G.T.; Writing—review & editing, N.M.P. and M.T.; Visualization, G.T.; Supervision, N.M.P. and M.T.; Project administration, N.M.P. and M.T.; Funding acquisition, M.T. All authors have read and agreed to the published version of the manuscript.

Funding: The work presented in this paper received funding from the European Union's Horizon 2020 research and innovation program under the Marie Skłodowska-Curie grant agreement no. 812997 and under the name CLEOPATRA (Cross-lingual Event-centric Open Analytics Research Academy).

Institutional Review Board Statement: Not applicable.

Informed Consent Statement: Not applicable.

Data Availability Statement: The original contributions presented in the study are included in the article, further inquiries can be directed to the corresponding authors.

Conflicts of Interest: The authors declare no conflicts of interest. The funders had no role in the design of the study; in the collection, analyses, or interpretation of data; in the writing of the manuscript; or in the decision to publish the results.

References

1. Goodfellow, I.; Bengio, Y.; Courville, A. *Deep Learning*; MIT Press: Cambridge, MA, USA, 2020.
2. Géron, A. *Hands-on Machine Learning with Scikit-Learn, Keras, and TensorFlow*; O'Reilly Media, Inc.: Sebastopol, CA, USA, 2022.
3. Bengio, Y. Practical Recommendations for Gradient-Based Training of Deep Architectures. *Neural Netw. Tricks Trade Second. Ed.* **2012**, *7700*, 437–478. [CrossRef]
4. Halevy, A.; Norvig, P.; Pereira, F. The Unreasonable Effectiveness of Data. *IEEE Intell. Syst.* **2009**, *24*, 8–12. [CrossRef]
5. Schreiner, C.; Torkkola, K.; Gardner, M.; Zhang, K. Using Machine Learning Techniques to Reduce Data Annotation Time. In Proceedings of the Human Factors and Ergonomics Society Annual Meeting, Sydney, Australia, 20–22 November 2006; pp. 2438–2442. [CrossRef]
6. Wei, J.; Zou, K. EDA: Easy Data Augmentation Techniques for Boosting Performance on Text Classification Tasks. *arXiv* **2019**, arXiv:1901.11196.
7. Kobayashi, S. Contextual Augmentation: Data Augmentation by Words with Paradigmatic Relations. In Proceedings of the 2018 Conference of the North American Chapter of the Association for Computational Linguistics: Human Language Technologies, Volume 2 (Short Papers), New Orleans, LA, USA, 1–6 June 2018; pp. 452–457. [CrossRef]
8. Ribeiro, M.T.; Wu, T.; Guestrin, C.; Singh, S. Beyond Accuracy: Behavioral Testing of NLP models with CheckList. In Proceedings of the 58th Annual Meeting of the Association for Computational Linguistics, Online, 5–10 July 2020; pp. 4902–4912.
9. Wang, J.; Lan, C.; Liu, C.; Ouyang, Y.; Qin, T. Generalizing to Unseen Domains: A Survey on Domain Generalization. In Proceedings of the Thirtieth International Joint Conference on Artificial Intelligence, IJCAI-21, Montreal, QC, Canada, 19–27 August 2021; Zhou, Z.-H., Ed.; Survey Track, pp. 4627–4635. [CrossRef]
10. Zhang, X.; Zhao, J.; LeCun, Y. Character-Level Convolutional Networks for Text Classification. *Adv. Neural Inf. Process. Syst.* **2015**, *28*, 649–657. [CrossRef]
11. Goodfellow, I.J.; Shlens, J.; Szegedy, C. Explaining and Harnessing Adversarial Examples. *arXiv* **2014**, arXiv:1412.6572.
12. Zuo, X.; Chen, Y.; Liu, K.; Zhao, J. KnowDis: Knowledge Enhanced Data Augmentation for Event Causality Detection via Distant Supervision. In Proceedings of the 28th International Conference on Computational Linguistics, Barcelona, Spain (Online), 8–13 December 2020; pp. 1544–1550. [CrossRef]
13. Yang, Y.; Malaviya, C.; Fernandez, J.; Swayamdipta, S.; Bras, R.L.; Wang, J.; Bhagavatula, C.; Choi, Y.; Downey, D. Generative Data Augmentation for Commonsense Reasoning. In Proceedings of the Findings of the Association for Computational Linguistics: EMNLP 2020, Online, 16–20 November 2020; pp. 1008–1025. [CrossRef]
14. Miller, G.A. WordNet: A Lexical Database for English. *Commun. ACM* **1995**, *38*, 39–41. [CrossRef]
15. Brown, T.B.; Mann, B.; Ryder, N.; Subbiah, M.; Kaplan, J.; Dhariwal, P.; Neelakantan, A.; Shyam, P.; Sastry, G.; Askell, A.; et al. Language Models Are Few-Shot Learners. *arXiv* **2020**, arXiv:2005.14165.
16. Vaswani, A.; Shazeer, N.; Parmar, N.; Uszkoreit, J.; Jones, L.; Gomez, A.N.; Kaiser, Ł.; Polosukhin, I. Attention Is All You Need. In Proceedings of the 31st Conference on Neural Information Processing Systems (NIPS 2017), Long Beach, CA, USA, 4–9 December 2017; pp. 5998–6008.
17. Su, P.; Li, G.; Wu, C.; Vijay-Shanker, K. Using Distant Supervision to Augment Manually Annotated Data for Relation Extraction. *PLoS ONE* **2019**, *14*, e0216913. [CrossRef] [PubMed]
18. Mintz, M.; Bills, S.; Snow, R.; Jurafsky, D. Distant Supervision for Relation Extraction without Labeled Data. In Proceedings of the Joint Conference of the 47th Annual Meeting of the ACL and the 4th International Joint Conference on Natural Language Processing of the AFNLP, Suntec, Singapore, 2–7 August 2009; pp. 1003–1011.
19. Garg, S.; Ramakrishnan, G. BAE: BERT-based Adversarial Examples for Text Classification. In Proceedings of the 2020 Conference on Empirical Methods in Natural Language Processing (EMNLP), Online, 16–20 November 2020; pp. 6174–6181.
20. Li, L.; Ma, R.; Guo, Q.; Xue, X.; Qiu, X. BERT-ATTACK: Adversarial Attack Against BERT Using BERT. In Proceedings of the 2020 Conference on Empirical Methods in Natural Language Processing (EMNLP), Online, 16–20 November 2020; pp. 6193–6202.
21. Yoo, J.Y.; Qi, Y. Towards Improving Adversarial Training of NLP Models. In Proceedings of the Findings of the Association for Computational Linguistics: EMNLP 2021, Virtual, 16–20 November 2021; pp. 945–956.

22. Ren, S.; Deng, Y.; He, K.; Che, W. Generating Natural Language Adversarial Examples through Probability Weighted Word Saliency. In Proceedings of the 57th Annual Meeting of the Association for Computational Linguistics, Florence, Italy, 28 July–2 August 2019; pp. 1085–1097. [CrossRef]
23. Samanta, S.; Mehta, S. Towards Crafting Text Adversarial Samples. *arXiv* **2017**, arXiv:1707.02812.
24. Li, D.; Zhang, Y.; Peng, H.; Chen, L.; Brockett, C.; Sun, M.; Dolan, B. Contextualized Perturbation for Textual Adversarial Attack. In Proceedings of the 2021 Conference of the North American Chapter of the Association for Computational Linguistics: Human Language Technologies, Online, 6–11 June 2021; pp. 5053–5069. [CrossRef]
25. Sennrich, R.; Haddow, B.; Birch, A. Improving Neural Machine Translation Models with Monolingual Data. In Proceedings of the 54th Annual Meeting of the Association for Computational Linguistics (Volume 1: Long Papers), Berlin, Germany, 7–12 August 2016; pp. 86–96. [CrossRef]
26. Edunov, S.; Ott, M.; Auli, M.; Grangier, D. Understanding Back-Translation at Scale. In Proceedings of the 2018 Conference on Empirical Methods in Natural Language Processing, Brussels, Belgium, 31 October–4 November 2018; pp. 489–500. [CrossRef]
27. Longpre, S.; Wang, Y.; DuBois, C. How Effective is Task-Agnostic Data Augmentation for Pretrained Transformers? In Proceedings of the Findings of the Association for Computational Linguistics: EMNLP 2020, Online, 16–20 November 2020; pp. 4401–4411.
28. Xie, Q.; Dai, Z.; Hovy, E.; Luong, T.; Le, Q. Unsupervised Data Augmentation for Consistency Training. *Adv. Neural Inf. Process. Syst.* **2020**, *33*, 6256–6268.
29. Go, A.; Bhayani, R.; Huang, L. Twitter sentiment classification using distant supervision. *CS224N Proj. Rep. Stanf.* **2009**, *1*, 2009.
30. Martinc, M.; Montariol, S.; Pivovarova, L.; Zosa, E. Effectiveness of Data Augmentation and Pretraining for Improving Neural Headline Generation in Low-Resource Settings. In Proceedings of the LREC 2022, Marseille, France, 20–25 June 2022.
31. Cheung, T.-H.; Yeung, D.-Y. {MODALS}: Modality-agnostic Automated Data Augmentation in the Latent Space. In Proceedings of the International Conference on Learning Representations, Online, 3–7 May 2021 .
32. Goldsmith, J.; Riggle, J.; Alan, C.L. *The Handbook of Phonological Theory*; Wiley Online Library: Hoboken, NJ, USA, 1995.
33. Glavaš, G.; Korenčić, D.; Šnajder, J. Aspect-Oriented Opinion Mining from User Reviews in Croatian. In Proceedings of the 4th Biennial International Workshop on Balto-Slavic Natural Language Processing, Sofia, Bulgaria, 8–9 August 2013; pp. 18–23.
34. Kapukaranov, B.; Nakov, P. Fine-Grained Sentiment Analysis for Movie Reviews in Bulgarian. In Proceedings of the International Conference Recent Advances in Natural Language Processing, Hissar, Bulgaria, 7–9 September 2015; pp. 266–274.
35. Pecar, S.; Simko, M.; Bielikova, M. Improving Sentiment Classification in Slovak Language. In Proceedings of the 7th Workshop on Balto-Slavic Natural Language Processing, Florence, Italy, 2 August 2019.
36. Kadunc, K.; Robnik-Šikonja, M. Opinion Corpus of Slovene Web Commentaries KKS 1.001; Slovenian Language Resource Repository CLARIN.SI. 2017. Available online: http://hdl.handle.net/11356/1115 (accessed on 19 July 2022).
37. Anaby-Tavor, A.; Carmeli, B.; Goldbraich, E.; Kantor, A.; Kour, G.; Shlomov, S.; Tepper, N.; Zwerdling, N. Do Not Have Enough Data? Deep Learning to the Rescue! In Proceedings of the Thirty-Fourth AAAI Conference on Artificial Intelligence, AAAI 2020, The Thirty-Second Innovative Applications of Artificial Intelligence Conference, IAAI 2020, the Tenth AAAI Symposium on Educational Advances in Artificial Intelligence, EAAI 2020, New York, NY, USA, 7–12 February 2020; pp. 7383–7390. [CrossRef]
38. Bollegala, D.; Weir, D.; Carroll, J.A. Using Multiple Sources to Construct a Sentiment Sensitive Thesaurus for Cross-Domain Sentiment Classification. In Proceedings of the 49th Annual Meeting of the Association for Computational Linguistics: Human Language Technologies, Portland, OR, USA, 19–24 June 2011; pp. 132–141.
39. Gamon, M. Sentiment Classification on Customer Feedback Data: Noisy data, large feature vectors, and the role of linguistic analysis. In Proceedings of the COLING 2004: 20th International Conference on Computational Linguistics, Geneva, Switzerland, 23–27 August 2004; pp. 841–847.
40. Itertools Combinations. 2022. Available online: https://docs.python.org/3/library/itertools.html#itertools.combinations (accessed on 26 July 2022).
41. Itertools Permutations. 2022. Available online: https://docs.python.org/3/library/itertools.html#itertools.permutations (accessed on 26 July 2022).
42. Erjavec, T.; Fišer, D. Building Slovene Wordnet. In Proceedings of the Fifth International Conference on Language Resources and Evaluation (LREC'06), Genoa, Italy, 22–28 May 2006.
43. Koeva, S.; Genov, A.; Totkov, G. Towards Bulgarian Wordnet. *Rom. J. Inf. Sci. Technol.* **2004**, *7*, 45–60.
44. Raffaelli, I.; Tadic, M.; Bekavac, B.; Agic, Ž. Building croatian wordnet. In Proceedings of the GWC, Szeged, Hungary, 22–25 January 2008; pp. 349–360.
45. Baeza-Yates, R.; Ribeiro-Neto, B. *Modern Information Retrieval*; ACM Press: New York, NY, USA, 1999; Volume 463.
46. Gemini Team Google. Gemini: A Family of Highly Capable Multimodal Models. *arXiv* **2023**, arXiv:2312.11805. [CrossRef]
47. Yeh, A. More Accurate Tests for the Statistical Significance of Result Differences. In Proceedings of the COLING 2000 Volume 2: The 18th International Conference on Computational Linguistics, Saarbrücken, Germany, 31 July–4 August 2000.
48. Kolesnichenko, L.; Velldal, E.; Øvrelid, L. Word Substitution with Masked Language Models as Data Augmentation for Sentiment Analysis. In Proceedings of the Second Workshop on Resources and Representations for Under-Resourced Languages and Domains (RESOURCEFUL-2023), Torshavn, Danmark, 22 May 2023; pp. 42–47.
49. Pang, B.; Lee, L. Opinion Mining and Sentiment Analysis. In *Foundations and Trends in Information Retrieval*; Now Publishers Inc.: Norwell, MA, USA, 2008; pp. 1–135. [CrossRef]
50. Liu, B. *Sentiment Analysis: Mining Opinions, Sentiments, and Emotions*; Cambridge University Press: Cambridge, UK, 2020.

51. Khandelwal, A.; Sawant, S. NegBERT: A Transfer Learning Approach for Negation Detection and Scope Resolution. In Proceedings of the Twelfth Language Resources and Evaluation Conference, Marseille, France, 11–16 May 2020; pp. 5739–5748.
52. Askalidis, G.; Kim, S.J.; Malthouse, E.C. Understanding and Overcoming Biases in Online Review Systems. *Decis. Support Syst.* **2017**, *97*, 23–30. [CrossRef]
53. Utsab, B.; Das, A.; Joachim, W.; Foster, J. Code Mixing: A Challenge for Language Identification in the Language of Social Media. In Proceedings of the First Workshop on Computational Approaches to Code Switching, Doha, Qatar, 25 October 2014; pp. 13–23.

eng MDPI

Article

Enhanced Skin Lesion Segmentation and Classification Through Ensemble Models

Su Myat Thwin and Hyun-Seok Park *

Department of Computer Science and Engineering, Ewha Womans University, Seoul 03760, Republic of Korea; sumyatthwin@ewhain.net
* Correspondence: sw4everyone@gmail.com or neo@ewha.ac.kr

Abstract: This study addresses challenges in skin cancer detection, particularly issues like class imbalance and the varied appearance of lesions, which complicate segmentation and classification tasks. The research employs deep learning ensemble models for both segmentation (using U-Net, SegNet, and DeepLabV3) and classification (using VGG16, ResNet-50, and Inception-V3). The ISIC dataset is balanced through oversampling in classification, and preprocessing techniques such as data augmentation and post-processing are applied in segmentation to increase robustness. The ensemble model outperformed individual models, achieving a Dice Coefficient of 0.93, an IoU of 0.90, and an accuracy of 0.95 for segmentation, with 90% accuracy on the original dataset and 99% on the balanced dataset for classification. The use of ensemble models and balanced datasets proved highly effective in improving the accuracy and reliability of automated skin lesion analysis, supporting dermatologists in early detection efforts.

Keywords: ensemble model; segmentation; classification; oversampling

Citation: Thwin, S.M.; Park, H.-S. Enhanced Skin Lesion Segmentation and Classification Through Ensemble Models. *Eng* **2024**, *5*, 2805–2820. https://doi.org/10.3390/eng5040146

Academic Editor: Antonio Gil Bravo

Received: 29 September 2024
Revised: 26 October 2024
Accepted: 30 October 2024
Published: 31 October 2024

1. Introduction

The early detection and accurate classification of skin lesions are critical for effective skin cancer treatment. Dermoscopic analysis has improved diagnostic accuracy, but manual interpretation remains time-consuming and subjective, highlighting the need for automated diagnostic tools [1,2]. Deep learning, particularly with CNNs like VGG16, ResNet-50, and Inception-V3, has proven effective in medical image interpretation [3–5]. Skin cancer, especially melanoma, BCC, and SCC, is challenging to diagnose early, and the ISIC dataset provides valuable resources for algorithm development. Deep learning models like U-Net, SegNet, and DeepLabv3 excel in segmentation tasks, capturing spatial details. This study proposes a deep ensemble model combining these architectures to enhance skin lesion segmentation accuracy. For classification, ensemble models of VGG16, ResNet-50, and Inception-V3 are used to improve recognition of BCC, SCC, and melanoma. To address imbalanced datasets, oversampling techniques are employed, aiming to develop a reliable framework for skin cancer diagnosis, contributing to better outcomes and early intervention.

This work [6] focuses on segmenting skin lesions using preprocessed dermoscopic images, removing noise through a fusion of six image-processing techniques. Modified U-Net architectures, particularly U-Net 46, are used, yielding 93% accuracy, 97% specificity, and 91% sensitivity on the ISIC 2018 dataset. Another study [7] employs atrous convolutions in a CNN to improve lesion segmentation by expanding the receptive field without lowering resolution, showing precise lesion region identification. A modified cGAN with Factorized Channel Attention (FCA) [8] enhances segmentation, reducing computational complexity. A multi-task deep neural network [9] achieves high AUCs for lesion classification and segmentation accuracy. A novel CNN architecture [10] uses auxiliary edge prediction and multi-scale feature aggregation, yielding superior segmentation performance. Research [11] reviews 177 deep learning models, highlighting challenges in skin lesion segmentation due to low contrast and fuzzy borders. An attention-based U-Net model with DenseNet [12] improves segmentation via adaptive gamma correction. Another method [13] combines deep and classical learning, achieving high segmentation and classification accuracy using a cubic support vector machine. A deep learning approach [14] with ResUNet++ and post-processing techniques like CRF and TTA improves segmentation, achieving high Jaccard Index scores. Finally, a CNN-based technique [15] recovers lesion locations with enhanced accuracy through pixel-level segmentation and post-processing.

This collection of studies explores various deep learning approaches for skin lesion classification. Study [16] utilized Inception-V3 and DenseNet to categorize seven skin lesion types, with DenseNet outperforming in classification accuracy and using visualizations to enhance interpretability. In study [17], an AI-based system won first place in the ISIC 2019 challenge, addressing challenges like class imbalance by integrating patient data. Study [18] introduced a Deep Focused Sub-network (DFS) and Information-guided Weights Rebalancing (IWR) to highlight significant skin lesion areas and mitigate class disparity, achieving a state-of-the-art performance on ISIC datasets. In study [19], a system using attention mechanisms and SMOTE for data balancing demonstrated improved accuracy on the HAM10000 dataset. The authors [20] presented an advanced deep residual network designed to enhance skin lesion classification accuracy. The model uses multilevel feature extraction and cross-channel correlation to capture complex lesion features and patterns effectively. Additionally, it incorporates an outlier detection mechanism to improve robustness by filtering out data that could negatively impact model performance. This approach demonstrates significant promise for improving diagnostic accuracy in skin lesion classification by capturing subtle lesion characteristics and addressing data anomalies. It introduces a sophisticated residual network that employs multilevel feature extraction and cross-channel correlation, proving effective in capturing complex lesion patterns. However, while Skin-Net offers a robust approach to feature extraction, it may face limitations in handling highly imbalanced datasets without targeted data balancing strategies.

This paper reviews [21] the latest advancements in deep learning and optimization-based techniques for skin lesion segmentation, focusing on the integration of machine learning, neural networks, and optimization algorithms to enhance segmentation accuracy and clinical relevance. It covers prominent deep learning architectures like U-Net, FCN, and GANs, alongside optimization strategies that improve model efficiency and generalization. The review addresses challenges such as class imbalance, variability in lesion appearance, and the need for robust models in real-world applications, providing insights into current trends and future directions for skin lesion segmentation in dermatological diagnostics. It provides a thorough review of segmentation techniques, offering valuable insights into the optimization of segmentation performance. However, as a review, it lacks experimental validation, leaving practical performance considerations in real-world clinical contexts unexplored.

The proposed method enhances skin lesion segmentation and classification by introducing an innovative ensemble model that integrates U-Net, SegNet, and DeepLabv3 for segmentation, and VGG16, ResNet-50, and Inception-V3 for classification. This unique combination leverages the strengths of each model—U-Net's precision in boundary detection, SegNet's efficiency in semantic segmentation, and DeepLabv3's ability to capture multi-scale context—to deliver highly accurate and robust lesion segmentation. For classification, the ensemble of VGG16, ResNet-50, and Inception-V3 captures a broader range of feature representations, boosting diagnostic accuracy for complex lesion patterns. Unlike conventional single-model approaches, our ensemble method addresses critical challenges in differentiating melanoma, BCC, and SCC, especially with a balanced dataset and data augmentation techniques that mitigate the impact of class imbalance. By demonstrating improved precision, recall, and F1-scores across lesion types, this method offers a precise and impactful advancement in automated skin lesion analysis, underscoring its potential for real-world clinical applications. Using the ISIC dataset, the system demonstrates superior diagnostic accuracy, offering a reliable clinical tool for early skin cancer detection.

Our work presents several key contributions aimed at enhancing skin lesion analysis.

- A novel ensemble architecture for segmentation: the proposed method integrates U-Net, SegNet, and DeepLabv3 with augmentation, each contributing unique strengths—boundary detection, semantic segmentation, and multi-scale context capture—for precise and efficient skin lesion segmentation.
- An ensemble of deep models for classification: through combining VGG16, ResNet-50, and Inception-V3, the model captures diverse feature representations, improving classification accuracy for melanoma, basal cell carcinoma (BCC), and squamous cell carcinoma (SCC) lesions.
- Balanced dataset and data augmentation: employing oversampling and augmentation techniques, the model effectively addresses class imbalance, enhancing reliability across lesion categories.
- Enhanced performance metrics: the ensemble method achieves superior precision, recall, and F1-scores compared to single-model approaches, demonstrating its robustness for clinical applications.
- Real-world clinical relevance: the method's improved diagnostic accuracy and robustness make it a promising tool for real-world skin lesion analysis and potential integration into clinical workflows.

The format of this document is as follows: The proposed deep ensemble model for skin lesion segmentation and the suggested deep ensemble model for skin lesion classification are offered in Section 2. Performance results are discussed in Section 3, Section 4 compares these results with the existing literature, and the last section concludes with future directions for this paper.

2. Materials and Methods

This section outlines proposed skin lesion segmentation and proposed skin lesion classification performance using a dermoscopic image (ISIC) dataset. Algorithm 1 describes the proposed segmentation and classification algorithm.

Algorithm 1: Proposed Method for Skin Lesion Segmentation and Classification

Input:
Dermoscopic images of skin lesions
Corresponding labels for each image (melanoma, BCC, or SCC)
Hyperparameters (e.g., learning rate, batch size, epochs, etc.)

Output:
Segmentation masks and classification labels

Step 1: Data Preparation
Load the dataset of dermoscopic images and their corresponding labels.
Split the dataset into training (80%) and validation (20%) sets for segmentation, and training (75%) and validation (25%) sets for classification.
Resize images to appropriate dimensions:
Segmentation models: 256 × 256 pixels
Classification models: 224 × 224 pixels (VGG16, ResNet50) or 299 × 299 pixels (InceptionV3).
Normalize pixel values of images to the range [0, 1].
Apply data augmentation techniques to the training set:
Random rotations
Horizontal and vertical flips
Painting.

Algorithm 1 *Cont.*

Step 2: Class Balancing
Identify class distribution within the training dataset.
Use random oversampling to duplicate instances of minority classes (BCC, SCC) until balanced with the majority class (melanoma).

Step 3: Model Training
Segmentation Models:
Initialize U-Net, DeepLabV3, and SegNet models.
Compile each model using an appropriate optimizer (e.g., Adam) and loss function (e.g., Binary Crossentropy, unet3p_hybrid_loss).
Train each segmentation model on the augmented and balanced training set for a specified number of epochs.
Monitor training and validation loss/accuracy.

Classification Models:
Initialize VGG16, ResNet50, and InceptionV3 models
Compile each model using an appropriate optimizer and loss function (e.g., Categorical Crossentropy).
Train each classification model on the augmented and balanced training set for a specified number of epochs.
Monitor training and validation metrics (accuracy, precision, recall).

Step 4: Ensemble Prediction
Generate predictions from each trained segmentation model for the validation set.
Combine segmentation outputs using a voting or averaging mechanism to obtain the final segmentation mask.
Generate predictions from each trained classification model for the validation set.
Combine classification outputs using an ensemble technique (e.g., weighted average or majority voting) to obtain the final class label.

Step 5: Model Evaluation
Calculate evaluation metrics for segmentation:
Intersection over Union (IoU)
Dice Coefficient
Accuracy.

Calculate evaluation metrics for classification:
Precision
Recall
F1-Score
AUC
Accuracy.

End Algorithm

2.1. Proposed Skin Lesion Segmentation

The system leverages an ensemble of U-Net [22], SegNet [23], and DeepLabV3 [24] to improve segmentation accuracy and robustness. U-Net excels at capturing fine details for precise lesion boundaries, SegNet maintains spatial resolution during upsampling, and DeepLabv3 handles multi-scale lesions with atrous convolutions. This ensemble outperforms individual models on the challenging ISIC dataset, making it applicable in real-world clinical scenarios. The segmentation model's ability to accurately identify lesion borders aids early detection and the diagnosis of skin malignancies, improving outcomes for patients. Figure 1 describes the segmentation design. The ISIC dataset [25], used for model training, includes 2357 images of various benign and malignant skin conditions, with a focus on three key cancers: melanoma (438 images), basal cell carcinoma (376 images), and squamous cell carcinoma (181 images). The ISIC 2018 dataset is a large and comprehensive dataset used for the analysis of skin lesions, particularly for tasks such as segmentation, classification, and the detection of skin cancer. It was made available as part of the International Skin Imaging Collaboration (ISIC) challenge, aimed at improving the early detection of skin cancer. The dataset contains 10,015 high-resolution dermoscopic images, each labeled with diagnostic information confirmed by expert dermatologists. The images cover various types of skin lesions, with a focus on conditions such as melanoma, BCC, and SCC, among others. The dataset is diverse, including images from different skin types, lesion shapes, sizes, and locations, providing a robust foundation for developing machine learning models that can generalize well to real-world scenarios. Each image is provided in JPG format with varying resolutions, making it essential to preprocess the images before feeding them into models.

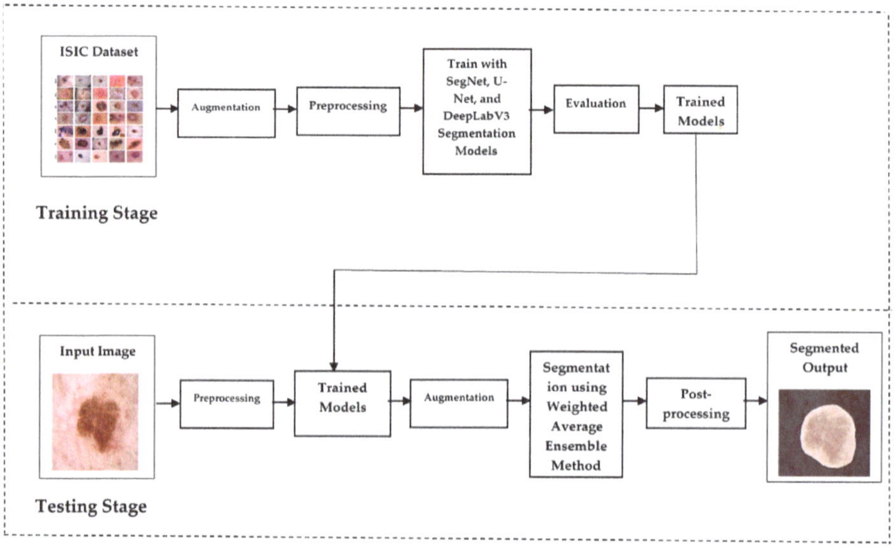

Figure 1. Skin lesion segmentation design.

The data is split into 80% for training and 20% for validation. Annotation is performed using LabelMe tools, and the annotated JSON files are converted into mask images. Images are resized to 256 × 256 pixels, and pixel intensities are normalized. Data augmentation techniques, such as flipping, in-painting, and rotation, are applied to create five variations of each image to improve model robustness.

This process involves two types of transformations. In flipping (2 images), the original image is flipped to create two additional variations. This might include horizontal and vertical flips. In-painting (2 images) is a technique which involves removing a portion of the image (in this case, one lung) and filling in the missing part with surrounding pixels or predicted values. In medical imaging, this can help models learn to identify structures even when parts of them are missing. Rotation (1 image) involves rotating the original image to create variations. This can involve rotations by angles between $-10°$ to $+10°$.

Each of the five augmented images is independently fed into the segmentation model, producing a binary or probabilistic mask for each. The masks are averaged pixel-wise across the five augmentations to create a more robust final prediction. The resulting masks from the three models—SegNet, DeepLabv3, and U-Net—are then combined using a weighted average based on each model's performance (e.g., Dice Coefficient, IoU, and accuracy). This combined mask, containing continuous values representing the probability of each pixel belonging to a skin lesion, is then thresholded at 0.5 to produce a final binary mask, classifying pixels as lesion or non-lesion. This approach leverages the strengths of both augmentations and the model diversity, resulting in a more accurate and reliable skin lesion segmentation.

$$\text{Binary Mask (i, j)} = \begin{cases} 1 \text{ if Combined Mask (i, j)} > 0.5 \\ 0 \text{ if Combined Mask (i, j)} \leq 0.5 \end{cases} \quad (1)$$

In the post-processing stage of skin lesion segmentation, a series of morphological operations is applied to refine the segmentation mask. Morphological Opening removes small noise and smooths lesion boundaries, and is followed by Erosion to reduce small white pixels and shrink the lesion edges. Dilation then restores lesion size while maintaining noise-free regions. Morphological Closing fills small holes within the lesion areas, and an additional Dilation step closes gaps. A final Erosion ensures the lesion regions are continuous and hole-free. These operations enhance the combined mask, leading to more precise and reliable skin lesion segmentation. By applying morphological operations, such as opening and closing, the post-processing stage effectively refines the segmentation masks produced by the ensemble models. Morphological Opening removes small artifacts and noise, leading to smoother contours and more defined boundaries of the lesions. This is particularly important in medical imaging, where the accurate delineation of lesions is critical for diagnosis and treatment planning. Additionally, Morphological Closing fills in gaps and connects fragmented structures, ensuring that the segmented lesions are represented uniformly. These improvements in the quality of segmentation masks facilitate better feature extraction and analysis, enabling subsequent classification tasks to be performed more effectively. Furthermore, post-processing can help mitigate issues arising from class imbalance by ensuring that the detected lesions are consistently represented across different images. Overall, incorporating post-processing techniques significantly enhances the robustness of skin lesion segmentation, ultimately contributing to more reliable and effective medical diagnostics.

2.2. Proposed Skin Lesion Classification

The proposed system combines VGG16 [26], ResNet-50 [27], and Inception-V3 [28] architectures to enhance skin lesion detection and classification by addressing key challenges in dermatological image analysis, such as class imbalance, low intra-class variability, and high inter-class similarity. The ISIC dataset, a comprehensive collection of dermoscopic images, is used for this study. To mitigate class imbalance, oversampling strategies are applied, particularly to underrepresented classes [29]. Figure 2 shows the architecture design for skin lesion classification. The dataset is split into 75% for training and 25% for testing, with images resized to 224 × 224 pixels for VGG16 and ResNet-50, and 299 × 299 pixels for Inception-V3. Image preprocessing includes normalization using the min-max technique [30], scaling pixel values between 0 and 1 to ensure consistent feature contribution. Additional preprocessing steps, such as data augmentation (e.g., rotation, flipping, and zooming), further enhance model performance by increasing data diversity [31]. These include Center Crop to focus on central features, Random Rotate, Grid Distortion, Flips (Horizontal/Vertical), Optical Distortion, and Affine transformations, all applied with a 0.1 probability to introduce controlled distortions and variations. These augmentations helped the model learn robust features by increasing the diversity of the dataset.

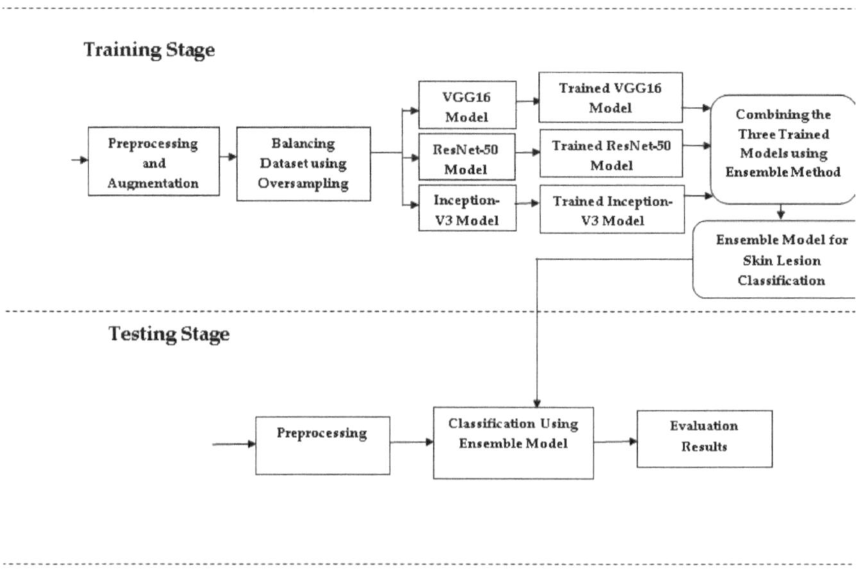

Figure 2. Skin lesion classification design.

This system classifies skin cancer into three categories: melanoma, BCC, and SCC, using 438 images of melanoma, 376 images of BCC, and 181 images of SCC. The significant disparity in class sizes could bias the model's performance, making it favor the majority class. To address this imbalance, the dataset was balanced using random oversampling [32]. This technique involves duplicating samples from the minority class, ensuring a more even class distribution. Specifically, simple random sampling with replacement (SRSWR) was employed, allowing the same sample to be chosen multiple times. This method effectively increases the representation of the minority class, mitigating the effects of class imbalance. This technique involves randomly duplicating instances from the minority classes—specifically, basal cell carcinoma (BCC) and squamous cell carcinoma (SCC)—until the number of samples in each class matches that of the majority class, melanoma. The goal of this approach is to prevent the model from being biased toward the overrepresented majority class by ensuring that all classes contribute equally to the training process. In addition to random oversampling, we also applied data augmentation techniques, such as rotation, scaling, and flipping, to introduce more variability into the dataset. This not only enriched the oversampled data but also mitigated the risk of overfitting by providing the model with diverse training samples. Together, random oversampling and data augmentation allowed us to create a balanced and robust dataset for training, improving the model's ability to classify skin lesions across all categories.

The balanced dataset was then used to train the model over 150 epochs, each representing a full pass through the training data, allowing the model to progressively improve. Once the desired accuracy was reached, the model was saved for potential reuse or further fine-tuning. The final classification system was constructed as a deep ensemble, combining predictions from three models—ResNet-50, VGG16, and Inception-V3—using a weighted average technique.

During testing, input images were first resized according to the model architecture: 224×224 pixels for VGG16 and ResNet-50, and 299×299 pixels for Inception-V3. These images were then normalized to ensure consistent pixel intensities across the dataset, and converted to tensor format for deep learning compatibility. The classification was performed using the ensemble model, which integrates predictions from multiple architectures to boost accuracy and robustness. Initially, testing was conducted on the original, imbalanced dataset to identify performance issues caused by class disparity. Further testing was carried out on a balanced dataset, achieved through oversampling and data augmentation techniques such as rotations, flips, and color adjustments. Finally, the model was evaluated on both the original and balanced datasets to assess its overall effectiveness and robustness.

3. Results

Performance evaluation for segmentation and performance evaluation for classification are the two sections that make up this section.

3.1. Performance Evaluation for Segmentation

The suggested system is trained and tested using the ISIC dataset, which has an 80% training to 20% testing partition ratio. Using the original dataset, a detailed assessment of the system's effectiveness and performance is carried out, with a particular focus on important metrics like Dice Coefficient, IoU, and accuracy. In the ensemble model segmentation, performance metrics are evaluated for individual models and the ensemble model. Table 1 describes hyperparameters for segmentation and Table 2 shows the performance results of ensemble model segmentation.

Table 1. Hyperparameters for segmentation.

Parameter	U-Net	DeepLab-V3	SegNet
Input Image Size	256 × 256	256 × 256	256 × 256
Batch Size	16	16	16
Learning Rate	0.001	0.001	0.001
Optimizer	Adam	Adam	Adam
Epochs	100	100	100
Loss Function	unet3p_hybrid_loss	unet3p_hybrid_loss	Binary Crossentropy Loss
Dropout Rate	0.5	0.5	0.5
Data Augmentation	Random Rotation, Flip, Painting	Random Rotation, Flip, Painting	Random Rotation, Flip, Painting
Training/Validation Split	80%/20%	80%/20%	80%/20%

Table 2. Segmentation results.

	Dice Coefficient	IoU	Accuracy
Model 1 (SegNet)	0.82	0.75	0.88
Model 2 (DeepLabV3)	0.86	0.79	0.90
Model 3 (U-Net)	0.84	0.81	0.91
Ensemble Model	0.93	0.90	0.95

The segmentation results reveal that the ensemble model of SegNet, U-Net, and DeepLabV3 significantly outperforms the individual models. Specifically, SegNet achieved a Dice Coefficient of 0.82, an IoU of 0.75, and an accuracy of 0.88, while DeepLabV3 delivered higher results with a Dice Coefficient of 0.86, an IoU of 0.79, and an accuracy of 0.90. U-Net also performed well, with a Dice Coefficient of 0.84, an IoU of 0.81, and an accuracy of 0.91. The ensemble model, however, achieved the highest scores with a Dice Coefficient of 0.93, an IoU of 0.90, and an accuracy of 0.95, demonstrating a substantial improvement in segmentation performance by leveraging the strengths of all three models. This enhancement demonstrates how the ensemble approach can effectively improve segmentation performance by utilizing the advantages of many models to produce a more reliable and accurate result.

3.2. Performance Evaluation for Classification

In this section, the performance of the suggested deep ensemble model is compared to that of the three separate models—ResNet-50, Inception-V3, and VGG16—with an emphasis on accuracy metrics. Specifically, the evaluation employs the Kaggle ISIC dataset to classify melanoma, SCC, and BCC using pictures. The initial, unbalanced dataset is used to test the system's performance. After that, the assessment is carried out once again using a balanced dataset that was produced using oversampling methods. TensorFlow serves as the backend, and Keras is used to build and train these models. In performance evaluation, 25% of the samples are used for testing and 75% are used for training. We use common performance measures including accuracy, recall, precision, and F1-score to assess the effectiveness of our suggested strategy. The model's ability to distinguish between melanoma, SCC, and BCC lesions may be quantified with the use of these metrics, which provide an extensive assessment of the model's classification performance across numerous classes. Table 3 describes hyperparameters for classification.

Table 3. Hyperparameters for classification.

Parameter	VGG16	ResNet-50	Inception-V3
Input Image Size	224 × 224	224 × 224	299 × 299
Batch Size	32	32	32
Learning Rate	0.001	0.001	0.001
Optimizer	Adam	Adam	Adam
Epochs	150	150	150
Loss Function	Categorical Crossentropy	Categorical Crossentropy	Categorical Crossentropy
Dropout Rate	0.5	0.5	0.5
Data Augmentation	Yes	Yes	Yes
Training/Validation Split	75%/25%	75%/25%	75%/25%

This study compares the performance of VGG16, ResNet-50, Inception-V3, and an ensemble model for skin lesion classification using both balanced and unbalanced datasets, with and without augmentation, as shown in Tables 4–7. Results show that balanced datasets consistently outperform unbalanced ones, particularly when combined with augmentation. VGG16 and ResNet-50 show significant improvements in precision, recall, F1-Score, and AUC when using balanced datasets with augmentation, while unbalanced datasets yield lower effectiveness. Similarly, Inception-V3 benefits from augmentation and balancing, achieving its highest accuracy on balanced datasets with augmentation. The ensemble model performs best with balanced datasets and augmentation, showing the highest precision and accuracy, underscoring the importance of these techniques in optimizing model performance.

Table 4. VGG16's results on balanced and unbalanced datasets.

Dataset Type	Class	Precision	Recall	F1-Score	AUC	Accuracy
Unbalanced (Without Augmentation)	Melanoma	0.75	0.70	0.72	0.80	0.71
	BCC	0.80	0.74	0.77	0.82	
	SCC	0.70	0.65	0.67	0.78	
Unbalanced (With Augmentation)	Melanoma	0.78	0.74	0.76	0.83	0.80
	BCC	0.84	0.78	0.81	0.85	
	SCC	0.73	0.69	0.71	0.80	
Balanced (Without Augmentation)	Melanoma	0.85	0.82	0.83	0.88	0.86
	BCC	0.87	0.85	0.86	0.90	
	SCC	0.80	0.78	0.79	0.86	

Table 4. *Cont.*

Dataset Type	Class	Precision	Recall	F1-Score	AUC	Accuracy
Balanced (With Augmentation)	Melanoma	0.90	0.87	0.88	0.92	0.94
	BCC	0.91	0.89	0.90	0.93	
	SCC	0.85	0.83	0.84	0.89	

Table 5. ResNet-50's results on balanced and unbalanced datasets.

Dataset Type	Class	Precision	Recall	F1-Score	AUC	Accuracy
Unbalanced (Without Augmentation)	Melanoma	0.77	0.72	0.74	0.78	0.74
	BCC	0.81	0.76	0.78	0.79	
	SCC	0.71	0.67	0.69	0.71	
Unbalanced (With Augmentation)	Melanoma	0.80	0.76	0.78	0.81	0.85
	BCC	0.84	0.79	0.81	0.83	
	SCC	0.74	0.71	0.72	0.74	
Balanced (Without Augmentation)	Melanoma	0.86	0.83	0.84	0.88	0.88
	BCC	0.88	0.86	0.87	0.89	
	SCC	0.81	0.80	0.80	0.82	
Balanced (With Augmentation)	Melanoma	0.98	0.87	0.88	0.91	0.98
	BCC	0.90	0.88	0.89	0.92	
	SCC	0.83	0.81	0.82	0.84	

Table 6. Inception-V3's results on balanced and unbalanced datasets.

Dataset Type	Class	Precision	Recall	F1-Score	AUC	Accuracy
Unbalanced (Without Augmentation)	Melanoma	0.75	0.71	0.73	0.76	0.70
	BCC	0.79	0.74	0.76	0.77	
	SCC	0.69	0.65	0.67	0.69	
Unbalanced (With Augmentation)	Melanoma	0.78	0.75	0.76	0.79	0.87
	BCC	0.81	0.77	0.79	0.80	
	SCC	0.72	0.68	0.70	0.72	
Balanced (Without Augmentation)	Melanoma	0.84	0.80	0.82	0.85	0.86
	BCC	0.87	0.83	0.85	0.86	
	SCC	0.78	0.76	0.77	0.79	
Balanced (With Augmentation)	Melanoma	0.86	0.83	0.84	0.87	0.98
	BCC	0.89	0.85	0.87	0.88	
	SCC	0.80	0.78	0.79	0.81	

Table 7. Deep ensemble model's results on balanced and unbalanced datasets.

Dataset Type	Class	Precision	Recall	F1-Score	AUC	Accuracy
Unbalanced (Without Augmentation)	Melanoma	0.80	0.75	0.77	0.82	0.78
	BCC	0.83	0.78	0.80	0.83	
	SCC	0.71	0.67	0.69	0.71	
Unbalanced (With Augmentation)	Melanoma	0.84	0.80	0.82	0.85	0.90
	BCC	0.86	0.82	0.84	0.87	
	SCC	0.74	0.70	0.72	0.74	
Balanced (Without Augmentation)	Melanoma	0.87	0.83	0.85	0.88	0.92
	BCC	0.89	0.85	0.87	0.89	
	SCC	0.81	0.78	0.79	0.82	
Balanced (With Augmentation)	Melanoma	0.89	0.86	0.87	0.90	0.99
	BCC	0.91	0.88	0.89	0.91	
	SCC	0.83	0.81	0.82	0.84	

The ensemble model with augmentation significantly outperforms individual models like VGG16, ResNet-50, and Inception-V3 in skin lesion classification for both the original and balanced ISIC datasets. On unbalanced datasets without augmentation, the ensemble model achieves the highest accuracy at 0.78 and, with augmentation, it reaches 0.90. On balanced datasets, the ensemble model's accuracy improves to 0.92 without augmentation and 0.99 with augmentation, consistently outperforming the individual models. These results highlight the effectiveness of combining multiple models with balanced datasets and augmentation, making skin lesion classification more accurate and reliable. This approach enhances diagnostic precision and holds potential for broader use in medical image analysis. Table 8 describes the confusion matrix. The ROC curves for VGG16, ResNet-50, Inception-V3, and the ensemble model for skin lesion classification using both balanced and unbalanced datasets, with and without augmentation, are shown in Figures 3–6.

Table 8. Confusion matrix.

	Predicted BCC	Predicted SCC	Predicted Melanoma
True Melanoma	300	30	8
True BCC	10	280	12
True SCC	5	15	150

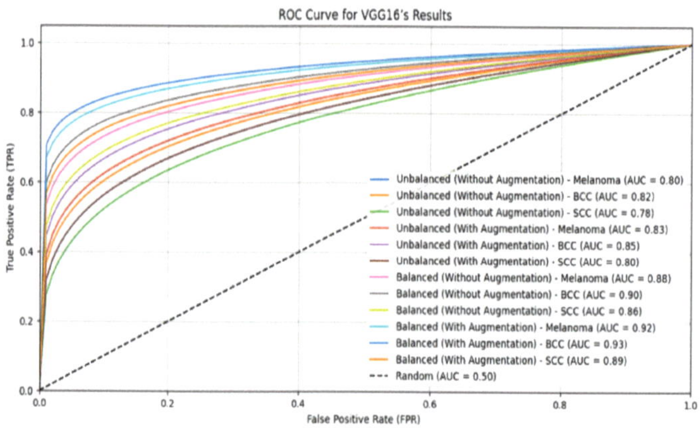

Figure 3. ROC curve for VGG16.

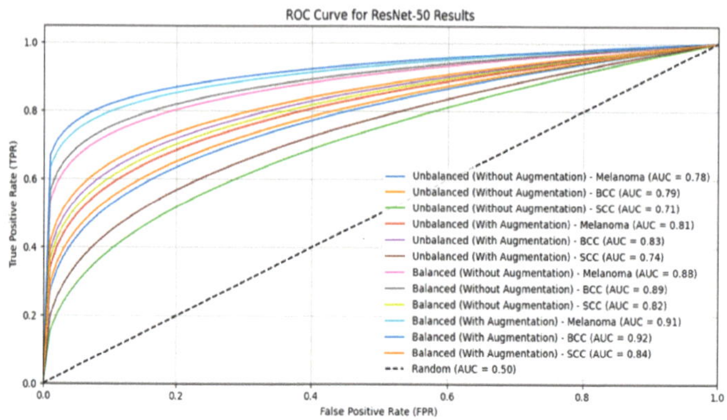

Figure 4. ROC curve for ResNet-50.

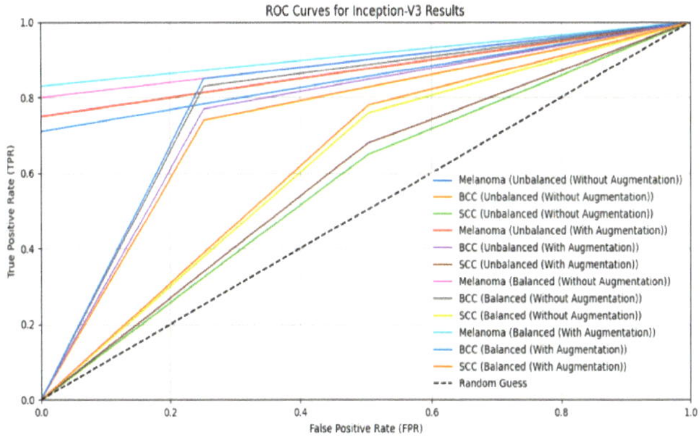

Figure 5. ROC curve for Inception-V3.

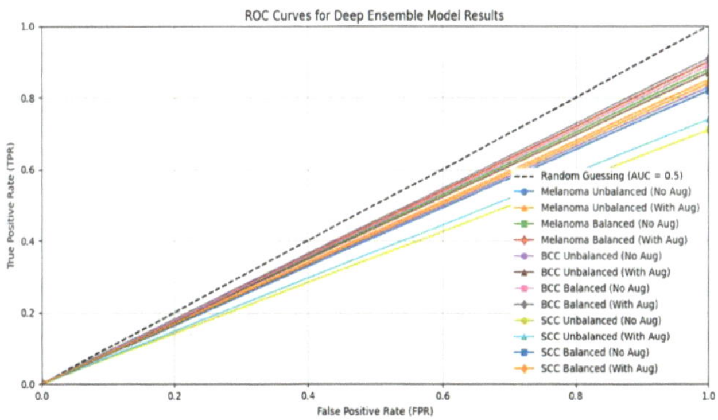

Figure 6. ROC curve for deep ensemble model.

Then, the system is tested using the ISIC 2019 dataset. The results of the ensemble model for skin lesion classification using both balanced and unbalanced datasets, with and without augmentation, are shown in Table 9. These findings underscore the advantages of integrating various models with balanced datasets and augmentation, improving the accuracy and reliability of skin lesion classification. This strategy increases diagnostic accuracy and shows promise for wider applications in medical image analysis.

Table 9. Deep ensemble model's results on balanced and unbalanced datasets of ISIC 2019.

Dataset Type	Class	Precision	Recall	F1-Score	AUC	Accuracy
Unbalanced (Without Augmentation)	Melanoma	0.70	0.65	0.67	0.76	0.80
	BCC	0.75	0.80	0.77	0.79	
	SCC	0.68	0.60	0.64	0.70	
Unbalanced (With Augmentation)	Melanoma	0.73	0.70	0.71	0.78	0.92
	BCC	0.78	0.82	0.80	0.81	
	SCC	0.70	0.65	0.67	0.72	
Balanced (Without Augmentation)	Melanoma	0.85	0.80	0.82	0.90	0.94
	BCC	0.88	0.85	0.86	0.92	
	SCC	0.82	0.78	0.80	0.88	
Balanced (With Augmentation)	Melanoma	0.87	0.85	0.86	0.91	0.99
	BCC	0.90	0.88	0.89	0.93	
	SCC	0.84	0.82	0.83	0.89	

4. Discussion

In this work, we investigated the segmentation and classification of skin lesions using cutting-edge deep learning methods. By leveraging a deep ensemble model combining U-Net, SegNet, and DeepLabv3 for segmentation, and VGG16, ResNet-50, and Inception-V3 for classification, we aimed to enhance the accuracy and reliability of skin lesion detection. The ISIC dataset was balanced using oversampling approaches to address class imbalances and offer a more resilient training process. The results demonstrate that the ensemble models effectively segmented skin lesions and classified them into basal cell carcinoma (BCC), squamous cell carcinoma (SCC), and melanoma, with notable performance improvements when using balanced datasets through oversampling. The segmentation models achieved high accuracy across different types of skin lesions, while the classification models showed significant enhancements in precision, recall, and F1-scores with balanced datasets. These improvements highlight the effectiveness of using ensemble methods and data balancing techniques in medical image analysis.

Comparison with Existing Literature

Table 10 shows comparative results from other existing research in the literature for skin lesion segmentation and classification systems. Our suggested approach yields more accurate findings compared to alternative methods. In comparison with those in the literature, the proposed ensemble model for segmentation outperforms previous methods, achieving a Dice Coefficient of 93 and an IoU of 90 on the ISIC 2018 dataset, surpassing the highest results reported for U-Net (89.3) and DenseUNet (92.23). For classification, the proposed ensemble model also shows superior performance, with an accuracy of 99 on the ISIC 2018 dataset, exceeding the accuracy of other methods such as DenseUNet (97.88) and CNN-based approaches (up to 98.5), and demonstrating its effectiveness in both segmentation and classification tasks.

Table 10. Comparison with existing literature for segmentation and classification.

	Existing Works	Methods	Dataset	Dice Coefficient	IoU	Accuracy
Segmentation	[7]	U-Net	ISIC 2018	89.3		
	[8]	FCA-Net	ISIC 2018		77.2	
	[12]	DenseUNet	ISIC 2016	85.64		
			ISIC 2017	86.61		
			ISIC 2018	92.23		
	Proposed System	Ensemble Model	ISIC 2018	93	90	
Classification	[6]	CNN	ISIC 2018			83.1
		ReNet-50				83.6
		Resnet50-Inception				84.1
		Inception V3				85.7
	[10]	CNN	ISBI2017			94.32
	[12]	DenseUNet	ISIC 2016			98.03
			ISIC 2017			96.19
			ISIC 2018			97.88
	[13]	Cubic SVM	ISIC 2017			96.7
	[15]	CNN	Dermquest database			98.5
	[16]	DenseNet	ISIC 2018			81
	[19]	CNN	HAM10000			95.94
	Proposed System	Ensemble Model	ISIC 2018			99

5. Conclusions

This study explored advanced deep learning techniques for skin lesion segmentation and classification. Using a deep ensemble approach with U-Net, SegNet, and DeepLabV3 for segmentation and VGG16, ResNet-50, and Inception-V3 for classification, we aimed to improve skin lesion detection. Balancing the ISIC dataset with oversampling addressed class imbalances, leading to enhanced model performance. The significant performance improvements achieved through ensemble methods and data balancing suggest that these techniques are highly effective for skin lesion analysis. The proposed method has several strengths, including improved accuracy for both segmentation and classification through the use of ensemble models, effectively addressing challenges like class imbalance with oversampling. Additionally, data augmentation enhances the model's robustness, making it more reliable for detecting different types of skin lesions. However, the method has some weaknesses, such as increased computational complexity due to ensemble learning, which demands more resources and time. It also relies heavily on large, well-annotated datasets and requires further clinical validation to assess its practical applicability in real-world settings. Future work could explore additional ensemble configurations or integrate other advanced techniques such as attention mechanisms to further enhance model performance. Additionally, expanding the dataset with more diverse samples could help in validating the robustness of the proposed models across different populations.

Limitations

While our study demonstrated promising results, there are limitations to consider. The oversampling technique, while effective, may introduce some noise in the data, which could potentially impact model generalizability. Future research should address these limitations by exploring alternative balancing techniques or incorporating advanced preprocessing methods. Additionally, we will explore the implications of these limitations in real-world clinical settings, considering factors such as variability in image quality, diversity of skin types, and the integration of our approach into existing clinical workflows. By providing a more nuanced discussion, we aim to better inform readers about the practical considerations of implementing our ensemble models in clinical practice and highlight areas for future research to address these challenges effectively.

Author Contributions: Conceptualization, S.M.T. and H.-S.P.; methodology, S.M.T. and H.-S.P.; software, S.M.T. and H.-S.P.; validation, S.M.T. and H.-S.P.; formal analysis, S.M.T. and H.-S.P.; investigation, S.M.T. and H.-S.P.; resources, H.-S.P. and S.M.T.; data curation, H.-S.P.; writing—original draft preparation, S.M.T.; writing—review and editing, S.M.T. and H.-S.P.; visualization, S.M.T. and H.-S.P.; supervision, H.-S.P.; funding acquisition, H.-S.P. All authors have read and agreed to the published version of the manuscript.

Funding: This research received no external funding.

Institutional Review Board Statement: Not Applicable.

Informed Consent Statement: Not applicable.

Data Availability Statement: Data are contained within the article.

Conflicts of Interest: The authors declare no conflict of interest.

References

1. Alkarakatly, T.; Eidhah, S.; Sarawani, M.A.; Sobhi, A.A.; Bilal, M. Skin Lesions Identification Using Deep Convolutional Neural Network. In Proceedings of the 2019 International Conference on Advances in the Emerging Computing Technologies (AECT), Al Madinah Al Munawwarah, Saudi Arabia, 10 February 2020; IEEE: Piscataway, NJ, USA; pp. 209–213. [CrossRef]
2. Murugan, A.; Nair, S.A.H.; Preethi, A.A.P.; Kumar, K.S. Diagnosis of skin cancer using machine learning techniques. *Microprocess. Microsyst.* **2021**, *81*, 103727. [CrossRef]
3. Salian, A.C.; Vaze, S.; Singh, P. Skin Lesion Classification using Deep Learning Architectures. In Proceedings of the 2020 3rd International Conference on Communication System, Computing and IT Applications (CSCITA), Mumbai, India, 3–4 April 2020; IEEE: Piscataway, NJ, USA; pp. 168–173. [CrossRef]

4. Ali, M.S.; Miah, M.S.; Haque, J.; Rahman, M.M.; Islam, M.K. An enhanced technique of skin cancer classification using deep convolutional neural network with transfer learning models. *Mach. Learn. Appl.* **2021**, *5*, 100036. [CrossRef]
5. Filali, Y.; Khoukhi, H.E.; Sabri, M.A. Texture Classification of skin lesion using convolutional neural network. In Proceedings of the 2019 International Conference on Wireless Technologies, Embedded and Intelligent Systems (WITS), Fez, Morocco, 3–4 April 2019. [CrossRef]
6. Gouda, W.; Sama, N.U.; Waakid, G.A. Detection of Skin Cancer Based on Skin Lesion Images Using Deep Learning. *Healthcare* **2022**, *10*, 1183. [CrossRef] [PubMed]
7. Araujo, R.L.; Rabelo, R.A.; Rodrigues, J.P.C.; Silva, R.V. Automatic Segmentation of Melanoma Skin Cancer Using Deep Learning. In Proceedings of the 2021 IEEE International Conference on E-Health Networking, Application & Services (HEALTHCOM), Shenzhen, China, 1–2 March 2021; IEEE: Piscataway, NJ, USA; pp. 1–6. [CrossRef]
8. Singh, V.K.; Abdel-Nasser, M.; Rashwan, H.A.; Akram, F.; Pandey, N.; Lalande, A.; Presles, B.; Romani, S.; Puig, D. FCA-Net: Adversarial learning for skin lesion segmentation based on multi-scale features and factorized channel attention. *IEEE Access* **2019**, *7*, 130552–130565. [CrossRef]
9. Yang, X.; Zeng, Z.; Yeo, S.Y.; Tan, C.; Tey, H.L.; Su, Y. A novel multi-task deep learning model for skin lesion segmentation and classification. *arXiv* **2017**, arXiv:1703.01025. [CrossRef]
10. Liu, L.; Tsui, Y.Y.; Mandal, M. Skin Lesion Segmentation Using Deep Learning with Auxiliary Task. *J. Imaging* **2021**, *7*, 67. [CrossRef]
11. Mirikharaji, Z.; Abhishek, K.; Bissoto, A.; Barata, C.; Avila, S.; Valle, E.; Celebi, M.E.; Hamarneh, G. A survey on deep learning for skin lesion segmentation. *Med. Image Anal.* **2023**, *88*, 102863. [CrossRef] [PubMed]
12. Jimi, A.; Abouche, H.; Zrira, N.; Benmiloud, I. Skin Lesion Segmentation Using Attention-Based DenseUNet. In Proceedings of the 16th International Joint Conference on Biomedical Engineering Systems and Technologies (BIOSTEC 2023), Lisbon, Portugal, 16–18 February 2023; BIOINFORMATICS. Volume 3, pp. 91–100. [CrossRef]
13. Bibi, A.; Khan, M.A.; Javed, M.Y.; Tariq, U.; Kang, B.G.; Nam, Y.; Mostafa, R.R.; Sakr, R.H. Skin Lesion Segmentation and Classification Using Conventional and Deep Learning Based Framework. *Comput. Mater. Contin.* **2022**, *71*, 2477–2495. [CrossRef]
14. Ashraf, H.; Waris, A.; Ghafoor, M.F.; Gilani, S.O.; Niazi, I.K. Melanoma segmentation using deep learning with test-time augmentations and conditional random fields. *Sci. Rep.* **2022**, *12*, 3948. [CrossRef]
15. Jafari, M.H.; Karimi, N.; Nasr-Esfahani, E.; Samavi, S.; Soroushmehr, S.M.R.; Ward, K.; Najarian, K. Skin Lesion Segmentation in Clinical Images Using Deep Learning. In Proceedings of the 2016 23rd International Conference on Pattern Recognition (ICPR), Cancún Center, Cancún, México, 4–8 December 2016. [CrossRef]
16. Chandra, R.; Hajiarbabi, M. Skin Lesion Detection Using Deep Learning. *J. Autom. Mob. Robot. Intell. Syst.* **2022**, *16*, 56–64. [CrossRef]
17. Gessert, N.; Nielsen, M.; Shaikh, M.; Werner, R.; Schlaefer, A. Skin lesion classification using ensembles of multi-resolution EfficientNets with meta data. *MethodsX* **2020**, *7*, 100864. [CrossRef] [PubMed]
18. Ding, S.; Wu, Z.; Zheng, Y.; Liu, Z.; Yang, X.X.; Yuan, G.; Xie, J. Deep attention branch networks for skin lesion classification. *Comput. Methods Programs Biomed.* **2021**, *212*, 106447. [CrossRef] [PubMed]
19. Alhudhaif, A.; Almaslukh, B.; Aseeri, A.O.; Guler, O.; Polat, K. A novel nonlinear automated multi-class skin lesion detection system using soft-attention based convolutional neural networks. *Chaos Solitons Fractals* **2023**, *170*, 113409. [CrossRef]
20. Alsahafi, Y.S.; Kassem, M.A.; Hosny, K.M. Skin-Net: A novel deep residual network for skin lesions classification using multilevel feature extraction and cross-channel correlation with detection of outlier. *J. Big Data* **2023**, *10*, 105. [CrossRef]
21. Hosny, K.M.; Elshoura, D.; Mohamed, E.R.; Vrochidou, E.; Papakostas, G.A. Deep Learning and Optimization-Based Methods for Skin Lesions Segmentation: A Review. *IEEE Access* **2023**, *11*, 85467–85488. [CrossRef]
22. Ronneberger, O.; Fischer, P.; Brox, T. U-Net: Convolutional Networks for Biomedical Image Segmentation. In Proceedings of the Medical Image Computing and Computer-Assisted Intervention–MICCAI 2015: 18th International Conference, Munich, Germany, 5–9 October 2015. [CrossRef]
23. Badrinarayanan, V.; Kendall, A.; Cipolla, R. SegNet: A Deep Convolutional Encoder-Decoder Architecture for Image Segmentation. *IEEE Trans. Pattern Anal. Mach. Intell.* **2017**, *39*, 2481–2495. [CrossRef] [PubMed]
24. Chen, L.; Zhu, Y.; Papandreou, G.; Schroff, F.; Adam, H. DeepLabv3: Rethinking Atrous Convolution for Semantic Image Segmentation. *arXiv* **2018**, arXiv:1706.05587. [CrossRef]
25. MNOWAK061. Skin Lesion Dataset. *ISIC2018 Kaggle Repository*. 2021. Available online: https://www.kaggle.com/datasets/mnowak061/isic2018-and-ph2-384x384-jpg (accessed on 10 April 2022).
26. Simonyan, K.; Zisserman, A. Very Deep Convolutional Networks for Large-Scale Image Recognition. *arXiv* **2014**, arXiv:1409.1556. [CrossRef]
27. He, K.; Zhang, X.; Ren, S.; Sun, J. Deep Residual Learning for Image Recognition. In Proceedings of the IEEE Conference on Computer Vision and Pattern Recognition (CVPR), Honolulu, Hawaii, 21–26 July 2016. [CrossRef]
28. Sayyad, J.; Patil, P.; Gurav, S. Skin Disease Detection Using VGG16 and InceptionV3. *Int. J. Intell. Syst. Appl. Eng.* **2024**, *12*, 148–155.
29. Barua, S.; Islam, M.M.; Murase, K. A novel synthetic minority oversampling technique for imbalanced data set learning. *Lect. Notes Comput. Sci.* **2011**, *7063*, 735–744. [CrossRef]

30. Han, J.; Kamber, M.; Pei, J. *Data Mining: Concepts and Techniques (The Morgan Kaufmann Series in Data Management Systems)*, 3rd ed.; Elsevier Science Ltd.: Amsterdam, The Netherlands, 2011; pp. 1–703. [CrossRef]
31. Harangi, B.; Baran, A.; Hajdu, A. Assisted deep learning framework for multi-class skin lesion classification considering a binary classification support. *Biomed. Signal Process. Control* **2020**, *62*, 102041. [CrossRef]
32. Mariani, G.; Scheidegger, F.; Istrate, R.; Bekas, C.; Malossi, C. BAGAN: Data Augmentation with Balancing GAN. *arXiv* **2018**, arXiv:1803.09655.

eng

MDPI

Article

Optical Fiber Technology for Efficient Daylighting and Thermal Control: A Sustainable Approach for Buildings

Lokesh Udhwani [1,*], Archana Soni [1], Erdem Cuce [2,3,4] and Sudhakar Kumarasamy [5,6,1,*]

1 Energy Centre, Maulana Azad National Institute of Technology, Bhopal 462003, Madhya Pradesh, India; soniarchana@gmail.com
2 Department of Mechanical Engineering, Faculty of Engineering and Architecture, Recep Tayyip Erdogan University, Zihni Derin Campus, Rize 53100, Turkey
3 Department of Mechanical Engineering, College of Engineering, Birmingham City University, Birmingham B4 7XG, UK
4 University Centre for Research and Development, Chandigarh University, Mohali 140413, Punjab, India
5 Faculty of Mechanical and Automotive Engineering Technology, Universiti Malaysia Pahang Al Sultan Abdullah, Pekan 26600, Pahang, Malaysia
6 Centre for Automotive Engineering (Automotive Centre), Universiti Malaysia Pahang Al Sultan Abdullah, Pekan 26600, Pahang, Malaysia
* Correspondence: udhwanilokeshmanit@gmail.com (L.U.); sudhakar@umpsa.edu.my (S.K.)

Abstract: Different direct solar harvesting systems for daylighting are being explored to achieve high uniform illumination deep within buildings at minimal cost. A promising solution to make these systems cost-effective is the use of plastic optical fibers (POFs). However, heat-related issues with low-cost POFs need to be addressed for the widespread adoption of efficient daylighting technologies. Previous studies have explored solutions for this overheating problem, but their effectiveness remains uncertain. This study proposes a low-cost fiber optic daylighting system integrated with a newly patented mechanical component designed to secure the fiber optic bundle at the focal point, providing three levels of heat filtration while ensuring uniform illumination. Our methodology involves selecting a small area, installing the setup, and measuring both heat and light readings, followed by validation through software simulations. The operational principle of this technology is explained, and experimental tests using lux meters and infrared thermometers were conducted to investigate the system's characteristics. The three-level heat filtration device reduces temperature by approximately 35 °C at the surface of the optical fiber, and the average illumination of the room is around 400 lux. These results were further verified using RELUX simulation software. The findings demonstrate the promising potential of this new device in solar heat filtration and achieving uniform illumination. Recommendations for mitigating overheating damage and exploring heat filtering possibilities in new parabolic solar daylighting systems for further research are also provided.

Keywords: plastic optical fibers (POFs); uniform illumination; daylighting; RELUX simulation; solar harvesting

Citation: Udhwani, L.; Soni, A.; Cuce, E.; Kumarasamy, S. Optical Fiber Technology for Efficient Daylighting and Thermal Control: A Sustainable Approach for Buildings. *Eng* **2024**, *5*, 2680–2694. https://doi.org/10.3390/eng5040140

Academic Editors: Antonio Gil Bravo and Maria Founti

Received: 6 June 2024
Revised: 2 September 2024
Accepted: 6 September 2024
Published: 18 October 2024

1. Introduction

There is a keen interest in developing renewable energy resources to tackle the ever-increasing need for electricity and the resulting global warming [1]. Non-domestic buildings are a major aspect to consider when calculating total energy consumption [2]. Lighting is the most energy-intensive application, which holds a major share of the average electricity consumption. About 25–30% of total energy usage results from lighting applications in most developing countries, which may increase further to 40% in developed countries [3]. Luminaries account for 31% of household and 44% of office electricity consumption in the Indian scenario [4,5]. The finances involved in using the different resources are studied in recent research works [6]. SBTool has been presented as a useful evaluation of buildings for energy efficiency [7]. Experimental methods for cost assessment and performance enhancement strategies have been in focus for a long time [8]. Also, automatic systems for data collection have been demonstrated by researchers [9]. Smart strategies, like automatic switching, dimming, and controlled lighting, have been gaining popularity for efficient energy usage [10]. Since non-domestic buildings like offices and laboratories operate during the day, smart lighting and efficient daylight utilization can help reduce the burden on energy resources. Daylighting can be employed to reduce power consumption by 20–40% [11]. Daylight is the best source of light to replace the energy-intensive artificial lighting [12,13]. Research studies also suggest that daylight with artificial lighting can save 10.8 to 44% energy savings [14]. In addition, employing proper control systems can help further reduce energy consumption for indoor lighting [15]. Efficient use of daylight can also reduce the need for cooling appliances as it has lesser heating effects on the building walls [16–20]. Special-purpose, smart energy-efficient lighting by experimental analysis has been proposed previously in a few reports [21,22]. Efficient utilization of daylight is desirable to avoid unnecessary wastage of energy. However, in many cases, it may not be possible to harness natural light, so there is a dependence on artificial lighting. While various daylighting techniques exist and are commercially available, including systems like Himawari (Japan), Parans (Sweden), and Hybrid solar lighting (USA) [23,24], their widespread adoption is hindered by factors such as low efficiency, high initial costs, and technical complexities. Among the systems utilized, the parabolic dish concentrator coupled with optical fiber guiding wires stands out as one of the most cost-effective options. Plastic optical fibers (POFs) present a viable solution for achieving uniform illumination deep within buildings without significantly impacting costs. However, a notable drawback of POFs is their susceptibility to heat, which becomes apparent when attempting to enhance efficiency parameters such as enlarging the capturing device, resulting in the melting of POF wires [25–27]. This study centers on monitoring the temporal temperature at the input of POF wires.

Numerous studies conducted over the past decades have explored the utilization of optical fiber guiding systems [3–26]. Previous studies have explored the performance and characteristics of solar concentrator systems integrated with various fiber optic cables using simulations and experiments. Daylighting systems using optical fiber wires (DSvOF) can be categorized based on solar concentrating technologies and fiber optic types.

In the past, the discussion surrounding Daylighting Systems via Optical Fiber (DSvOF) did not emphasize the detrimental effects of overheating on fiber optic wires. This problem, however, has emerged as a significant obstacle in the advancement of optical-fiber-based daylighting systems. The primary challenge lies in safeguarding plastic optical fibers (POFs) from the deleterious components of solar radiation. What follows is a critical analysis of proposed systems concerning their susceptibility to heat damage affecting POFs.

Kandilli et al. [28,29] evaluated a cost-effective fiber optic daylighting system. This system used a dual sun-tracking offset parabolic dish to collect sunlight and transmit it via Polymethyl Methacrylate (PMMA) plastic optical fibers positioned at the focal point. The researchers achieved an average annual exergy efficiency of 15% and an overall system efficiency of 59%, with a calculated power output of 1041.6 kW per square meter from the dish. While the study explored methods to improve energy and exergy efficiencies, it lacked a detailed analysis of overheating, a critical issue for PMMA fibers. The authors suggested using optical filters to mitigate heat damage but did not specify filter type or filtration degree or test their effectiveness. Kaiyan et al. [30] proposed a concentrator for use in a solar fiber lamp, comprising multiple components including a compound curved surface concentrator and a secondary compound parabolic concentrator. Experimental results indicated that the solar fiber lamp could produce brightness equivalent to that of a 6–8 W electrical energy-saving lamp. However, the study did not delve into the effects of heat on the POFs used, despite their known sensitivity to high temperatures. Wang et al. [31] introduced a fluorescent fiber solar concentrator (FFSC) system consisting of a plate device with 150 fluorescent fibers embedded in a PMMA plate, each connected to a 10 m long PMMA clear optical fiber. While the system demonstrated reasonable light-to-light efficiency and lighting effects, there was minimal risk of the PMMA fibers melting due to the absence of a solar concentrator. Han et al. [32] developed a fiber optic solar concentrator system utilizing a small dish concentrator, flat mirror, and homogenizer. Photometric characteristics were presented for various lens options, highlighting the efficacy of concave lenses in dispersing light. However, the use of PMMA fibers in conjunction with a second reflecting mirror posed challenges to uniform illumination due to the overshadowing of central optical fibers. Sapia [33] explored the potential of a hybrid lighting system comprising a primary parabolic collector and a secondary flat optical reflector. The system, employing synthetic optical fibers, was assessed for its potential cost savings in solar daylighting. The author suggested using cold mirrors to mitigate overheating of optical fibers but acknowledged potential drawbacks including the impact on the uniformity of illumination. Xue et al. [34] investigated a solar optic fiber illumination system incorporating a novel multi-surface co-focus compound parabolic concentrator (CPC). Their experimental results indicated transmittance rates of up to 17.5% for the solar concentrator and up to 11% for the novel illumination system. Factors influencing transmittance included reflectivity, tracking accuracy of the solar concentrator, and characteristics of the optical fiber.

Lingfors et al. [35] acknowledged the limitations of plastic optical fibers (POFs) in high-temperature environments and proposed several solutions to mitigate the heat problem. They suggested limiting the concentrator with a concentration ratio of 278 for the novel sunlight guiding system, coupled with an acceptance angle of 3.4°. This design not only facilitates the connection of optical fibers but also aids in reducing heat damage. However, these solutions do not necessarily correlate with higher efficiencies and uniform illumination, as mentioned previously [3–7]. Song et al. devised a model based on a parallel mechanism, comprising 48 concentrating cells arranged in a 7×7 array. Through a series of tests, they demonstrated that the system could achieve a light transmission of 25% at a distance of 10 m, providing an average illumination of 122 lx for a 36 m^2 area. However, they noted that exceeding a daylight flux concentration level of 2500 suns could lead to the overheating of PMMA fibers. They proposed adopting silica fibers for their higher temperature resistance but acknowledged their higher cost and lack of flexibility. Ullah et al. [25–27] addressed the issue of overheating damage affecting fiber optic wires in recent studies. They proposed two efficient approaches utilizing plastic optical fibers, parabolic mirrors, and Fresnel lenses to achieve collimated light. They emphasized the importance of uniform illumination for enhancing efficiency and reducing heat effects. Silica optical fibers (SOFs) were introduced as a solution to the overheating problem due to their higher resistance, with POFs used for most of the transmission parts for their low cost and flexibility. Index matching gel was applied between SOFs and POFs to minimize losses. Sedki et al. [15–26] presented a study of a low-cost system utilizing mirrored simple parabolic reflectors and POFs. The system, which reflected direct non-diffuse sunlight into a focal point corresponding to the input area of a PMMA fiber optic bundle, demonstrated an overall energy efficiency of 69% in optimal solar conditions. They proposed using an extended hot mirror to protect the fiber bundle from overheating, with further studies planned to investigate the characteristics, field testing, and efficiency of such a device. In conclusion, addressing the overheating problem is crucial for the development of fiber optic daylighting systems. Proposed solutions, including limiting concentration ratios, adopting silica fibers, and implementing efficient optical designs, aim to mitigate heat damage while maximizing system performance and efficiency.

Architects and building developers have benefited greatly from the development of several simulation software programs. According to research carried out in 2004, about 80% of lighting optimization could be performed using simulation and analysis [36]. Simulations can follow diverse kinds of algorithms. They all run on mathematical and analytical programs that may depend on the choice of environmental conditions, type of luminaries, design schemes, etc. [37]. A few software programs consider the internal properties of the room, such as the paints on the walls, their reflection, the presence of room furniture, etc. [38]. Many researchers have demonstrated the effect of these interior components on the simulation results [39,40]. Reliability is thus a major concern with such simulation software, as estimating the errors is a tedious task, which often may not have a fixed algorithm [41]. Relux is a reliable software that has recently gained popularity and can be used for lighting simulations, estimations, and analysis. Due to different luminaries, Relux can be used to estimate illuminance at a standard height from the floor. Recent research uses Relux as a lighting simulation tool to optimize the placement and patterning of laminar schemes within a room or building. It was employed for optimizing daylighting and integrated artificial luminaries [42]. Previous studies on different geographical regions and specific applications also employ Relux as a reliable software program [43–45].

Daylighting systems using plastic optical fibers (POFs) offer a promising solution for bringing natural light deep into buildings. However, a critical challenge hindering their wider adoption is heat damage to POFs. While various designs have been proposed, effectively mitigating heat remains an area requiring innovative solutions. To address the heat damage issue in POF-based daylighting systems, this study had the following aims:

1. To analyze the previous studies to understand the extent of heat-induced damage on POFs and evaluate proposed solutions in existing systems.
2. To design a customized parabolic reflector to efficiently capture daylight and evaluate the thermal performance of POF daylighting systems.
3. To assess the effectiveness of our proposed solution in mitigating heat damage and achieving desired illumination levels and validate it with RELUX simulations.

This paper addresses the gap in the literature concerning the integration of effective heat filtration with existing optical fiber lighting systems. While existing systems struggle with heat management, our research introduces a solution that reduces heat by 35 degrees, allowing for the safe transfer of daylight to interior spaces. The uniqueness of this paper lies in its novel design of a heat filtration device specifically tailored for optical fiber lighting systems. Unlike previous approaches, our design effectively manages heat while maintaining sufficient daylight transmission, offering a practical and innovative solution to a common challenge in the field. The most significant contribution of this paper is the development of a heat filtration device that not only significantly reduces the heat by 35 degrees but also ensures the effective transfer of daylight to interior spaces. This innovation enhances the practicality and safety of optical fiber lighting systems, potentially expanding their application in environments where heat management is critical.

2. Description of the Parabolic Solar Daylighting Design

The proposed parabolic solar daylighting system is designed to maximize the efficiency of daylight harvesting and distribution within buildings using plastic optical fibers (POFs). However, the heat issues associated with these plastic optical fibers must be addressed to facilitate their widespread adoption in new, efficient daylighting technologies. To investigate this issue, our focus is on monitoring the temporal temperature at the focal point. Additionally, we introduce a novel mechanical component designed to secure the fiber optic bundle at the focal point while ensuring three levels of heat filtration and uniform illumination. The current parabolic solar daylighting system is installed on the roof of the Energy Centre, MANIT-Bhopal, India. The apparatus consists of a light-capturing device equipped with a Single-axis tracking system, a mechanical support structure containing a filter, a light fiber, and a light guide terminal. The key components of the system include the following:

2.1. Parabolic Reflector

A high-reflectivity parabolic mirror is used to concentrate sunlight onto a focal point. The reflector is designed to capture and focus as much sunlight as possible throughout the day. The operational principle of the system involves the utilization of solar radiation, which is captured by the parabolic dish surface and concentrated, then reflected onto the focal area. Subsequently, the concentrated sunlight is transmitted through a fiber optic bundle (FOB) to convey it to the designated destination as shown in Figure 1a. The primary component of the system is the parabolic solar concentrator area, comprising polyester film. This high-quality aluminum-paint-coated film increases the intensity and spreads light more evenly with a high specular reflectivity index of 95%, illustrated in Figure 1d.

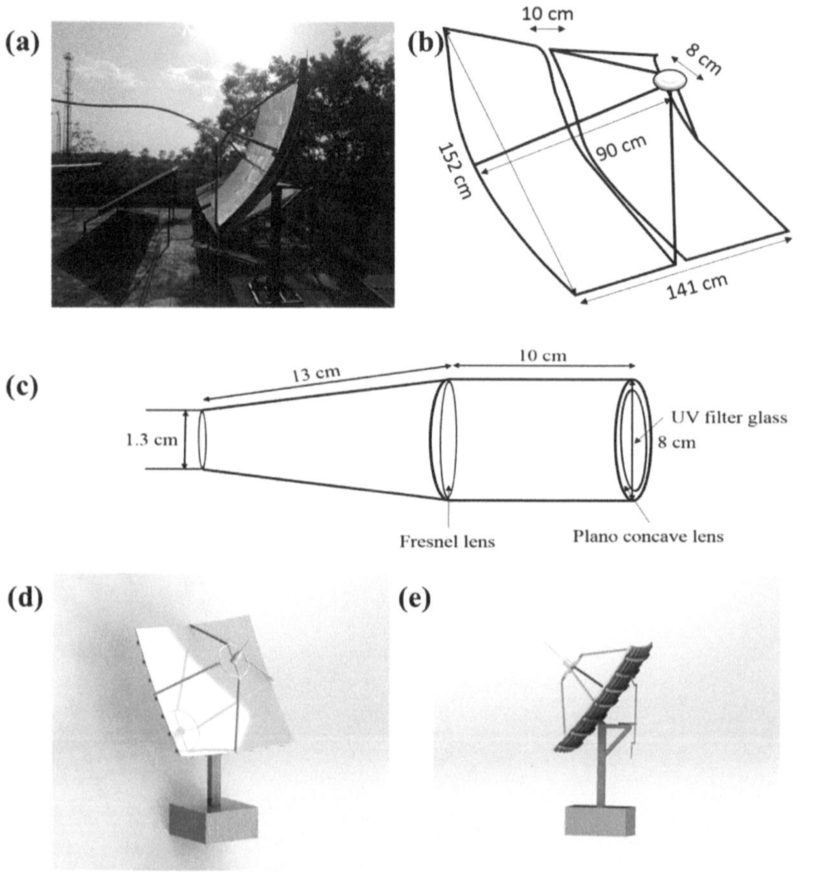

Figure 1. (**a**) Installed solar daylighting system on the roof of an energy center. (**b**) Dimensions of the installed apparatus. (**c**) Receiver cavity. (**d**,**e**) Three-dimensional drawings of the design.

2.2. *Plastic Optical Fibers (POFs)*

Low-cost POFs are selected for their flexibility and cost-effectiveness. However, their susceptibility to heat damage necessitates careful design considerations to mitigate this issue. To achieve the objective of providing uniform illumination deep within the building at a low cost, the utilization of plastic optical fibers (POFs) emerges as one of the most viable solutions, ensuring cost-effectiveness. A bundle of Polymethyl Methacrylate fiber optic wires (PMMA) was utilized, comprising 60 pieces of 1 mm fibers positioned at the center of the bundle, surrounded by 60 pieces of 1 mm fibers. The mathematical relationship between the energy at the inlet of the fiber optic bundle (Q_{in}) (Equation (1)) and the energy at the end of the bundle (Q_{out}) (Equation (2)) is described by the following equations [25,26].

$$Q_{out} = 10^{-\left(\frac{LdB_{loss}}{10}\right)} Q_{in} \tag{1}$$

$$Q_{in} = \pi F^2 \rho_m G_b \left(\sin^2 \varphi_r\right) \tag{2}$$

Assuming that the focal distance is F. The polyester film's reflection index is ρ_m. The radiation of the normal beam is Gb. φ_r is the parabolic dish's rim angle, and dBloss is attenuation. Apart from their cost-effectiveness, PMMA fibers also provide the advantage of flexibility within the building environment.

2.3. Mechanical Support for Heat Filtration

Initially, the innovative mechanical support was devised to solely bear the weight of the fiber bundle and maintain its position on the focal surface during solar tracking. This support mechanism was primarily constructed with a single level of filtration, featuring a housing for an optical filter (refer to Figure 1d,e). The Figure 1 depict the industrial design and actual implementation of the innovative mechanical component. An extended hot mirror was employed within the filter to mitigate heat within the optical system without compromising its visible output. Despite the implementation of this filter, which extends the reflection range to approximately 1750 nm, the operational experience revealed that prolonged system usage still led to temperature-induced damage to the optical fiber. To address this issue, we enhanced the mechanical component by incorporating additional filtration levels. We introduced a new filtration combination where two lenses and one glass are used for heat reduction.

2.4. Heat Filtration Mechanism

The heat filtration mechanism consists of three layers of filters designed to reduce the thermal load on the POFs. The three components are UV filter glass, plano-concave lens, and Fresnel lens. A newly patented mechanical component is integrated at the focal point to secure the fiber optic bundle and provide three levels of heat filtration. This component is crucial for protecting the POFs from overheating while maintaining high optical transmission efficiency.

Primary Heat Filter: This layer is composed of a UV-absorbing material that blocks a significant portion of ultraviolet radiation, which contributes to the degradation of POFs. UV filter glass for daylighting functions by selectively blocking ultraviolet (UV) radiation while allowing visible light to pass through. This filtration process helps reduce UV-induced glare and heat, creating a more comfortable and visually appealing indoor environment. By minimizing UV penetration, the filter helps protect interior furnishings and occupants from UV-related damage and health risks. Additionally, it maintains the natural color balance of daylight, enhancing the quality of interior illumination.

Secondary Heat Filter: A plano-concave lens for daylighting works by diverging incoming light rays and spreading them out across a wider area. Made from a heat-resistant polymer, this layer further reduces infrared radiation, preventing excessive heat from reaching the POFs. This lens design helps to evenly distribute natural daylight, reducing glare and shadows in indoor spaces. By dispersing light, it enhances overall illumination levels while maintaining uniformity. Placed strategically in daylighting systems, such lenses optimize the utilization of natural light, promoting energy efficiency in buildings.

Tertiary Heat Filter: The final layer consists of a reflective lens that deflects any remaining heat while allowing visible light to pass through efficiently. A Fresnel lens for daylighting functions by refracting incoming light, focusing it toward a specific area or direction. Its unique design features concentric grooves that bend light, concentrating it onto a focal point. This lens efficiently captures and redirects natural daylight, maximizing its intensity within a space. By concentrating sunlight, it enhances interior illumination levels, reducing the need for artificial lighting. Installed in daylighting systems, Fresnel lenses help optimize energy usage in buildings while providing ample natural light for occupants.

Additionally, we strategically created a calculated space between the filters to facilitate natural ventilation within the system.

As the direct beam from the concentrator converges on the focal point (refer to Figure 1c), it encounters the first UV filter that selectively blocks ultraviolet (UV) radiation while allowing visible light to pass through, and maximum heat should be reduced. After passing through this initial filtration, the beam traverses through a plano-concave lens which evenly distributes natural daylight in a chamber after the vacuum chamber for natural heat ventilation induced by rays, constituting the second level of heat reduction. Subsequently, the light beam encountered through a Fresnel lens refracts incoming light, focusing it toward a specific direction. Finally, the beam enters the fiber optic inlet, transmitting light to the terminal via the light fiber. This refined approach ensures uniform light distribution across each optical fiber, leading to improved efficiency in terms of illumination.

3. Materials and Methods

This study investigated the performance of a novel fiber optic daylighting system designed for efficient light transmission and heat filtration.

3.1. Experimental Setup

The system utilizes a parabolic collector to concentrate sunlight and a newly developed mechanical component to secure the fiber optic bundle at the focal point. For the manually adjustable tracking system utilized in the experiment, a one-axis system was employed. A parabolic collector with a specific focal length was fabricated using the material specification, e.g., a reflective aluminum sheet. The dimensions and surface properties of the collector were optimized for efficient light capture based on reference design. Plastic optical fibers (POFs) were chosen for their cost-effectiveness and light transmission properties. The newly developed mechanical component, designed to secure the fiber optic bundle at the focal point, incorporates three levels of heat filtration. The complete daylighting system, including the parabolic collector, fiber optic bundle with the securing component, and light diffuser (optional), was assembled. The system was positioned outdoors under direct sunlight with the collector facing south (or appropriate direction based on location).

3.2. Instrumentation and Testing

The system's performance was evaluated through a series of experiments designed to measure the illumination levels and temperature at various points within the system. Measurements were taken at different times of the day to assess the system's performance under varying sunlight conditions. The experiment was conducted between 1 March 2023 and 31 May 2023. The data for each month represent the conditions on the first day of the month. On 1 March, in the morning, Bhopal experienced essentially clear sky conditions, which became partly cloudy around 10 a.m., followed by a return to clear skies. The average temperature that day was 25 °C, with a wind speed of 6 km/h. On 1 April 2023, there were clear sky conditions throughout the day, with an average temperature of 29 °C and a wind speed of 5 km/h. On 1 May 2023, there were scattered clouds initially, followed by partly cloudy conditions later in the day. The average temperature was 23 °C, with a wind speed of 8 km/h. Illuminance measurements were taken at various distances within the building using a lux meter with multiple data points to assess light distribution and uniformity. It is worth noting that the results encompass both direct lighting and inter-reflection within the measurement environment. The room dimensions, including the height, are as follows: 2.5 m × 2.2 m × 3.2 m. Lux meters were placed at different locations within the building to measure the intensity and uniformity of the distributed light. The uniformity of illumination was calculated using standard metrics like uniformity ratio (U0) or coefficient of variation (CV). An infrared thermometer was used to measure the temperature at different points within the system, including the parabolic collector, securing component, and emerging light from the fiber optic bundle. The experimental setup is illustrated in Figure S1 of the supporting information. Infrared thermometer data were used to assess the effectiveness of the heat filtration mechanism in reducing heat transfer through the fiber optic bundle. Infrared thermometers were used to monitor the temperature of the POFs and other critical components. These measurements helped assess the effectiveness of the heat filtration mechanism in protecting the POFs from overheating. The temperature measurements were compared with the ambient temperature to determine the system's thermal performance.

3.3. Simulation with RELUX Software

RELUX simulation software was used to model the daylighting system and its light distribution characteristics. The simulated illumination map was compared with the experimental data to verify the performance of the system and the accuracy of the design. To verify and complement the experimental results, the system was modeled and simulated using RELUX, a lighting simulation software (version 2019.1) program. The physical dimensions and reflective properties of the parabolic reflector, the optical characteristics of the POFs, and the heat filtration mechanism were modeled in RELUX. Simulations were run under various lighting conditions, including different sun angles and intensities, to predict the system's performance in real-world scenarios. The simulated illumination levels and temperature profiles were compared with the experimental data to validate the accuracy and reliability of the simulation model.

Figure 2 shows the flow chart detailing the design and implementation of the research methodology. This chart provides a comprehensive overview of the sequential steps and processes involved in our study. It illustrates the various stages of the research, from initial planning and experimental setup to data collection and analysis. The flow chart also highlights the key components and their interrelationships, offering a visual representation of how each step integrates into the overall research framework. This aims to enhance understanding and facilitate replication of our methodology by clearly mapping out the entire research process.

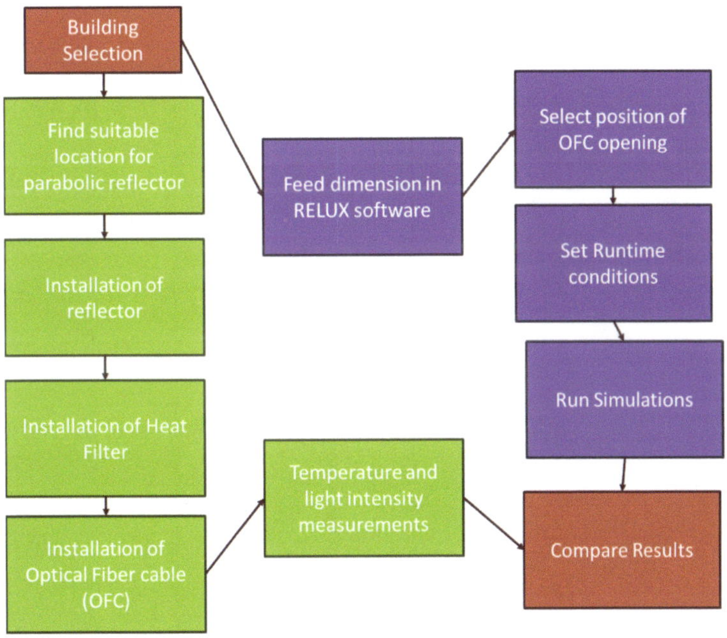

Figure 2. Flow chart of research methodology.

4. Results

4.1. Temperature Control

The infrared thermometer readings demonstrated that the heat filtration mechanism effectively reduced the temperature of the POFs. The maximum recorded temperature of the POFs was consistently below 40 °C, well within the safe operating range for these materials. Figure 3 depicting the temperature difference between ambient conditions and the input of the optical fiber after installing the component showcases a remarkable reduction in temperature, particularly evident during May, March, and April. The multi-layer heat filtration system was particularly effective in reducing the thermal load. The primary filter reduced UV radiation by approximately 70%, while the secondary and tertiary filters collectively reduced infrared radiation by an additional 50%. This substantial decrease, averaging around 35 degrees Celsius, signifies an exceptional improvement in heat management within the system.

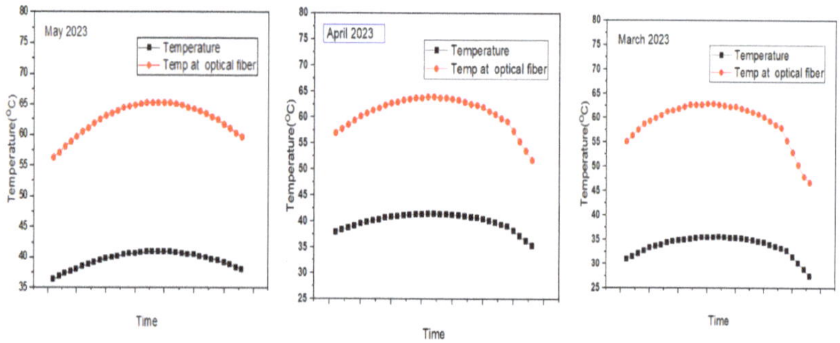

Figure 3. Temperature comparison of ambient vs. internal at optical fiber for different months.

Such a significant reduction in temperature can be attributed to the effectiveness of the installed component in mitigating heat transfer to the optical fiber. Overall, the substantial decrease in temperature observed highlights the effectiveness of the installed component in managing heat within the optical fiber system. By implementing measures to reduce heat transfer and enhance thermal management, the component successfully maintains lower temperatures at the fiber input, ensuring optimal performance and longevity of the optical fiber in various environmental conditions.

4.2. Optimum Illumination Distribution

The effectiveness of the component in carrying illumination from outdoor ambient conditions to indoor spaces is crucial for evaluating its performance. The distribution of light was found to be highly uniform, with a standard deviation of less than 10% across the measured points. This uniformity is critical for maintaining consistent lighting conditions throughout the day. The lux meter readings indicated that the system achieved high levels of illumination deep within the building. The measured illuminance varied between 500 lux to 350 lux at peak sunlight hours, ensuring sufficient lighting for typical indoor activities. Figure 4 showcases the difference in outdoor and indoor illumination levels providing valuable insight into the component's efficacy. The availability of approximately 400–450 lux of illumination indoors, as measured using the setup, indicates a significant success in achieving the desired outcome.

Achieving indoor illumination levels of 400–450 lux using the setup demonstrates the high effectiveness of the component in carrying illumination from outdoor ambient conditions to indoor spaces. This success not only enhances visual comfort and productivity but also promotes energy efficiency and occupant well-being, highlighting the significant benefits of utilizing natural light in indoor environments. The ability to achieve 400–450 lux of indoor illumination demonstrates effective daylighting, where natural light is harnessed to illuminate indoor environments. This is particularly important for reducing reliance on artificial lighting sources during daylight hours, leading to energy savings and environmental benefits. Adequate indoor illumination levels contribute to improved visual comfort and productivity for occupants. By providing sufficient natural light, the component enhances the overall indoor environment, creating a more pleasant and conducive atmosphere for various activities. The ability of the component to deliver substantial indoor illumination levels ensures that occupants can reap these health-related advantages. By reducing the need for artificial lighting, the effective transmission of natural light indoors results in cost savings associated with energy consumption.

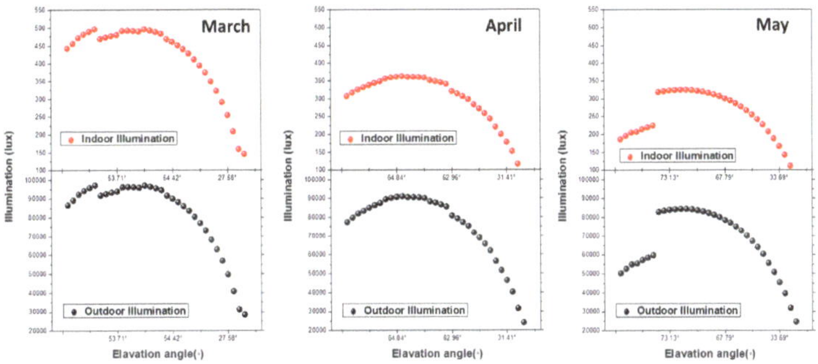

Figure 4. Illumination comparison ambient vs. indoor at optical fiber for different months.

4.3. Simulation Results

Additionally, as a proof of concept, simulations using Relux software were carried out to demonstrate the effectiveness of the setup. The RELUX simulations closely matched the experimental results, with predicted illuminance levels within 5% of the measured values. This high level of accuracy confirms the reliability of the simulation model in predicting the system's performance. The temperature profiles obtained from the simulation were also in good agreement with the experimental data, further validating the effectiveness of the heat filtration mechanism.

Figure 5a is the 3D model that showcases a demo room meticulously designed to serve as a controlled environment for simulating various light sources. Its purpose revolves around studying the behavior and impact of different lighting conditions, encompassing natural sunlight. The design emphasizes flexibility and precision, featuring adjustable walls, ceilings, and floors to modify the space's dimensions, while surfaces and materials are chosen for their reflective properties. The room houses sophisticated simulation equipment to accurately reproduce diverse lighting scenarios, enabling researchers and designers to analyze the effects of illumination on spaces, objects, and human perception. Figure 5b shows the pseudo-color model depicting light distribution in the room illustrating a uniform spread of illumination across its entirety. Through meticulous calibration and analysis, this model accurately represents the consistent distribution of light intensity, showcasing minimal variation or discrepancies throughout the space. Each color gradient corresponds to specific light intensity levels, with a cohesive and balanced spectrum pervading the room. This further asserts the usefulness of the proposed system. The pseudo-reference plane depicted in the 2D cross-sectional representation Figure 5b illustrates a comprehensive view of light distribution within space. This visual tool effectively demonstrates the consistency and evenness of light intensity across different planes within the room. By showcasing how light propagates and spreads through the environment, it offers crucial insights into the efficacy of the new lighting system. The uniformity and balance observed in the distribution of light along the reference plane confirm the effectiveness of the system in achieving its intended objective of providing consistent illumination throughout the designated area. This validation supports the system's suitability for various applications, affirming its capability to meet lighting requirements with precision and reliability.

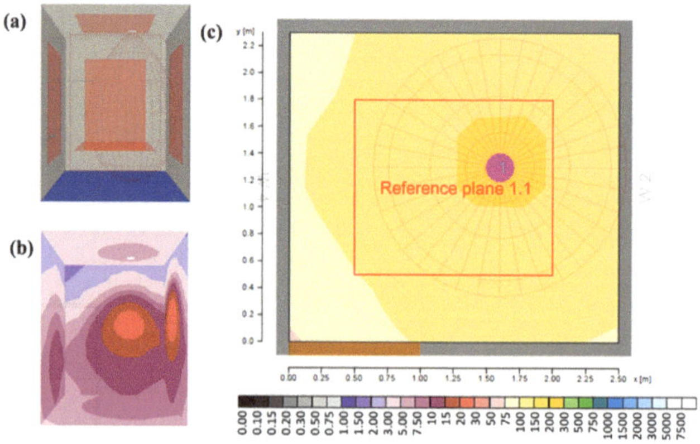

Figure 5. Simulation results: (**a**) 3D view of the demo room, (**b**) pseudo-color model showing light distribution (lux), and (**c**) pseudo-reference plane.

4.4. Discussion

The primary objective of optical fibers is indeed to transfer light efficiently, rather than heat. Optical fibers are specifically designed to guide and transmit light signals over long distances with minimal loss and distortion. However, it is essential to acknowledge that in practical applications, optical fibers can still be subjected to heat due to various factors, including environmental conditions, nearby heat sources, and the intensity of light they are transmitting. While the focus is on light transmission, heat management remains crucial for ensuring the optimal performance and longevity of optical fibers. Excessive heat can degrade the optical properties of the fiber, leading to signal attenuation, increased signal noise, and even physical damage in extreme cases. Effective heat management is essential to ensure their reliable operation and longevity. By minimizing heat-related effects, optical fibers can maintain their efficiency and integrity, providing reliable transmission of light signals over extended periods. In this study, the addition of a heat filtration device resulted in a significant reduction in the temperature input to the optical fiber. This improvement can be scientifically explained by considering the following factors:

The heat filtration device may act as a heat sink, absorbing and dissipating heat away from the optical fiber. When optical fiber is exposed to sunlight, it absorbs radiant energy, leading to an increase in temperature. By introducing a heat sink, the excess heat is transferred away from the fiber, reducing its temperature. The filtration device may also provide thermal insulation, creating a barrier between the optical fiber and external heat sources. This insulation prevents the direct transfer of heat to the fiber, thereby reducing its temperature. The design of the filtration device may incorporate features that increase its surface area. A larger surface area facilitates better heat dissipation, allowing more efficient cooling of the optical fiber. The choice of materials for the filtration device is crucial in determining its thermal conductivity and heat dissipation capabilities. Materials with high thermal conductivity, such as metals or heat-conductive polymers, are effective in rapidly transferring heat away from the optical fiber. High temperatures can cause thermal degradation of the fiber's materials, leading to changes in its optical properties. For example, prolonged exposure to heat can cause the fiber's core and cladding materials to expand or contract, altering the refractive index and affecting light transmission. The precise positioning of the filtration device relative to the optical fiber is essential for optimal heat reduction. Placing the part near the fiber ensures efficient heat transfer and minimizes temperature buildup.

The overall cost of the project was USD 281.75 with the heat filtration device accounting for USD 15.49 of that total (Table 1). When calculated as a percentage, the cost of the heat filtration device represents 5.5% of the entire project expenditure. This indicates that a small portion of the project's budget was allocated to the heat filtration device, highlighting its relative cost-effectiveness within the broader scope of the project.

Table 1. Cost analysis of the system.

S No	Item	USD Cost
1	Parabolic reflector	220.42
2	Optical fiber cable	28.0
3	Heat filtration device Lens Base Reflective sheet Total	 9.53 3.57 2.38 15.49
4	Other charges (installation, transportation, etc.)	17.87
5	Total setup charges	281.75

Overall, the introduction of a heat filtration device serves to enhance the thermal management of the optical fiber within the solar daylighting system. By effectively dissipating heat and providing insulation, this component contributes to maintaining lower temperatures, thereby improving the performance and longevity of the fiber.

5. Conclusions

The combined experimental and simulation results highlight the system's potential to provide efficient and uniform daylighting while mitigating the risk of POF overheating. The design demonstrated its ability to effectively manage heat, improve light transmission, and offer numerous benefits to indoor environments, a significant advancement in enhancing its efficacy and overall performance. Firstly, the component's capability to mitigate heat transfer to the optical fiber contributes to the system's efficiency and longevity. By reducing the temperature input to the fiber, potential heat-related damage is minimized, ensuring prolonged operational life and durability of the optical fiber. Secondly, the improved light transmission achieved with the component leads to enhanced indoor illumination levels. This not only promotes visual comfort and productivity but also reduces reliance on artificial lighting sources, resulting in energy savings and environmental sustainability. Furthermore, the component's ability to reduce indoor temperatures underscores its role in contributing to thermal comfort within indoor spaces. By effectively managing heat, occupants can enjoy a more comfortable environment, particularly during periods of intense sunlight and high temperatures.

Overall, the addition of the filtration device component to the optical fiber daylighting system represents a holistic approach to optimizing the optical fiber daylighting system. As we continue to explore innovative solutions for sustainable lighting and environmental comfort, the integration of such components underscores our commitment to creating healthier, more efficient indoor spaces for the well-being of occupants and the planet alike. Although the three levels of heat filtration reduce the lighting efficiency, the guidelines of the minimum requirement of light for a living area require around 300 to 500 lux. The setup can efficiently deliver an average light in the daytime of around 400 lux which is well within acceptable limits. This is useful for basements or other areas where daylight cannot reach through windows. The filtration device helps in employing optical fibers with lower temperature ratings. These findings suggest that the proposed parabolic solar daylighting system, with its innovative heat filtration mechanism, is a viable solution for sustainable and efficient daylighting in modern buildings

Future scope: Recommendations for future research include exploring advanced materials for even more effective heat filtration and optimizing the geometry of the parabolic reflector to enhance light concentration and distribution. The relationship between the distance from the light source and the illuminance level can be analyzed to understand the light attenuation characteristics of the fiber optic bundle. The impact of different heat filtration levels on light transmission efficiency can be investigated by comparing the performance with and without specific filtration components. The optimal balance between heat dissipation and lighting effects can be explored. This includes investigating advanced heat filtration techniques that enable maximum light transmission while preserving the integrity of the optical fibers. By optimizing this balance, the system's efficiency could be improved, allowing for broader applications in areas where both high-intensity lighting and effective thermal management are essential. Further research and development could lead to widespread adoption of this technology, significantly reducing the energy consumption and environmental impact of artificial lighting.

Supplementary Materials: The following supporting information can be downloaded at: https://www.mdpi.com/article/10.3390/eng5040140/s1, Figure S1: Experimental set-up of the system; Table S1: Cost analysis of the system. Figure S2: Flow chart of research methodology.

Author Contributions: Conceptualization, A.S. and S.K.; methodology, E.C.; software, L.U.; validation, L.U., A.S. and S.K.; formal analysis, E.C.; investigation, A.S.; resources, L.U.; data curation, L.U.; writing—original draft preparation, A.S.; writing—review and editing, E.C.; visualization, S.K.; supervision, A.S.; project administration, E.C.; funding acquisition, S.K. All authors have read and agreed to the published version of the manuscript.

Funding: This research received no external funding.

Institutional Review Board Statement: Not applicable.

Informed Consent Statement: Not applicable.

Data Availability Statement: No new data were created or developed. The original contributions presented in the study are included in the article, further inquiries can be directed to the corresponding authors.

Acknowledgments: The technical support of MANIT-Bhopal, India, and Universiti Malaysia Pahang Al Sultan Abdullah (www.umpsa.edu.my, accessed on 12 April 2024) is greatly acknowledged. We are also thankful to our Alumni student Vipin Vipinraj Sugathan, Nanyang Technological University Singapore for his assistance.

Conflicts of Interest: The authors declare no conflicts of interest.

References

1. Yi, R.; Shao, L.; Su, Y.; Riffat, S. Daylighting performance of atriums in subtropical climate. *Int. J. Low-Carbon Technol.* **2009**, *4*, 230–237. [CrossRef]
2. Krarti, M.; Erickson, P.M.; Hillman, T.C. A simplified method to estimate energy savings of artificial lighting use from daylighting. *Build. Environ.* **2005**, *40*, 747–754. [CrossRef]
3. Li, D.H.; Wong, S. Daylighting and energy implications due to shading effects from nearby buildings. *Appl. Energy* **2007**, *84*, 1199–1209. [CrossRef]
4. Barton, J.; Thomson, M.; Sandwell, P.; Mellor, A. A Domestic Demand Model for India. In *Advances in Energy Research*; Springer: Singapore, 2020; pp. 743–753. [CrossRef]
5. The Official Website of Ministry of Environment, Forest and Climate Change, Government of India. Available online: http://www.moef.nic.in/en/downloads/public-information/Residentialpowe (accessed on 11 April 2022).
6. Rubab, S.; Kandpal, T.C. Resource-Technology Combinations for Domestic Lighting in Rural India: A Comparative Financial Evaluation. *Energy Sources* **1997**, *19*, 813–831. [CrossRef]
7. Bahadiroğlu, A.; Koç, A.Y.; Parlak, E.; Larsson, N.; Kujawski, W.; Unver, U. Sustainable building evaluation: A case study. *Energy Sources Part A Recover. Util. Environ. Eff.* **2022**, *44*, 3149–3163. [CrossRef]
8. Tabrizi, A.; Sanguinetti, P. Life-cycle cost assessment and energy performance evaluation of NZEB enhancement for LEED-rated educational facilities. *Adv. Build. Energy Res.* **2015**, *9*, 267–279. [CrossRef]
9. Terrill, T.J.; Bay, C.J.; Rasmussen, B.P. Autonomous lighting assessments in buildings: Part 2—Light identification and energy analysis. *Adv. Build. Energy Res.* **2016**, *11*, 227–244. [CrossRef]
10. Ihm, P.; Nemri, A.; Krarti, M. Estimation of lighting energy savings from daylighting. *Build. Environ.* **2009**, *44*, 509–514. [CrossRef]
11. A Critical Review of Simulation Techniques for Daylight Responsive Systems | L Doulos—Academia.edu. Available online: https://www.academia.edu/25457051/A_critical_review_of_simulation_techniques_for_daylight_responsive_systems (accessed on 11 April 2022).

12. Udhwani, L.; Soni, A. Evaluation of daylighting performance in an office building: A case study. *Mater. Today Proc.* **2021**, *46*, 5626–5631. [CrossRef]
13. Sayigh, A. Sustainability, Energy and Architecture Case Studies in Realizing Green Buildings. 2014. Available online: http://wrenuk.co.uk (accessed on 11 April 2022).
14. Ghisi, E.; Tinker, J.A. Evaluating the potential for energy savings on lighting by integrating fibre optics in buildings. *Build. Environ.* **2006**, *41*, 1611–1621. [CrossRef]
15. Doulos, L.; Tsangrassoulis, A.; Topalis, F. Quantifying energy savings in daylight responsive systems: The role of dimming electronic ballasts. *Energy Build.* **2008**, *40*, 36–50. [CrossRef]
16. Kurian, C.; Aithal, R.; Bhat, J.; George, V. Robust control and optimisation of energy consumption in daylight—Artificial light integrated schemes. *Light. Res. Technol.* **2008**, *40*, 7–24. [CrossRef]
17. Loutzenhiser, P.G.; Maxwell, G.M.; Manz, H. An empirical validation of the daylighting algorithms and associated interactions in building energy simulation programs using various shading devices and windows. *Energy* **2007**, *32*, 1855–1870. [CrossRef]
18. Liu, G.; Liu, H. Using Insulation in China's Buildings: Potential for Significant Energy Savings and Carbon Emission Reductions. *Low Carbon Econ.* **2011**, *02*, 220–223. [CrossRef]
19. Rosen, M.A. Net-Zero Energy Buildings and Communities: Potential and the Role of Energy Storage. *J. Power Energy Eng.* **2015**, *03*, 470–474. [CrossRef]
20. Öztürk, L.D. Determination of Energy Losses in Lighting in Terms of Good Vision Efficiency. *Arch. Sci. Rev.* **2008**, *51*, 39–47. [CrossRef]
21. Shankar, A.; Krishnasamy, V.; Babu, B.C. Smart LED lighting system with occupants' preference and daylight harvesting in office buildings. *Energy Sources Part A Recover. Util. Environ. Eff.* **2020**, *18*, 1–21. [CrossRef]
22. Akrasakis, S.; Tsikalakis, A.G. Corridor lighting retrofit based on occupancy and daylight sensors: Implementation and energy savings compared to LED lighting. *Adv. Build. Energy Res.* **2018**, *12*, 274–288. [CrossRef]
23. Mayhoub, M.S. Innovative Daylighting Systems' Challenges: A Critical Study. *Energy Build.* **2014**, *80*, 394–405. Available online: https://www.sciencedirect.com/science/article/pii/S0378778814003193 (accessed on 20 April 2024). [CrossRef]
24. Chadichal, S.S.; Misra, S.; Vashishta, D.S. Circular Model/Economy and Sustainability Strategies for Long-Term Prosperity for a Retail Sector in India: A Study on Clothing Consumerism. May 2022. Available online: http://gnanaganga.inflibnet.ac.in:8080/jspui/handle/123456789/573 (accessed on 20 April 2024).
25. Ullah, I.; Shin, S.-Y. Development of optical fiber-based daylighting system with uniform illumination. *J. Opt. Soc. Korea* **2012**, *16*, 247–255. Available online: https://opg.optica.org/abstract.cfm?uri=josk-16-3-247 (accessed on 20 April 2024). [CrossRef]
26. Ullah, I.; Whang, A.J.-W. Development of Optical Fiber-Based Daylighting System and Its Comparison. *Energies* **2015**, *8*, 7185–7201. [CrossRef]
27. Ullah, I.; Shin, S. Uniformly Illuminated Efficient Daylighting System. *Smart Grid Renew. Energy* **2013**, *04*, 161–166. [CrossRef]
28. Kandilli, C.; Ulgen, K. Review and modelling the systems of transmission concentrated solar energy via optical fibres. *Renew. Sustain. Energy Rev.* **2007**, *13*, 67–84. [CrossRef]
29. Kandilli, C.; Ulgen, K.; Hepbasli, A. Exergetic assessment of transmission concentrated solar energy systems via optical fibres for building applications. *Energy Build.* **2008**, *40*, 1505–1512. [CrossRef]
30. He, K.; Zheng, H.; Li, Z.; Taotao; Dai, J. Design and investigation of a novel concentrator used in solar fiber lamp. *Sol. Energy* **2009**, *83*, 2086–2091. [CrossRef]
31. Wang, C.; Abdul-Rahman, H.; Rao, S. Daylighting can be fluorescent: Development of a fiber solar concentrator and test for its indoor illumination. *Energy Build.* **2009**, *42*, 717–727. [CrossRef]
32. Han, H.; Kim, J.T. Application of high-density daylight for indoor illumination. *Energy* **2010**, *35*, 2654–2666. Available online: https://www.sciencedirect.com/science/article/pii/S0360544209001935 (accessed on 20 April 2024). [CrossRef]
33. Sapia, C. Daylighting in buildings: Developments of sunlight addressing by optical fiber. *Sol. Energy* **2013**, *89*, 113–121. [CrossRef]
34. Xue, X.; Zheng, H.; Su, Y.; Kang, H. Study of a novel sunlight concentrating and optical fibre guiding system. *Sol. Energy* **2011**, *85*, 1364–1370. [CrossRef]
35. Lingfors, D.; Volotinen, T. Illumination performance and energy saving of a solar fiber optic lighting system. *Opt. Express* **2013**, *21*, A642–A655. Available online: https://opg.optica.org/abstract.cfm?uri=oe-21-S4-A642 (accessed on 20 April 2024). [CrossRef]
36. Reinhart, C.; Fitz, A. Findings from a survey on the current use of daylight simulations in building design. *Energy Build.* **2006**, *38*, 824–835. [CrossRef]
37. Kim, C.-S.; Chung, S.-J. Daylighting simulation as an architectural design process in museums installed with toplights. *Build. Environ.* **2010**, *46*, 210–222. [CrossRef]
38. Ng, E.Y.-Y.; Poh, L.K.; Wei, W.; Nagakura, T. Advanced lighting simulation in architectural design in the tropics. *Autom. Constr.* **2001**, *10*, 365–379. [CrossRef]
39. Li, D.; Lau, C.; Lam, J. Predicting daylight illuminance by computer simulation techniques. *Light. Res. Technol.* **2004**, *36*, 113–128. [CrossRef]
40. Mardaljevic, J. Verification of program accuracy for illuminance modelling: Assumptions, methodology and an examination of conflicting findings. *Light. Res. Technol.* **2004**, *36*, 217–239. [CrossRef]
41. Maamari, F.; Fontoynont, M.; Adra, N. Application of the CIE test cases to assess the accuracy of lighting computer programs. *Energy Build.* **2006**, *38*, 869–877. [CrossRef]

42. Bhavani, R.G.; Khan, M.A. An intelligent simulation model for blind position control in daylighting schemes in buildings. *Build. Simul.* **2009**, *2*, 253–262. [CrossRef]
43. Bhavani, R.; Khan, M. Advanced Lighting Simulation Tools for Daylighting Purpose: Powerful Features and Related Issues. *Trends Appl. Sci. Res.* **2011**, *6*, 345–363. [CrossRef]
44. Bhavani, R.G.; Khan, M.A. Present Trends and Future Direction of Lighting Control in Dubai. In Proceedings of the 4th IEEE GCC Conf 2007, Manama, Kingdom of Bahrain, 11–14 November 2007; pp. 1–5.
45. Bhavani, R.G.; Khan, M. Prevalence and Penetration of Lighting Control Systems in Dubai Buildings: A Pointer to Future Measures. *J. Appl. Sci.* **2008**, *8*, 3460–3466. [CrossRef]

eng

MDPI

Article

Development of a Smart Material Resource Planning System in the Context of Warehouse 4.0

Oleksandr Sokolov [1,2,*], Angelina Iakovets [1], Vladyslav Andrusyshyn [1,2] and Justyna Trojanowska [3]

1 Faculty of Manufacturing Technologies with the Seat in Prešov, Technical University of Košice, Bayerova 1, 080 01 Prešov, Slovakia; angelina.iakovets@tuke.sk (A.I.); vladyslav.andrusyshyn@tuke.sk (V.A.)

2 Department of Manufacturing Engineering, Machines and Tools of the Sumy State University, Kharkivska 116, 40007 Sumy, Ukraine

3 Department of Production Engineering, Poznan University of Technology, M. Skłodowska-Curie Square 5, 60-965 Poznan, Poland; justyna.trojanowska@put.poznan.pl

* Correspondence: oleksandr.sokolov@tuke.sk

Abstract: This study explores enhancing decision-making processes in inventory management and production operations by integrating a developed system. The proposed solution improves the decision-making process, managing the material supply of the product and inventory management in general. Based on the researched issues, the shortcomings of modern enterprise resource planning systems (ERPs) were considered in the context of Warehouse 4.0. Based on the problematic areas of material accounting in manufacturing enterprises, a typical workplace was taken as a basis, which creates a gray area for warehouse systems and does not provide the opportunity of quality-managing the company's inventory. The main tool for collecting and processing data from the workplace was the neural network. A mobile application was proposed for processing and converting the collected data for the decision-maker on material management. The YOLOv8 convolutional neural network was used to identify materials and production parts. A laboratory experiment was conducted using 3D-printed models of commercially available products at the SmartTechLab laboratory of the Technical University of Košice to evaluate the system's effectiveness. The data from the network evaluation was obtained with the help of the ONNX format of the network for further use in conjunction with the C++ OpenCV library. The results were normalized and illustrated by diagrams. The designed system works on the principle of client–server communication; it can be easily integrated into the enterprise resource planning system. The proposed system has potential for further development, such as the expansion of the product database, facilitating efficient interaction with production systems in accordance with the circular economy, Warehouse 4.0, and lean manufacturing principles.

Keywords: warehouse 4.0; machine vision; neural network; material management

Citation: Sokolov, O.; Iakovets, A.; Andrusyshyn, V.; Trojanowska, J. Development of a Smart Material Resource Planning System in the Context of Warehouse 4.0. *Eng* **2024**, *5*, 2588–2609. https://doi.org/10.3390/eng5040136

Academic Editor: Antonio Gil Bravo

Received: 31 August 2024
Revised: 9 October 2024
Accepted: 10 October 2024
Published: 12 October 2024

1. Introduction

In the context of lean production [1], tracking material balances at workplaces is becoming increasingly challenging. Most enterprises make material reserves so customers can receive warranty or post-warranty service. The desire to produce a unique product affects the number of excess product components, which increase in enterprises and often do not go into the production cycle. It is essential not only to maintain flexibility in production and meet the needs of the customer but also to be able to put surplus or residual components into production to reduce the costs of servicing warehouse balances [2–4].

The constant development of production also affects the enterprise's non-production activities, such as warehouses. Warehouse management has become subject to the basic principle of loading and unloading materials and an intelligent way of managing it, which should provide transparent information on the movement of materials and their accounting. Modern requirements for warehouse systems are mentioned in more detail in the concept of Warehouse 4.0, which implies the use of modern material labeling, IoT systems, and even digital twins. Full integration of all Warehouse 4.0 principles is a challenge for some companies, as it requires a readjustment of the company's information system and even its reorganization. Processes of this type are very labor-intensive and costly. The problems of implementing Warehouse 4.0 principles are highlighted by the scientific articles of Agnieszka A. Tubis and Juni Rohman [5], K. Aravindaraaj and P. Rajan Chinna [6], Lihle N. Tikwayo and Tebello N. D. Mathaba [7], Walaa Hamdy and team [8], and Arso Vukicevic and team [9].

Agnieszka A. Tubis and Juni Rohman. [5], in their study, found that bottlenecks and the detailed research of Warehouse 4.0 were given the greatest share of attention in such research areas as Automation and Control Systems (23.3% of articles) and Business and Economics (22.1% of articles) [5]. These are 113 articles with practical and laboratory research, considering the fact that both research areas are inseparably connected with enterprise resource planning (ERP) systems and enterprise information systems (EIS). In the article by S.O. Ongbali et al. [10], various variables of manufacturing bottlenecks that limit production capacity are analyzed. The study identifies key factors such as equipment failure and material unavailability critical to prioritizing efforts to improve manufacturing processes. The study results indicate that eliminating these bottlenecks is essential to optimize warehouse operations in the context of Industry 4.0. S.O. Ongbali et al. propose using simulations to solve the warehouse bottleneck problem. Such a proposal has a place, but identifying new bottlenecks will be challenging without using a digital twin.

For example, Nicole Franco-Silvera and her team [11] propose the 5S concept for optimal warehouse management. Their concept makes sense but still puts complete information about the material in the "under-processing" stage. Material of this type is most often removed from the process of sending to the supplier because it is a component of a product that, for some reason, did not meet the requirements set in the order and was released in reserve quantities, and so remained on the records of the manufacturing company. Material flow control (MFC) is essential to production planning and management [4]. In the practical conditions of a manufacturing enterprise, such materials or components remain in the so-called "gray zone" of inventory systems.

The emergence of "gray zones" is directly related to the bottlenecks of ERP and other accounting systems. Summarizing some of the studies [12–16], we can highlight the main areas of ERP system bottlenecks in production:

- Common problems that arise in inventory management when using ERP systems;
- Inaccurate data tracking, poor integration with other software, and inefficient processes;
- Bottlenecks in material accounting at certain workstations, which can lead to discrepancies in the stock of the entire warehouse;
- Problems with updating the balance in inter-warehouses, which affects delays in order fulfilment and an increase in operating costs, which ultimately affects the overall efficiency of the business;
- Problems that arise in inventory management associated with manual processes;
- Lack of real-time visibility and, as a result, poor production planning.

In inventory management and ERP systems, materials are classified into six main statuses that determine their availability and usability for various operational activities. When inventory is labeled as available, it is ready for immediate use and can be allocated to production tasks or sales orders. In contrast, items marked as unavailable cannot be used in any transactions, often due to being damaged, defective, or otherwise unsuitable. Inventory given a blocked status is physically present but restricted from use, generally due to the need for inspection or further action. Meanwhile, materials placed on hold are temporarily inaccessible, often awaiting quality checks or administrative procedures. Additionally, items categorized as in transit are those currently being transported between locations and are not accessible for use until they arrive at their intended destination. Lastly, items with a reserved status have been allocated to specific orders but have not yet been picked up or shipped, thus remaining set aside until needed [11].

The "gray zone" is an inter-warehouse of a department of the enterprise, where the material has been stacked and is in unavailable, blocked, or on hold status. The problem with such workplaces is that the material on record in this part of the enterprise is often suitable for use and can be used to launch a new batch of products [17]. Reusing such material would align with the principles of the circular economy [18], which entails reducing waste disposal costs, warranty and post-warranty repair costs, and product costs, and a reduction in warehouse maintenance costs [19]. Removing quality material from the "gray zone" will make it possible to reduce production downtime and reduce the amount of production waste. Tools for achieving the concept of reuse can provide a system built according to the principle of Warehouse 4.0. The idea of Warehouse 4.0 is based on eliminating errors, real-time updates, scalability and flexibility, automation, convenience, and security [20]. Medium-sized and big enterprises in Slovakia already have enterprise resource planning programs and inventory mobile scanners for warehouse management. However, this system has bottlenecks that create barriers to effective real-time planning processes. Managers of the work shift need to have relevant information about the amount of material in the disassembly workplaces to provide enough material for the product and thus reduce downtime [21].

The proposed study was conducted in the SmartTechLab laboratory [22], where the main idea was to improve the inventory management processes at the enterprise based on the trends of lean manufacturing [23], Warehouse 4.0, DFMA (Design for Manufacturing, Assembly, and Disassembly) [24], and the circular economy using a client–server software platform based on REST API architecture with an integrated convolutional neural network [25]. The following parts of the proposed study are presented in the following order: Section 1 discusses the challenges in managing material balances in flexible production environments and introduces the concept of Warehouse 4.0 for improving material flow control and reducing costs. Section 2 describes DFMA principles and a smart integrated system combining ERP and machine vision technologies with convolutional neural networks to optimize inventory management and decision-making. Section 3 presents the performance and accuracy evaluation of the YOLOv8 convolutional neural network model in the detection task of production components, demonstrating high precision and recall in various scenarios. Section 4 summarizes the development and potential benefits of the smart material resource planning system, highlighting its impact on reducing waste, improving decision-making, and supporting circular economy principles.

2. Materials and Methods

Since this research is about improving the decision-making process [26], the DFMA principle was chosen as the basic principle. DFMA (Design for Manufacturing, Assembly, and Disassembly) is a research methodology focused on optimizing design and processes in manufacturing and construction to reduce costs, improve efficiency, and increase sustainability. The approach integrates Design for Facilitated Manufacturing and Assembly (DFM and DFA) [27] with the ability to easily disassemble and reuse components to help minimize material waste and product lifecycle costs. The DFMA methodology identifies and eliminates unnecessary complexity during the design phase, ultimately improving productivity, reducing risks, and making processes more environmentally responsible. By integrating enterprise material planning automation solutions, DFMA can significantly improve inventory management and decision-making in manufacturing operations.

As part of this study, an integrated quality and logistics management system was developed and implemented. Our system block diagram is presented in Figure 1. This system integrates an ERP system, a decision-making module, a mobile application, a neural network, and a camera system to provide an integrated real-time management of production processes and inventories. The system's backbone is the ERP server, responsible for collecting and managing data on the materials and spare parts required for production. The essential functions of the ERP system are requesting information on the availability and composition of spare parts for Turbine 1, estimating material requirements, and controlling stock levels, allowing timely replenishment decision-making. The Decision Module, which performs coordination and analysis functions, uses data from the ERP system and other sources to determine further actions, such as allocating or moving parts based on availability. The mobile application serves as the interface for users and provides access to information and critical operations, such as confirming the availability of parts at the correct location. An important aspect is that all the management processes should go through the mobile app, which provides centralized control and flexibility in management. A convolutional neural network integrated into the system uses machine learning algorithms to analyze data and automate decision-making, minimizing the reaction time to changes. The machine vision system module, which supports the convolutional neural networks, determines the availability and quantity of parts and materials at the workplace at the specified frequency. This module makes it possible to speed up decision-making through better awareness of the actual amount of parts and materials. The system's flexible architecture, including various selection nodes and event scenarios, allows the control to be adapted to current operational conditions, ensuring the high accuracy and responsiveness of quality management.

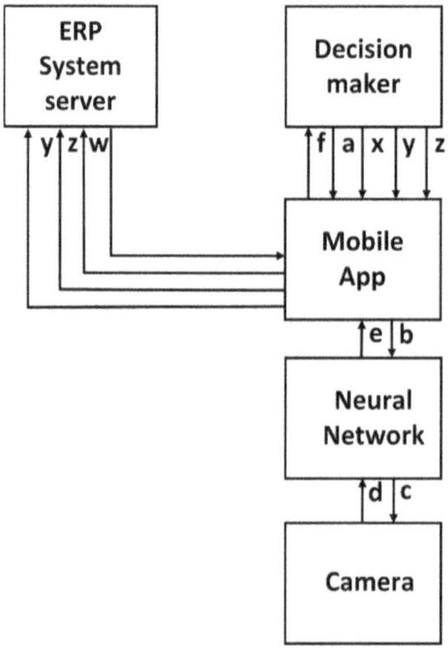

Figure 1. Block diagram of the proposed system.

The ERP system server is an enterprise management program server that contains information about all activities and the enterprise warehouse. The system operation begins with a camera installed at the operator's workplace. The camera sends a video stream with data to the neural network, which sends processed data to the mobile application, but only at the request of the Design Maker through the mobile application, where the information from the neural network is already converted into something suitable for the manager. The manager, in turn, uses the received data for the work order form for the warehouse or production operators to determine if this material is sufficient and/or is in demand in the production process.

Figure 1 shows the communication diagram of the system components at the manager's request:

- Data flow (a) is the user's request to receive data from his profile;
- Data stream (b) is a request stream for the neural network to send a signal to the video camera for further processing;
- Data stream (c)—request to neural network to receive a video stream;

- (d) is a reverse unprocessed video stream;
- (e) is a transformed video stream with data on detected objects. In the mobile application, (f) is the number of detected parts with data on them;
- (x) is a request to check the completeness of the turbine;
- (y) is a request to re-order missing components for a work order for production. This request is sent directly to the ERP system and the logistics department;
- (z) is the formulation of the production work order; therefore, this signal is sent through the application directly to the ERP system.

2.1. Workplace Configuration

This study utilized a workstation configuration, where an employee performs the disassembly of defective assemblies using a camera system to capture reusable parts, which is illustrated in Figure 2.

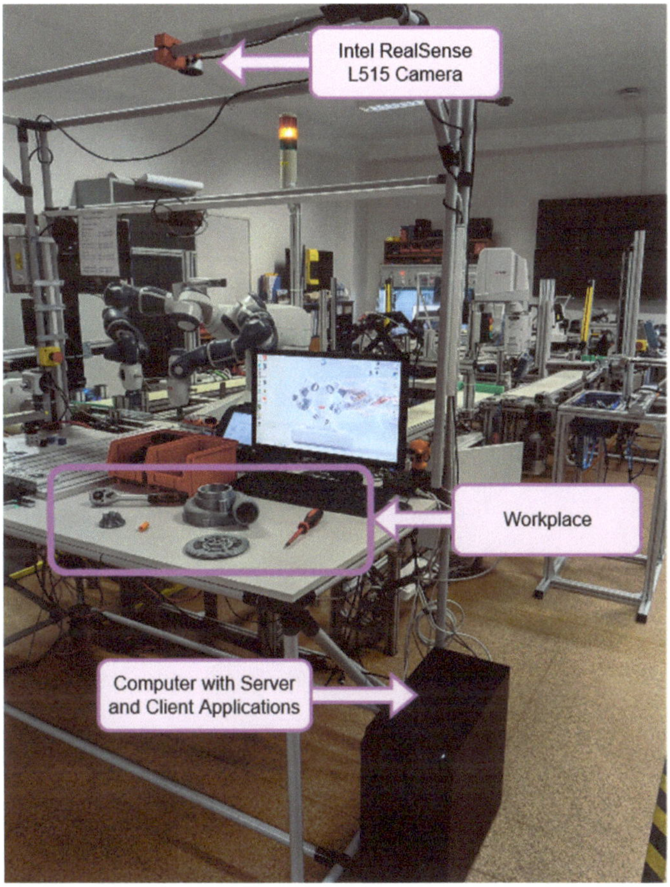

Figure 2. Workstation configuration.

Intel RealSense L515, capable of RGB images in 1920 × 1080 pixels, was chosen as the camera. The camera is connected to a computer running a client that is authorized as a worker and has permissions to send images to the server.

Technically, the client part for the workstation and the server part can be installed on different devices, but in the case under consideration, the client part of the workstation and the server part are installed on the same computer. The computer contains a discrete video card to accelerate the computation of the neural network. Technical characteristics of the computer on which the server solution was tested: CPU Intel Core i5-13600KF, RAM 64 GB, GPU NVIDIA GeForce RTX 3080 Ti 12 GB.

2.2. *Server Configuration*

The server application consists of the HTTP server, MySQL client, and machine vision module. The main element of the HTTP server is the listener, which creates HTTP sessions when clients connect.

When establishing a connection with a client, the server first reads the HTTP request header, which contains the user credentials (login and password) encoded in Base64 format. Then, the server decodes these data, extracts the login and password, checks for the presence of forbidden characters, and passes them to the MySQL server client for processing and executing an SQL query to the database to verify the authenticity of the user account. The server decides whether to allow or reject the request depending on the database response. The proposed authentication approach is compatible with modern web browsers, so it is possible to prepare a web version of the client for other devices in the future.

The considered variant uses an insecure primary authentication mechanism, but for industrial applications, it is recommended to use more secure authentication mechanisms such as OAuth 2.0.

After successful authentication, the following requests are available depending on the user rights:

- POST request to the storekeeper to add parts to the warehouse (/add);
- POST request for a storekeeper to remove parts from the warehouse (/remove);
- POST request to the assembler's workplace to send a photo of the current state of disassembly (/workplace);
- GET manager request for current warehouse status (/warehouse_status);
- GET manager request for the current state of disassembly (/workplace _status).

If the POST request is successful, the client receives an error code 200 OK; if errors occur during the request execution that are not related to authentication, the client receives an error 400 Bad Request (for example, removing more parts from the warehouse than there are in stock, or the presence of prohibited characters in the request).

Since the requests require access to a database, a MySQL client is integrated into the server application. The HTTP server is implemented using the C++ library Boost Beast, and the MySQL client uses the Boost MySQL library. For simplicity, the current version of the application uses an unsecured HTTP connection. However, in the future, for industrial applications, to improve security, we can use an encrypted HTTPS connection. The source code will require minor changes as the Boost Beast library [28] and MySQL [29] have built-in support for SSL-encrypted connections.

The server application uses a machine vision module written using the OpenCV library [30] to process the builder workstation requests. The OpenCV library allows image processing using convolutional neural networks for object detection tasks.

2.3. *Client Part Configuration*

The client part of the application is developed in the Mendix platform [26] and is a web interface through which users can interact with the system. The main components of the client application are a visual interface for working with warehouse data and those for managing assembly operations and monitoring the current state of operations (Figure 3).

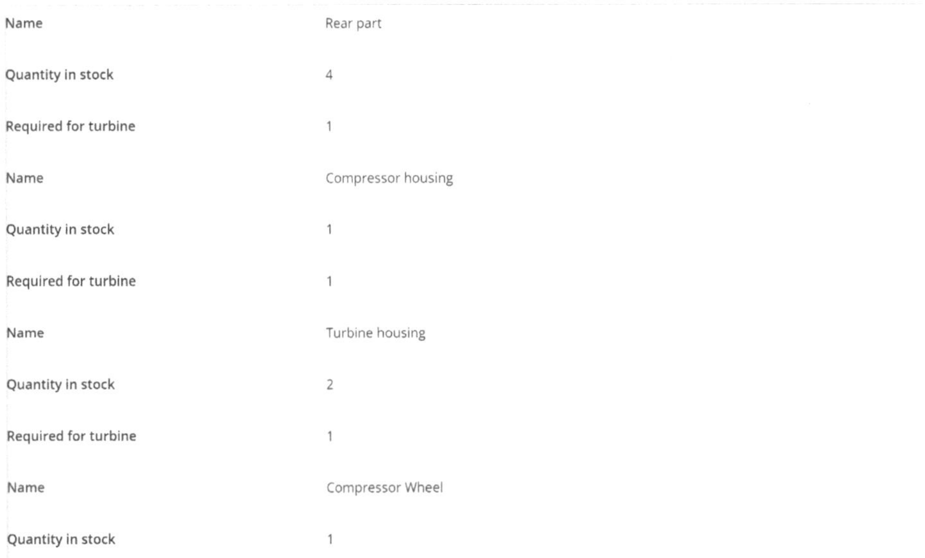

Figure 3. Mendix web interface for monitoring current state of operation in Warehouse 4.0.

When logging into the system, users go through an authentication process based on verified credentials. Authentication is accomplished through a secure mechanism integrated into Mendix, allowing the system to identify users and grant them access to different functions based on their roles.

After successful authentication, users have access to the following functions depending on their access rights: adding parts to the warehouse and removing them from the warehouse. Users responsible for inventory can use the form to enter data about new parts and their quantities. This operation is performed using the AddPart microflow, which sends the data to the server to update the database. A form is also provided to remove parts from the warehouse; the RemovePart microflow, illustrated in Figure 4, verifies that the parts are in stock and their quantities are correct before performing the operation. In addition, assemblers can send photos of the current disassembly status at their workstations using the UploadWorkplaceImage microflow, which leverages file-processing and server integration capabilities in Mendix. Managers can view the current warehouse status through an interface that is implemented using Data Grid or List View widgets to display data from the database. They can also retrieve the current status of the assembly process on the job site using Data View widgets that display updated data on a page.

For all requests, the client side uses a server side developed API server that supports REST API. The client side sends HTTP requests to the server and receives appropriate responses (e.g., 200 OK for successful requests or 400 Bad Requests when non-authentication errors occur).

The Mendix web interface allows extending and scaling the application, adding new features, and managing user access rights. The client side uses standard Mendix security elements, such as OAuth 2.0, in the current implementation for data protection and access control.

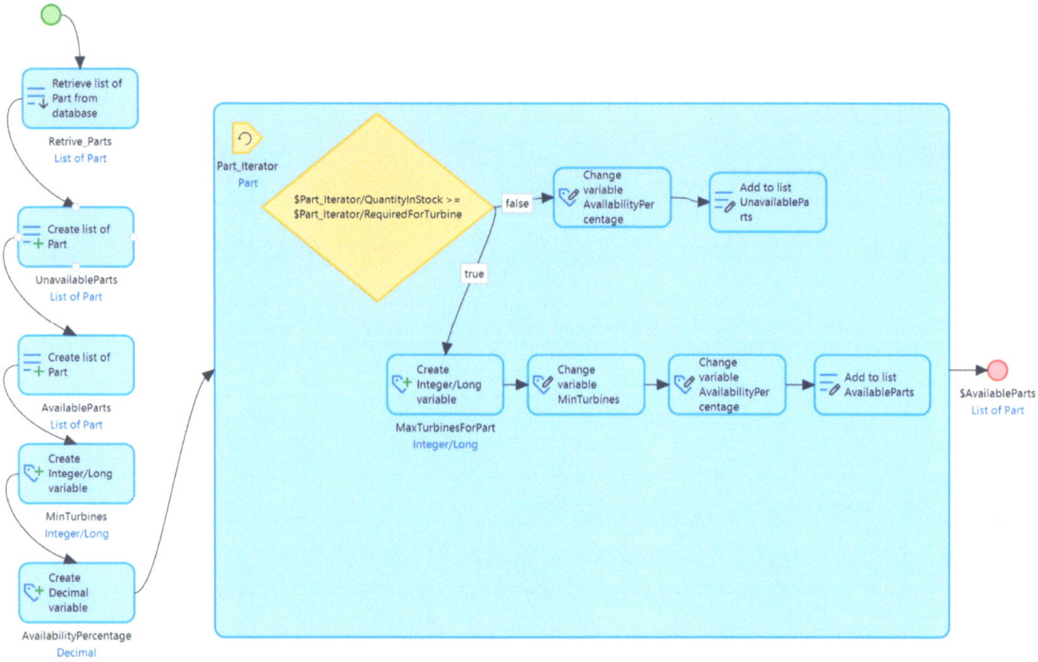

Figure 4. Microflow for decision-making process in Mendix database.

2.4. *YOLLOv8 Convolutional Neural Network Architecture*

Object detection models such as YOLO (You Only Look Once) have gained widespread popularity due to their exceptional image-processing speed, making them ideal for applications with critical real-time performance. Unlike two-stage detectors such as Faster R-CNN, which split the process into region suggestion generation and object classification, YOLO performs both tasks in a single pass through the network. This single-stage approach allows YOLO to process images much faster, often in real time, which is essential for systems requiring immediate object recognition, such as autonomous vehicles or surveillance cameras. Redmon and Farhadi [31] demonstrated in their work on YOLOv3 that this model can balance speed and accuracy, making it a powerful tool for high-speed applications.

While two-stage detectors such as the Faster R-CNN provide higher accuracy, especially in complex scenes with small or closed objects, they inherently suffer from slower processing times due to the need for region proposal networks (RPNs). Ren et al. [32] emphasized the significant accuracy improvements that Faster R-CNN provides, especially for more detailed tasks such as small object detection. However, this comes at the cost of speed. For example, Faster R-CNN is significantly slower than YOLO despite achieving an impressive accuracy, with a mean average precision (mAP) of 76.4% on the VOC2007 dataset [32]. In contrast, YOLOv4, as shown by Bochkovskiy et al. [33], achieves a mAP of 43.5% at 65 FPS, illustrating its optimal balance between speed and performance, which is crucial in time-sensitive environments.

Our application prioritizes fast image processing, so YOLO becomes the preferred solution. Although Faster R-CNN and similar two-stage detectors, such as Mask R-CNN [34], provide more detailed detection, their slower performance is unsuitable for our real-time needs. YOLOv3 and YOLOv4 have consistently demonstrated that they can handle object detection tasks with sufficient accuracy and at much higher speeds, making them more suitable for applications where latency cannot be compromised [31,33].

Since defining the parts to optimize the process is necessary, Ultralytics' YOLOv8 [35] object detection models were suggested. Figure 5 illustrates the structure of this neural network, which consists of a Focus Layer, Backbone, SPP Block, Neck (PANet), and Detection Head [36–38]. The values of these parameters are given in Table 1.

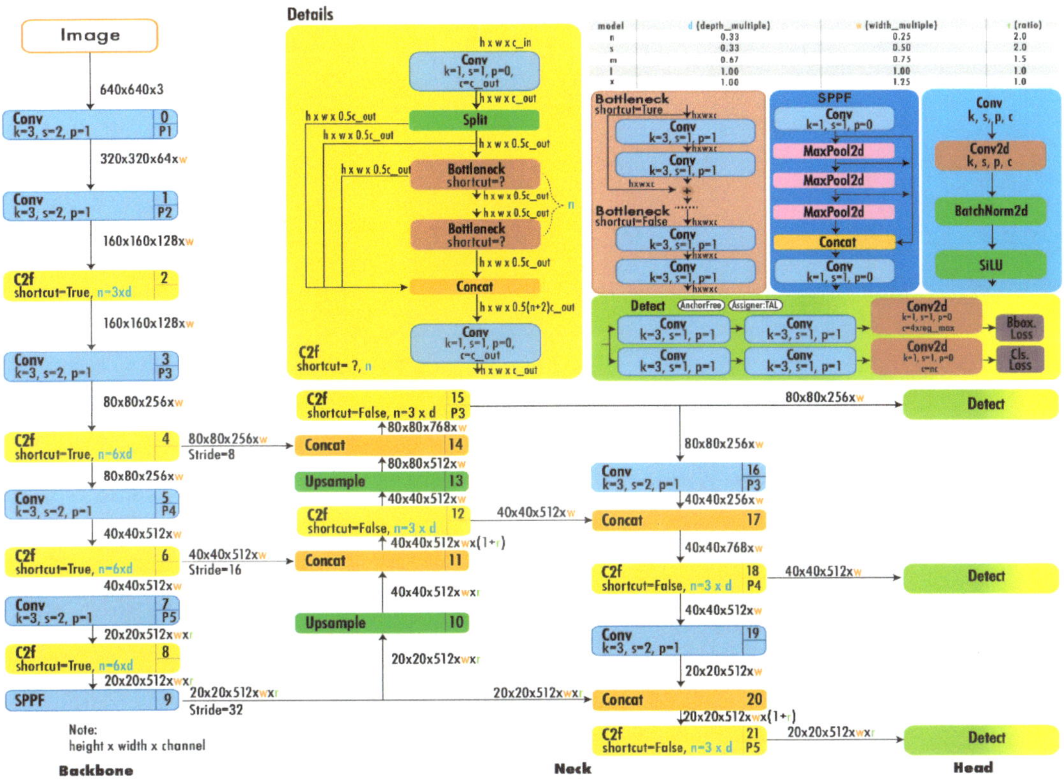

Figure 5. Structure of YOLOv8 [36].

The Focus Layer transforms the input image by splitting it into channels, reducing its size and highlighting essential features. The Backbone is responsible for extracting crucial features from the image by applying convolutional layers, creating a multi-level feature representation. The SPP Block (Spatial Pyramid Pooling) aggregates features at different scales, improving the network's ability to detect objects of various sizes. The Neck (PANet) combines features from different levels of abstraction, enhancing the detection of both small and large objects. Finally, the Detection Head performs the final detection and classification of objects, outputting the coordinates of bounding boxes and the corresponding class labels.

Table 1. YOLOv8 parameters.

Component	Parameter	Values
Focus Layer	Input image size	(3, 640, 640)
	Output size after convolution	(32, 320, 320)
	Convolution kernel size	3×3
	Stride	2
Backbone (CSPDarknet53)	Number of CSP Bottleneck Blocks	5
	Input size for CSP1	(32, 320, 320)
	Output size for CSP1	(64, 160, 160)
	Number of filters	64, 128, 256, 512, 1024
	Convolution kernel sizes	3×3
	Stride	2
SPP Block (Spatial Pyramid Pooling)	Input size	(1024, 10, 10)
	Pooling sizes	$5 \times 5, 9 \times 9, 13 \times 13$
	Output size	(1024, 10, 10)
Neck (PANet)	FPN Path input size	(1024, 10, 10)
	Output size after FPN Path	(256, 40, 40), (128, 80, 80)
	PANet Path Blocks (C3 Blocks) input size	(512, 20, 20), (256, 40, 40), (128, 80, 80)
	Output size after PANet Path Blocks	(256, 40, 40), (128, 80, 80), (64, 160, 160)

A more detailed description of the architecture of the YOLOv8 convolutional network is given in [37], and a detailed review of the evolution of the YOLO family of convolutional neural networks and a detailed description of their operation is given in [37,38].

The Focus Layer is used to pre-process the input image to reduce its size and increase the number of channels, and it functions according to the following algorithm.

The first step of the algorithm is to load the original image, whose dimensions can be described with Equation (1).

$$S = W \times H \tag{1}$$

In this equation, S is square of picture, W is the width and H is the height of the image. In our study, we use camera images with a resolution of 1920 × 1080 pixels, which corresponds to a 16:9 aspect ratio. However, for the correct functioning of the YOLOv8 model, the input images should have a square shape of 640 × 640 pixels with three color RGB channels. This necessity is determined by the architectural peculiarities of the YOLOv8 model, which requires inputting a strictly defined data size.

Further, the second step involves resizing the image while preserving its proportions. Such a step is critical because non-compliance with the proportions can lead to the distortion of objects in the image and, as a result, will negatively affect the model's accuracy in the detection task. Therefore, the resizing process is performed as follows: if the image width W is more significant than its height H, the width is reduced to 640 pixels, and the height is scaled proportionally by Equation (2).

$$H^I = \frac{640 \times H}{W} \tag{2}$$

In case the height of the image is greater than or equal to its width, similarly, the height is set to 640 pixels and the width is scaled proportionally, which can be expressed as Equation (3).

$$W^I = \frac{640 \times W}{H} \tag{3}$$

The third step is to augment the image to a square format with a resolution of 640 × 640 pixels. After resizing, the image may have a shape other than a square. To correct this, empty bars are added to the edges of the image—both top and bottom, as well as on the sides. If, for example, after resizing, the height of the image is less than 640 pixels, empty bars are added to the top and bottom of the image in equal proportions. The number of pixels added to the top and bottom is calculated by Equations (4) and (5).

$$\text{top\_pad} = \frac{640 - H^I}{2} \quad (4)$$

$$\text{bottom\_pad} = 640 - H^I - \text{top\_pad} \quad (5)$$

The remaining value of pixels is added to the bottom of the image. The same process is performed for the width complement if the image's width is less than 640 pixels.

The fourth step involves normalizing the pixel values. YOLOv8 requires the input data to be normalized into the range [0, 1] for using this model. This is achieved by dividing each pixel value by 255, which converts the input data (from the [0, 255] range typical of images) into a standard range that is convenient for processing by the model. Normalized pixel values improve the performance of convolutional neural networks, helping to stabilize the learning process and accelerate convergence. Mathematically, normalization is expressed by Equation (6).

$$\text{normalized\_pixel\_value} = \frac{\text{original\_pixel\_value}}{255} \quad (6)$$

Finally, the final step is to check that the image has three channels corresponding to the RGB color model.

The Focus Layer performs the function of splitting the input image into four parts and combining them into one tensor. This reduces the spatial dimensions of the image by a factor of two in width and height, while increasing the number of channels (depth) by a factor of four. This tensor is then passed through a convolution layer (Conv2D), which allows a better extraction of the initial features from the image. This operation reduces the computational overhead in the following steps and allows for a more efficient capture of fine details to be used for further processing.

The Backbone, based on CSPDarknet53, is the main part of the neural network responsible for extracting features at different depth levels. It consists of a number of CSP (Cross Stage Partial) blocks, such as CSP1, CSP2, CSP3, CSP4, and CSP5. Each of these blocks uses convolution layers that reduce the spatial dimensions of the image and increase the number of channels to produce more abstract and higher-level features. CSP blocks divide the input tensor into two parts: one part is processed by standard convolutional operations, and the other part is passed directly to the next layer, after which they are combined. This helps to reduce the number of computations and improve the flow of gradients through the network, making training more stable and efficient. In addition, Batch Normalization (Batch Normalization) and the Leaky ReLU activation function are applied after each convolution operation, which adds nonlinearity to the model and helps to avoid the problem of vanishing gradients.

The SPP Block (Spatial Pyramid Pooling) is designed to improve the capture of contextual features at different scales. This block applies multiscale pooling (pooling) with fixed core sizes (e.g., 5×5, 9×9, 13×13) to the same input. This allows the model to better capture the features of objects of different sizes and improves its ability to handle objects that may vary in size in the image. The SPP Block retains the original spatial dimensions of the output, but significantly increases the perceptual domain, allowing the model to work with more context, which is critical for object detection tasks in different environments.

The Neck is the part of the neural network that is responsible for aggregating and combining features from different depth levels derived from the Backbone to provide a better ability to detect objects of different sizes. YOLOv8 utilizes a modified Path Aggregation Network (PANet) architecture that includes Feature Pyramid Network (FPN) blocks and additional processing paths (PANet Path Blocks). The FPN performs an upsampling (size increase) operation for features extracted from deep layers and combines them with features from shallower layers to provide a denser representation of features at different layers. PANet Path Blocks then perform a downsampling (size reduction) operation and combine features from different layers to further improve the model's ability to localize and classify objects. This allows the network to capture multi-layer information and improves its accuracy in image processing.

The Detection Head is responsible for the final prediction of objects in the image, including their classes, the coordinates of the bounding boxes, and the degree of confidence the model has in these predictions. YOLOv8 uses a hybrid system that incorporates both anchor-free and anchor-based prediction heads. The anchor-free approach predicts the location of bounding boxes directly, without using predefined anchors, which simplifies the process and improves performance in detecting objects of different sizes and shapes. At the same time, the anchor-based approach uses predefined anchor points and dimensions to improve prediction accuracy when localizing objects. Combining these two approaches allows YOLOv8 to achieve a high level of accuracy and reliability in a wide range of object detection scenarios.

2.5. Training and Validation of YOLOv8

For our system, we used 174 photos for neural network training and 20 photos for validation. The number of object classes is 8. An example of a used training photo is shown in Figure 6.

Figure 6. Used photo for training neural network.

The proposed concept was tested using mock-up turbine components, which are characterized by complex shapes, varying sizes, and being made from different materials. Due to the lack of access to the full range of turbine parts, it was decided to use 3D-printed replicas for validation. The mock-up parts were printed using the Fused Filament Fabrication (FFF) method with PLA plastic as the material, and they are shown in Figure 7. The dimensions of the printed components ranged from 15 × 15 × 30 mm to 210 × 180 × 80 mm, allowing for a diverse representation of the actual turbine parts.

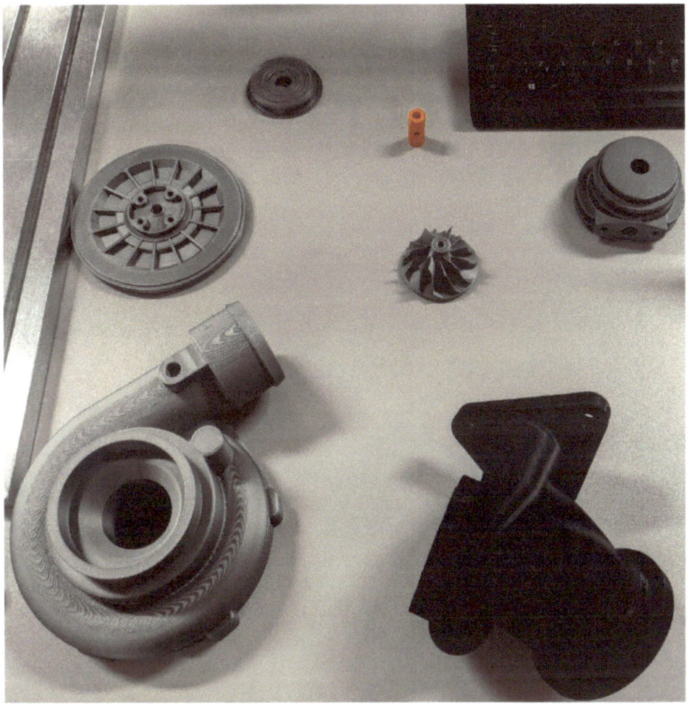

Figure 7. Some 3D-printed turbine components.

In the experiment, we evaluated the frame-processing speed of a system utilizing the YOLOv8 convolutional neural network. Three image resolutions were selected for testing: 416 × 416, 640 × 640, and 768 × 768. The models used for the tests included three configurations of YOLOv8: YOLOv8s (11.2 M parameters), YOLOv8m (25.9 M parameters), and YOLOv8l (43.7 M parameters). Technical characteristics of the laptop on which the convolutional neural network was trained: CPU Intel Core i7-13700HX, RAM 16 GB, GPU Nvidia GeForce RTX 4060 8 GB.

The system's performance was measured in terms of frames processed per minute. Specifically, we calculated how many frames the system could handle within one minute, translating that into frames per second (fps). For each resolution and model configuration, we recorded the training time and frame-processing time. Table 2 contains the microparameters that were used in the training.

To measure the system's real-time performance, the test aimed to determine how quickly the model could process frames while maintaining the accurate detection of objects. The key parameters used during the inference are shown in Table 3.

Table 2. Training hyperparameters.

Hyperparameter	Value	Description
Epochs	100	The number of times the entire dataset is passed through the model during training
Batch	16	The number of images processed in one training iteration
Iou	0.7	The threshold for determining whether overlapping bounding boxes should be merged during non-maximum suppression
Max_det	300	The maximum number of objects the model can predict in one image
lr0	0.01	The starting rate at which the model's weights are updated during training
lrf	0.01	The learning rate maintained during the final phase of training
momentum	0.937	Controls the amount of influence past updates have on the current weight updates
weight_decay	0.0005	A regularization parameter that helps prevent the model from overfitting by penalizing large weights
warmup_epochs	3.0	The number of epochs during which the learning rate gradually increases from a very low value to the set initial learning rate
warmup_momentum	0.8	The starting momentum value during the warmup phase, which gradually increases as training progresses
warmup_bias_lr:	0.1	The initial learning rate for bias parameters during the warmup phase, helping them converge faster in the early epochs

Table 3. Parameters for running the neural network.

Micrometer	Value	Description
Confidence Threshold	0.25	This is the minimum probability at which the model considers that the detected region contains an object
Score Threshold	0.45	The score threshold takes into account both the model's confidence in the presence of the object and its classification
Non-Maximum Suppression (NMS) Threshold	0.50	This parameter defines the overlap threshold between predicted bounding boxes

The resulting Python Package YOLOv8 model can be imported into the ONNX format for further use in conjunction with the C++ OpenCV library [36].

3. Results and Discussion

This section presents the results of a performance analysis of the YOLOv8 neural network trained on the task of industrial component classification. Test results for frame-processing speed depending on the model configuration and image resolution are given in Table 4. Therefore, YOLOv8m with image resolution 640 × 640 was chosen for our research.

Table 4. Test results based on model configuration and image resolution.

Image Resolution	YOLOv8s (11.2 M Parameters)	YOLOv8m (25.9 M Parameters)	YOLOv8l (43.7 M Parameters)
416 × 416	Training Time: 4 m 2 s Processing Speed: max: 0.185 s min: 0.007 s avg: 0.008 s Note: Detected object (bearing) showed false positives near edges	Training Time: 9 m 30 s Processing Speed: max: 0.225 s min: 0.009 s avg: 0.013 s Note: Bearing detection had false positives near edges	Training Time: 16 m 1 s Processing Speed: max: 0.248 smin: 0.014 savg: 0.016 s
640 × 640	Training Time: 5 m 23 s Processing Speed: max: 0.200 s min: 0.011 s avg: 0.013 s Note: Compressor housing not always detected, some false positives	Training Time: 13 m 37 s Processing Speed: max: 0.235 s min: 0.016 s avg: 0.018 s	Out of Memory
768 × 768	Training Time: 9 m 14 s Processing Speed: max: 0.225 s min: 0.017 s avg: 0.019 s Noti: Bearing false positives still present	Out of Memory	Out of Memory

To evaluate the quality of the model, we constructed confusion matrices, curves of F1 metrics, accuracy, and completeness versus confidence level, and examined graphs of loss during training.

The results presented in the error matrix (Figure 8a) show that the YOLOv8 model demonstrates high accuracy in classifying most classes. The diagonal elements of the matrix indicate that classes such as "compressor wheel", "compressor housing", "turbine housing", "turbine cover", "compressor holder", and "bearing" were correctly classified 100% of the time, indicating that there were no errors for these classes. However, for the "Center housing" class, the model made an error in one case by predicting it as a "rear part" and conversely, one case of a "rear part" was incorrectly classified as a "center housing". This indicates that there may be confusion between the two classes, probably due to their visual similarity or overlapping features.

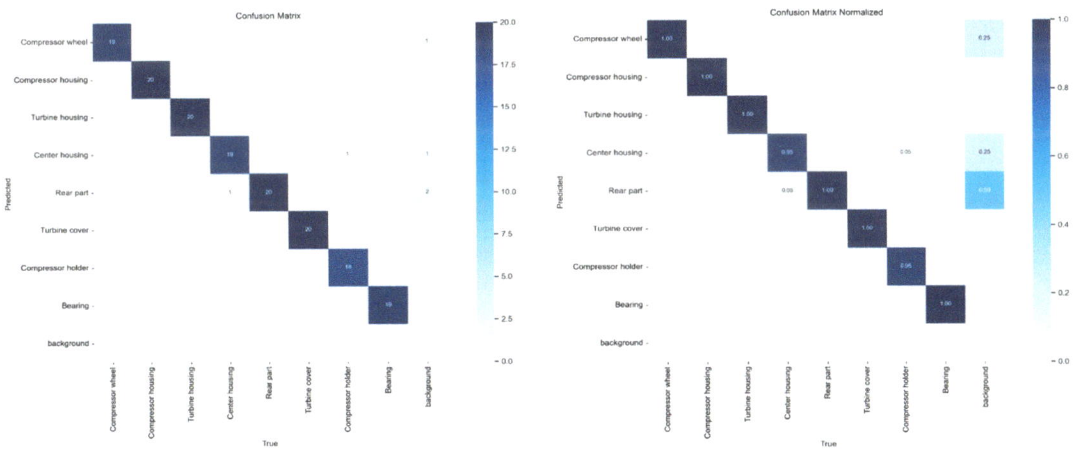

(a) (b)

Figure 8. (**a**) YOLOv8 confusion matrix; (**b**) YOLOv8 confusion matrix normalized.

The normalized error matrix provides similar information in relative values, allowing a better understanding of the error rate for each class. The results in Figure 8b show that the classification accuracy for most classes is 1.0, confirming that there are no errors. However, for the center housing class, the classification accuracy is 95%, and for the rear part it is 100%, with a 5% error in classifying a center housing as rear part. This indicates the need for additional model optimization or improved data quality to distinguish complex classes more accurately.

Analysis of the F1 metrics versus the confidence threshold curve, illustrated in Figure 9, shows that at a confidence level of about 0.635, the model reaches a maximum F1 score of 0.98 for all classes. This means that the best balance between precision (Precision) and completeness (Recall) is achieved at this confidence level. This is especially important for practical applications where minimizing false positives and misses is critical. The optimal F1 score value also confirms that the YOLOv8 model performs effectively on the classification task for the given parameters.

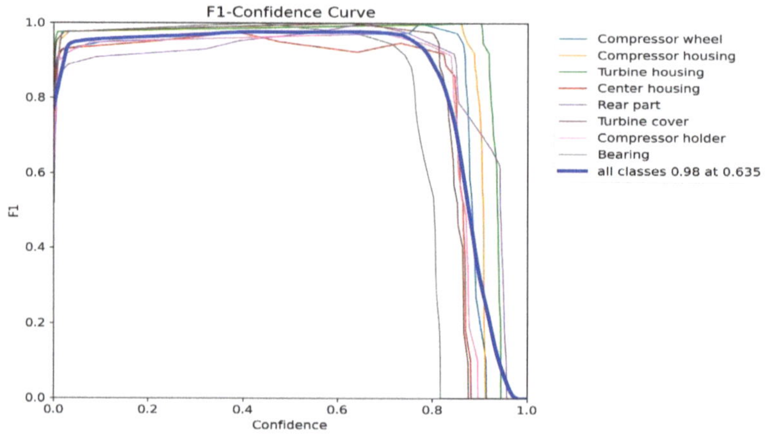

Figure 9. F1–confidence curve.

The label distribution histogram, which is presented in Figure 10, demonstrates that the data are relatively balanced for most of the classes, such as "compressor housing", "turbine housing", "bearing", and "compressor wheel", which have about 150–175 instances. This contributes to the stability of the model training and reduces the likelihood of a prediction bias in favor of classes with more data. The scatter plots, which are illustrated in Figure 11, show various object sizes and positions, confirming the model's ability to generalize and recognize objects of different shapes and scales.

Analyzing the precision–confidence (Figure 12a) and completeness curves (Figure 12b) shows that the Yolov8 model achieves a precision value of 1.0 at a confidence level of 0.929, confirming its ability to classify without false positives at high confidence levels accurately. Completeness reaches 1.0 at low confidence levels, allowing the model to find all possible objects, although some false positives may accompany this. These results allow the model to be customized for specific applications where finding the optimal balance between accuracy and completeness is essential.

The Precision–Recall curve (Figure 13) shows the high performance of the YOLOv8 model in object classification. The mean accuracy value (mAP) reaches 0.994 at a threshold of 0.5, demonstrating the model's ability for high-accuracy prediction and efficient object detection. This makes the model suitable for application in tasks that require high accuracy and confidence in the results, such as industrial equipment and automated surveillance systems.

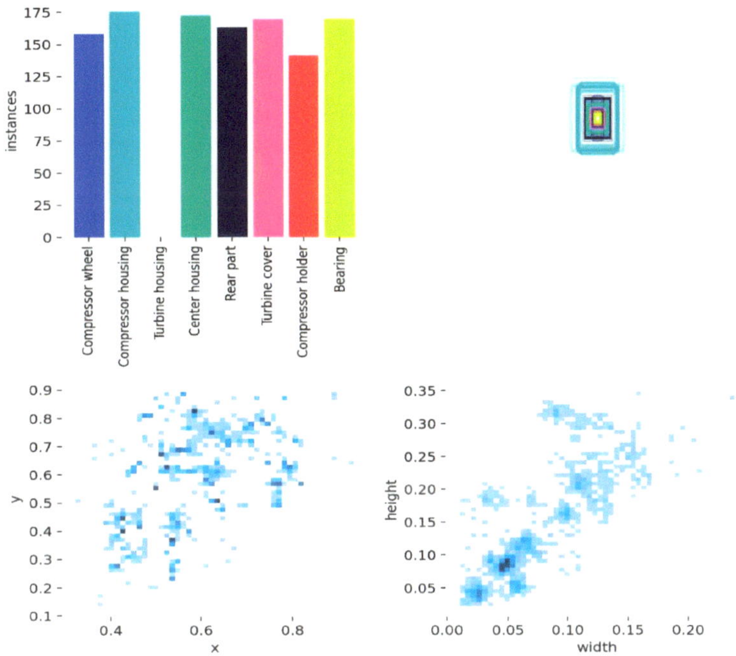

Figure 10. Label distribution histogram.

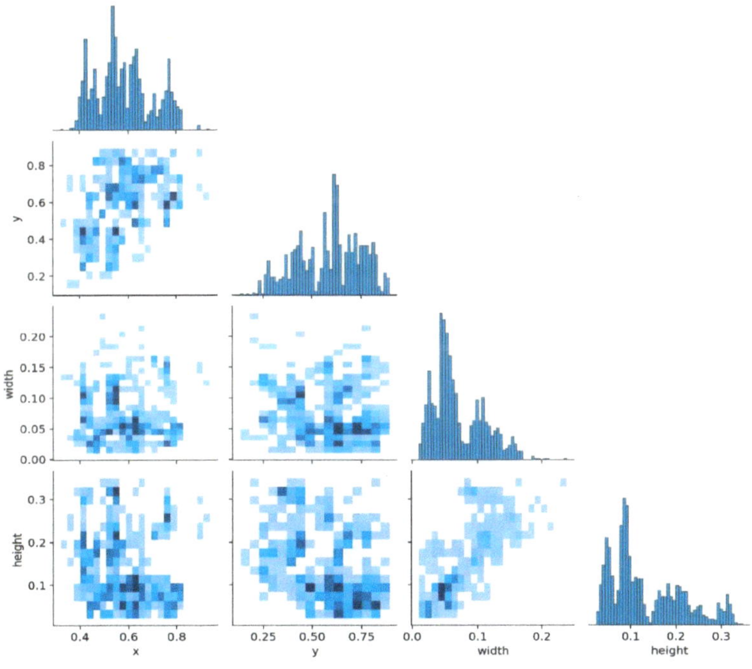

Figure 11. Scatter plots.

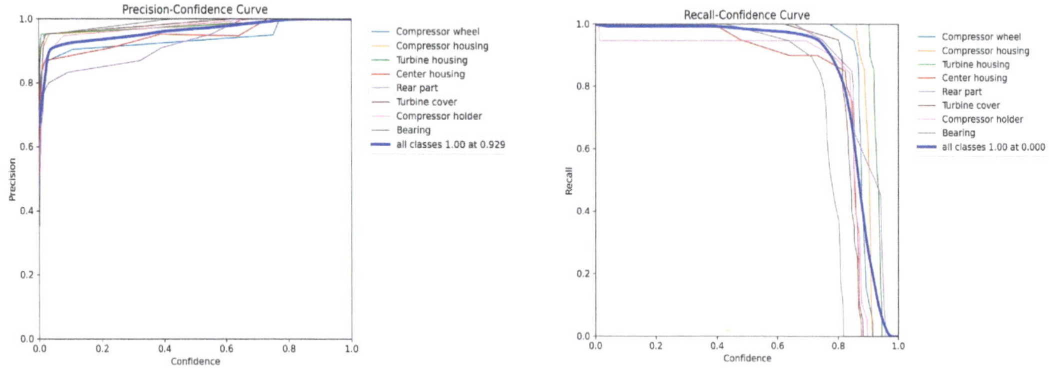

(**a**) (**b**)

Figure 12. (**a**) Precision–confidence curve; (**b**) completeness curve.

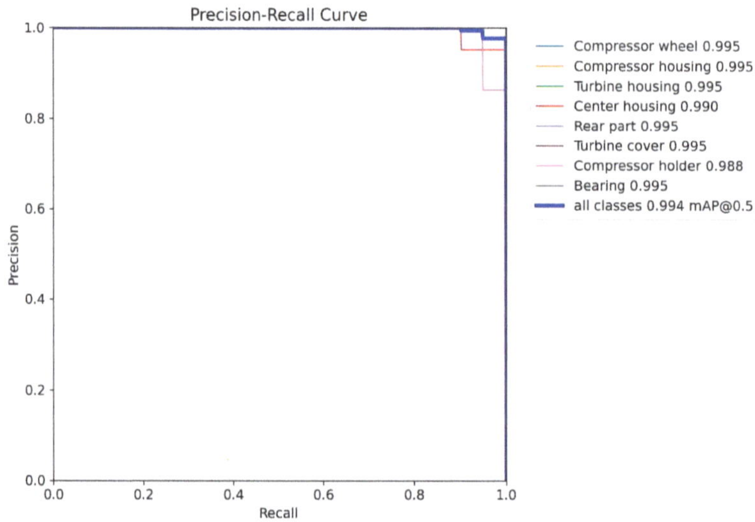

Figure 13. Precision–Recall curve.

The loss plots, illustrated in Figure 14, show that the losses for the training and validation data steadily decrease as the number of epochs increases, indicating the good convergence of the Yolov8 model. The minimal differences between training and validation losses indicate no overtraining, allowing the model to generalize well to new data. The losses eventually stabilize, confirming that the training process is complete and the model is ready for practical application.

Future research directions include expanding the material and component database to improve the identification accuracy, integrating with advanced manufacturing systems and the IoT to improve collaboration and real-time control, developing predictive and prescriptive analytics to optimize decision-making, improving the user interface to speed up workflows, implementing support for flexible and adaptive manufacturing, and improving the scalability and resiliency of client–server architecture for large-scale applications.

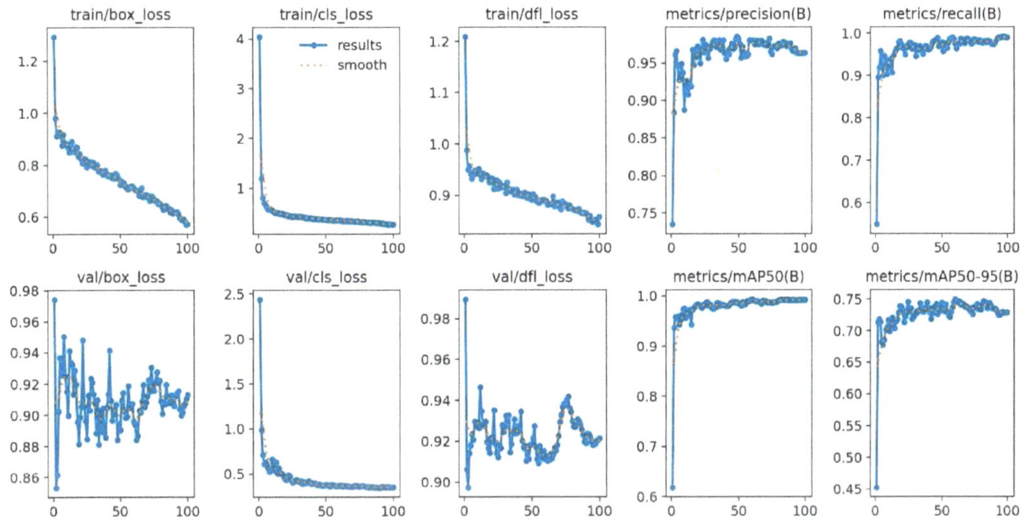

Figure 14. Loss plots.

4. Conclusions

The article is devoted to developing a material resource planning system in the context of the Warehouse 4.0 concept, which aims to automate and improve inventory and production operations management efficiency. The developed software solution utilizes the YOLOv8 neural network to accurately identify materials and production parts. Laboratory experiments have shown that the proposed system improves the decision-making process for material management, production planning, and launching new batches of products. The system is easily integrated into the enterprise's existing ERP systems, providing centralized management and planning flexibility.

This article pays special attention to optimizing material management processes, which is especially important in flexible production conditions and the implementation of circular economy principles.

The proposed system covers the "gray areas" of material accounting systems, where it is quite difficult to ensure the accounting for each component and their dispatch to the main warehouse may be untimely. Time losses associated with transfers to the main warehouse would be automatically eliminated, since the manager making the decision remotely can ensure the transfer of this product directly to the production process and create a work order for the production employees. This approach reduces the workforce downtime associated with waiting for material for production.

The proposed system covers bottlenecks of warehouse systems through the efficient utilization of excess or unclaimed components, thereby reducing the cost of maintaining stock balances, shortening production downtime, and reducing production waste. Applying the DFMA (Design for Manufacturing and Assembly) approach helps optimize the design and processes at the manufacturing and assembly stages, reducing product lifecycle costs.

5. Discussion

The proposed study showed that the proposed concept is workable. Practical efficiency should be calculated in the conditions of a real enterprise, where the internal costs of the enterprise associated with untimely decision-making and the liquidation and provision of warehouse balances that were not transferred to the production process would be considered. At this stage, it can be stated that the system works and, in the future, can reduce the time and increase the efficiency of operational planning based on the data collected by the neural network.

The faculty where the experiment was conducted closely cooperates with the production sector and, based on the collected data, can state that at this stage, the problems of operational planning mentioned in the Introduction have not been eliminated, and neural systems are used mainly only for quality control or the detection of certain objects. Our concept is innovative, as it opens up prospects for using neural networks for remote decision-making.

Further research should cover such important aspects as:

- Expanding the range of components;
- Training the system to detect external defects, since at this stage, the system is not able to do this;
- Identifying factors that can interfere with the quality of camera operation in real enterprise conditions;
- Integrating the proposed neural network into warehouse facilities in order to reduce the costs of holding stale material and offering several decision-making options for this category of material in accordance with the principles of the circular economy and Warehouse 4.0;
- Expanding the areas of use of the neural network by mobile applications, since the studied workplaces are important for operational planning and quality control. From this, it is possible to determine further areas of application for the proposed system: personnel management, quality control, logistics, material supply for productionm and the process of creating a cost chain.

Author Contributions: Conceptualization, O.S.; methodology, A.I.; formal analysis, O.S; software, V.A.; validation, V.A. and O.S.; writing—original draft preparation O.S.; writing—review and editing, all authors; supervision, J.T. All authors have read and agreed to the published version of the manuscript.

Funding: This research was funded by the EU NextGenerationEU through the Recovery and Resilience Plan for Slovakia under projects No. 09I03-03-V01-00095 and 09I03-03-V01-00102, and was also carried out within the project "Intensification of production processes and development of intelligent product quality control systems in smart manufacturing" (State reg. no. 0122U200875, Ministry of Education and Science of Ukraine).

Data Availability Statement: The original contributions presented in the study are included in the article, further inquiries can be directed to the corresponding author.

Acknowledgments: This work was also supported by the Slovak Research and Development Agency under contract No. APVV-23-0591, and by projects VEGA 1/0700/24 and KEGA 022TUKE-4/2023 supported by the Ministry of Education, Research, Development, and Youth of the Slovak Republic.

Conflicts of Interest: The authors declare no conflicts of interest.

References

1. Husár, J.; Hrehova, S.; Trojanowski, P.; Wojciechowski, S.; Kolos, V. Perspectives of Lean Management Using the Poka Yoke Method. In *Lecture Notes in Mechanical Engineering, Proceedings of the Advances in Design, Simulation and Manufacturing VI, High Tatras, Slovakia, 6–9 June 2023*; Ivanov, V., Trojanowska, J., Pavlenko, I., Rauch, E., Piteľ, J., Eds.; DSMIE 2023; Springer: Cham, Switzerland, 2023. [CrossRef]
2. Wang, H. Customer Needs Assessment and Screening for Transmission Solution Selection. In *Lecture Notes in Mechanical Engineering, Proceedings of the 8th International Conference on Advances in Construction Machinery and Vehicle Engineering, Shanghai, China, 13–16 October 2023*; Halgamuge, S.K., Zhang, H., Zhao, D., Bian, Y., Eds.; ICACMVE 2023; Springer: Singapore, 2024. [CrossRef]
3. Kaya, B.; Karabağ, O.; Çekiç, F.R.; Torun, B.C.; Başay, A.Ö.; Işıklı, Z.E.; Çakır, Ç. Inventory Management Optimization for Intermittent Demand. In *Lecture Notes in Mechanical Engineering, Proceedings of the Industrial Engineering in the Industry 4.0 Era, Antalya, Türkiye, 5–7 October 2023*; Durakbasa, N.M., Gençyılmaz, M.G., Eds.; ISPR 2023; Springer: Cham, Switzerland, 2024. [CrossRef]

4. Thürer, M.; Fernandes, N.O.; Lödding, H.; Stevenson, M. Material flow control in make-to-stock production systems: An assessment of order generation, order release and production authorization by simulation. *Flex. Serv. Manuf. J.* **2024**. [CrossRef]
5. Tubis, A.A.; Rohman, J. Intelligent Warehouse in Industry 4.0—Systematic Literature Review. *Sensors* **2023**, *23*, 4105. [CrossRef] [PubMed]
6. Aravindaraj, K.; Chinna, P.R. A systematic literature review of integration of industry 4.0 and warehouse management to achieve Sustainable Development Goals (SDGs). *Clean. Logist. Supply Chain.* **2022**, *5*, 100072. [CrossRef]
7. Tikwayo, L.N.; Mathaba, T.N.D. Applications of Industry 4.0 Technologies in Warehouse Management: A Systematic Literature Review. *Logistics* **2023**, *7*, 24. [CrossRef]
8. Hamdy, W.; Al-Awamry, A.; Mostafa, N. Warehousing 4.0: A proposed system of using node-red for applying internet of things in warehousing. *Sustain. Futures* **2022**, *4*, 100069. [CrossRef]
9. Vukicevic, A.; Mladineo, M.; Banduka, N.; Macuzic, I. A smart Warehouse 4.0 approach for the pallet management using machine vision and Internet of Things (IoT): A real industrial case study. *Adv. Prod. Eng. Manag.* **2021**, *16*, 297–306. [CrossRef]
10. Ongbali, S.O.; Afolalu, S.A.; Oyedepo, S.A. Aworinde AK, Fajobi MA. A study on the factors causing bottleneck problems in the manufacturing industry using principal component analysis. *Heliyon* **2021**, *7*, e07020. [CrossRef] [PubMed] [PubMed Central]
11. Franco-Silvera, N.; Valdez-Yrigoen, A.; Quiroz-Flores, J.C. Warehouse Management Model under the Lean Warehousing Approach to Increase the Order Fill Rate in Glass Marketing SMEs. In Proceedings of the 2023 9th International Conference on Industrial and Business Engineering (ICIBE '23), Beijing, China, 22–24 September 2023; Association for Computing Machinery: New York, NY, USA, 2023; pp. 379–387. [CrossRef]
12. Manufacturing ERP Software Development: All You Need to Know. Available online: https://appinventiv.com/blog/manufacturing-erp-software-development/ (accessed on 26 September 2024).
13. ERP Implementation Challenges Manufacturers Need to Be On Top Of. Available online: https://dwr.com.au/erp-implementation-challenges-manufacturers-need-to-be-on-top-of/ (accessed on 26 September 2024).
14. TechTarget. Common Problems with Inventory Management. Available online: https://www.techtarget.com/searcherp/tip/Common-problems-with-inventory-management (accessed on 26 September 2024).
15. NetSuite. Inventory Management Challenges: Overcoming Obstacles to Success. Available online: https://www.netsuite.com/portal/resource/articles/inventory-management/inventory-management-challenges.shtml (accessed on 26 September 2024).
16. Aptean. Challenges of Inventory Management: How to Overcome Them. Available online: https://www.aptean.com/fr/insights/blog/challenges-of-inventory-management (accessed on 26 September 2024).
17. Battaïa, O.; Dolgui, A.; Heragu, S.S.; Meerkov, S.M.; Tiwari, M.K. Design for manufacturing and assembly/disassembly: Joint design of products and production systems. *Int. J. Prod. Res.* **2018**, *56*, 7181–7189. [CrossRef]
18. Arakawa, M.; Park, W.Y.; Abe, T.; Tasaki, K.; Tamaki, K. Development of Service and Product Design Processes Considering Product Life Cycle Management for a Circular Economy. In *Lecture Notes in Mechanical Engineering, Proceedings of the Industrial Engineering and Management, Chengdu, China, 17–19 November 2023*; Chien, C.F., Dou, R., Luo, L., Eds.; SMILE 2023; Springer: Singapore, 2024. [CrossRef]
19. Cebi, S.; Baki, B.; Ozcelik, G. Overcoming Barriers in Circular Economy Implementation with Industry 4.0 Technologies: The Case of Defense Industry. In *Lecture Notes in Mechanical Engineering, Proceedings of the Industrial Engineering in the Industry 4.0 Era, Antalya, Türkiye, 5–7 October 2023*; Durakbasa, N.M., Gençyılmaz, M.G., Eds.; ISPR 2023; Springer: Cham, Switzerland, 2024. [CrossRef]
20. Plakantara, S.P.; Karakitsiou, A.; Mantzou, T. Managing Risks in Smart Warehouses from the Perspective of Industry 4.0. In *Disruptive Technologies and Optimization Towards Industry 4.0 Logistics*; Karakitsiou, A., Migdalas, A., Pardalos, P.M., Eds.; Springer Optimization and Its Applications, vol 214; Springer: Cham, Switzerland, 2024. [CrossRef]
21. Favi, C.; Mandolini, M.; Campi, F.; Cicconi, P.; Raffaeli, R.; Germani, M. Design for Manufacturing and Assembly: A Method for Rules Classification. In *Lecture Notes in Mechanical Engineering, Proceedings of the Advances on Mechanics, Design Engineering and Manufacturing III, Aix-en-Provence, France, 2–4 June 2020*; Roucoules, L., Paredes, M., Eynard, B., Morer Camo, P., Rizzi, C., Eds.; JCM 2020; Springer: Cham, Switzerland, 2021. [CrossRef]
22. Demčák, J.; Lishchenko, N.; Pavlenko, I.; Pitel', J.; Židek, K. The Experimental SMART Manufacturing System in SmartTechLab. In *Lecture Notes in Mechanical Engineering, Proceedings of the Advances in Manufacturing II, Poznan, Poland, 19–22 May 2019*; Springer International Publishing: Cham, Switzerland, 2022; pp. 228–238. [CrossRef]
23. Sá, J.C.; Dinis-Carvalho, J.; Costa, B.; Silva, F.J.G.; Silva, O.; Lima, V. Implementation of Lean Tools in Internal Logistic Improvement. In *Lean Thinking in Industry 4.0 and Services for Society*; Antosz, K., Carlos Sa, J., Jasiulewicz-Kaczmarek, M., Machado, J., Eds.; IGI Global: Hershey, PA, USA, 2023; pp. 125–137. [CrossRef]
24. Widanage, C.; Kim, K.P. Integrating Design for Manufacture and Assembly (DfMA) with BIM for infrastructure. *Autom. Constr.* **2024**, *167*, 105705. [CrossRef]
25. Pathan, M.S.; Richardson, E.; Galvan, E.; Mooney, P. The Role of Artificial Intelligence within Circular Economy Activities—A View from Ireland. *Sustainability* **2023**, *15*, 9451. [CrossRef]
26. Iakovets, A.; Andrusyshyn, V. Design of a Decision-Making Model for Engineering Education. In *EAI/Springer Innovations in Communication and Computing, Proceedings of the 2nd EAI International Conference on Automation and Control in Theory and Practice, Orechová Potôň, Slovakia 7–9 February 2024*; Balog, M., Iakovets, A., Hrehová, S., Berladir, K., Eds.; EAI ARTEP 2024; Springer: Cham, Switzerland, 2024. [CrossRef]

27. Quality-One. (n.d.). Design for Manufacturing/Assembly (DFM/DFA). Available online: https://quality-one.com/dfm-dfa/#:~:text=What%20is%20Design%20for%20Manufacturing,assembled%20with%20minimum%20labor%20cost (accessed on 2 October 2024).
28. GitHub—Boostorg/Beast: HTTP and WebSocket built on Boost.Asio in C++11. Available online: https://github.com/boostorg/beast (accessed on 11 August 2024).
29. GitHub—Boostorg/Mysql: MySQL C++ Client Based on Boost.Asio. Available online: https://github.com/boostorg/mysql (accessed on 11 August 2024).
30. OpenCV—Open Computer Vision Library. Available online: https://opencv.org/ (accessed on 11 August 2024).
31. Redmon, J.; Farhadi, A. YOLOv3: An Incremental Improvement. *arXiv* **2018**, arXiv:1804.02767. Available online: https://arxiv.org/abs/1804.02767 (accessed on 2 October 2024).
32. Bochkovskiy, A.; Wang, C.-Y.; Liao, H.-Y.M. YOLOv4: Optimal Speed and Accuracy of Object Detection. Computer Science. Computer Vision and Pattern Recognition. *arXiv* **2020**, arXiv:2004.10934. [CrossRef]
33. He, K.; Gkioxari, G.; Dollar, P.; Girshick, R. Mask R-CNN. In Proceedings of the IEEE International Conference on Computer Vision (ICCV) 2017, Venice, Italy, 22–29 October 2017; pp. 2961–2969.
34. Ultralytics | Revolutionizing the World of Vision AI. Available online: https://www.ultralytics.com/ (accessed on 11 August 2024).
35. YOLOv8. (n.d.). YOLOv8 Architecture; Deep Dive into its Architecture. Available online: https://yolov8.org/yolov8-architecture/ (accessed on 11 August 2024).
36. Terven, J.; Córdova-Esparza, D.-M.; Romero-González, J.-A. A Comprehensive Review of YOLO Architectures in Computer Vision: From YOLOv1 to YOLOv8 and YOLO-NAS. *Mach. Learn. Knowl. Extr.* **2023**, *5*, 1680–1716. [CrossRef]
37. Lalinia, M.; Sahafi, A. Colorectal polyp detection in colonoscopy images using YOLO-V8 network. *Signal Image Video Process.* **2024**, *18*, 2047–2058. [CrossRef]
38. Ultralytics. Ultralytics YOLO. Available online: https://github.com/ultralytics/ultralytics (accessed on 2 October 2024).

eng

Article

Vibration Analysis of a Centrifugal Pump with Healthy and Defective Impellers and Fault Detection Using Multi-Layer Perceptron

Masoud Hatami Garousi [1], Mahdi Karimi [2], Paolo Casoli [1], Massimo Rundo [3] and Rasoul Fallahzadeh [1,*]

1 Department of Engineering for Industrial Systems and Technologies, University of Parma, 43124 Parma, Italy; masoud.hatamigarousi@unipr.it (M.H.G.); paolo.casoli@unipr.it (P.C.)
2 Department of Engineering, Bu-Ali Sina University, Hamedan 65175-38695, Iran; m_karimi@basu.ac.ir
3 Department of Energy, Politecnico di Torino, 10129 Turin, Italy; massimo.rundo@polito.it
* Correspondence: rasoul.fallahzadeh@unipr.it

Abstract: Centrifugal pumps (CPs) are widely utilized in many different industries, and their operations are maintained by their reliable performance. CPs' most common faults can be categorized as mechanical or flow-related faults: the first ones are often associated with damage at the impeller, while the second ones are associated with cavitation. It is possible to use computational algorithms to monitor both failures in CPs. In this study, two different problems in pumps, the defective impeller and cavitation, have been considered. When a CP is working in a faulty condition, it generates vibrations that can be measured using piezoelectric sensors. Collected data can be analyzed to extract time- and frequency-domain data. Interpreting the time-domain data showed that distinguishing the type of defect is not possible. However, indicators like kurtosis, skewness, mean, and variance can be used as input for the multi-layer perceptron (MLP) algorithm to classify pump faults. This study presents a detailed discussion of the vibration-based method outcomes, emphasizing the benefits and drawbacks of the multi-layer perceptron method. The results show that the suggested algorithm can identify the occurrence of different faults and quantify their severity during pump operation in real time.

Keywords: vibration analysis; multi-layer perceptron; centrifugal pump; cavitation; fault detection

Citation: Garousi, M.H.; Karimi, M.; Casoli, P.; Rundo, M.; Fallahzadeh, R. Vibration Analysis of a Centrifugal Pump with Healthy and Defective Impellers and Fault Detection Using Multi-Layer Perceptron. *Eng* **2024**, *5*, 2511–2530. https://doi.org/10.3390/eng5040131

Academic Editor: Antonio Gil Bravo

Received: 21 August 2024
Revised: 3 October 2024
Accepted: 4 October 2024
Published: 8 October 2024

1. Introduction

Centrifugal pumps are important in oil, gas, and other industries. Therefore, knowledge and understanding of the behavior of these types of pumps are essential. Experimental tests are a suitable method for troubleshooting devices. For example, measuring and identifying different vibration frequencies related to the pump impeller, rotating shafts, and bearings is a crucial test in diagnosing potential issues and ensuring the efficient and reliable operation of the pump. By identifying specific vibration patterns, it is possible to detect early signs of wear or damage, allowing for timely maintenance and reducing the risk of unexpected failures. Vibration signals in the time, frequency, or time-frequency domains provide information about the status of equipment and devices [1–3]. In CPs, various problems, such as impeller failure and cavitation, can occur. Detecting and fixing these defects requires a detailed understanding of how the device works and the reasons for the defect. Predictive maintenance involves monitoring the vibration of rotating machines to discover defects in the early stages and prevent the progression of failure. Predictive maintenance helps to determine the conditions of the working equipment to define the appropriate maintenance time [4–7]. In other words, predictive maintenance uses the actual operating conditions of factory equipment and devices to optimize the overall operation of the equipment [8–10].

Mousmoulis et al. [11] investigated cavitation in centrifugal pumps, concentrating on its emergence and growth in three distinct impellers. They also demonstrated that impeller geometry significantly influences cavitation behavior, affecting inception points and behavior. Using flow visualization, acoustic emissions, and vibration monitoring, they found that accelerometers and acoustic emission sensors were effective in detecting cavitation onset well before the total head drop. Cavitation inception occurs at higher flow rates, consistently appearing at the suction side of leading blade edges. Al-Obaidi [12] investigated the effect of pump rotational speed on the performance and detection of cavitation using vibration signals. The results showed that the amount of cavitation has a direct relationship with the rotational speed and the flow rate. Mousmoulis et al. [13] investigated the vibration analysis, flow observation technique, and wave propagation technique on the plexiglass impeller pump to detect cavitation. They showed that the cavitation values increased with the increase in flow rate. Sun et al. [13] used the Hilbert spectrum to detect cavitation in a CP. They concluded that the Hilbert transform is suitable for processing transient and unsteady signals, and time-frequency domain characteristics can be extracted. Murovec et al. [14] used the acoustic diffusion method to detect cavitation in a radial CP. Using acoustic criteria and mathematical parameters, they provided an effective method for predicting and classifying cavitation in a CP. Kumar et al. [15] used time-frequency domain analysis and support vector machines to detect faults in a CP. At first, they trained the machine using the obtained data, and then they used it to identify defects such as cavitation and bearing defects. Azizi et al. [16] employed the hybrid feature selection technique to identify the degree of cavitation. Using this method, it is possible to differentiate between three states: no bubbling, limited bubbling, and cavitation that occurred at the outflow. Birajdar et al. [17] studied the sources of vibration and noise in CPs and the methods of troubleshooting this type of pump. They listed the sources of vibration in CPs as unbalanced rotors, defective impellers, broken shafts, damaged bearings, and cavitation at critical speeds. They concluded that cavitation is created randomly and in a high-frequency energy range, but it has sometimes been mistaken for blade passing frequency. They presented a diagram for the range of the frequency spectrum. Sakthivel et al. [18] investigated fault diagnosis and fault prediction in CPs with the help of a soft computing approach. They used different classification patterns, such as support vector machines, genetic algorithms, and wavelet analysis, to classify the defects. Six parameters, including the bearing defect, impeller defect, leakage defect, impeller and bearing defect together, and cavitation, were considered, and an algorithmic decision tree was used to select the defect. They analyzed the results using a genetic algorithm (GA), support vector machine (SVM), and wavelet analysis (WA) and finally observed that both the GA and SVM have better performance compared to other classifications for defects. Askari et al. [19] investigated the troubleshooting and diagnosis of the cavitation phenomenon by using the vibration analysis method and provided solutions to solve it. They investigated the abnormal performance of the CPs of a refining unit with the modal analysis method and finally determined the vibration sources of the pump according to the frequency range determined for cavitation, predicting the existence of this phenomenon in the pump. Al-Braik et al. [20] investigated the troubleshooting of CPs to detect impeller defects. They conducted a test on a healthy impeller and five impellers with varying degrees of defects on the vane tips. Vibration data were collected at different flow rates, and they obtained the frequency spectrum for these conditions. They showed that discrete spectral components at vane-passing frequencies and higher-order harmonics of the shaft frequency are effective for diagnosing impeller faults and that primary spectral components of the turbulent flow occur above 1 kHz. Spadafor et al. [21] conducted dynamic system simulations and optimizations of a CP to investigate impeller failure. They aimed to determine whether the failure was due to metallurgical issues, such as improper heat treatment, corrosion, or mechanical factors. After replacing the pump with another impeller with the same geometry but made from a stainless steel alloy, they observed the same failure results. Vibration analysis revealed that torsional vibrations introduced by motor oscillations were affecting the pump. Their experimental research, using a hydraulic response system, demonstrated that pressure fluctuations were closely related to shaft speed variations and changes in engine torque, ultimately causing damage to the pump impeller. Wang and Cheng [22] used wavelet transmission (WT) and a partially linearized neural network (PNN) to extract the characteristics of cavitation from the vibration signals and then used non-dimensional symptom parameters like the mean and standard deviations as input for the artificial neural network (ANN) to determine whether the pump was healthy or defective. The results obtained showed that WT successfully processes cavitation vibration signals, and other processing methods are also effective. Barrio et al. [23] investigated radial loads on centrifugal pump impellers and found significant unsteady components that were challenging to estimate. They showed that the unsteady component (the fluctuating part of the radial load on the impeller) could represent 40% to 70% of the average load when operating at off-design conditions (deviations from the optimal flow rate and pressure for which the centrifugal pump was designed). The CFD simulations demonstrated strong agreement with the experimental data in terms of both global performance and unsteady pressure distribution. Their findings highlight the importance of accounting for unsteady loads in pump design to prevent fatigue failure. Casoli et al. [24,25] studied a vibration signal-based method for fault identification and classification in hydraulic axial piston pumps. Based on the vibration signals and the use of SVM and ANN, they proposed an algorithm to detect the health state of a variable displacement axial piston pump. By using the time-sampling raw signals, they achieved a satisfying accuracy. They showed that the proposed algorithm can identify the faults in the axial piston pump for each working condition.

Following the studies in the literature, vibration analysis and fault detection in CPs are presented in this paper. Most of the researchers explored the utilization of statistical parameters in the frequency domain as input for an ANN. Moreover, there has been a noticeable absence of emphasis on investigating the time-domain, frequency-domain, and classification methods, such as MLP, within the scope of centrifugal pumps. The main objective of this work is to present the practical vibration analysis method for a CP and use statistical parameters of the time domain, including mean, kurtosis, variance, RMS, and skewness in the MLP algorithm to identify the faults in CPs. This study offers superior advantages of MLP over fast Fourier transform (FFT). Unlike FFT, MLP automates the feature extraction process from time-domain signals, eliminating the need for manual interpretation and reducing reliance on expert knowledge. Furthermore, MLP's ability to handle non-stationary and transient signals enhances its effectiveness for real-time fault detection.

2. Methodology

In this section, the test method is explained, and then the multi-layer perceptron neural network method is introduced.

2.1. Test Method

The test bench used, as seen in Figure 1, is made of several parts, such as a DC motor, a voltage converter, suction and discharge tanks, a centrifugal pump (radial flow), a pressure sensor, and a flow and pressure adjustment valve. Vibration signals in our experiment were measured using a piezoelectric accelerometer under the trade name 2224C (Endevco, Halifax, NC, USA), which was mounted horizontally and parallel to the pump axis. Specifically, the accelerometer was installed on the body of the centrifugal pump, near the impeller housing, and on the horizontal ventilation bolt, as this area is highly sensitive to vibrations induced by impeller defects or cavitation. Equipment such as a piezoelectric accelerometer, an amplifier, and a DS20080A two-channel oscilloscope card (OC) -(TNM Electronics, Tehran, Iran) were used to collect the signals. The specifications of the Endevco 2224C accelerometer are presented in Table 1.

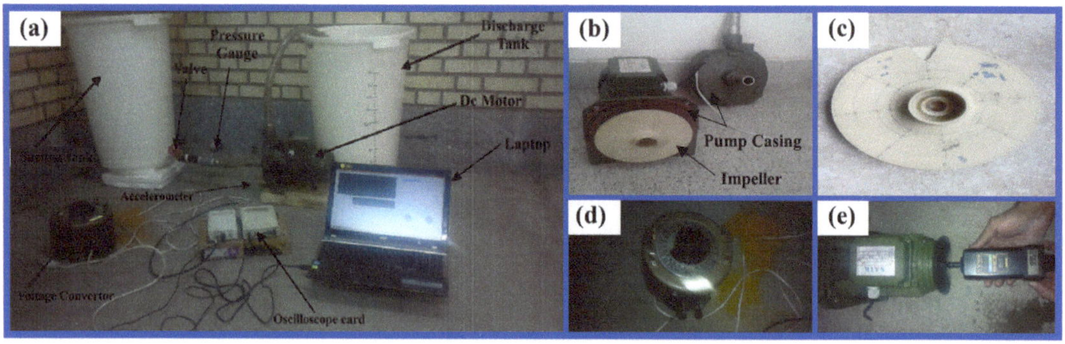

Figure 1. (**a**) Laboratory test rig, (**b**) pump with healthy impeller, (**c**) defective impeller, (**d**) voltage converter, and (**e**) tachometer.

Table 1. Key parameters of the Endevco 2224C accelerometer.

Attribute	Testing Conditions/Remarks	Measurement Unit
Sensitivity	±12	pC/g
Sinusoidal limit	1000	g
Shock limit	2000	g
Operating temperature	−55–+177	°C (°F)
Frequency response	0.1–10,000	±1 dB Hz

The pump used in this experiment is the SAER CMP76, which is used to pump the water with an impeller made of polyethylene. The pump's and impeller's specifications are shown in Table 2.

Table 2. Main parameters of SAER CMP76 centrifugal pump.

Name	Value
Flow rate	0–6 m^3/h
Head	21–29 m
Efficiency	75%
Impeller inlet diameter	36 mm
Impeller outlet diameter	148 mm
Impeller outlet width	2 mm
Power	0.75 kW
Blade number	6
Specific speed	58.45
Flow coefficient	0.00165
Head coefficient	0.1233

To test a case representing faulty mechanical conditions, the impeller was damaged by removing a triangular piece from it, as shown in Figure 1c. This type of defect in pump impellers, known as a notch defect, usually occurs at the outlet diameter because of the higher speed. The size of the notch defect can indeed vary depending on several factors, such as the operational conditions of the pump (e.g., speed and flow rate), the material of the impeller, and the duration of exposure to mechanical stress or cavitation. In our study, we removed a triangular piece of the impeller to simulate a typical notch defect often observed at the outlet diameter due to high-speed and tension operations. However, in real-world scenarios, the size and shape of the defect can differ. Notch defects may grow over time as the pump continues to operate under faulty conditions, especially if cavitation or abrasive particles are present in the fluid, further aggravating the damage. Thus, our experiment used a controlled defect size for repeatability, but the same methodology can be applied to varying sizes of impeller damage. Indeed, changing the size of the notch will change the amplitude of the peaks, but the frequency range will remain constant. The defect notch was created specifically on the impeller shroud, not the diffuser or volute. Regarding the dimensions, the notch was a triangular cut with approximately 10 mm sides and a 15 mm height, located at the impeller outlet where the tension is highest. The notch was positioned facing the shroud. Figure 1 depicts a test bench with both healthy and defective impellers, while Figure 2 shows a schematic representation of the test bench. "AS" in Figure 2 refers to the accelerometer sensor used to measure vibration signals.

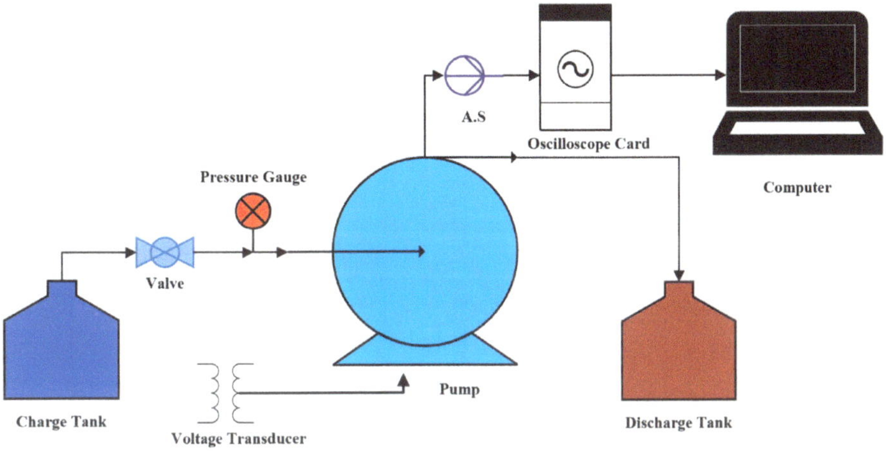

Figure 2. Schematic diagram of the experimental setup.

A single-phase induction motor is directly coupled to the pump and drives it. For changing the rotational speed of the motor, a voltage converter was used, and the speed was measured with a portable tachometer. The pump speed ranges from 500 to 2880 r/min. The tank that supplies water for the pump was positioned to ensure sufficient head at the inlet and to prevent any inherent cavitation. Vibration signals were acquired by the accelerometer sensor and transferred to the OC and computer to save the data. The sampling frequency of the OC was 100,000 samples per second. When the pump is operating, the procedure consists of gradually closing the valve on the inlet pipe to create cavitation until the bubbles are visible through the transparent suction pipe. The cavitation bubbles were visually detected through the transparent suction pipe, and no additional optical devices were used for this purpose. The transparent pipe provided sufficient visibility for the manual observation of cavitation onset and progression during the experiments. The procedure was first carried out with a healthy impeller and then with the damaged one.

2.2. Multi-Layer Perceptrons

An artificial neural network (ANN) is characterized by its internal connectivity and associative connections used to process information. In most cases, an ANN is a flexible system that adjusts its structure in response to input from either the outside or inside domain [26,27]. One of the main types of ANN is the feedforward neural network (FNN), characterized by the unidirectional flow of information between its layers. A significant sub-branch of the FNN is the multi-layer perceptron (MLP). MLP neurons typically use nonlinear activation functions, allowing the network to identify complex patterns in the data. An output signal is generated by every unit (neuron), and this signal is a function of the sum of its inputs. The output as a function of the input and weights is expressed as:

$$y_i = f\left(\sum x_i w_i\right) \quad (1)$$

The activation function can be any function; however, sigmoid functions (also known as the hyperbolic tangent) are most frequently employed. An MLP consists of layers stacked one after another, with varying numbers of processing units in each layer. The units in the first layer are fully connected to the units in the hidden layer and receive input from the external environment [27]. Meanwhile, the units in the hidden layer are fully connected to the units in the output layer, and the units in the output layer produce the MLP's output, as shown in Figure 3.

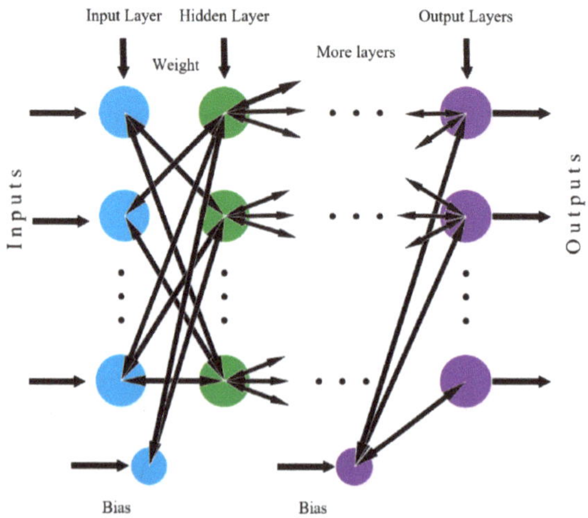

Figure 3. Architecture of the MLP network.

2.2.1. Training Method

To achieve a desired task, an MLP must first be trained. This means that the network's connection weight values need to be defined so that it can produce the correct output for each input pattern. A training algorithm calculates these appropriate weights. There are numerous training methods and their variations. For instance, the backpropagation algorithm is a commonly used method where the error is propagated backward through the network to iteratively adjust the weights. This process involves passing an input forward through the network to obtain an output, calculating the error by comparing this output to the expected result, and then updating the weights to minimize this error over many iterations. It should be noted that another goal of training a neural network is to achieve a good generalization ability rather than merely memorizing the training set. In other words, the network should generate accurate output values for inputs that were not seen during training [28]. During training, the early stopping technique [29–31] is employed to enhance the network's generalization performance and prevent overtraining. This method involves selecting a validation set that is distinct from the training set. The validation error serves as the stopping condition during the training process. In this study, early stopping criteria are implemented to achieve optimal performance. The training algorithm adopted in this study is the Levenberg–Marquardt algorithm.

2.2.2. Levenberg–Marquart Algorithm

The Levenberg–Marquardt (LM) algorithm is an optimization method that effectively combines the gradient descent and Gauss–Newton algorithms to solve nonlinear least squares problems. It aims to minimize the sum of squared errors between predicted and actual outputs. The algorithm updates parameters iteratively using the following formula:

$$P_{k+1} = P_k - \left(J_k^T J_k + \lambda I\right)^{-1} J_k^T e_k \tag{2}$$

where J_k is the Jacobian matrix of partial derivatives, e_k is the error vector, λ is a damping factor, and I is the identity matrix. The damping factor λ adjusts the blend between gradient descent and the Gauss–Newton method. For large λ, the algorithm behaves like gradient descent, ensuring stability, while for small λ, it behaves like the Gauss–Newton method, ensuring fast convergence. This adaptive approach makes the LM algorithm particularly suitable for training neural networks and curve fitting, providing a balance between efficiency and robustness, although it can be memory-intensive due to the computation of the Jacobian matrix [32].

2.3. *Statical Parameters as Input Data*

Statistical parameters like mean and variance are primarily used for data normalization and understanding the spread of data. Skewness and kurtosis help in understanding the asymmetry and tailenders of data, guiding data transformation processes, while root mean square (RMS) provides an accurate measure of the magnitude of varying quantities, making it essential for assessing prediction errors and fluctuations in data. The accuracy and reliability of an MLP can be significantly improved by ensuring that the data input into the model is well-prepared by utilizing these statistical properties [33].

In the following, statistical parameters like mean, kurtosis, variance, skewness, and RMS are described.

Calculating the average of measured data is done by:

$$\overline{Z} = \frac{Z_1 + Z_2 + \ldots Z_n}{n} \tag{3}$$

where $\overline{Z}$ is the mean value, $Z_1, Z_2, \ldots Z_n$ are measured data at each time, and n is the time of data recorded.

Variance is the parameter used to measure the distribution and dispersion of data around their average value:

$$Var_x = \frac{1}{N}\left[\sum_{i=1}^{N}(x_i - \overline{x})^2\right] \tag{4}$$

where N is the number of samples, x is the time signal sample, and $\overline{x}$ is the mean value.

Standard deviation is the amount of variation or dispersion in a set of data values. It indicates how much individual data points differ from the mean (average) of the dataset.

$$\sigma = \sqrt{Var_x} = \sqrt{\frac{1}{N}\left[\sum_{i=1}^{N}(x_i - \overline{x})^2\right]} \tag{5}$$

Skewness measures the distortion or asymmetry of a signal, which can be visually interpreted by examining the signal's distributional shape, and is given as:

$$\chi = \frac{\sum_{i=1}^{N}(x_i - \overline{x})^3}{N\sigma^3} \tag{6}$$

where $\overline{x}$ is the mean value, and σ is the standard deviation of the data [33].

Kurtosis is defined as the fourth-order moment of data and shows the degree of a peak in a set of data [33]. In other words, kurtosis is an index to detect larger peaks among the data, or it can be said that it determines the shape of a data distribution and is given as:

$$K = \frac{N\sum_{i=1}^{N}(x_i - \overline{x})^4}{\left[\sum_{i=1}^{N}(x_i - \overline{x})^2\right]^2} \tag{7}$$

Our interpretation of the time-domain data also included the use of the root mean square (RMS) parameter. The RMS value is a statistical measure of the magnitude of a varying quantity. It is especially useful in contexts where the values can be both positive and negative [33]. RMS is given as:

$$\mathrm{RMS_x} = \sqrt{\frac{1}{\mathrm{N}}\left[\sum_{\mathrm{i}=1}^{\mathrm{N}} \mathrm{x_i}^2\right]} \quad (8)$$

where x is the time signal sample, N is the number of samples, and i is the sample index [33].

The statistical convergence of skewness and kurtosis was ensured by employing a sufficiently large sample size during data collection and analysis. Additionally, to measure the stability of the statistical assessment, a test was conducted across multiple trials to confirm their reliability.

2.4. Training Procedure

The training procedure for an MLP network involves feeding the network with input data and corresponding target outputs and then adjusting the network's weights and biases through backpropagation to minimize the error between predicted and actual values. The neural network employed here consists of three layers: the input, middle, and output layers. The input layer has neurons that represent the normalized features of the extracted vibration signals. To minimize the mean square error or functional function between the network's output and the target, networks are alternately trained. The application also automatically generates the network's initial weights and biases. This is the reason why the network has undergone several iterations of training. Only 1 and 0, which denote the healthy and faulty states of the pump, respectively, can be the target values for the two output neurons. However, since the combination of the intermediate and output transfer functions is considered a mix of hyperbolic and linear transfer functions, the neural network's output can range from zero to one and may even exceed one. Extracted statistical parameters from the time domain were collected in the form of a 5 × 20 matrix. The number 5 represents the statistical features used, and 20 is equivalent to the data in four different functional states and tested at five rotational speeds (500, 1000, 2000, 2500, and 2880 rpm). From the 5 × 20 matrix, a 5 × 16 matrix was used for training the algorithm, while a 5 × 4 matrix was reserved for validation to determine if the algorithm can correctly anticipate faults. The target matrix of the neural network was chosen as 2 × 1. The data used for training and testing the ANN is shown in Table 3.

Table 3. Input data of neural network.

$$\begin{bmatrix} \mathrm{M, h500} & & \mathrm{M, h2880} & \mathrm{M, hc500} & & \mathrm{M, hc2880} & \mathrm{M, d500} & & \mathrm{M, d2880} & \mathrm{M, dc, 500} & & \mathrm{M, dc, 2880} \\ \mathrm{K, h500} & & \mathrm{K, h2880} & \mathrm{K, hc500} & & \mathrm{S, hc2880} & \mathrm{K, d500} & & \mathrm{S, d2880} & \mathrm{K, dc, 500} & & \mathrm{K, dc, 2880} \\ \mathrm{S, h500} & \cdots & \mathrm{S, h2880} & \mathrm{S, hc500} & \cdots & \mathrm{S, hc2880} & \mathrm{S, d500} & \cdots & \mathrm{S, d2880} & \mathrm{S, dc, 500} & \cdots & \mathrm{S, dc, 2880} \\ \mathrm{V, h500} & & \mathrm{V, h2880} & \mathrm{V, hc500} & & \mathrm{V, hc2880} & \mathrm{V, d500} & & \mathrm{V, d2880} & \mathrm{V, dc, 500} & & \mathrm{V, dc, 2880} \\ \mathrm{RMS, h500} & & \mathrm{RMS, h2880} & \mathrm{RMS, hc500} & & \mathrm{RMS, hc2880} & \mathrm{RMS, d500} & & \mathrm{RMS, d2880} & \mathrm{RMS, dc, 500} & & \mathrm{RMS, dc, 2880} \end{bmatrix}$$

In this matrix, M is the mean, K is the kurtosis, S is the skewness, V stands for the variance, and RMS shows the root mean square value. Also, h represents healthy status, hc is for health with a cavitation status, d is for the defect status, and dc is for a defect with cavitation status. The numbers after the letters show the rotational speed. For example, M, hc2880 means the mean value of the health with a cavitation status at 2880 rpm.

There are ones and zeros in the target matrix. In this selected target matrix, if the output of the neural network is closer to the value of one, it is a sign of the health of the pump, and if it is closer to the value of zero, it indicates the presence of a defect.

The flowchart of the process for training ANNs is shown in Figure 4.

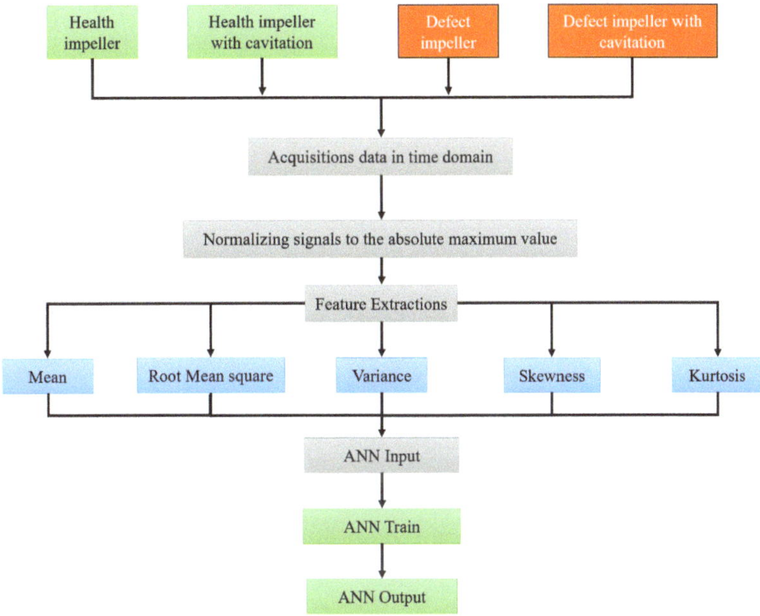

Figure 4. Flowchart of the neural network training process.

3. Results and Discussion

Time-domain analysis was conducted, and graphs were generated using the data obtained from the accelerometer. In the next step, statistical factors such as the mean, kurtosis, skewness, variance, and RMS were extracted from the normalized time signals. Subsequently, frequency-domain analysis was performed using Fourier transform to diagnose and monitor the faults. Finally, an MLP neural network was employed to predict the defects based on the statistical data.

3.1. Time-Domain Analysis

The time domain displays the range value of signals and vibrations measured by the accelerometer at moments during sampling. Basic troubleshooting techniques were performed to analyze the signals in the time domain. Since the maximum input frequency is 48 Hz (the maximum rotational speed of the pump is 2880 rpm), the periodicity of the signals is 0.0208 s. The interval of the horizontal axis can be selected from 0 to 208 samples, and since the sampling rate is equal to 10,000 samples per second, this interval shows the periodicity of the signal with the input frequency. However, due to the nature of the randomness of the signal, this interval was chosen from 0 to 416. It should be noted that the value of the data collection frequency has been chosen to prevent the occurrence of an aliasing phenomenon. It is important to make sure that the sampling rate for transient monitoring is high enough to capture the parameters of the system dynamics. Data were normalized to fix the data range between −1 and 1 based on this formula:

$$\widetilde{x} = 2 \times \left(\frac{x - \min(x)}{\max(x) - \min(x)} \right) - 1 \tag{9}$$

In this formula, x is the original value, $\widetilde{x}$ is normalized value, and min(x) and max(x) are the minimum and maximum values in the dataset, respectively. The normalized graphs in the time domain for different states of the pump at 2880 rpm are shown in Figure 5.

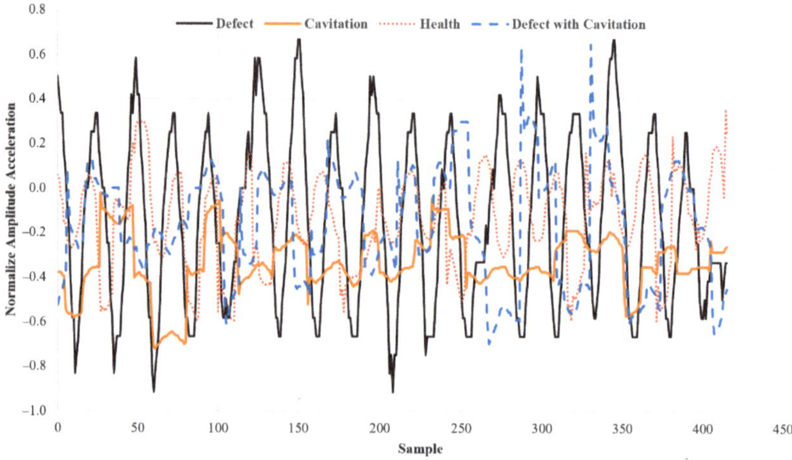

Figure 5. Time domain diagrams for 4 different statuses of the pump (2880 rpm).

The negative values in Figure 5 indicate that the measured vibration signals exhibit a shift in their amplitude relative to the baseline level. This could suggest that the vibrations are oscillating around a mean value, which may occur due to factors like noise in the system or the inherent characteristics of the signal processing method used. As it is well known, detecting pump defects using this diagram is challenging. Consequently, variations in the instantaneous time signals do not provide a comprehensive understanding of identifying the pump defects. Therefore, further analysis is needed to analyze the signals in the time domain and detect the defects using various statistical features. For better diagnosis and analysis of the time-domain graphs, statistical parameters such as the mean, variance, and kurtosis could be extracted. The mean value is presented in Figure 6.

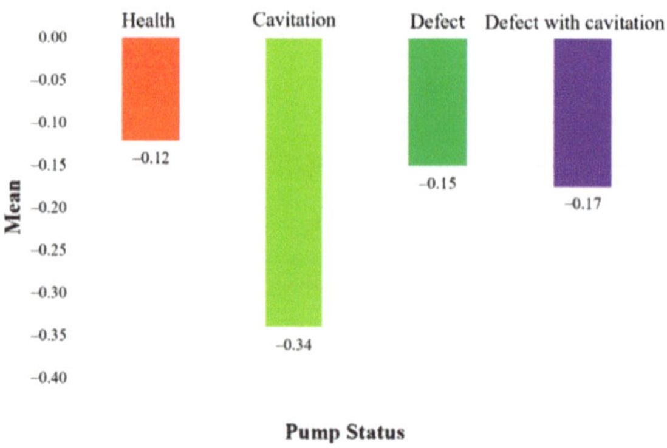

Figure 6. Mean values for different pump statuses.

This analysis shows that the average signal level varies across different pump states, with the cavitation state showing the most significant decrease, followed by the defect with cavitation state, the defect state, and finally, the healthy state with the smallest decrease.

The kurtosis diagram for the extracted data for the 2880 rotational speed is shown in Figure 7.

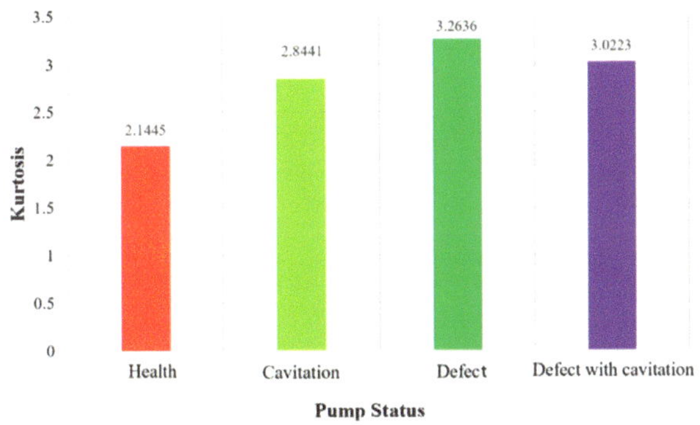

Figure 7. Kurtosis values for different pump statuses.

The kurtosis value indicates a distribution with higher peaks and heavier tails than a normal distribution. The kurtosis value for the cavitation state is 2.84; this kurtosis value is more than that of the healthy state, indicating that the cavitation state has a higher peak and heavier tails compared to the healthy state. The kurtosis value for the defect state is 3.26; this value shows that the defect state has the highest kurtosis value, suggesting a spike distribution with heavier tails compared to all other states. This analysis shows that the defect with and without cavitation states exhibit higher peaks and heavier tails in their distributions, indicating more extreme values. In contrast, the healthy state has a flatter distribution, suggesting fewer extreme variations.

The unequal distribution of a signal about its mean value is measured by its skewness. Therefore, in the next step, the value of the signal's skewness was calculated, as shown in Figure 8.

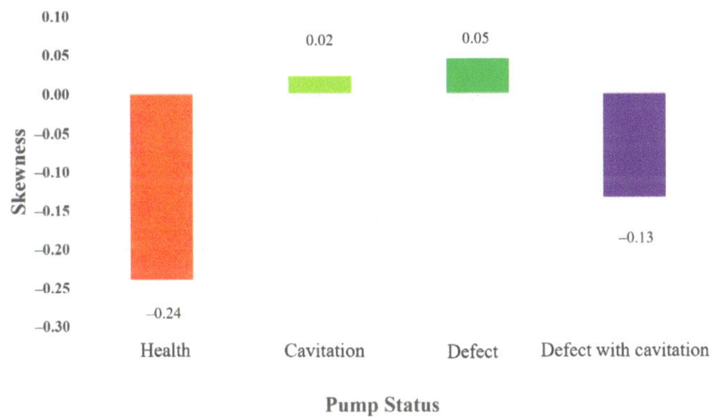

Figure 8. Skewness values for different pump statuses.

The negative skewness value indicates that the signal for the healthy pump status shows a significant left skew distribution, suggesting that the data points are spread out more on the left side of the mean. A skewness value close to zero indicates that the signal distribution for the cavitation state is almost symmetrical. A slightly positive skewness value indicates that the signal for the defect state has a longer tail on the right side of the distribution, suggesting that the data points are spread out more on the right side of the mean. The negative skewness value indicates that the signal for the defect with cavitation state has a longer tail on the left side of the distribution, like the healthy state, but less than that. This analysis suggests that the skewness of the signal varies depending on the pump status, with the healthy and defect with cavitation states showing left-skewed distributions and the cavitation and defect states showing near-symmetrical or slightly right-skewed distributions. Figure 9 shows the RMS values for 2880 rpm.

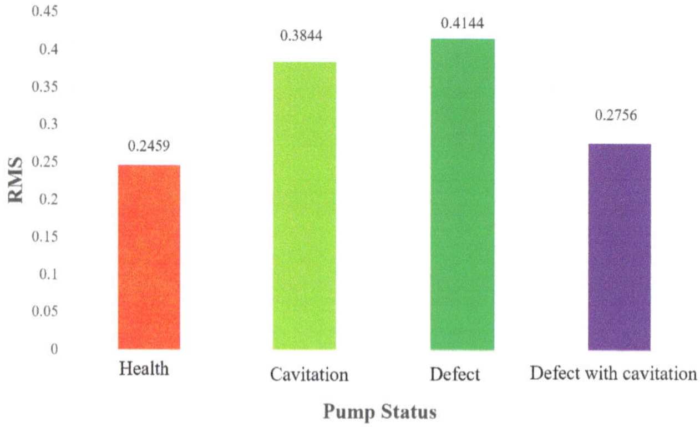

Figure 9. RMS values for different pump statuses.

The RMS value for the healthy pump status indicates a relatively lower value of the signal compared to the other states. The defect with a cavitation state has a higher RMS value. The RMS values in the chart provide a clear indication of the pump's condition. Lower RMS values suggest stable and healthy operation, while higher RMS values indicate increased signal power or intensity due to cavitation or defects.

Figure 10 shows the variance values for the obtained data.

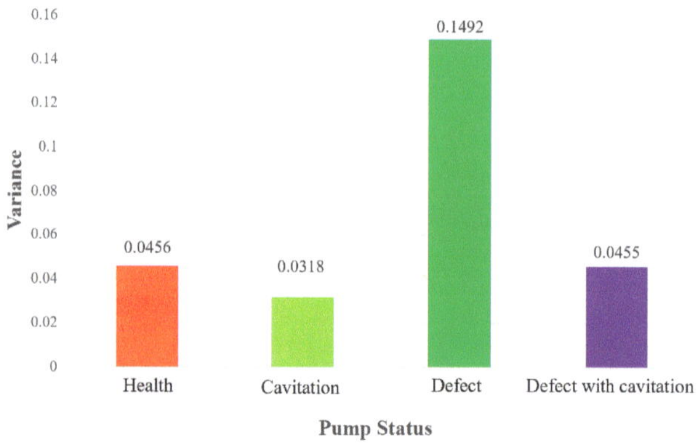

Figure 10. Variance diagram.

The variance value for the healthy pump status indicates a moderate level of spread in the signal values. The cavitation state shows a lower variance value, indicating less spread in the signal values compared to the healthy state. The defect state has the highest variance value, indicating a significant level of spread in the signal values. The defect with the cavitation state has a variance value like the healthy state, indicating a moderate level of spread in the signal values. This analysis suggests that the diffusion of signal values varies across different pump states, with the defect state showing the most significant variation, followed by the healthy and defect with cavitation states, and finally, the cavitation state with the least variation.

In general, the results of time-domain analysis are challenging, revealing that different pump states exhibit distinct statistical features. Specifically, the defect state shows the highest kurtosis and variance, indicating extreme and widely spread values, while the healthy state maintains the lowest RMS and moderate variance, suggesting stable operation. These variations underscore the necessity of using multiple statistical parameters to effectively diagnose pump conditions. While time-domain analysis alone may not provide comprehensive insights into complex signal patterns, the statistical parameters calculated in this domain can be used to train artificial intelligence networks, such as ANNs, to establish suitable criteria for fault diagnosis in devices. It should be noted that one of the effective methods for fault diagnosis is through frequency-domain analysis. The next section will explore the analysis of data in the frequency domain.

3.2. Frequency Domain Analysis

In the frequency domain, the range of signals is shown in terms of amplitude and frequency. Therefore, with the help of fast Fourier transform, time-domain signals can be converted into a frequency spectrum. Fourier transform is given as:

$$F(\omega) = \int_{-\infty}^{+\infty} f(t)e^{-j\omega t}dt \tag{10}$$

where f(t) is the time-domain data and F(ω) is Fourier transforms f(t) in the time domain.

Figure 11 shows the frequency spectrum of the installed sensor in a healthy state for a rotational speed of 2880 rpm. In this figure, the horizontal axis shows the frequency range in Hz, and the vertical axis shows the amplitude of the signal in decibel voltage (dbv).

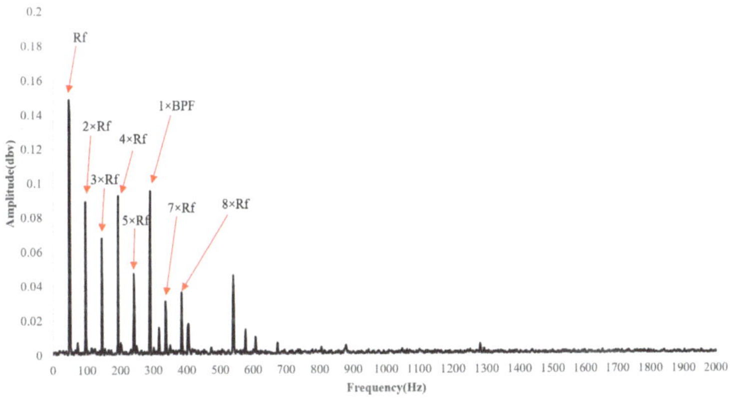

Figure 11. Frequency spectrum of the pump in a healthy state.

Pressure fluctuations can be detected at discrete frequencies that are multiples of the rotation frequency and the number of blades. These frequencies are also called blade passage frequencies (BPF). They are generated by the interaction of the rotating blades with a stationary component. Each blade passing a fixed point produces a distinct frequency, known as the blade pass frequency:

$$BPF = N \times Rf \tag{11}$$

In this equation, N is the number of blades, and Rf is the rotational speed in revolutions per second.

The amplitude of such pressure fluctuations depends on the number of blades, diffuser design parameters, and operational parameters. Figure 12 shows the periodicity in the centrifugal pump. It can be observed that two main dominant frequencies are present in the frequency range; the dominant frequency in this range is associated with the shaft rotating frequency (Rf), the blade passing frequency (BPF), and their harmonics. These effects result from the interaction between the impeller blades and the pumped flow, as well as the significant interaction between the impeller blades and the stationary components. Blade passage frequencies appear at the lower end of the frequency spectrum. However, they are not evident for a specific number of blades in most spectra, which can be attributed to the unique vibration characteristics and design of the pump. Blade passage frequencies typically appear at the lower end of the spectrum, as they correspond to the fundamental mechanical vibrations of the rotating blades. Figure 12 shows the frequency spectrum of the pump in the state of cavitation.

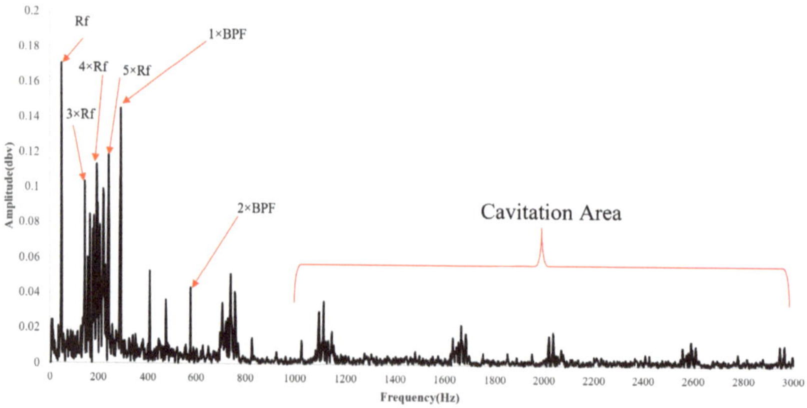

Figure 12. Frequency spectrum of the pump in a cavitation state.

From Figure 12, the first rotational frequency occurs at 48 Hz with an amplitude of 0.17 dbv. The first harmonic amplitude in healthy status was 0.148, so the amplitude of harmonics increased in this status. The amplitude of the first and second BPF in this status increased, and the amplitude of the other harmonics increased. In fact, more bubbles form and burst in the pump as cavitation progresses, which causes the pump to vibrate more intensely, leading to increased fluctuations in amplitude. Cavitation occurs in the high-frequency range and can be seen; the frequency peaks in the high-frequency range increase with the start of cavitation. In Figure 13, the frequency spectrum for the defective impeller is shown.

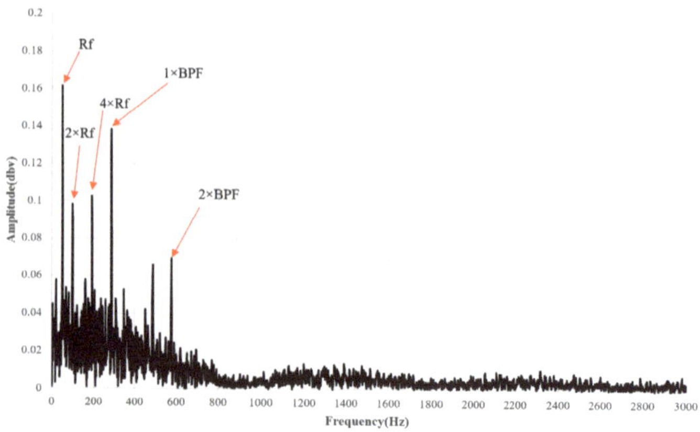

Figure 13. Frequency spectrum of the pump under a faulty impeller condition.

In Figure 13, the amplitude of $1\times$Rf and $2\times$Rf, compared to the healthy status, has increased. The BPF also increased in this status from 0.092 to 0.138 dbv. Additionally, the frequency range of the other harmonics has increased. A defective impeller introduces additional mechanical vibrations and irregularities in the flow, and it is clear that many other frequencies are produced by various sources, including the driving motor and the bearing misalignment, as observed. It can be concluded that in a pump with a defective impeller, the maximum peak corresponding to the Rf and blade passing frequencies increases.

Figure 14 shows the frequency spectrum for the state of the defective impeller with cavitation. In this spectrum, the frequency harmonic is shown, along with the frequency range of cavitation.

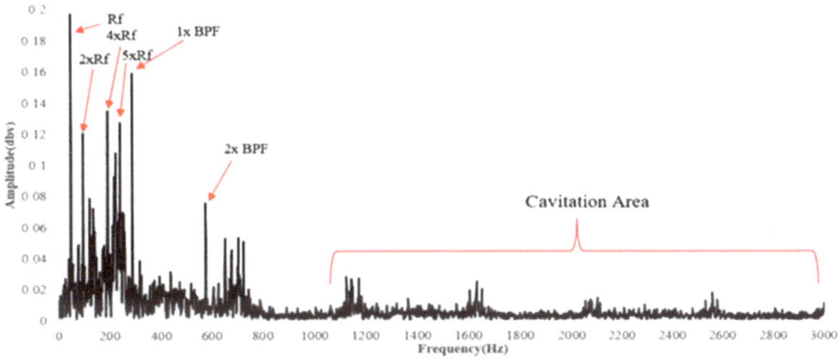

Figure 14. Frequency spectrum of the pump with cavitation-induced impeller damage.

In Figure 14, the amplitudes of the harmonics of 1× Rf, 2× Rf, 4× Rf, and 5× Rf have increased, and these harmonics have higher amplitudes compared to the cavitation and defect statuses. The BPF also increases compared to the defective and cavitation statuses. The combined effects of cavitation and a defective impeller significantly amplify across all of the vibration levels. The interaction between impeller defects and cavitation bubbles generates complex vibration patterns, leading to increased amplitude in both the low-frequency and high-frequency ranges. In fact, defects in the impeller cause turbulence, which alters the pressure contours on both the impeller and the pump body. This results in an increase in amplitude in the frequency domain.

3.3. Fault Diagnosis Using ANN

While the analysis of vibration signals in the frequency domain is effective for detecting faults in pumps, interpreting these signals accurately requires a comprehensive understanding of signal processing techniques and the operational dynamics of pumps. This interpretation becomes even more complex when considering signals across multiple domains, such as time, frequency, and time-frequency, which demand experience and expertise. To address the challenge of accurately diagnosing pump faults, an automatic, fast, and reliable troubleshooting method has been developed. ANNs can analyze vast amounts of vibration data, learn from patterns, and provide predictive maintenance insights, thereby reducing downtime and maintenance costs. By automating the signal interpretation process, these AI-based methods enhance the efficiency and reliability of pump condition monitoring, making it accessible, even to those with limited expertise in signal processing.

After trying different layers and making many try-and-error processes, the number of hidden layers was finally chosen to be ten layers. A linear transfer function was used for the output layer, and a hyperbolic tangent transfer function was used for the intermediate layer to spread the data regarding zeros and ones. To predict the output value from the primary matrix, which is a 5 × 20 matrix, a 5 × 16 matrix was assigned for training the neural network, and a 5 × 4 matrix that included the five statistical parameters for four states of the pump at specific speeds was assigned as a test matrix. The first column of this matrix includes a healthy impeller with cavitation at a rotational speed of 1000 rpm; the second column includes a defective impeller at a rotational speed of 2000 rpm; the third column includes a defective impeller with cavitation at a rotational speed of 2500 rpm; and the fourth column includes a healthy impeller at a rotational speed of 2880 rpm.

Figure 15 shows the regression percentages for eleven iterations of the neural network.

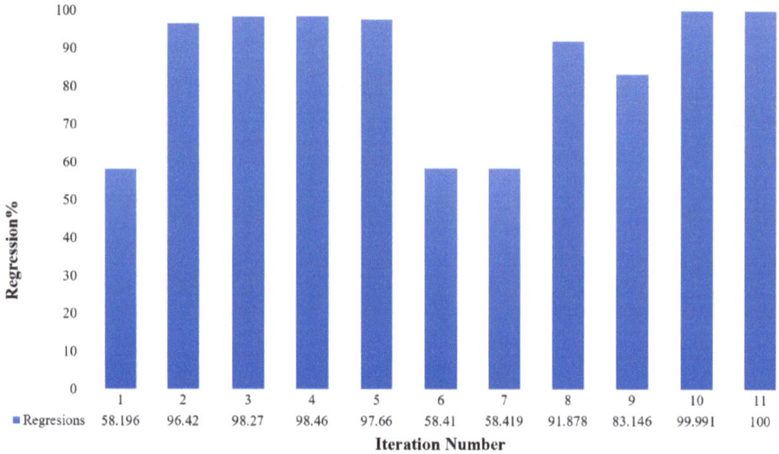

Figure 15. Percent accuracy of the trained network.

In Figure 15, the best regression results are observed for iterations 10 and 11. To ensure the accuracy of network prediction, the columns have been randomly extracted at different speeds. In Figure 16, the regression percentage is shown for seven and eleven repetitions.

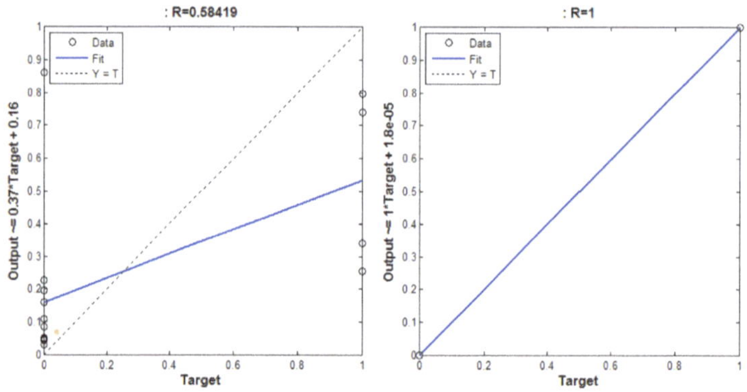

(a) (b)

Figure 16. Regression graph: (**a**) seventh iteration and (**b**) eleventh iteration.

The charts represent the performance of an ANN in predicting target values. The horizontal axis shows the target values (T), while the vertical axis shows the output values produced by the ANN (Y). The open circular symbols in Figure 16 represent the target values during the neural network's training process. These symbols indicate the expected outputs for the various states of the pump being analyzed. This value represents the correlation coefficient, indicating the strength and direction of the linear relationship between the target and output values. An R-value of 0.58419 suggests a moderate positive correlation. The fit line equation is given as Output $\approx 0.37 \times$ Target + 0.16. This indicates that the ANN's output is only about 37% of the target value plus a small constant offset (0.16). The slope of 0.37 suggests the ANN is under-predicting the target values. In this figure, the dashed line represents the ideal scenario where the output perfectly matches the target (Output = Target). As is clear from Figure 15, the regression percentage for the seventh iteration is 58.419%, and for the eleventh iteration, it is 100%. The eleventh iteration has the best performance, so it has been used to train data and predict faults in the network simulation part. The performance of the ANN based on the mean squared error is shown in Figure 17.

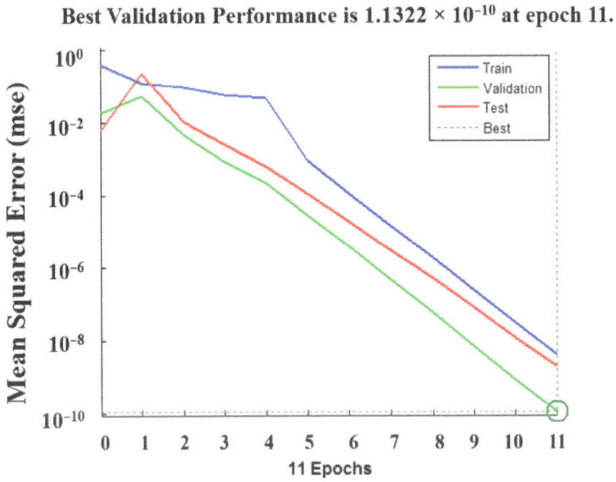

Figure 17. Mean square error for the training artificial neural networks.

In Figure 17, the horizontal axis (epochs) represents the number of training epochs, which is the number of times the learning algorithm has processed the entire training dataset. The mean squared error (MSE) in the vertical axis represents a measure of the difference between the predicted and actual values. A lower MSE indicates better model performance. In this chart, the blue line represents the MSE for the training dataset over the epochs, the green line represents the MSE for the validation dataset over the epochs, the red line (test) represents the MSE for the test dataset over the epochs, and the dashed green line represents the best validation performance achieved during the training process. The chart demonstrates effective training of the ANN, with the MSE decreasing consistently across training, validation, and test datasets. The model shows initial overfitting but quickly improves, achieving optimal performance by epoch eleven. As it is clear, the best validation performance is 1.1322×10^{-10} in epoch eleven. This interpretation indicates that the ANN training process is highly effective, with the model achieving excellent performance and minimal error by the final epoch.

To accurately predict faults in the pump, a neural network model was employed, consisting of three layers, ten neurons per layer, and trained over eleven iterations. The results of this neural network's prediction were then calculated. Subsequently, the final prediction matrix, sized 5×4, as mentioned earlier, was input into the algorithm to analyze and validate the outcomes. This approach not only ensured precise fault detection but also demonstrated the robustness and efficiency of using neural networks for predictive maintenance in pump systems. The results are shown in Figure 18.

In this diagram, the first column is a number close to zero, which indicates that the pump is defective (a healthy impeller with cavitation). The second column is a number close to zero, which indicates that the pump is faulty (a defective impeller). The third column is a number close to zero, which indicates that the pump is defective (a defective impeller with cavitation), but the fourth column is a number close to one, which indicates that the pump is healthy. As is evident, the neural network used has worked well. Therefore, it can be said that the designed neural network is successful in detecting the fault.

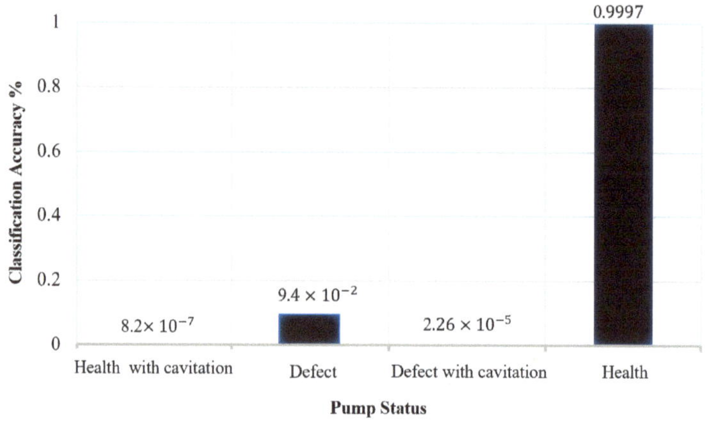

Figure 18. The result prediction of ANN.

4. Conclusions

In this study, the vibration analysis of a CP with a healthy and defective impeller, with and without cavitation, was performed at different rotational speeds in two domains of time and frequency. The data obtained from the time domain were used as input for the neural network to predict faults in the pump. The results prove that the data obtained from the normalized time domain do not provide interpretable information. Therefore, statistical factors such as the kurtosis, mean, and variance were calculated using these data to interpret the results. The calculated data reveal that the results are difficult to interpret and that, in certain situations, different pump statuses exhibit diverse statistical patterns. To proceed, the FFT diagrams of the frequency domain were used to identify the defect, and it was shown that all defects in the CP can be detected and diagnosed using these diagrams. Finally, the results of this research demonstrate that the MLP properties of vibration and current signals could indicate pump function under faulty conditions. The application of the MLP algorithm in this study has demonstrated its superiority. The key advantage of using MLP is its ability to process and learn from large volumes of time-domain data and automatically extract meaningful patterns. This reduces dependency on expert knowledge for signal interpretation and allows for real-time monitoring and diagnosis. The MLP's robustness in handling non-stationary and transient signals further emphasizes its potential as a powerful tool for predictive maintenance, enabling more accurate and timely interventions to prevent pump failures. To make the model more sensitive to the type of issue, we will attempt to investigate this in subsequent research. The current work's next phase is concentrated on creating an MLP that can classify all the defects into different groups. To enhance the algorithm's capability to differentiate between various faulty states, we plan to expand our training dataset to include a broader range of defect types. This can be achieved by modifying the ANN architecture to support multi-class output that allows the model to recognize and differentiate among multiple faulty conditions. The proposed method can also be used for frequency-domain analysis to investigate pump fault detection. This study focused on a specific notch defect in the impeller. Next, research could also focus on a wider range of defects, such as blade cracks, impeller pitting, and other mechanical failures, or changing the dimensions of defect on the impeller to assess how well the ANN generalizes to different fault conditions.

Author Contributions: Conceptualization, M.H.G. and M.K.; Writing—original draft preparation, M.R. and P.C.; Writing—review and editing, M.K. and R.F.; Investigation, M.H.G.; Visualization, M.R. and P.C.; Supervision, M.K. All authors have read and agreed to the published version of the manuscript.

Funding: This research received no external funding.

Institutional Review Board Statement: Not applicable.

Informed Consent Statement: Not applicable.

Data Availability Statement: The original contributions presented in the study are included in the article, further inquiries can be directed to the corresponding author.

Conflicts of Interest: The authors declare no conflicts of interest.

Nomenclature

ANN	Artificial neural network
BPF	Blade passage frequency
CP	Centrifugal pump
FNN	Feedforward neural network
GA	Genetic algorithm
LM	Levenberg–Marquardt
MLP	Multi-layer perceptron
OC	Oscilloscope card
PNN	Partially linearized neural network
RMS	Root mean square
SVM	Support vector machine
RF	shaft rotating frequency
WA	Wavelet analysis

References

1. Gülich, J.F. Pump Types and Performance Data. In *Centrifugal Pumps*; Springer: Cham, Switzerland, 2014; pp. 43–78.
2. Muralidharan, V.; Sugumaran, V.; Indira, V. Fault diagnosis of monoblock centrifugal pump using svm. *Eng. Sci. Technol. Int. J.* **2014**, *17*, 152–157. [CrossRef]
3. Muralidharan, V.; Sugumaran, V. A comparative study of naïve bayes classifier and bayes net classifier for fault diagnosis of monoblock centrifugal pump using wavelet analysis. *Appl. Soft Comput.* **2012**, *12*, 2023–2029. [CrossRef]
4. Flett, J.; Bone, G.M. Fault detection and diagnosis of diesel engine valve trains. *Mech. Syst. Signal Process.* **2016**, *72–73*, 316–327. [CrossRef]
5. Moosavian, A.; Najafi, G.; Ghobadian, B.; Mirsalim, M.; Jafari, S.M.; Sharghi, P. Piston scuffing fault and its identification in an ic engine by vibration analysis. *Appl. Acoust.* **2016**, *102*, 40–48. [CrossRef]
6. Dolenc, B.; Boškoski, P.; Juričić, Đ. Distributed bearing fault diagnosis based on vibration analysis. *Mech. Syst. Signal Process.* **2016**, *66–67*, 521–532. [CrossRef]
7. Orhan, S.; Aktürk, N.; Çelik, V. Vibration monitoring for defect diagnosis of rolling element bearings as a predictive maintenance tool: Comprehensive case studies. *NDT E Int.* **2006**, *39*, 293–298. [CrossRef]
8. Rizal, M.; Ghani, J.A.; Nuawi, M.Z.; Haron, C.H.C. Online tool wear prediction system in the turning process using an adaptive neuro-fuzzy inference system. *Appl. Soft Comput.* **2013**, *13*, 1960–1968. [CrossRef]
9. Xiang, J.; Zhong, Y.; Gao, H. Rolling element bearing fault detection using ppca and spectral kurtosis. *Measurement* **2015**, *75*, 180–191. [CrossRef]
10. Gan, M.; Wang, C.; Zhu, C. Multiple-domain manifold for feature extraction in machinery fault diagnosis. *Measurement* **2015**, *75*, 76–91. [CrossRef]
11. Mousmoulis, G.; Karlsen-Davies, N.; Aggidis, G.; Anagnostopoulos, I.; Papantonis, D. Experimental Analysis of Cavitation in a Centrifugal Pump Using Acoustic Emission, Vibration Measurements, and Flow Visualization. *Eur. J. Mech.-B/Fluids* **2019**, *75*, 300–311. [CrossRef]
12. Al-Obaidi, A.R. Investigation of effect of pump rotational speed on performance and detection of cavitation within a centrifugal pump using vibration analysis. *Heliyon* **2019**, *5*, e01989. [CrossRef] [PubMed]
13. Sun, H.; Luo, Y.; Yuan, S.; Yin, J. Hilbert spectrum analysis of unsteady characteristics in centrifugal pump operation under cavitation status. *Ann. Nucl. Energy* **2018**, *114*, 607–615. [CrossRef]
14. Murovec, J.; Čurović, L.; Novaković, T.; Prezelj, J. Psychoacoustic approach for cavitation detection in centrifugal pumps. *Appl. Acoust.* **2020**, *165*, 107323. [CrossRef]
15. Kumar, A.; Kumar, R. Time-frequency analysis and support vector machine in automatic detection of defect from vibration signal of centrifugal pump. *Measurement* **2017**, *108*, 119–133. [CrossRef]
16. Azizi, R.; Attaran, B.; Hajnayeb, A.; Ghanbarzadeh, A.; Changizian, M. Improving accuracy of cavitation severity detection in centrifugal pumps using a hybrid feature selection technique. *Measurement* **2017**, *108*, 9–17. [CrossRef]

17. Birajdar, R.; Patil, R.; Khanzode, K. Vibration and Noise in Centrifugal Pumps: Sources and Diagnosis Methods. In Proceedings of the 3rd International Conference on Integrity, Reliability and Failure, Porto, Portugal, 20–24 July 2009; Volume 81, pp. 2631–2644.
18. Sakthivel, N.R.; Nair, B.; Sugumaran, V. Soft computing approach to fault diagnosis of centrifugal pump. *Appl. Soft Comput.* **2012**, *12*, 1574–1581. [CrossRef]
19. Askari, A.; Alekathir, J.; Mator, F.; Rezazade, A. Troubleshooting and diagnosis of cavitation in centrifugal pump using vibration analysis and providing solution to fix it. In Proceedings of the Second Technical Conference of Monitoring and Troubleshooting, Tehran, Iran, 24 February 2008. (In Persian).
20. Al-Braik, A.; Hamomd, O.; Gu, F.; Ball, A. Diagnosis of impeller faults in a centrifugal pump using vibration signals. *J. Sound Vib.* **2014**, *78*, 695–712.
21. Espadafor, F.J.; Villanueva, J.B.; García, M.T.; Trujillo, E.C. Experimental and dynamic system simulation and optimization of a centrifugal pump-coupling-engine system. part 1: Failure identification. *Eng. Fail. Anal.* **2011**, *18*, 1–11. [CrossRef]
22. Wang, H.; Chen, P. Intelligent diagnosis method for a centrifugal pump using features of vibration signals. *Neural Comput. Appl.* **2009**, *18*, 397–405. [CrossRef]
23. Barrio, R.; Fernandez, J.; Blanco, E.; Parrondo, J. Estimation of radial load in centrifugal pumps using computational fluid dynamics. *Eur. J. Mech.-B/Fluids* **2011**, *30*, 316–324. [CrossRef]
24. Casoli, P.; Pastori, M.; Scolari, F.; Rundo, M. A vibration signal-based method for fault identification and classification in hydraulic axial piston pumps. *Energies* **2019**, *12*, 953. [CrossRef]
25. Casoli, P.; Bedotti, A.; Campanini, F.; Pastori, M. A methodology based on cyclostationary analysis for fault detection of hydraulic axial piston pumps. *Energies* **2018**, *11*, 1874. [CrossRef]
26. Paliwal, M.; Kumar, U.A. Neural networks and statistical techniques: A review of applications. *Expert Syst. Appl.* **2009**, *36*, 2–17. [CrossRef]
27. Kutlu, Y.; Kuntalp, M.; Kuntalp, D. Optimizing the performance of an mlp classifier for the automatic detection of epileptic spikes. *Expert Syst. Appl.* **2009**, *36*, 7567–7575. [CrossRef]
28. Haykin, S. *Neural Networks*; Prentice Hall: Hoboken, NJ, USA, 1999.
29. Amari, S.I. Training error, generalization error, and learning curves in neural learning. In Proceedings of the 1995 Second New Zealand International Two-Stream Conference on Artificial Neural Networks and Expert Systems, Dunedin, New Zealand, 20–23 November 1995; pp. 4–5.
30. Demuth, H.; Beale, M. *Neural Network Toolbox for Use with MATLAB*; The MathWorks Inc.: Natick, MA, USA, 1998; pp. 10–30.
31. Hagiwara, K.; Kuno, K. Regularization Learning and Early Stopping in Linear Networks. In Proceedings of the IEEE-INNS-ENNS International Joint Conference on Neural Networks. IJCNN 2000: Neural Computing: New Challenges and Perspectives for the New Millennium, Como, Italy, 27 July 2000; pp. 511–516.
32. Ranganathan, A. The Levenberg-Marquardt Algorithm. *Tutor. LM Algorithm* **2004**, *11*, 101–110.
33. Shankar, P.M. *Probability, Random Variables, and Data Analytics with Engineering Applications*; Springer: Berlin/Heidelberg, Germany, 2021; Volume 473.

eng

Article

A Computationally Time-Efficient Method for Implementing Pressure Load to FE Models with Lagrangian Elements

Adnan Shahriar [1,*], Arsalan Majlesi [2] and Arturo Montoya [2]

1 Department of Mechanical Engineering, The University of Texas at San Antonio, San Antonio, TX 78249, USA
2 Department of Civil Engineering, The University of Texas at San Antonio, San Antonio, TX 78249, USA; arsalan.majlesi@my.utsa.edu (A.M.); arturo.montoya@utsa.edu (A.M.)
* Correspondence: adnan.shahriar@my.utsa.edu; Tel.: +1-210-8035-230

Abstract: A computationally time-efficient method is introduced to implement pressure load to a Finite element model. Hexahedron elements of the Lagrangian family with Gauss–Lobatto nodes and integration quadrature are utilized, where the integration points follow the same sequence as the nodes. This method calculates the equivalent nodal force due to pressure load using a single Hadamard multiplication. The arithmetic operations of this method are determined, which affirms its computational efficiency. Finally, the method is tested with finite element implementation and observed to increase the runtime ratio compared to the conventional method by over 20 times. This method can benefit the implementation of finite element models in fields where computational time is crucial, such as real-time and cyber–physical testbed implementation.

Keywords: equivalent pressure load; Lagrangian; Gauss–Lobatto nodes; computational time efficient; real-time FEM

Citation: Shahriar, A.; Majlesi, A.; Montoya, A. A Computationally Time-Efficient Method for Implementing Pressure Load to FE Models with Lagrangian Elements. *Eng* **2024**, *5*, 2379–2394. https://doi.org/10.3390/eng5030124

Academic Editor: Antonio Gil Bravo

Received: 15 July 2024
Revised: 16 September 2024
Accepted: 19 September 2024
Published: 22 September 2024

1. Introduction

Surface traction and body forces are the two main forms of external loadings that cause solids to deform. Body forces act on the inner, dispersed mass of the solid, whereas surface tractions act by applying normal and shearing stresses to the surface of the solid [1–3]. In the context of the finite element method (FEM), these two loading conditions are used to solve for the displacement of solids [3]. Other properties, such as stress and strain, can be computed from the displacement to analyze and design components of solid objects, i.e., structures. If multiple solid bodies are involved, contact models are crucial for enforcing continuity and compatibility conditions between the interacting surfaces [3–5]. There are several contact modeling techniques, such as surface-to-surface or node-to-surface models, where surface tractions are applied to model interfaces for coupling. However, the surface tractions cannot be applied to a FEM directly; instead, an equivalent nodal force vector is computed and applied. The computation of the equivalent nodal force consists of several steps: determination of (1) traction force, (2) equivalent nodal force for each of the integration points on the surface, and (3) assembly. This procedure involves several matrix–matrix multiplications. The total FE simulation consists of two phases: (1) precomputation and (2) runtime. For a dynamic case, during the runtime phase, if there is varying pressure loading or the model is in contact with another model, force vector assembly, followed by the computation of an equivalent nodal force vector for surface traction, needs to be carried out at each time step. Such computation requires substantial computational effort.

In fields where real-time computation is involved, computational time becomes very crucial. Such fields include cyber–physical testing, which involves real-time communication between an experimental and computational model [6]. Also, in a system of systems (SoS) [7,8] framework, there is a growing need to incorporate multiple solid or structural bodies to achieve accurate and holistic modeling in real time [9]. Recently, such testing has been implemented to design resilient space habitats [10,11] that require high-fidelity two- and three-dimensional dynamic FE models. These models can be subjected to pressure loading, and the interaction between the cyber and physical components (cyber–physical testing [12]) can also employ such loading that needs to be computed in real time [13]. Hence, implementing a computational time-efficient method to calculate equivalent nodal force due to pressure load during runtime is necessary.

Efforts have been made to reduce computational time, specifically for dynamic applications. For example, diagonal mass matrices [13] have been implemented in the field of spectral element methods [14–17]. Because it is efficient to invert a mass matrix, this approach can expedite the explicit dynamic simulation. Here, the diagonalization of the mass matrices has resulted from the hexahedral elements of the Lagrangian family with Gauss–Lobatto nodes [18] and integration quadrature, which is different from the Gauss–Legendre quadrature [19]. However, for FE application, equivalent nodal force vector calculation from pressure load is computationally demanding, and minimizing the computational time associated with it has not been approached.

In this paper, a method is developed to reduce the computational cost associated with the implementation of the pressure load to a three-dimensional FEM during the runtime phase of a dynamic simulation. According to this method, the computational overhead for the total procedure to compute equivalent nodal force can be split into two sub-procedures, where the majority of the computation is performed during the precomputation phase. For this purpose, the hexahedral elements of the Lagrangian family with Gauss–Lobatto nodes and integration quadrature are implemented for surface integration. For such an integration quadrature, the integration points overlap with the nodal coordinates. As a result, the value of a shape function evaluated at the integration points is one at one integration point and zero at all others. This property is then utilized to reduce the equivalent nodal force vector computations to a single Hadamard multiplication [20] for the runtime phase. Such multiplication is also known as elementwise multiplication and is implemented in many languages, such as PYTHON (3.12.6) and MATLAB®(2023b). This method will come with a limitation as it cannot be implemented for large deformation when the surface area or surface normal vector changes.

The remainder of this paper is organized as follows. The methodology section first describes the implementation of the Lagrangian elements of the hexahedral family and integration quadrature. Then, the properties of such elements are utilized to develop a computationally time-efficient algorithm that utilizes a single Hadamard multiplication to compute the equivalent nodal force vector from the pressure load. Next, arithmetic operations are determined for the new algorithm and compared against the existing one. Then, the algorithms are implemented into FE models, and the CPU time is computed, which shows the efficiency of the current method compared to the conventional method. Next, a convergence study is conducted, and a dynamic analysis is carried out using an FE model of a space habitat subjected to a dynamic pressure load. Then, the limitations of the current method are discussed in detail. Finally, the conclusion section concludes this current work with some remarks.

2. Materials and Methods

This section first discusses the computation of the equivalent nodal force vector due to surface pressure. Next, the element that will be incorporated into this method is illustrated. Finally, the procedure to implement Hadamard multiplication to compute equivalent nodal force based on the framework discussed priorly is presented.

2.1. Conventional Method to Compute the Equivalent Nodal Force from Pressure Load

For an element, the equation of motion of the model in the coordinate $\boldsymbol{x}$ (x, y, z) is as follows:

$$[m]\{\ddot{u}\}^T + [c]\{\dot{u}\}^T + [k]\{u\}^T = \{r_e\}^T \tag{1}$$

where $[m]$, $[c]$, $[k]$ are the mass, damping, and stiffness matrix; $\{u\}$, $\{\dot{u}\}$, and $\{\ddot{u}\}$ are the displacement, velocity, and acceleration, respectively, for an element. On the right-hand side of Equation (1), $\{r_e\}$ is the equivalent nodal force due to external forces on element e. The matrices $[m]$, $[c]$, $[k]$, and $\{r_e\}$ are computed following the isoparametric formulation that can be found in the literature [4]. The vector of shape functions for an element in the isoparametric coordinate $\boldsymbol{\xi}(\xi, \eta, \zeta)$ is expressed as follows:

$$\{\chi\} = \{N_1 \quad N_2 \quad \ldots \quad N_{n_e}\} \tag{2}$$

where n_e is the number of nodes of an element and N_i is the value of i shape function at the isoparametric coordinate $\boldsymbol{\xi}$. Following Equation (3), a shape function matrix is defined as

$$[N] = \begin{bmatrix} N_1 & 0 & 0 & N_2 & 0 & 0 & \ldots & N_{n_e} & 0 & 0 \\ 0 & N_1 & 0 & 0 & N_2 & 0 & \ldots & 0 & N_{n_e} & 0 \\ 0 & 0 & N_1 & 0 & 0 & N_2 & \ldots & 0 & 0 & N_{n_e} \end{bmatrix} \tag{3}$$

Different types of forces, such as surface traction, $\{f_{trac}\}$ and body force, $\{f_{body}\}$, contribute to the equivalent nodal force vector. Considering the normal surface traction $\{\varphi\}$, the $\{f_{trac}\}$ can be determined following isoparametric formulation as follows:

$$\{f_{trac}\}^T = \int_{-1}^{1}\int_{-1}^{1} [N]^T_{.} P\{d\Gamma\}^T d\eta d\xi = \sum_{i=1}^{i=n_g} [N]_i^T P_i \{d\Gamma\}_i^T w_i \tag{4}$$

Here, the normal surface traction $\{\varphi\} = P\{\hat{n}\}$, where P is the traction magnitude and $\{\hat{n}\}$ is the normal unit vector on the surface of an element pointing outward from the element. P_i is the normal traction magnitude pointing outward from the element surface, w_i is the weight at integration point i, and n_g is the number of integration points of the element on the surface of the element subjected to the traction load. Note: pressure is the negative of the P_i. $\{d\Gamma\}$ is a vector in the global coordinate perpendicular to the surface the pressure is acting on, with $\|d\Gamma\|$ being the ratio of a differential area of the element between global and local coordinates. There are six surfaces for hexahedral elements. For demonstration purposes in this paper, considering $[J]$ as the Jacobian matrix, only the case of the $\zeta = 1$ plane of a hexahedral element is presented, as follows:

$$\{d\Gamma\}^T = \begin{Bmatrix} J_{22}J_{33} - J_{23}J_{32} \\ J_{23}J_{31} - J_{21}J_{33} \\ J_{21}J_{32} - J_{22}J_{31} \end{Bmatrix} \tag{5}$$

Evaluation of the traction value at the integration point i can be simplified by assigning nodal traction value. If the tractions of all nodes of the element e are $\{P\}_e$, following the isoparametric formulation

$$P_i = \{\chi\}_i \{p\}_e^T \tag{6}$$

The $\{f_{trac}\}$ is the equivalent nodal force vector for a single surface of an element. A single element can have multiple surfaces, i.e., hexahedral elements have six surfaces. Hence, $\{f_{trac}\}$ needs to be computed for all the surfaces of an element. This paper considers just one surface of each element subjected to pressure loading. Hence, the total number of surfaces of the whole model equals the number of elements. Considering n_E as the number of elements of the model, the computation of the global force vector can be determined by summing through elements. The equation of motion (EOM) for the assembled system is as follows:

$$[M]\{\ddot{U}\}^T + [C]\{\ddot{U}\}^T + [K]\{U\}^T = \{F_{trac}\}^T + \{F_{body}\}^T \tag{7}$$

where $\{\ddot{U}\}$, $\{\ddot{U}\}$, and $\{U\}$ are the acceleration, velocity, and displacement vectors, $[M]$, $[C]$, and $[K]$ are the mass, damping, and stiffness matrices, and $\{F_{body}\}$ is the body force vector for the assembled model. Several time-stepping algorithms exist to solve the EOM. For this study, the implicit Newmark-Beta method [4] is implemented, which is an implicit time-stepping method for solving the displacement at discrete time $t + \Delta t$ as follows:

$$[\overline{K}]\left\{{}^{t+\Delta t}_{\cdot}U\right\}^T = \{{}^{t}_{\cdot}\overline{R}\}^T \tag{8}$$

where $[\overline{K}]$ is the equivalent stiffness matrix of the form

$$[\overline{K}] = [K] + a_0[M] + a_1[C] \tag{9}$$

And $\{{}^{t}_{\cdot}\overline{R}\}$ is the equivalent nodal force vector

$$\{{}^{t}_{\cdot}\overline{R}\}^T = [M]\left(a_0\{{}^{t}_{\cdot}U\}^T + a_2\left\{{}^{t}_{\cdot}\dot{U}\right\}^T + a_3\left\{{}^{t}_{\cdot}\ddot{U}\right\}^T\right) + [C]\left(a_1\{{}^{t}_{\cdot}U\}^T + a_4\left\{{}^{t}_{\cdot}\dot{U}\right\}^T + a_5\left\{{}^{t}_{\cdot}\ddot{U}\right\}^T\right) + \{F_{trac}\}^T + \left\{F_{body}\right\}^T \tag{10}$$

Upon obtaining the displacement vector, the acceleration and velocity can be obtained, respectively, as follows:

$$\left\{{}^{t+\Delta t}_{\cdot}\ddot{U}\right\} = a_0\left(\left\{{}^{t+\Delta t}_{\cdot}U\right\} - \{{}^{t}_{\cdot}U\}\right) - a_2\left\{{}^{t}_{\cdot}\dot{U}\right\} - a_3\left\{{}^{t}_{\cdot}\ddot{U}\right\} \tag{11}$$

$$\left\{{}^{t+\Delta t}_{\cdot}\dot{U}\right\} = \left\{{}^{t}_{\cdot}\dot{U}\right\} + a_7\left\{{}^{t+\Delta t}_{\cdot}\ddot{U}\right\} + a_6\left\{{}^{t}_{\cdot}\ddot{U}\right\} \tag{12}$$

where the definition of $a_0 - a_7$ in Equations (8)–(12) can be found in [4].

2.2. *Lagrangian Element with Legendre-Gauss-Lobatto Nodes and Integration Quadrature*

The hexahedron elements used for FE analysis can be divided into two families: (1) serendipity, i.e., a 20-node brick quadratic widely used in commercial software such as Abaqus 6.9 [21] and (2) Lagrangian, i.e., a 27-node brick quadratic [22]. The Lagrangian element consists of Legendre–Gauss–Lobatto (LGL) nodes and integration quadrature. For a one-dimensional n-point LGL quadrature between -1 and 1, along the ξ axis, the nodal coordinates are the solution of

$$\left(1 - \xi^2\right)\frac{d}{d\xi}L_{n-1} = 0 \tag{13}$$

where L_{n-1} indicates the Legendre polynomial of the degree $n - 1$. The weight corresponding to a node i, with coordinate ξ_i, can be calculated as follows:

$$w_i = \frac{2}{n(n-1)[L_n(\xi_i)]^2} \tag{14}$$

This formulation can be extended to 3D elements with a different order along the ξ, η, ζ axis, as n_ξ, n_η, and n_ζ, respectively. A schematic of such 3D elements with $n_\xi = 6$, $n_\eta = 6$, and $n_\zeta = 3$ utilized by [23] to simulate wave propagation is presented in Figure 1. One of the shape functions is plotted, and the 0 value is presented as void. The integration points of the element overlap with the nodal coordinates; hence, the shape function is 1 at one integration point and 0 at all others.

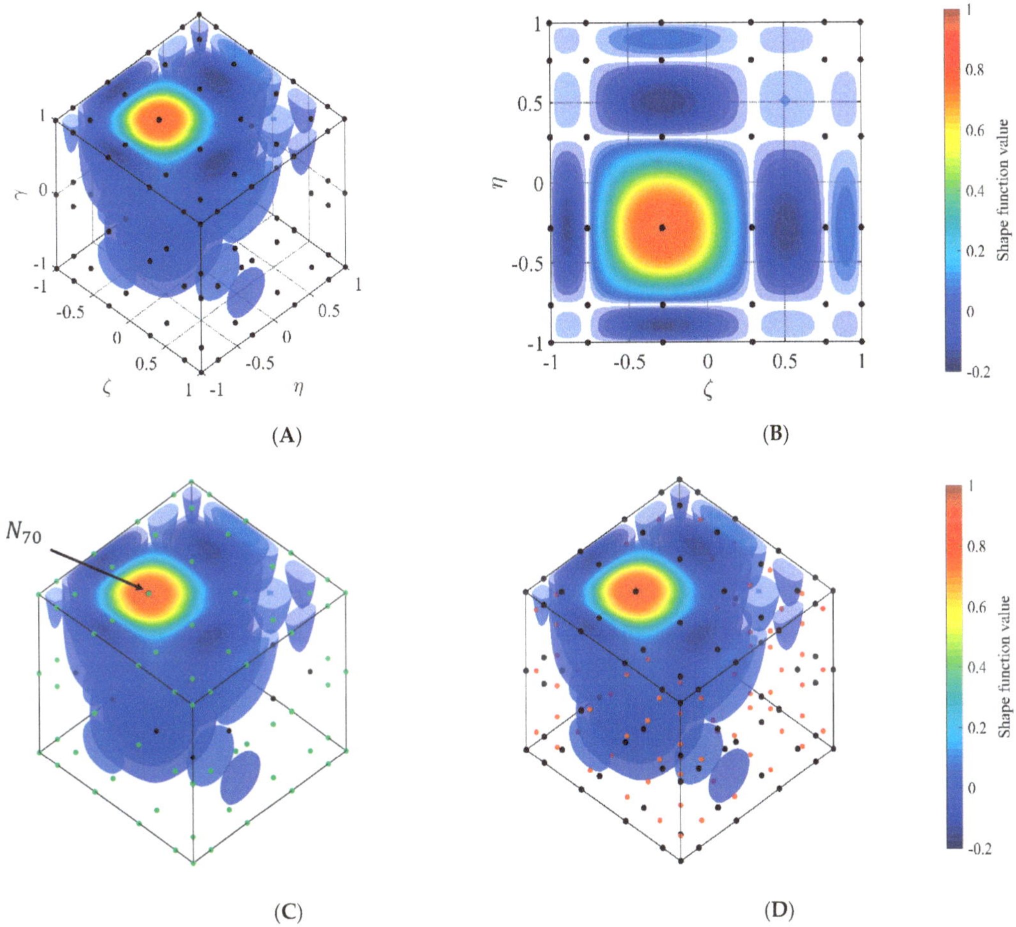

Figure 1. Schematic of a Lagrangian element with Legendre–Gauss–Lobatto nodes and integration quadrature. $n_\xi = 6$, $n_\eta = 6$, and $n_\zeta = 3$, (**A**) isometric, (**B**) top view, (**C**) Legendre–Gauss–Lobatto integration quadrature on the surface (green), (**D**) Gauss–Legendre integration quadrature for the region (red).

2.3. *Development of a Computationally Time-Efficient Algorithm to Compute Equivalent Nodal Force from Pressure Load*

The procedure to implement Hadamard multiplication to compute equivalent nodal force is presented in this section. First, the implementation of the Lagrangian element is discussed. In this paper, the Legendre–Gauss–Lobatto integration quadrature for a Lagrangian element is computed such that the sequence of node numbers is the same as the sequence of the integration points. A schematic of such an element is presented in Figure 1C, with all integration points on the surface (green) shown to coincide with the nodes on the surface. Also, for a shape function N_i ($i = 70$ in the figure), the i indicates both the node number and the integration point number of the element. Note: the Gauss–Legendre integration quadrature can be used for region integration, as shown in Figure 1D. As the node numbers and integration points follow the same coordinate and same sequence, for an integration point number i, Equations (2) and (3) are as follows:

$$\{\chi\}_i = \{0 \quad \ldots \quad N_i = 1 \quad \ldots \quad 0\} \tag{15}$$

$$[N]_i = \begin{bmatrix} 0 & 0 & 0 & \dots & N_i = 1 & 0 & 0 & \dots & 0 & 0 & 0 \\ 0 & 0 & 0 & \dots & 0 & N_i = 1 & 0 & \dots & 0 & 0 & 0 \\ 0 & 0 & 0 & \dots & 0 & 0 & N_i = 1 & \dots & 0 & 0 & 0 \end{bmatrix} \quad (16)$$

Next, the algorithm that reduces the equivalent nodal force calculation into a single Hadamard multiplication utilizing Equations (15) and (16) is developed. Following Equation (15), Equation (6) is reduced into

$$P_i = p_i \quad (17)$$

Similarly, following Equation (16) and replacing P_i with p_i, the components of $\{f_{trac}\}$ are as follows:

$$\begin{Bmatrix} f_{trac_{3i-2}} \\ f_{trac_{3i-1}} \\ f_{trac_{3i}} \end{Bmatrix} = [N]_i^T w_i \{d\Gamma\}_i^T p_i \quad (18)$$

Equation (18) can be written in terms of Hadamard multiplication as follows:

$$\begin{Bmatrix} f_{trac_{3i-2}} \\ f_{trac_{3i-1}} \\ f_{trac_{3i}} \end{Bmatrix} = \left\{ [N]_i^T w_i \{d\Gamma\}_i^T \right\} \circ \begin{Bmatrix} p_i \\ p_i \\ p_i \end{Bmatrix}^T \quad (19)$$

Due to the linear independence of the components of $\{f_{trac}\}$ from $3i-2$ to $3i$, the force vector for an element can be written as follows:

$$\{f_{trac}\}^T = \left(\sum_{i=1}^{i=n_g} [N]_i^T w_i \{d\Gamma\}_i^T \right) \circ \{f_p\}^T \quad (20)$$

During assembly, if node i of element e_1 and node j of the element e_2 have the same coordinate and are denoted as node k in the global coordinate, the k components of the force vector of the global coordinate are as follows:

$$\begin{Bmatrix} F_{trac_{3k-2}} \\ F_{trac_{3k-1}} \\ F_{trac_{3k}} \end{Bmatrix} = \begin{Bmatrix} f_{trac_{3i-2}} \\ f_{trac_{3i-1}} \\ f_{trac_{3i}} \end{Bmatrix}_{e_1} + \begin{Bmatrix} f_{trac_{3j-2}} \\ f_{trac_{3j-1}} \\ f_{trac_{3j}} \end{Bmatrix}_{e_2} \quad (21)$$

Following Equation (19), Equation (21) can be written as

$$\begin{Bmatrix} F_{trac_{3k-2}} \\ F_{trac_{3k-1}} \\ F_{trac_{3k}} \end{Bmatrix} = \left\{ \left\{ [N]_i^T w_i \{d\Gamma\}_i^T \right\}_{e_1} + \left\{ [N]_j^T w_j \{d\Gamma\}_j^T \right\}_{e_2} \right\} \circ \begin{Bmatrix} p_k \\ p_k \\ p_k \end{Bmatrix}^T \quad (22)$$

Hence, the total force vector is as follows:

$$\{F_{trac}\}^T = \left(\sum_{j=1}^{j=e} \left(\sum_{i=1}^{i=n_g} [N]_i^T w_i \{d\Gamma\}_i^T \right) \right) \circ \{F_p\}^T \quad (23)$$

where

$$\{F_p\}^T = \{\{p_1 \;\; p_1 \;\; p_1\} \;\; \{p_2 \;\; p_2 \;\; p_2\} \;\; \dots \;\; \{p_{n_n} \;\; p_{n_n} \;\; p_{n_n}\}\} \quad (24)$$

where n_n is the total number of nodes of the assembled model. The summation term of Equation (24) is defined as $\{F_{P\_unity}\}$, which follows

$$\{F_{P\_unity}\}^T = \left(\sum_{j=1}^{j=e}\left(\sum_{i=1}^{i=n_g}[N]_i^T w_i \{d\Gamma\}_i^T\right)\right) \quad (25)$$

Although the Gauss–Legendre quadrature is widely used in finite element applications and requires two fewer integration points than Gauss–Lobatto to integrate a polynomial exactly, Equation (17) cannot be implemented as the nodal coordinates do not coincide with the integration points. Hence, Equation (6) needs to be used, restricting the development towards Equation (23).

If large deformation is not involved, $\{F_{P\_unity}\}$ is constant for a model; hence, it can be precomputed before the simulation, and only the Hadamard multiplication needs to be carried out between 2 vectors according to Equation (25). The procedure and the number of arithmetic operations are determined for the conventional and current methods in Tables 1 and 2, respectively. In Table 2, the operations are computed for the precomputed and runtime phases.

Table 1. Arithmetic operations of the conventional method.

Operation	Operation Description	No of Arithmetic Operations
Extract $\{p\}_e^{\cdot}$ from $\{p\}$ of Equation (6)	1: Elementwise addition $\underset{n_e\times 1}{\{p\}_e^{\cdot}} = \underset{n_e\times 1}{\{p\}_e^{\cdot}} + \underset{n_n\times 1}{\{p\}}$	n_e
Compute P_i, Equation (6)	2: Matrix vector multiplication $\underset{1\times 1}{P_i} = \underset{1\times n_e}{\{\chi\}_i} \times \underset{n_e\times 1}{\{p\}_e^T}$	$2n_e$
Compute $\{f_{trac}\}$, Equation (4)	3.1: Matrix–matrix multiplication $\underset{3n_e\times 1}{\{a\}} = \underset{3n_e\times 3}{[N]_i^T} \times \underset{3\times 1}{\{d\Gamma\}_i^T}$	$12n_e$
	3.2: Scalar–scalar multiplication $\underset{1\times 1}{b} = \underset{1\times 1}{P_i} \times \underset{1\times 1}{w_i}$	1
	3.3: Scalar–vector multiplication $\underset{3n_e\times 1}{\{c\}} = \underset{3n_e\times 1}{\{a\}} \times \underset{1\times 1}{b}$	$3n_e$
	3.4: Sum to $\{f_{trac}\}$ $\underset{3n_e\times 1}{\{f_{trac}\}} = \underset{3n_e\times 1}{\{f_{trac}\}} + \underset{3n_e\times 1}{\{c\}}$	$3n_e$
Assembly of the force vector for one surface	4.1: Operation 1 once and 2 to 3, n_g times	$(20n_e+1)n_g+n_e$
	4.2: Elementwise addition $\underset{3n_n\times 1}{\{F_{trac}\}} = \underset{3n_n\times 1}{\{F_{trac}\}} + \underset{3n_e\times 1}{\{f_{trac}\}}$	$3n_n$
Total for the whole model	5: Operation 4, n_E times	$((20n_e+1)n_g+n_e+3n_n)n_E$

The algorithm flowchart for both methods is presented in Figure 2, with the precomputation and runtime phases marked as blue and yellow, respectively. The pressure load in an input to the FE models is used for the equivalent nodal force computation. It is evident from Figure 2B that the precomputation phase handles most of the operations, leaving only $6nn$ operations during the runtime phase for such computation. Note that the conventional method involves $((20n_e+1)n_g+n_g+3n_n)n_E \gg 6n_n$ operations that have to be carried out during runtime (Figure 2A).

Table 2. Arithmetic operations of the current method.

Operation	Operation Description	No of Arithmetic Operations for Precomputation	No of Arithmetic Operation Runtimes
Compute $\left\{F_{P\_unity}\right\}$, Equation (25)	1.1: Matrix–matrix multiplication $\underset{3n_e \times 1}{\{a\}} = \underset{3n_e \times 3}{[N]_i^T} \times \underset{3 \times 1}{\{d\Gamma\}_i^T}$	$12n_e$	
	1.2: Scalar–vector multiplication $\underset{3n_e \times 1}{\{c\}} = \underset{3n_e \times 1}{\{a\}} \times \underset{1 \times 1}{w_i}$	$3n_e$	
	1.3: Sum to $\{d\}$ $\underset{3n_e \times 1}{\{d\}} = \underset{3n_e \times 1}{\{d\}} + \underset{3n_e \times 1}{\{c\}}$	$3n_e$	
	1.4 : Operations 1.1 to 1.3, n_g times	$18n_e \times n_g$	
	1.5: Elementwise addition $\underset{3n_n \times 1}{\left\{F_{P\_unity}\right\}} = \underset{3n_n \times 1}{\left\{F_{P\_unity}\right\}} + \underset{3n_e \times 1}{\{d\}}$	$3n_n$	
	1.6 : Operations 1.4 and 1.5, n_E times	$(18n_e \times n_g + 3n_n)n_E$	
Compute $\{F_P\}$, Equation (24)	2: Elementwise addition $\underset{3n_n \times 1}{\{F_P\}} = \underset{3n_n \times 1}{\{F_P\}} + \underset{n_n \times 1}{\{p\}}$		$3n_n$
Compute $\{F_{trac}\}$, Equation (23)	3: Hammard multiplication $\underset{3n_n \times 1}{\{F_{trac}\}} = \underset{3n_n \times 1}{\left\{F_{P\_unity}\right\}} \circ \underset{3n_n \times 1}{\{F_P\}}$		$3n_n$
Total for the whole model		$(18n_e \times n_g + 3n_n)n_E$	$6n_n$

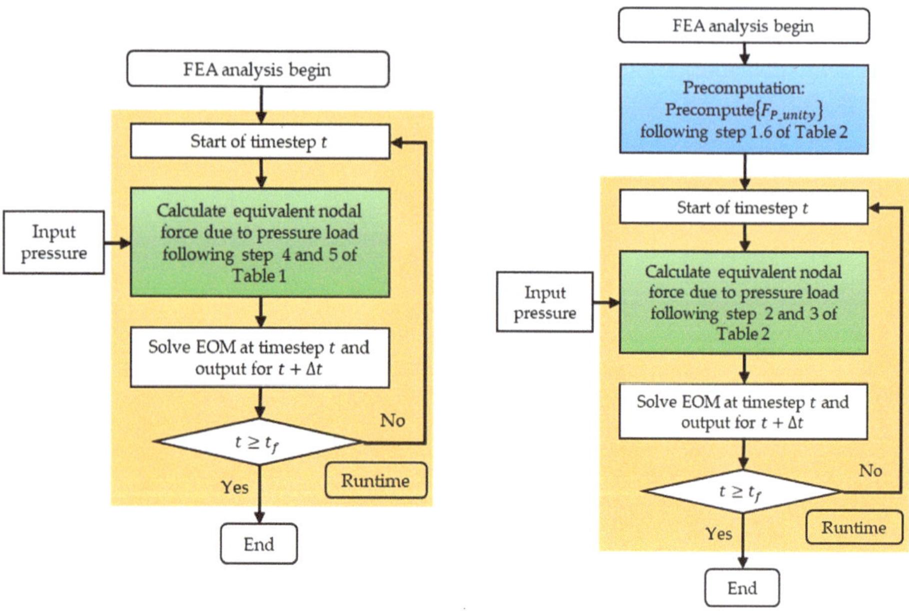

(A) **(B)**

Figure 2. Flowchart for (**A**) conventional method using Table 1 and (**B**) current method using Table 2.

Additionally, as the $\{F_{P\_unity}\}$ is a one-dimensional vector of size $3n \times 1$, the memory requirement for storing the vector is the same as the force vector for the whole model computed during the assembly operation. Hence, this current method does not increase the memory overhead with respect to the conventional method.

2.4. Application: Model Details

First, the efficiency in terms of computational time is tested with a plate model. Although, for 3D implementation, a single layer of elements is not recommended, the objective is to prove computational efficiency. Hence, the algorithm is tested to a simple model: a one-element layer thick plate of dimension 30 m × 30 m × 3 m, where pressure load is applied on the top surface (yellow surface of Figure 3). The Gauss–Legendre quadrature is used for region integration, whereas the Gauss–Lobatto is used for equivalent nodal force calculation from the pressure load.

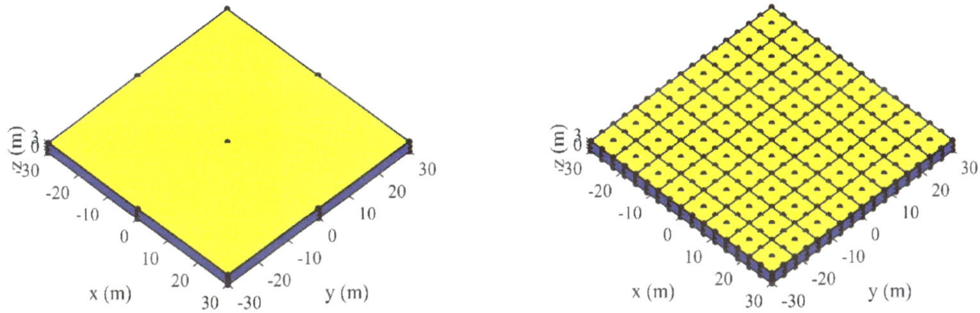

(A) (B)

Figure 3. Plate model with (**A**) 1 element and (**B**) 64 elements with 27-node elements.

As only one of the surfaces for each element is involved, the total number of surfaces is equal to the number of elements. The computational time with respect to the conventional method is measured as the runtime ratio, which is the ratio of CPU time following the current and conventional method during runtime.

Two cases are carried out. First, the number of elements increases from 1 to 121 as 1, 4, 9, 16, 25, 36, 49, 64, 81, 100, and 121, by changing the number of elements along x and y and keeping just 1 element along z. The equivalent nodal forces are calculated for a pressure load of 1 *Pa*.

Next, for the case of 36 elements, the nodes of a single element are increased from 2 to 9 as n_η and n_ξ and n_ζ is kept as 3. A schematic of one of the cases, $n_\xi = n_\eta = 7$, is presented in Figure 4.

Next, a convergence study is carried out using the plate model. The static displacement with the material property described in Table 3 was utilized for convergence analysis. The edge of the plate on the $z = 0$ plane was considered fixed for the boundary condition, and a pressure load of 0.1 MPa was implemented. The finest mesh of all of these test cases is with 121 elements. Although increasing the number of elements is expected to reduce error, the convergence behavior can be shown with the 121-element mesh. Hence, the displacement of the fine mesh with 121 elements was computed and set as the reference.

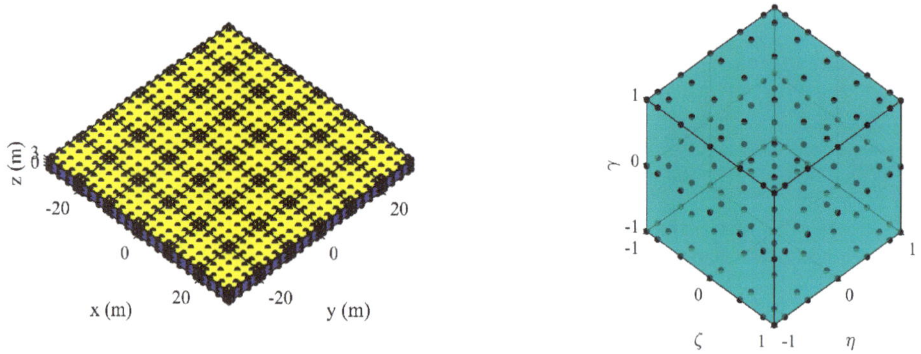

Figure 4. Plate model (**Left**) of 36 elements with the 147-node element (**Right**) $n_\xi = n_\eta = 7$ and $n_\zeta = 3$.

Table 3. Material properties.

Material Properties	Value
Modulus of elasticity	68 GPa
Poission's ratio	0.3
Density	2703 kg/m^3
Damping [C]	0.00001[K]

With u_{121} determined, the difference in displacement or error for other models Δu is computed as follows:

$$\Delta u = \frac{(u - u_{121})}{u_{121}} \times 100\% \tag{26}$$

Finally, this method is implemented on a space habitat model. A space habitat dome model with inner and outer radii of 2.5 and 2.9 m is modeled using 64 27-node quadratic elements, 819 nodes, and a 2457 DOF, as shown in Figure 5. The habitat has two surfaces: red indicates the inner, and blue indicates the outer surface, subjected to pressure load P_{inner} and P_{outer}, respectively. The yellow region from $z = 0$ to -1 is developed for ground consideration, and the habitat will be connected to the ground. For the analysis, the nodes at the yellow region with $z < 0$ are considered fixed-end boundary conditions.

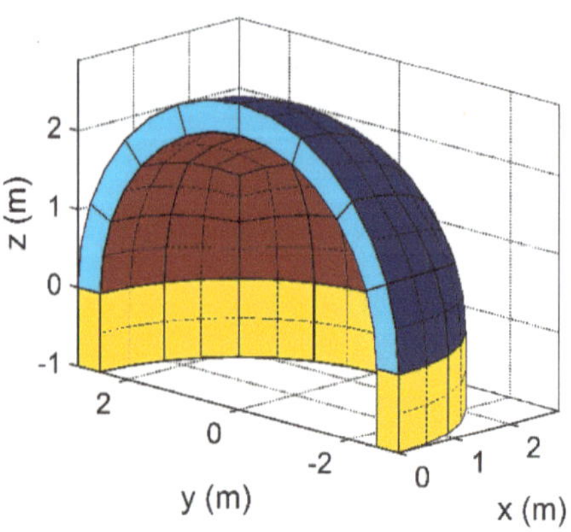

Figure 5. The space habitat FE model.

Dynamic analyses are carried out using the traction load following both current and conventional methods. The time-varying pressure load is set as follows:

$$P_{outer} = 10^4 y \times \sin(2\pi t) \tag{27}$$

$$P_{inner} = 101325 + 10^2 \sin(20\pi t) \tag{28}$$

For analysis, first, during the precomputation stage, a static analysis with $P_{inner} = 101,325$ and $P_{outer} = 0$ is performed. Next, the inverse of $\lfloor K \rceil$ is precomputed following Equation (9). Finally, the dynamic simulation is performed as the runtime stage, following the NB integration scheme (Equations (8)–(12)) with timestep $\Delta t = 0.0001$ s. Besides displacement, velocity, and acceleration, the stress is calculated for the integration points following the procedure described in [6]. Finally, the stress is extrapolated to the nodal points.

3. Results and Discussion

3.1. Computational Efficiency and Convergence Test

The runtime ratio is obtained following both conventional and current methods for the plate subjected to pressure load. The obtained equivalent nodal force for both procedures matches and verifies the procedure. For the study with the increase in element number, the results are presented in Figure 6A, where the current method is 5.3 times faster for one element. As the number of elements increases, the ratio reaches 27.5 asymptotically. The runtime ratio for the increase in element order is illustrated in Figure 6B, where it is evident that the current method is at least > 20 times faster than the conventional method, but the change in element order does not affect the computational performance predictably.

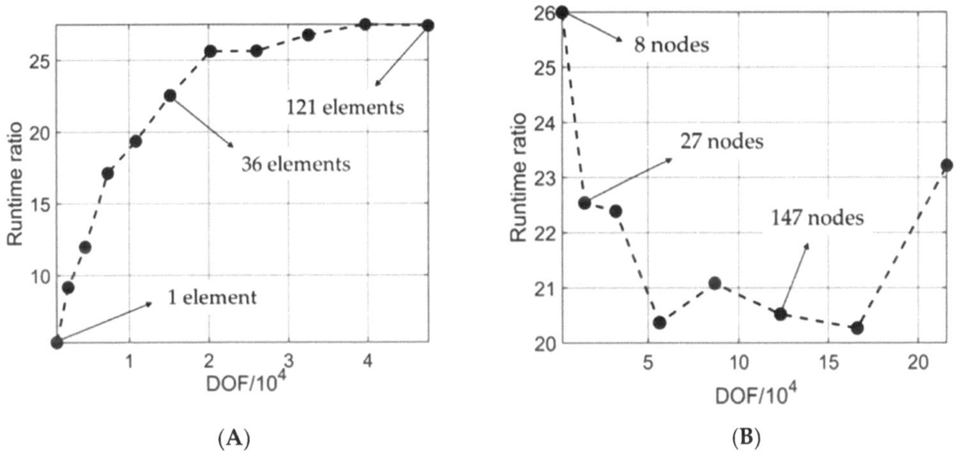

Figure 6. The runtime ratio for (**A**) an increase in element number with 27-node elements and (**B**) an increase in element order with 36 elements.

The displacement for the 121-element model is computed as $u_{121} = 16.12$ mm. A 3D overview of the displacement profile is shown in Figure 7. Equation (26) is used to obtain the error for the increase in the number of elements and the increase in the element order, presented in Figures 8A and 8B, respectively. It is obvious from the plot that the error decreases with an increase in both the number of elements and element order.

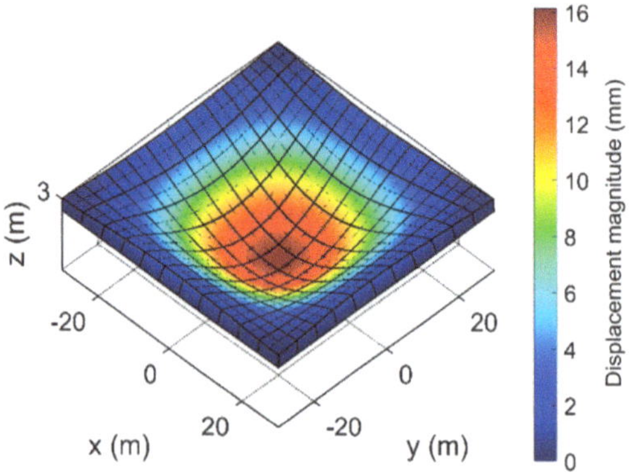

Figure 7. Plate model with 121 27-node brick elements.

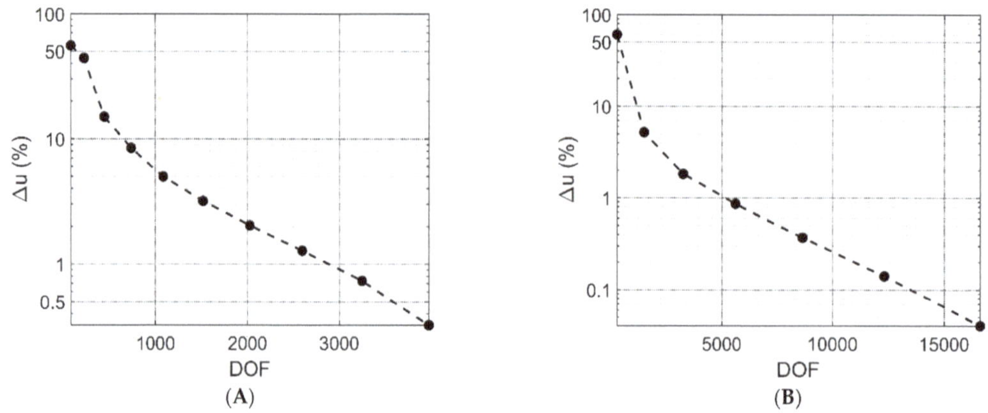

Figure 8. Convergence for (**A**) an increase in element number with 27-node elements and (**B**) an increase in element order with 36 elements.

3.2. *Application to Space Habitat Models*

The displacement profile for the dynamic scheme for the topmost node at z = 2.9 is shown in Figure 9A along y and Figure 9B along the *z*-axis, which shows that the frequency of the sinusoidal behavior of the displacement matches the loading frequency. A 3D profile of the displacement and von Mises stress are presented in Figures 10A and 10B at 0.75 s, respectively.

The time required for the pressure calculation following the current and conventional method, as well as the NB integration with precomputed $[K]$, is presented in Figure 11. It is obvious from the figure that the current method (around 6 s) takes significantly less time compared to the conventional method (around 175 s), which is higher than the NB integration scheme (around 55 s). The calculated runtime ratio is 26.25.

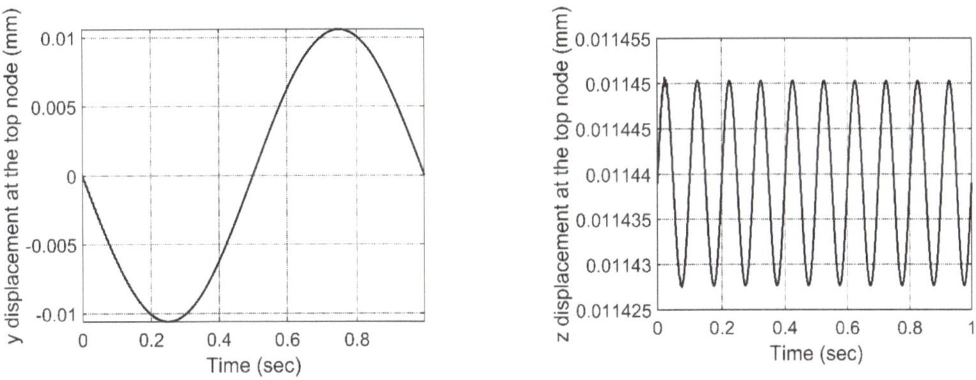

(**A**) (**B**)

Figure 9. Displacement at the top node at z = 2.9 m along (**A**) y- and (**B**) *z*-axis.

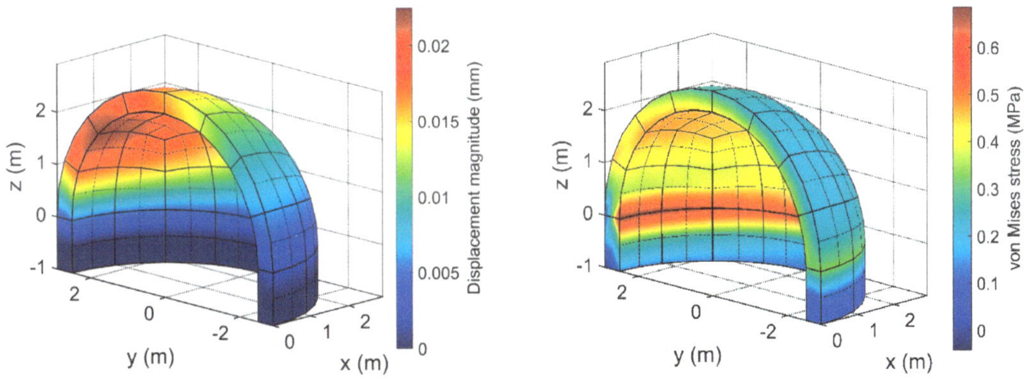

(**A**) (**B**)

Figure 10. (**A**) Displacement and (**B**) von Mises stress profile at 0.75 s.

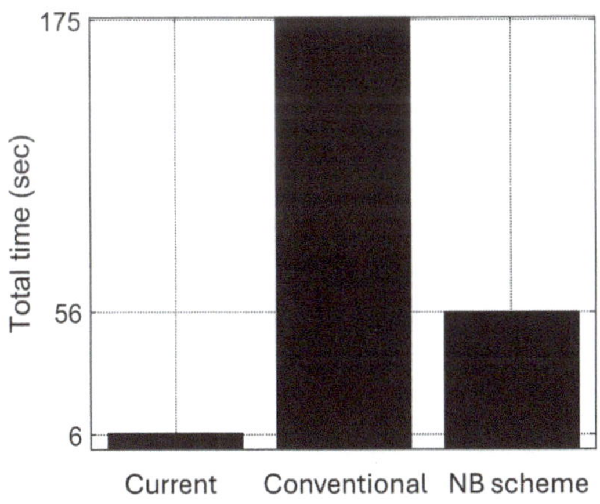

Figure 11. Total time taken by the current and conventional methods for equivalent nodal force calculation from pressure load and NB integration during the dynamic simulation.

3.3. Limitations

This work has three limitations: First, only the Lagrangian element can benefit from this method, as nodes of the serendipity family elements do not coincide with the LGL integration quadrature. Second, time-invariant $\{d\Gamma\}$ is required to precompute Equation (25). Hence, this method may not be applicable to large deformations [24]. Finally, it primarily considers elements with a single surface subjected to traction loading. If multiple surfaces of an element experience unequal pressure loads, the method presented cannot be directly applied. While the detailed implementation of such cases is beyond the scope of this paper, a procedure to address this scenario is outlined here. Figure 12 illustrates an example where elements e_1 and e_2 each have two faces subjected to traction loads. For element e_1, node i experiences equal pressure loading on both surfaces, denoted as a. Conversely, for the element e_2, node j is subjected to differing traction loads from two distinct surfaces. To handle this, the method involves partitioning all surfaces into two groups, Γ_1 and Γ_2, represented by red and blue, respectively. Subsequently, Equation (23) is applied independently for each group and calculates the $\{F_p\}$ vector. Finally, a Hadamard product is employed to compute the global force vector, as shown in Equation (29).

$$\{F_{trac}\}^T = \left[\left\{ F_{P_{unity}} \right\} \Big|_{\Gamma_1} \quad \left\{ F_{P_{unity}} \right\} \Big|_{\Gamma_2} \quad \cdots \right]^T \circ \left[\{F_p\} |_{\Gamma_1} \quad \{F_p\} |_{\Gamma_2} \quad \cdots \right]^T \tag{29}$$

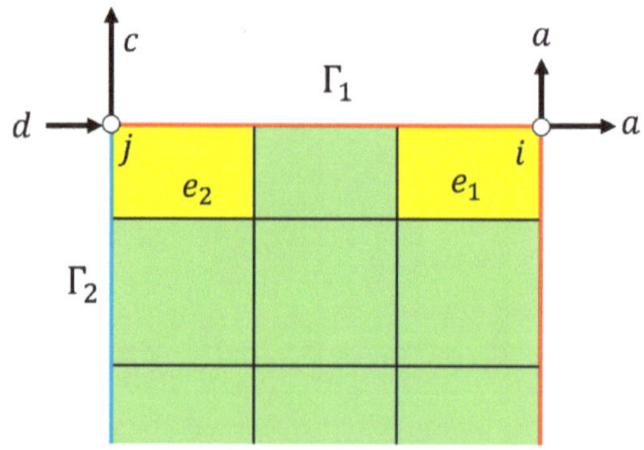

Figure 12. Surface partitioning to account for corner nodes subjected to two different traction loads from two different surfaces.

4. Conclusions

A computationally time-efficient method was developed to compute equivalent nodal force due to pressure load. Elements of the Lagrangian family with Gauss–Lobatto nodes and integration quadrature were implemented in such a way that the integration points follow the same sequence as the nodes. Through the implementation of such an element, the computation of the equivalent nodal force is reduced into a single Hadamard multiplication. Computational efficiency was established by computing the arithmetic complexity of this method. This method is implemented in a single-element thick plate model with different element densities. As the number of elements increases, the runtime ratio increases, surpassing 20 for 36 elements for a one-element thick plate model. It is also observed that the increase in element order decreases the runtime ratio by around 22% for the 36-element density but still outperforms the conventional method. Finally, a habitat model is developed, and dynamic analysis is carried out with a time-varying pressure load on two different surfaces that showed a runtime ratio over 20 by the current method to compute the equivalent nodal force from the pressure load.

Author Contributions: Conceptualization, A.S.; methodology, A.S.; software, A.S.; validation, A.S., and A.M. (Arsalan Majlesi); formal analysis, A.S.; investigation, A.S. and A.M. (Arsalan Majlesi); resources, A.S. and A.M. (Arturo Montoya); data curation, A.S. and A.M. (Arsalan Majlesi); writing—original draft preparation, A.S. and A.M. (Arsalan Majlesi); writing—review and editing, A.M. (Arturo Montoya); visualization, A.S.; supervision, A.M. (Arturo Montoya); project administration, A.M. (Arturo Montoya); funding acquisition, A.M. (Arturo Montoya). All authors have read and agreed to the published version of the manuscript.

Funding: This material was based on work carried out under the Resilient Extra-Terrestrial Habitat Institute (RETHi) supported by a Space Technology Research Institute grant (No. 80NSSC19K1076) from NASA's Space Technology Research Grants Program. This paper does not have any conflicts of interest.

Data Availability Statement: Data is contained within the article.

Conflicts of Interest: The authors declare no conflict of interest.

Nomenclature

$[m]$	Mass matrix for an element
$[c]$	Damping matrix for an element
$[k]$	Stiffness matrix for an element
$\{\ddot{u}\}$	Acceleration vector for an element
$\{\dot{u}\}$	Velocity vector for an element
$\{u\}$	Displacement vector for an element
$\{r_e\}$	Equivalent nodal force due to external forces on element e
$\{f_{body}\}$	Body force vector of an element
$\{f_{trac}\}$	Tractor force vector of an element
$\{F_{trac}\}$	Traction vector of the assembled model
$\{F_{body}\}$	Body force vector of the assembled model
$[M]$	Mass matrix for the assembled model
$[K]$	Stiffness matrix for the assembled model
$[C]$	Damping matrix for the assembled model
$\{\ddot{U}\}$	Acceleration vector for the assembled model
$\{\dot{U}\}$	Velocity vector for the assembled model
$[\overline{K}]$	Equivalent stiffness matrix for the assembled model
$\{U\}$	Displacement vector for the assembled model
$\{\varphi\}$	Normal surface traction vector
N_i	Shape function corresponds to node i of an element
$[J]$	Jacobian matrix
P_i	Traction magnitude on an integration point i
w_i	Weight of an integration point i
$\{\chi\}$	vector of shape functions
n	Number of nodes
n_ξ	Number of nodes along ξ axis
n_η	Number of nodes along η axis
n_ζ	Number of nodes along ζ axis
p_i	Pressure at node i
n_e	Number of nodes of an element e
n_g	Number of integration points of an element e
n_E	Number of elements of the assembled model
n_n	Number of nodes of the assembled model
Abbreviations	
FE	Finite element
FEM	Finite element method
LGL	Legendre–Gauss–Lobatto
NB	Newmark-Beta
EOM	Equation of motion
SoS	System of systems

References

1. Sadd, M.H. Elasticity: Theory, Applications, and Numerics. Academic Press: Cambridge, MA, USA, 2009.
2. Reismann, H. On the forced motion of elastic solids. *Appl. Sci. Res.* **1968**, *18*, 156–165. [CrossRef]
3. Zienkiewicz, O.; Taylor, R.; Zhu, J.Z. *The Finite Element Method: Its Basis and Fundamentals: Seventh Edition*; Butterworth-Heinemann: Oxford, UK, 2005.
4. Bathe, K.-J. *Finite Element Procedures*; Prentice Hall Education: Upper Saddle River, NJ, USA, 1996.
5. Papadopoulos, P.; Solberg, J.M. A Lagrange multiplier method for the finite element solution of frictionless contact problems. *Math. Comput. Model.* **1998**, *28*, 373–384. [CrossRef]
6. Perić, D.; Owen, D.R.J. Computational model for 3-D contact problems with friction based on the penalty method. *Int. J. Numer. Methods Eng.* **1992**, *35*, 1289–1309. [CrossRef]
7. Fagcang, H.; Stobart, R.; Steffen, T. A review of component-in-the-loop: Cyber-physical experiments for rapid system development and integration. *Adv. Mech. Eng.* **2022**, *14*, 16878132221109969. [CrossRef]
8. Delp, C.; Cooney, L.; Dutenhoffer, C.; Gostelow, R.; Jackson, M.; Wilkerson, M.; Kahn, T.; Piggott, S. The Challenge of Model-based Systems Engineering for Space Systems, Year 2. *Insight* **2009**, *12*, 36–39. [CrossRef]
9. Szarazi, J.; Reichwein, A.; Bock, C. Integrating Finite Element Analysis with Systems Engineering Models. In Proceedings of the NAFEMS World Congress, Stockholm, Sweden, 11 June 2017.
10. Dyke, S.J.; Marais, K.; Bilionis, I.; Werfel, J.; Malla, R. Strategies for the design and operation of resilient extraterrestrial habitats. In Proceedings of the Sensors and Smart Structures Technologies for Civil, Mechanical, and Aerospace Systems, Online, 22–26 March 2021; p. 1159105. [CrossRef]
11. Shahriar, A.; Montoya, H.; Majlesi, A.; Avila, D.; Montoya, A. Coupling Independent Solid Mechanics-Based Systems in a System-of-Systems Modeling Framework. *AIAA J.* **2024**, *62*, 1–16. [CrossRef]
12. Maghareh, A.; Lenjani, A.; Krishnan, M.; Dyke, S.; Bilionis, I. Role of cyber-physical testing in developing resilient extraterrestrial habitats. *Earth Space* **2021**, 1059–1068.
13. Teukolsky, S.A. Short note on the mass matrix for Gauss–Lobatto grid points. *J. Comput. Phys.* **2015**, *283*, 408–413. [CrossRef]
14. Palacz, M.; Krawczuk, M.; Żak, A. Spectral Element Methods for Damage Detection and Condition Monitoring. In *Smart Innovation, Systems and Technologies*; Springer: Berlin/Heidelberg, Germany, 2020; pp. 549–558. [CrossRef]
15. Ostachowicz, W.; Kudela, P.; Krawczuk, M.; Zak, A. *Guided Waves in Structures for SHM: The Time-Domain Spectral Element Method*; John Wiley & Sons, Ltd.: Hoboken, NJ, USA, 2012.
16. Patera, A.T. A spectral element method for fluid dynamics: Laminar flow in a channel expansion. *J. Comput. Phys.* **1984**, *54*, 468–488. [CrossRef]
17. Komatitsch, D.; Tromp, J. Spectral-element simulations of global seismic wave propagation—I. Validation. *Geophys. J. Int.* **2002**, *149*, 390–412. [CrossRef]
18. Gautschi, W. High-order Gauss–Lobatto formulae. *Numer. Algorithms* **2000**, *25*, 213–222. [CrossRef]
19. Swarztrauber, P.N. On computing the points and weights for Gauss--Legendre quadrature. *SIAM J. Sci. Comput.* **2003**, *24*, 945–954. [CrossRef]
20. Million, E. The hadamard product. *Course Notes* **2007**, *3*, 1–7.
21. Smith, M. *ABAQUS/Standard User's Manual, Version 6.9*; Dassault Systèmes Simulia Corp.: Johnston, RI, USA, 2009.
22. Maździarz, M. Unified isoparametric 3D lagrangeFinite elements. *CMES Comput. Model. Eng. Sci.* **2010**, *66*, 1–24.
23. Soman, R.; Kudela, P.; Balasubramaniam, K.; Singh, S.K.; Malinowski, P. A study of sensor placement optimization problem for guided wave-based damage detection. *Sensors* **2019**, *19*, 1856. [CrossRef] [PubMed]
24. Ouyang, P.-F.; Li, D.M.; Xie, J.-X. Modeling nonlinear deformation of slender auxetic structures under follower loads with complex variable meshfree methods. *Mech. Adv. Mater. Struct.* **2024**, *31*, 4969–4983. [CrossRef]

eng

MDPI

Article

Improved Lift for Thick Flatback Airfoils in the Inboard Blades of Large Wind Turbines

Micol Pucci * and Stefania Zanforlin

Department of Energy, Systems, Territory and Constructions Engineering, University of Pisa, Largo Lucio Lazzarino, 56122 Pisa, Italy; stefania.zanforlin@unipi.it
* Correspondence: micol.pucci@phd.unipi.it

Abstract: Thick airfoils are often used in the inboard sections of blades in commercial wind turbines. The main reason for this is to give the blade greater structural strength, but it is well known that thick airfoils degrade aerodynamic performance by stalling at relatively small angles of attack. The adoption of flatback airfoils instead of sharp trailing edges allows high lift coefficient to be maintained in thick airfoils. In this paper, we propose a novel airfoil design based on a passive flap to further improve the lift coefficient. This new design was tested by numerical simulation on several airfoils with different maximum thickness and different TE thickness. The improved design for flatback airfoils yields a higher lift coefficient, while the drag behaviour is strictly related to the baseline airfoil shape: some airfoils show a decrease in drag at certain angles of attack, while others exhibit a drag increase. In conclusion, the practical implications of the flap's utilisation on a state-of-the-art blade designed for a 5 MW wind turbine are analysed. The findings demonstrate that, due to the enhanced lift coefficient, it is feasible to shorten the chord while maintaining the power output, thereby reducing material costs.

Keywords: flatback; 2D; URANS; NREL 5MW; BEM

Citation: Pucci, M.; Zanforlin, S. Improved Lift for Thick Flatback Airfoils in the Inboard Blades of Large Wind Turbines. *Eng* **2024**, *5*, 2345–2361. https://doi.org/10.3390/eng5030122

Academic Editor: Antonio Gil Bravo

Received: 12 August 2024
Revised: 17 September 2024
Accepted: 19 September 2024
Published: 20 September 2024

1. Introduction

Wind energy is a crucial component of clean energy production and reducing greenhouse gas emissions. To achieve the climate targets, the European Union, for example, aims to cover more than one third of its electricity demand with wind power by 2030 and over 40% by 2050. The trend in recent years has been for wind turbines to increase in size. In fact, to be economically viable, wind turbines should increase their power output, which is proportional to the swept area of the rotor. This requires turbines with increasingly longer blades. At the same time, the mass of the blade increases approximately as the cube of the blade length, resulting in greater gravitational loads, particularly torsional, edgewise bending and flapwise bending loads (as explained in [1]). The latter is the most significant load, and to prevent the turbine from structural problems, it is necessary to adopt thick airfoils in the area of the blade close to the root. A thick airfoil (i.e., airfoil with $t/c \geq 0.25$ [2], where t is the maximum thickness and c is the airfoil chord) can be utilised to increase the resistance of the blade to load due to the enhanced moment of inertia. The downside is that the increase in thickness reduces the aerodynamic performance of the sharp trailing edge (TE) airfoil, since it stalls at small angles of attack, thereby preventing the achievement of a high lift coefficient C_L. As a compromise between structural requirements and aerodynamic performance, blunt trailing edge or flatback (FB) airfoils offer a good solution: they increase the maximum lift coefficient and the lift coefficient slope [3]. This is due to the different pressure distribution along the airfoil compared to the sharp trailing edge airfoil, resulting in a lower adverse pressure gradient on the suction side, which prevents premature boundary layer separation. In addition, flatback airfoils offer some transport advantages [4,5]. They are used in the inboard area where the chord is at its highest, so they allow for smaller dimensions.

On the other hand, flatback airfoils show a higher drag coefficient (C_D) compared to sharp TE and a wake development characterised by vortex shedding. However, when looking at the turbine as a whole, one of the most important parameters is the torque coefficient. Since the inboard region of the blade has a lower moment arm and lower relative velocity, the contribution to the torque generated by the entire blade is relatively small. Moreover, the kinetic energy available in the wind in a ring of thickness δr is proportional to the area of the ring, thus exhibiting a decrease towards the root. However, increasing the aerodynamic performance of the inboard region could be beneficial.

Considering Figure 1, a generic blade element located at a radial position r is shown. The infinitesimal lift δL and drag δD forces are highlighted as well as the relative velocity to the blade W. The infinitesimal torque and thrust acting on the blade element can be obtained considering the total force along the tangential (t) and normal (n) directions, respectively:

$$\delta Q = (\delta L \cdot sin(\phi) - \delta D \cdot cos(\phi))r = \frac{1}{2}\rho c W^2 C_Q r \delta r \quad (1)$$

$$\delta T = \delta L \cdot cos(\phi) + \delta D \cdot sin(\phi) = \frac{1}{2}\rho c W^2 C_T \delta r \quad (2)$$

where ϕ is the inflow angle, the sum of the angle of attack α and the twist angle β, and C_Q and C_T are the torque and thrust coefficients, respectively, defined as follows:

$$C_Q = C_L \cdot sin(\phi) - C_D \cdot cos(\phi) \quad (3)$$

$$C_T = C_L \cdot cos(\phi) + C_D \cdot sin(\phi) \quad (4)$$

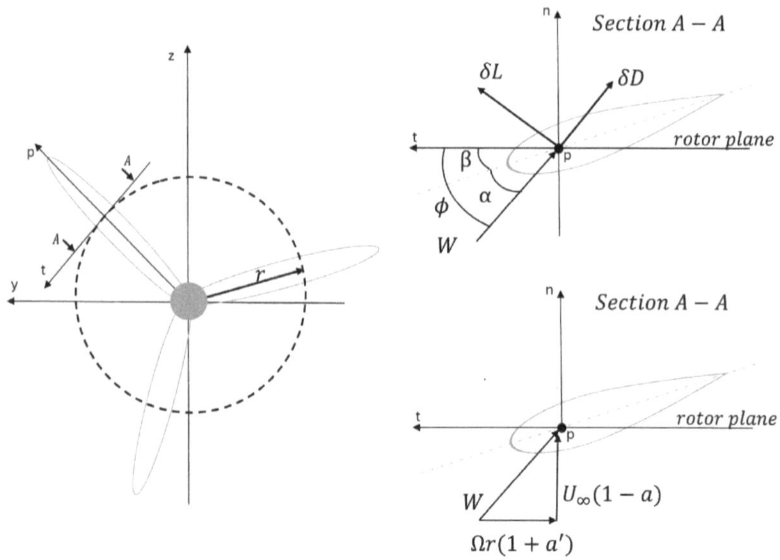

Figure 1. Qualitative representation of a blade element located at radial position r.

The main contribution to the torque in the inboard region of the blade is due to the lift coefficient C_L: near the root, the inflow angle ϕ is large, so the dominant term in the torque coefficient C_Q shown in Equation (3) is that which depends on C_L (as observed in [2,6,7]). A high C_D is therefore acceptable when combined with a high C_L, hence the adoption of flatback airfoils even when they have a high C_D.

The aim of this paper is to provide a solution to improve the lift coefficient of a flatback airfoil while keeping the drag coefficient almost the same. The latter, as mentioned, has little influence on the torque at the root of the blade but has a negative effect on the thrust coefficient C_T Equation (4). Consequently, it is necessary to identify novel geometries that will not substantially enhance the drag.

A review of the literature reveals numerous studies that have sought to reduce the drag coefficient of flatback airfoils. The primary objective of these studies was to reduce the drag coefficient and subsequently minimise noise emission due to vortex shedding. Various passive devices, including splitters and cavities (as shown in [8–10]), reveal a common trend in the aerodynamic performance resulting in a reduction in drag at relatively low angles of attack, accompanied by a concomitant reduction in the maximum lift coefficient achieved. Other works adopt a wavy TE such as in [11], where the drag reduction is accompanied by a lift loss. Another interesting solution is proposed in [4], where the swallow tail concept is studied. The trend shown in these works is that it is possible to attain a drag reduction paying the cost of a lift decrease. In the aforementioned works, it is not clear whether the lift losses are fully compensated for by the drag reduction or whether an increase in chord length may be necessary to maintain the same power output.

Other studies instead aim at increasing the lift coefficient; one example is given in [12] where the Gurney flap and vortex generators are analysed. The Gurney flap is a device that is typically set at a right angle at the TE on the pressure side of the airfoil. Its function is to enhance the camber of the airfoil, which results in a higher maximum lift coefficient and a reduction in the zero-lift angle. Vortex generators, on the other hand, delay the separation of the flow on the suction side of the airfoil, allowing for a higher maximum lift coefficient. These devices are beneficial for the lift, but a significant increase in drag is often linked to the increase in lift. The present paper has a similar objective: to enhance the lift coefficient of thick flatback airfoils. This may result in a slight increase in the drag coefficient, but given our focus on large-scale turbines, specifically those used in offshore applications, we will disregard the noise implications. The goal is to develop more efficient airfoils to achieve higher power production or to attain the same power output with a shorter chord, reducing plant costs. The techniques studied in the literature and listed above are not specifically designed for FB airfoils (except for the swallow tail), resulting in a poor overall performance (considering both the lift and drag coefficient). The present study proposes a flap designed specifically for FB airfoils. It consist of a passive device, which represents an economical solution to control the flow. Furthermore, the device is fixed to minimise maintenance costs.

The paper is structured as follows: Section 2 shows the methodology adopted and describes the geometry of the proposed flap. Section 3 describes the validation process, which encompasses the verification of calculation grids and the setup. To assess the robustness of the methodology, validation was conducted on multiple flatback airfoils, varying the airfoil under analysis. Subsequently, the novel geometry was implemented in a number of airfoils, as detailed in Section 4. In Section 5, a case study is presented for analysis. This case study involves the inboard blade of a turbine from the literature, which was analysed using conventional flatback airfoils. The results of this analysis are then compared with the new geometry proposed in this work. The paper concludes with a summary of its principal findings in Section 6.

2. Methodology

The objective of this study is to assess the potential for enhancing the aerodynamic performance of flatback airfoils. In particular, the use of passive devices has been explored as a means of increasing the C_L while minimising the impact of these devices on the C_D. Given the pivotal role of the C_L in determining the torque generated at the root of the blade, small differences and increments in C_D are deemed acceptable. The method used to enhance the lift is a flap of a specific design. The latter is depicted in grey in Figure 2. This kind of device can be added to any flatback airfoil; the slope of the flap is given by the line passing through points 1 and 2 (red in the figure), with point 1 located on the pressure side of the airfoil at a distance of 0.2%c from the TE, while point 2 is on the TE. The extent of the flap is 5% of the chord length in the x direction, and the thickness is 2% of the chord length.

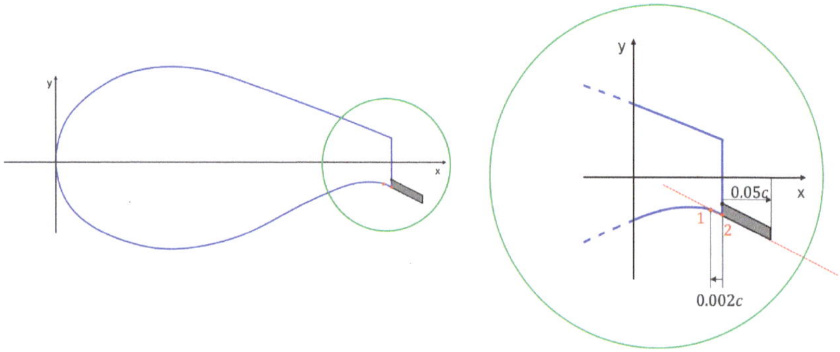

Figure 2. Qualitative representation of the flap proposed in this work. The zoom in the green circle highlights the geometric parameters of the flap (this is merely a qualitative representation; the image is not to scale in order to more clearly depict the flap details).

In order to conduct this study, 2D fluid dynamic simulations were employed. All computational grids were generated with the ANSYS ICEM v19.3 software and were structured multi-block grids. Such grids are a combination of the O-grid and the C-grid concept. This approach allows the advantages inherent to both grid types to be exploited. As documented in the literature [13], C-grids are well-suited for sharp TE airfoils, but they are less effective for finite thickness TE airfoils. Conversely, O-grids are adept at capturing the behaviour of flatback airfoils, but they tend to give poor predictions in the development of the wake and consequently in the C_D coefficient. Indeed, O-grid results in an excessively coarse grid in the wake region. Figure 3 shows a generic grid used in this work for flatback airfoils: a C-grid is visible with an O-grid surrounding the airfoil with a thickness of 2%c, which is enough to encompass the boundary layer thickness (the grid details in the red and violet boxes). The number of grid points on both the suction and pressure sides is 350, and there are 100 points on the TE. The wall distance from the first layer of cells is set to $1 \times 10^{-5}c$; this ensures that y^+ values are largely less than 1 throughout the airfoil wall. The computational domain extends 33c in the flow direction and 28c in the normal direction: the distance from the inlet to the airfoil nose is 14c as shown in Figure 3 on the left side. Figure 4 illustrates the grid employed in this study for the analysis of all FB airfoils with the flap. It differs from the grid utilised for flatback airfoils (as shown in Figure 3) in that an additional C-grid is incorporated at the TE in order to reproduce more complex geometry at the TE. The entire dominion extents are the same as those of Figure 3 on the left side. The boundary conditions used are a velocity inlet where the x velocity component is specified at the inlet and a pressure outlet at the exit where a free-stream pressure is set. The inlet has a C shape and also encompasses the top and bottom side; therefore, in order to set the free-stream velocity, we have to specify the velocity components: the x-component was set to $U_{\infty x}$, which is the free-stream velocity, while the y-component was zero. The numerical simulations were carried out with ANSYS Fluent v19.3 using URANS: a pressure-based approach was used with the SIMPLE scheme and default settings (also default controls). The turbulence model is $k - \omega$ SST, and the convergence criteria were 1×10^{-4} for all residuals. A total of 8000 time steps were used for each simulation, and the simulation time was such that the entire domain was traversed twice from the air flow. Therefore, the time step size was defined as $\Delta t = 2L_x/(U_{\infty x}8000)$, where L_x is the domain length in the x direction.

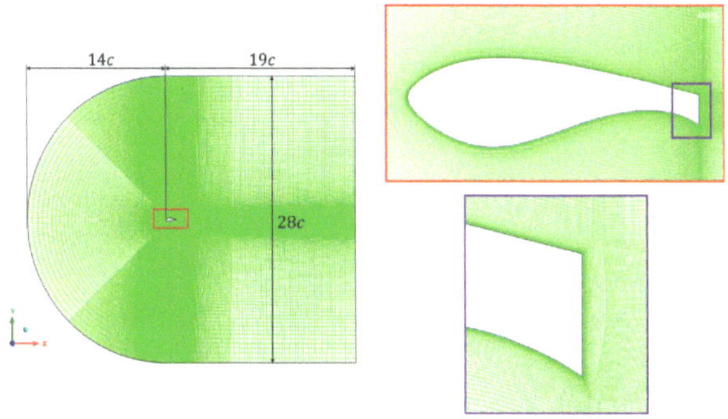

Figure 3. Example of a computational grid: (**left**) full domain extents and (**right**) detail of the grid near the blade (red rectangle) and near the TE (violet rectangle).

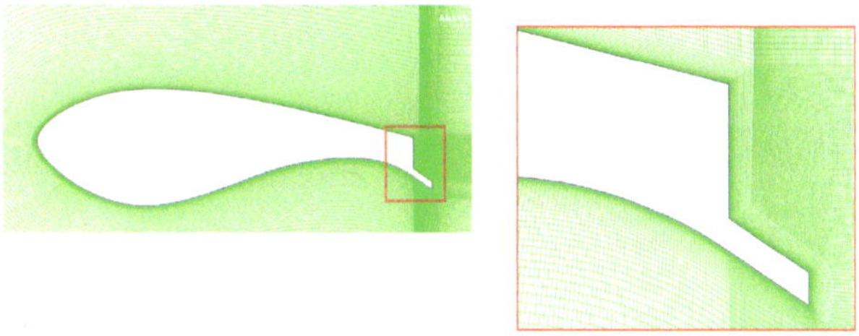

Figure 4. Example of the calculation grid used for FB airfoils with the flap (**left**), with the detail of the TE in the red rectangle (**right**).

3. Validation

Since the new flap proposed in this work is intended to be added to any flatback airfoil, the validation of the computational grid and the computational setup was carried out using three different airfoils to ensure a wide range of validity. The first airfoil is the one used in [14] called DU97FB (a flatback airfoil developed by modifying the DU97-W-300 airfoil); the second is called DU97ST and it is the airfoil used in [14] with the "swallow tail" (ST) concept (first appearance in [4]); while the third is the flatback airfoil proposed in [8], called FB-3500-1750. In Figure 5, the x-y coordinates of the profile are shown. The three airfoils are very different: DU97FB and DU97ST have a maximum thickness of 30%c and a TE thickness of 10%c, while the FB-3500-1750 airfoil has a maximum thickness of 35%c and a TE thickness of 17.5%c. Additional information about these validation cases is listed in Table 1. As previously stated, the simple flatback airfoils were evaluated using a grid type comparable to that illustrated in Figure 3, whereas FB airfoils with modifications at the TE (DU97ST in this case) were examined using a grid type analogous to that depicted in Figure 4.

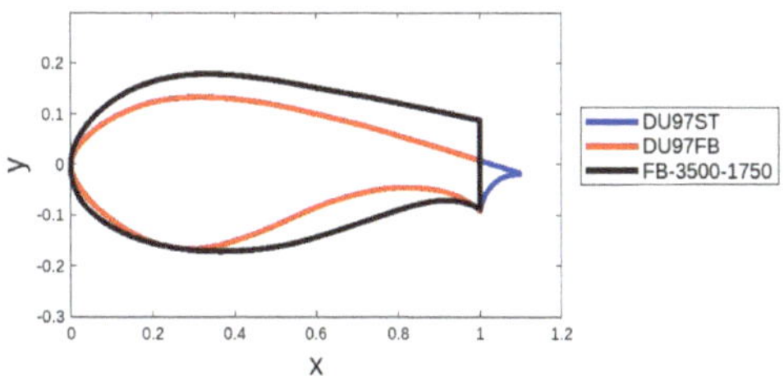

Figure 5. x-y coordinates of the three airfoils used for the validation task.

Table 1. Validation setup.

	DU97FB	DU97ST	FB-3500-1750
chord c [m]	0.65	0.65	0.66
$U_{\infty x}$ [m/s]	25	25	15
Re	1×10^6	1×10^6	6.66×10^5
Ma	0.07	0.07	0.04
transition	free	free	free

Figure 6 illustrates the lift and drag coefficients for all the airfoils analysed in the validation task. Comparisons are made between the 2D simulations of the present study and the corresponding experimental measurements. With regard to airfoils DU97FB and DU97ST, the experimental data are available in [14], while for FB-3500-1750, the free-transition measurements available in reference [15] were considered. The general trend of 2D URANS simulations is an overestimation of both the lift and drag coefficient (as also observed in [16]). Furthermore, it is evident that 2D results have difficulty in accurately capturing the occurrence of stall. A similar trend for FB-3500-1750 airfoil is shown in [17]. In particular, the linear trend is accurately represented with regard to the C_L, but the simulations overestimate both the maximum C_L and the alpha at which it occurs, resulting in a delayed stall. Therefore, when using 2D URANS it is important to operate at angles of attack far enough from the stall condition. For C_D, the simulated data consistently exceed the experimental data. These behaviours are consistent with those observed in other 2D numerical studies, as in [11,18,19]. Moreover, 2D URANS simulations have exhibit weakness in drag prediction as shown in [20], where a significant spanwise change in drag was measured.

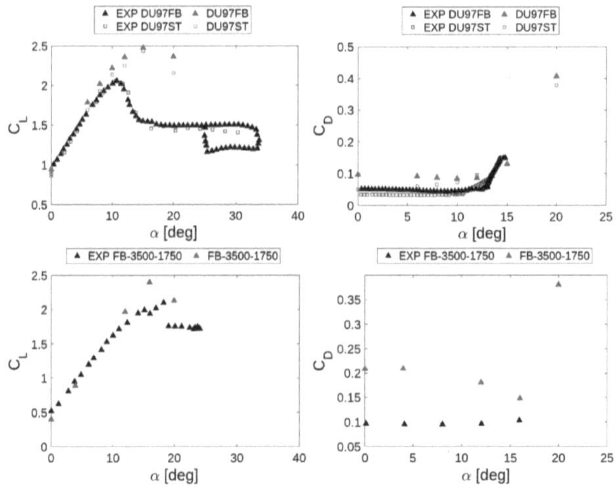

Figure 6. (**Top**) Lift and drag coefficient of the DU97FB and the DU97ST airfoils: comparison between the present 2D simulations and experimental measurements of [14]. (**Bottom**) Lift and drag coefficient of the FB-3500-1750 airfoil: comparison between the present 2D simulations and experimental measurements of [15].

4. Results

The flap geometry described in Figure 2 was applied to the test airfoils DU97FB and FB-3500-1750. Figure 7 illustrates the lift and drag coefficients obtained with and without the flap. It can be observed that the use of the flap allows for a higher lift coefficient across all angles of attack for both airfoils under consideration. With regard to the drag coefficient, the DU97FB airfoil exhibits a deterioration, manifested as an increase in drag across all considered angles of attack, with the only exception being for $\alpha = 0°$. Conversely, the FB-3500-1750 airfoil benefits from the use of the flap at low angles of attack (up to 12°). Figure 8 illustrates the pressure coefficient C_p for both airfoils with and without the flap, at two angles of attack, $\alpha = 0°$ and $\alpha = 12°$. The pressure coefficient is defined as follows:

$$C_p = \frac{p - p_\infty}{\frac{1}{2}\rho U_\infty^2} \tag{5}$$

where p is the static pressure on the blade, p_∞ is the undisturbed pressure and ρ is the air density. It is evident that the incorporation of the flap results in a higher overpressure on the pressure side of the airfoil, accompanied by a reduction in pressure on the suction side. The enhanced pressure difference across the airfoil is responsible for the observed lift enhancement. Conversely, the drag reduction observed at certain angles of attack can be attributed to the interference with the vortex shedding caused by the flap. The pathlines of the static pressure of DU97FB and FB-3500-1750 are plotted in Figures 9 and 10, respectively. The figures illustrate the situation with and without the flap at two different angles of attack, $\alpha = 0°$ and $\alpha = 12°$, and due to the oscillating trend of the C_L (and C_D) caused by the vortex shedding, the figures represent the instantaneous snapshot at the C_L peak. It is evident that the presence of the flap consistently confers an advantage to FB-3500-1750 across all angles of attack. The reduction in flow disturbance in the wake due to vortex shedding is a notable benefit of the flap. In the case of DU97FB, this remains true at $\alpha = 0°$, while the situation with the flap at $\alpha = 12°$ exhibits a slight deterioration. This is accompanied by an increase in drag, as illustrated in Figure 7. A similar pattern is evident in Figures 11 and 12, where the root-mean-square error (RMSE) of the mean velocity is plotted. It can be observed that the presence of the flap generally reduces velocity fluctuation in the wake, with the exception of DU97FB at $\alpha = 12°$. The preceding renderings were obtained through the utilisation of the data-sampling option in Fluent, given that vortex shedding is a periodic phenomenon, exhibiting varying frequencies across different scenarios, i.e., different airfoils under consideration. Hence, to ensure a fair comparison, the pressure coefficient of Figure 8 and the velocity fields of Figures 11 and 12 were calculated as the mean value of further 4000 time steps of simulation.

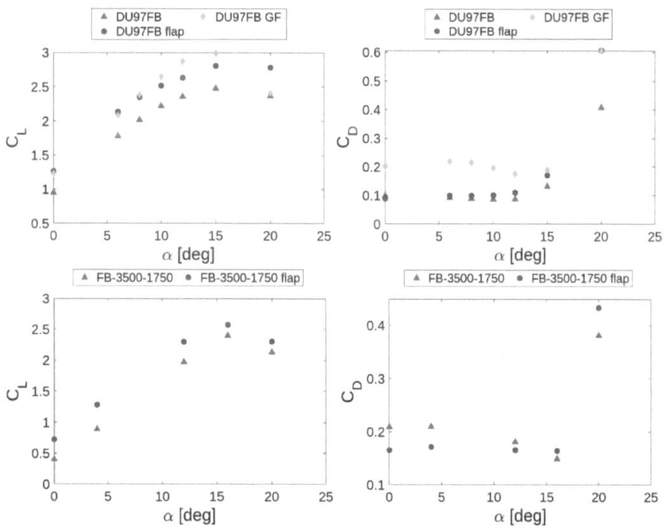

Figure 7. Coefficients of lift and drag: comparison between the simple flatback airfoil and the corresponding airfoil with the flap. DU97FB (**Top**) and FB-3500-1750 (**Bottom**). In case of DU97FB, the use of the Gurney flap (GF) is also analysed.

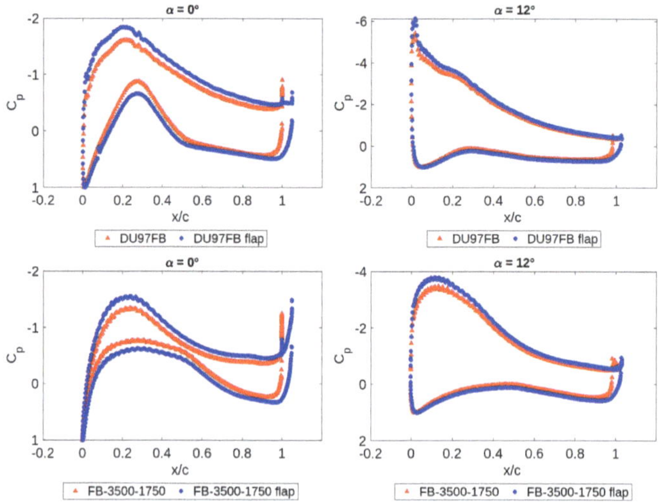

Figure 8. Pressure coefficient at $\alpha = 0°$ and $\alpha = 12°$ (**Top**) DU97FB with and without the flap (**Bottom**) FB-3500-1750 with and without the flap.

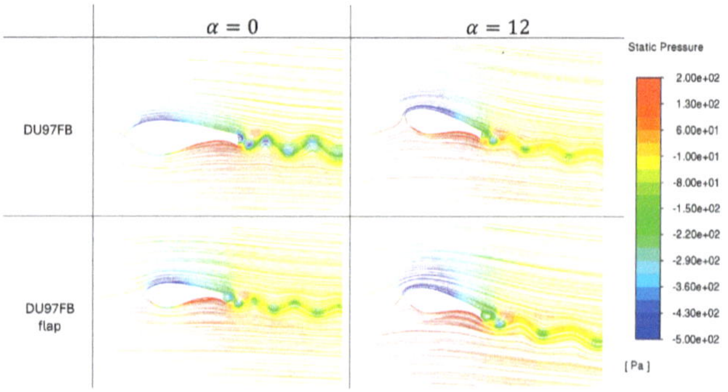

Figure 9. Pathlines coloured by static pressure at $\alpha = 0°$ and $\alpha = 12°$ for DU97FB with and without the flap.

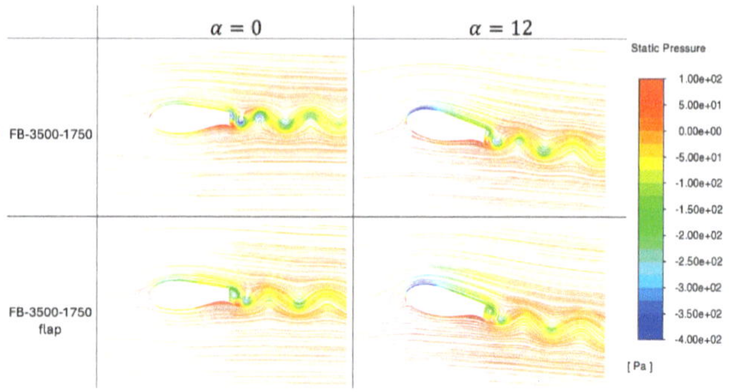

Figure 10. Pathlines coloured by static pressure at $\alpha = 0°$ and $\alpha = 12°$ for FB-3500-1750 with and without the flap.

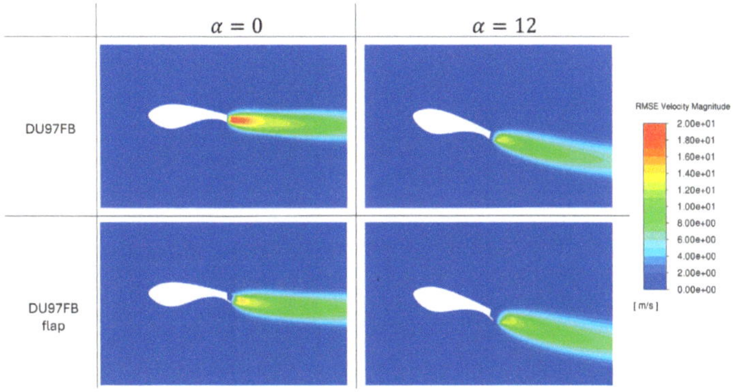

Figure 11. RMSE of the mean velocity magnitude at $\alpha = 0°$ and $\alpha = 12°$ for DU97FB with and without the flap.

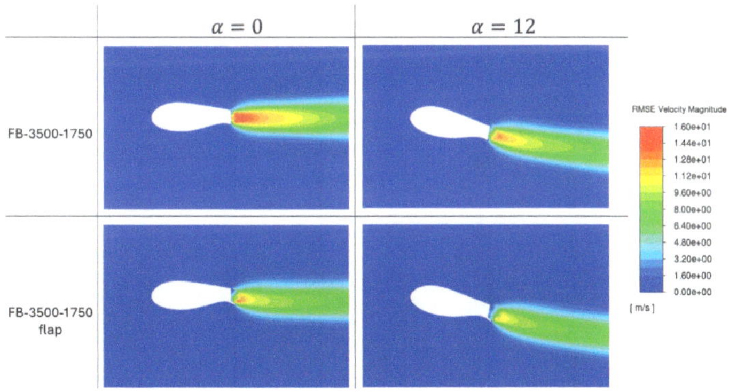

Figure 12. RMSE of the mean velocity magnitude at $\alpha = 0°$ and $\alpha = 12°$ for FB-3500-1750 with and without the flap.

In order to facilitate a comparative analysis between the novel flap geometry proposed in the present work and a widely utilised flap geometry, namely the Gurney flap (GF), the GF was applied to the simple DU97FB flatback airfoil. In this instance, the Gurney flap was scaled to the same dimensions as the proposed flap, with a thickness of 2%c and a vertical length of 5%c. Figure 7(top) presents a comparison of the lift and drag coefficients obtained with the addition of the Gurney flap and with our flap. The GF generates a notable increase in the drag coefficient at all alpha values, while the lift coefficient is comparable to that of our flap up to $\alpha = 8°$. For larger alpha values, the lift coefficient is higher when using the GF, with the exception of very high alpha values (e.g., $\alpha = 20°$). Figure 13 compares the streamlines colouring by static pressure at $\alpha = 10°$ for the simple flatback DU97FB airfoil, the airfoil with the flap and the airfoil with the GF. It is evident that the flap is capable of extending the overpressure region on the pressure side of the airfoil while maintaining relatively low drag values. The wake development in the case of the simple DU97FB (Figure 13a) and in the case with the flap (Figure 13b) are observed to be quite similar. In the case of the GF (Figure 13c), it is evident that the GF enhances the overpressure on the pressure side of the airfoil, but simultaneously increases the drag. The wake displays more pronounced depression regions. Therefore, the flap geometry proposed in this work, in contrast to the GF, allows for improvements in the lift coefficient while limiting the drag coefficient increment in comparison to the simple flatback airfoil case.

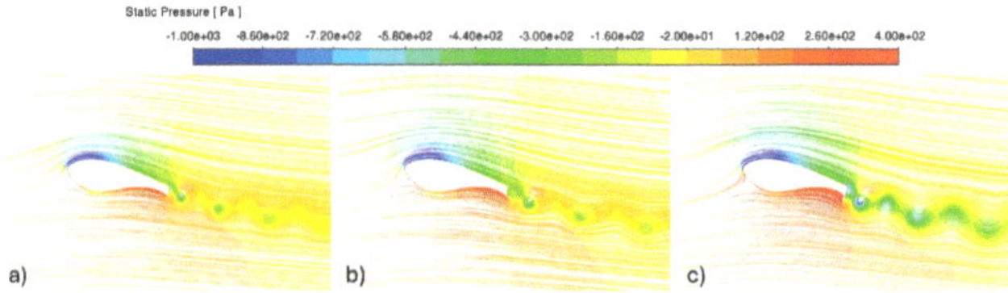

Figure 13. Streamlines coloured by static pressure of the DU97FB airfoil (**a**), DU97FB airfoil with a flap (**b**), and DU97FB airfoil with a Gurney flap (**c**) at $\alpha = 10°$.

The domain extent in the y direction was maintained at 28c for all simulations presented in this paper. To ensure that no significant boundary effects can occur with such a domain width, we conducted additional simulations, doubling the y extent. As illustrated in Figure 14, the new domain extends 47c in the flow direction and 56c in the normal direction. In particular, the new domain contains the old 28c domain, as illustrated in Figure 14, where the blue line encompasses the 28c domain. Consequently, the grid refinement in the calculation domain delimited by the blue line remains unchanged. The consideration of a larger domain has resulted in a slight increase in the number of grid cells, from 180 k to 200 k. In order to evaluate the influence of the domain y extent, the DU97FB airfoil was analysed at several angles of attack. The results demonstrate that enlarging the domain has no significant impact on the lift and drag coefficients, as illustrated in Figure 15. This is probably due to the fact that the grids used in this work have high refinement near the walls, so the grid surrounding the airfoil is very fine. Consequently, local effects are of greater consequence than far-field effects.

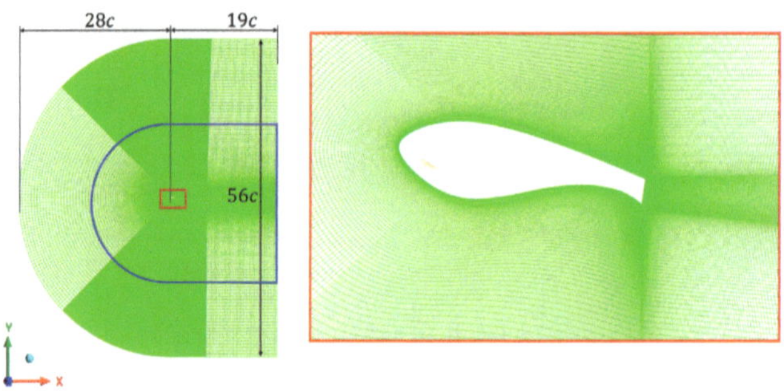

Figure 14. Enlarged calculation domain for sensitivity analysis. The new large domain includes the old domain, indicated by the blue line in the figure. The domain around the airfoil remains unchanged (red rectangle).

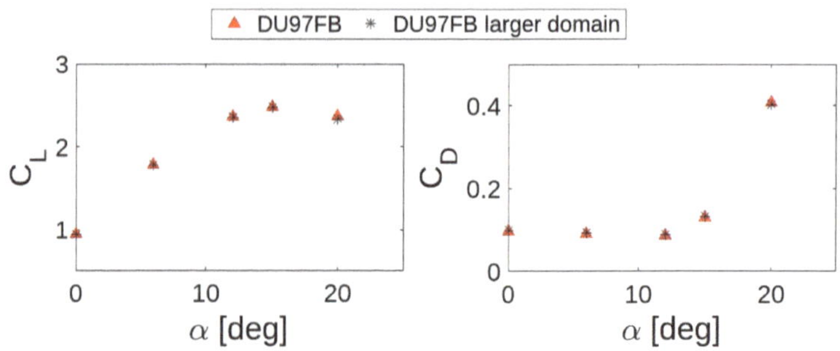

Figure 15. Coefficients of lift and drag: comparison between the "DU97FB" results (obtained with the 28c large domain) and the "DU97FB larger domain" results (obtained with the 56c large domain).

The flap geometry was then applied to two further airfoils: the DU40FB and the DU35FB airfoils. These airfoils are derived from the sharp trailing edge airfoils called DU40 and DU35, which are described in detail in [21,22] (coordinates available in [23] and in [24]). The DU40 and DU35 airfoils are obtained in turn from DU 99-W3-450 and DU 99-W3-350, respectively. All of the aforementioned airfoils are sharp TE airfoils, and the flatback airfoils were optimised in [25], where the x,y coordinates are provided for analysis. As would be expected, these airfoils have a maximum thickness of 40% and 35%, respectively, and a TE thickness of approximately 11% and 9%. The simulations for both airfoils were performed at Re of 7×10^6, with an undisturbed flow velocity of 34.4 m/s and a chord length of 3 m. All other settings are the same as the previous simulations. Figure 16 illustrates the lift and drag coefficients. For these airfoils, the flap exerts a beneficial influence on lift at all angles of attack, whereas the drag exhibits disparate behaviour. The DU40FB airfoil with the flap demonstrates a general increase in drag. In contrast, the DU35FB airfoil with the flap exhibits a drag reduction up to $\alpha = 10°$.

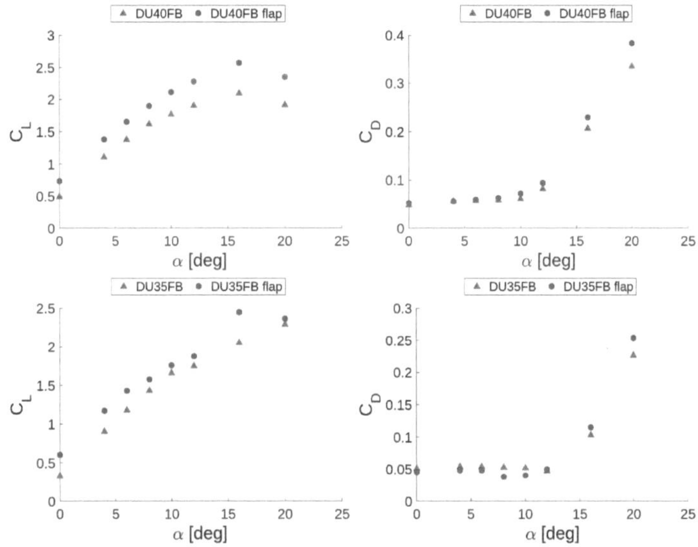

Figure 16. Coefficients of lift and drag: comparison between the simple flatback airfoil and the corresponding airfoil with the flap. DU40FB (**Top**) and DU35FB (**Bottom**).

A uniform conclusion regarding the flap's influence on drag cannot be drawn from the analysis of the results presented in Figures 7 and 16. The flap confers an advantage on some airfoils at relatively low angles of attack, while this is not the case for others. This behaviour appears to be independent of the thickness of the TE or the ratio between the maximum thickness of the airfoil and the TE thickness. Therefore, the limited number of airfoils tested precludes the formulation of a general rule for leveraging the flap solution in terms of drag reduction. However, the flap proposed in this work appears to offer a promising solution when compared to other devices. For instance, as discussed in reference [12], the adoption of the Gurney flap or vortex generators has been shown to significantly enhance lift, particularly in the case of Gurney flaps. However, this increase in lift is often accompanied by a significant increase in drag. In contrast, the flap proposed in this work has been observed to result in a reduction in drag at certain angles of attack.

Given the inherent limitations of 2D URANS analysis, namely the overestimation of drag and the delay in stall prediction, it is recommended that the proposed flap geometry be subjected to further analysis using a more sophisticated tool, such as 3D simulations or wind tunnel tests. However, without considering quantitative data, it is possible to make a qualitative assessment of the flap geometry proposed in this paper, which suggests that it has the potential to enhance the performance of the airfoil, particularly by increasing the lift coefficient.

In other words, a comprehensive campaign of 2D URANS fast simulations could serve as the initial phase of an airfoil optimisation methodology. However, the final verification should be conducted with more reliable CFD techniques, despite the increased computational cost.

5. Case Study

The DU40FB and DU35FB airfoils were selected for the purpose of analysing a portion of a realistic blade. The turbine in question is the NREL 5MW, for which the distribution of airfoils, chord and twist are available for reference in [21,26]. The focus of this analysis was on the portion of the blade from $\mu = 0.18$ to $\mu = 0.38$ (where μ is the dimensionless radius along the blade defined as $\mu = r/R$ and r is the local radius along the blade, while R is the turbine radius that in this case is 63 m). It should be noted that the original blade was constructed with sharp trailing edge airfoils; however, for the purposes of this analysis, these were substituted with FB airfoils. As previously stated, the DU40FB and DU35FB airfoils were optimised in [25], beginning with the sharp TE.

The objective of this section is to undertake a comparative analysis of the same portion of blade utilising both the flatback airfoils and the modified geometry with flaps. The analysis of the blade portion was based on blade element/momentum (BEM) theory. In particular, in order to calculate the torque Q and thrust T generated by the blade, it is necessary to calculate the relative velocity to the blade W (see Figure 1) as follows:

$$W = \sqrt{U_\infty^2(1-a)^2 + r^2\Omega^2(1+a')^2} \tag{6}$$

$$a = \frac{1}{3} \tag{7}$$

$$a' = \frac{a(1-a)}{\lambda^2\mu^2} \tag{8}$$

$$\lambda = \frac{R\Omega}{U_\infty} \tag{9}$$

where a is the axial induction factor, a' is the tangential induction factor, λ is the Tip Speed Ratio (TSR) and Ω is the rotation speed of the turbine. Equation (7) and consequently the a' value represent the condition of maximum power extraction. The angle of inflow ϕ and the angle of attack α are calculated as follows:

$$\phi = \arctan\left(\frac{1-a}{\lambda\mu(1+a')}\right) \tag{10}$$

$$\alpha = \phi - \beta \tag{11}$$

In order to evaluate and compare the torque and thrust of the "blade FB" and the "blade flap", where "blade FB" denotes the blade obtained using simple FB airfoils, while "blade flap" denotes the blade using FB airfoils with the flap, it is first necessary to ascertain the chord distribution. Given the analytical nature of this analysis, it is essential to pay close attention to certain aspects. To ensure a fair comparison in the torque and thrust calculation, it is necessary to use a different chord distribution when changing the airfoil under consideration. The theoretical blade design procedure proposed in [27] is based on the assumption of optimised operating conditions, which are represented by Equations (7) and (8). In this analysis, the fundamental parameter is the geometry parameter, defined as follows:

$$\frac{Bc}{2\pi R}\lambda C_L = \frac{4\lambda^2\mu^2 a'}{\sqrt{(1-a)^2 + (\lambda\mu(1+a'))^2}} \tag{12}$$

Equation (12) demonstrates that the chord distribution and the lift coefficient employed for the design are inversely proportional. Consequently, an airfoil with a higher lift coefficient will result in a shorter chord at each blade section. In light of the aforementioned concepts, modifying the airfoil entails modifying the chord distribution. In Equation (12), the chord distribution is a function of solely the C_L. The C_D is not considered, which is an acceptable simplification given that in [27] it was demonstrated that an almost equal chord distribution would have been obtained considering also the C_D. The drag effect will be evident only on the turbine performance, in particular, on the torque and thrust generated. If we assume that the value of C_L remains constant from the tip to the root, and thus the α, then the blade will exhibit a strongly non-linear tapering for $\mu \leq 0.6$. This is because the chord will result in very high values towards the root. Moreover, if a constant value of C_L is assumed, the blade will exhibit a high degree of twist, with a significant difference in the β value between the root and the tip. From an economic standpoint, the design procedure entails a linear tapering from the tip to the root, whereby the chord at the root is reduced and the loss is compensated for by an increase in the value of C_L and consequently in α. The left member of Equation (12) ($\frac{Bc}{2\pi R}\lambda C_L$) necessitates a specific product of $c \cdot C_L$ to achieve a given power extraction. Consequently, it is possible to reduce the chord when increasing the value of C_L, and vice versa. For the present analysis, we take R, the β distribution, λ and the rated wind speed from the real turbine proposed in [26]. Therefore, using equations from (6) to (11), it is possible to find the α distribution along the blade. The procedure is as follows:

1. Given R, λ, the β-distribution and the U_∞ of the NREL 5MW turbine ($R = 63$ m, $\lambda = 7$, and $U_\infty = 11.4$), we calculate α;
2. With the found α-distribution, we obtain the C_L-distribution;
3. With the C_L-distribution, we obtain the new chord distribution through Equation (12);

The ϕ-distribution is shown in Figure 17 (bottom right side) along with the α and β distributions. Using Equation (12), we found the chord distribution shown on the right side of Figure 17 (in the red box). The discontinuity at $\mu \approx 0.25$ is a consequence of the rapid change in the airfoil. For $\mu \leq 0.25$, the airfoil employed is DU40FB (or DU40FB with the flap), whereas otherwise the airfoil used is DU35FB (or DU35FB with the flap). In the case of a real blade, it would be advisable to incorporate a fitting to soften the impact of the airfoil change (see, for example, the airfoil distribution suggested in [28]). Figure 17 on the left illustrates the optimal chord distribution when utilising the DU35FB airfoil up to the tip of the blade. However, the present study focuses exclusively on the section of the blade between μ 0.18 and μ 0.38, as delineated by the red box.

The alpha value exhibits a notable decline as the μ value rises, which is responsible for the observed increase in the chord for $\mu \geq 0.25$.

Using Equations (3) and (4), it is possible to assess the torque and thrust coefficients obtained using the new designed blade section. Figure 18 illustrates the torque and thrust coefficients. As anticipated, the flap's deployment results in an elevated thrust coefficient, attributable to the enhanced lift coefficient and the concomitant increase in drag coefficient at specific angles of attack. With respect to the torque coefficient, the flap enables an increased value across all radial positions, substantiating the aforementioned assertion that at the inboard blade, the dominant influence on the torque coefficient is the lift coefficient, despite the rise in drag coefficient.

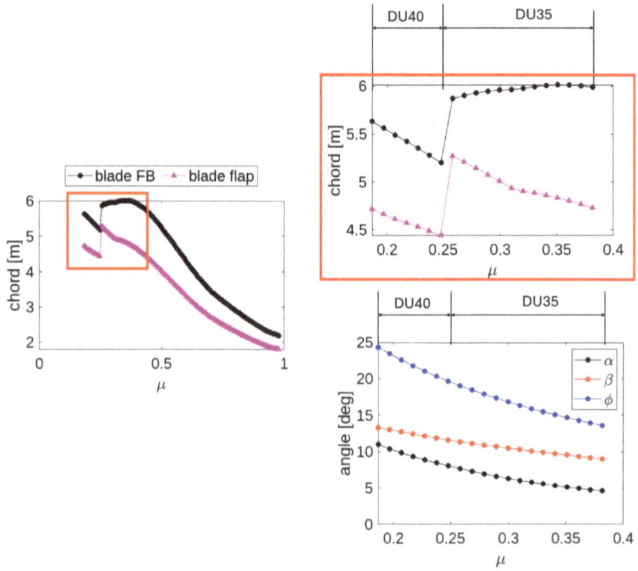

Figure 17. (**Left**) Chord distributions with zoom (red box) on the blade portion of interest. (**Right**) Chord distribution and ϕ, β and α distribution.

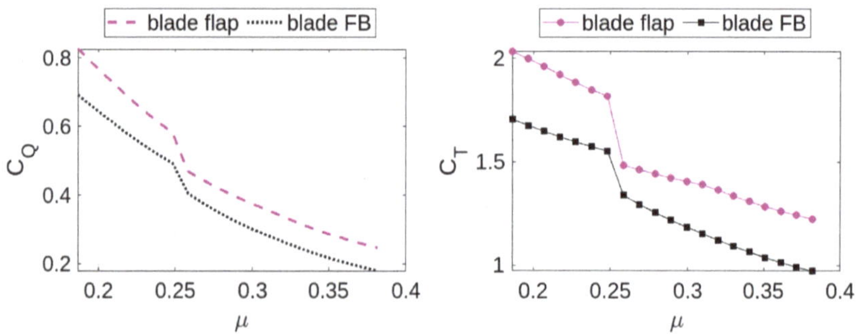

Figure 18. (**Left**) Torque coefficient along the inboard region of the blade and (**Right**) thrust coefficient.

By integrating Equations (1) and (2) along the blade section and multiplying by the number of blades, it is now feasible to obtain the torque and thrust, which are shown in Figure 19. The observed trend is consistent with the theoretical prediction, exhibiting a growth pattern from the root to the tip. The integration of the Q and T along the blade section yields an increase in total torque by 4% for the "blade flap" compared to the "blade FB" and at the same time a decrease in total thrust by 1%. It should be noted that all BEM calculations used to determine the Q and T values are based on the airfoil lift and drag coefficients presented in Figure 16, which were calculated at $Re = 7 \times 10^6$. The Reynolds number along the blade sections varies approximately between 6×10^6 and 10×10^6. Despite the simplifications employed in the analysis of this case study, the qualitative benefits of flap adoption are evident. It may be possible to construct the blade section with a shorter chord, reducing material requirements and costs while maintaining power generation and thrust on the structure. This can be achieved compatibly with structural requirements at the root.

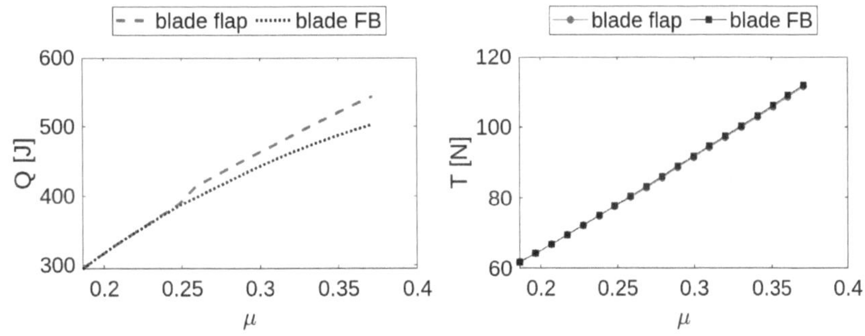

Figure 19. (**Left**) Torque along the inboard region of the blade and (**Right**) thrust.

6. Conclusions

In this study, we propose a novel flap geometry with the objective of enhancing the lift coefficient of flatback airfoils. The latter are employed at the inboard region of the blade in order to satisfy both structural and aerodynamic requirements. The flap can be added to any FB airfoil, and it typically results in an increase in the lift coefficient. With regard to the drag coefficient, no uniform trend is observed, as it is strictly dependent on the baseline airfoil shape. At specific angles of attack, some airfoils exhibit a reduction in drag when the flap is adopted, while other airfoils show a general increase in drag. The flap was evaluated on a number of airfoils, namely, DU97FB, FB-3500-1750, DU40FB, and DU35FB. However, due to the limited number of airfoils tested, no general conclusions can be drawn. Further analyses are required to assess the flap proposed in this work, as the selected airfoils exhibit significant variation in maximum thickness, thickness of the TE and camber.

The observed behaviour of the flap results in an increase in lift coefficient for all the airfoils, which coincides with the objective of this study. In the inboard region of the blade, where the torque coefficient is mainly influenced by the lift coefficient, enhancing the lift coefficient of the airfoil chosen for the blade design is of paramount importance. As a case study, we have attempted to evaluate the potential of adopting the flap in the inboard region of a realistic blade turbine. The method adopted is analytical, specifically based on BEM evaluations. The blade under consideration is the NREL 5MW turbine. The original blade features a sharp TE; however, a blade with FB airfoils ("blade FB") and a blade with the same airfoils but with the addition of the flap ("blade flap") were compared. In order to ensure a fair comparison through analytical evaluations, the chord distribution used for the two blades was modified. In particular, the airfoils with a higher C_L, i.e., airfoils with the flap, exhibit a shorter chord. The utilisation of the flap enables the attainment of a higher torque coefficient whilst concurrently a higher thrust coefficient. The total torque and thrust (Q and T) on the inboard region of the blade were evaluated, resulting in a +4% and −1% change, respectively. This is due to the fact that the BEM analysis enables the redesign of the blade, in particular, the chord distribution, as a result of the change in airfoil performance. The chord length exerts an influence on both Q and T; consequently, despite the fact that the thrust coefficient of the "blade flap" case is higher, the total thrust is lower due to the shorter chord. Neglecting quantitative results, the important thing in this case study was to evaluate the potential of adopting the flap in the inboard region of a realistic blade. What we can conclude is that it is possible to obtain approximately the same power output using fewer materials because of a shorter chord.

Future development of this work could be to further analyse the flap behaviour with more sophisticated tools, for instance, LES simulations or testing a couple of airfoil geometries with the flap addition in a wind tunnel. For a realistic blade inboard region, it will be useful to simulate an entire blade with 3D fluid dynamic simulations in order to assess the chord reduction. With 3D simulations, neglecting the BEM analytical approach, it will be possible also to analyse and compare airfoils with and without the flap using the same chord distribution, evaluating whether the flap could lead to higher power output.

Author Contributions: Conceptualisation, M.P. and S.Z.; methodology, M.P. and S.Z.; software, M.P. and S.Z.; validation, M.P. and S.Z.; formal analysis, M.P.; investigation, M.P. and S.Z.; data curation, M.P.; writing—original draft preparation, M.P.; writing—review and editing, M.P.; supervision, S.Z.; project administration, S.Z.; funding acquisition, S.Z. All authors have read and agreed to the published version of the manuscript.

Funding: This research was funded by the National Recovery and Resilience Plan (NRRP), CN1, Centro Nazionale di Ricerca in High-Performance Computing, Big Data e Quantum Computing, spoke 6: multiscale modelling and engineering applications.

Institutional Review Board Statement: Not applicable.

Informed Consent Statement: Not applicable.

Data Availability Statement: The raw data supporting the conclusions of this article will be made available by the authors on request.

Conflicts of Interest: The authors declare no conflicts of interest.

Abbreviations

The following abbreviations are used in this manuscript:

FB	Flatback
GF	Gurney flap
TE	Trailing edge
TSR	Tip Speed Ratio

References

1. Cox, K.; Echtermeyer, A. Structural design and analysis of a 10 MW wind turbine blade. *Energy Procedia* **2012**, *24*, 194–201. [CrossRef]
2. Standish, K.; Van Dam, C. Aerodynamic analysis of blunt trailing edge airfoils. *J. Sol. Energy Eng.* **2003**, *125*, 479–487. [CrossRef]
3. Baker, J.P.; Mayda, E.A.; van Dam, C.P. Experimental Analysis of Thick Blunt Trailing-Edge Wind Turbine Airfoils. *J. Sol. Energy Eng.* **2006**, *128*, 422–431. [CrossRef]
4. Grasso, F.; Ceyhan, O. Non-conventional flat back thick airfoils for very large offshore wind turbines. In Proceedings of the 33rd Wind Energy Symposium, Kissimmee, FL, USA, 5–9 January 2015; p. 0494. [CrossRef]
5. Barone, M.; Berg, D. Aerodynamic and aeroacoustic properties of a flatback airfoil an update. In Proceedings of the 47th AIAA Aerospace Sciences Meeting including The New Horizons Forum and Aerospace Exposition, Orlando, FL, USA, 5–8 January 2009; p. 271. [CrossRef]
6. Jackson, K.; Zuteck, M.V.; Van Dam, C.; Standish, K.; Berry, D. Innovative design approaches for large wind turbine blades. *Wind. Energy Int. J. Prog. Appl. Wind. Power Convers. Technol.* **2005**, *8*, 141–171. [CrossRef]
7. Law, S.; Gregorek, G. *Wind Tunnel Evaluation of a Truncated NACA 64-621 Airfoil for Wind Turbine Applications*; Technical Report; NASA Lewis Research Center: Cleveland, OH, USA, 1987. [CrossRef]
8. Baker, J.P. Drag Reduction of a blUnt Trailing-Edge Airfoil. Ph.D. Thesis, University of California, Davis, CA, USA, 2009.
9. Van Dam, C.; Cooperman, A.; McLennan, A.; Chow, R.; Baker, J. Thick airfoils with blunt trailing edge for wind turbine blades. In Proceedings of the Turbo Expo: Power for Land, Sea, and Air, Glasgow, UK, 14–18 June 2010; Volume 44007, pp. 923–931. [CrossRef]
10. Kahn, D.L.; van Dam, C.; Berg, D.E. *Trailing Edge Modifications for Flatback Airfoils*; Technical Report; Sandia National Laboratories (SNL): Albuquerque, NM, USA, 2008. [CrossRef]
11. Yang, S.J.; Baeder, J.D. Aerodynamic drag and aeroacoustic noise mitigation of flatback airfoil with spanwise wavy trailing edge. In Proceedings of the 33rd Wind Energy Symposium, Kissimmee, FL, USA, 5–9 January 2015; p. 0993. [CrossRef]
12. Cene, A.; Manolesos, M.; Grasso, F. Aerodynamic and Aeroacoustic Measurements of the Flow Past a Very Thick Flatback Airfoil with Passive Flow Control Devices. In Proceedings of the AIAA SCITECH 2022 Forum, San Diego, CA, USA, 3–7 January 2022; p. 0279. [CrossRef]
13. TPI Composites, Inc. *Innovative Design Approaches for Large Wind Turbine Blades*; Technical Report, SAND2004-0074; Sandia National Lab.: Albuquerque, NM, USA, 2004.
14. Ceyhan, O.; Timmer, W. Experimental evaluation of a non-conventional flat back thick airfoil concept for large offshore wind turbines. In Proceedings of the 2018 Applied Aerodynamics Conference, Atlanta, GA, USA, 25–29 June 2018; p. 3827. [CrossRef]

15. Mayda, E.A.; van Dam, C.P.; Chao, D.D.; Berg, D.E. *Flatback Airfoil Wind Tunnel Experiment*; Technical Report; Sandia National Laboratories (SNL): Albuquerque, NM, USA, 2008. [CrossRef]
16. Bangga, G.; Sasongko, H. Dynamic stall prediction of a pitching airfoil using an adjusted two-equation URANS turbulence model. *J. Appl. Fluid Mech.* **2017**, *10*, 1–10. [CrossRef]
17. Bangga, G.; Lutz, T.; Krämer, E. Numerical investigation of unsteady aerodynamic effects on thick flatback airfoils. *arXiv* **2017**. [CrossRef]
18. Sarlak, H.; Nishino, T.; Sørensen, J.N. URANS simulations of separated flow with stall cells over an NREL S826 airfoil. *AIP Conf. Proc.* **2016**, *1738*, 030039.
19. Szydlowski, J.; Costes, M. Simulation of flow around a static and oscillating in pitch NACA 0015 airfoil using URANS and DES. In Proceedings of the Heat Transfer Summer Conference, Charlotte, NC, USA, 11–15 July 2004; Volume 46911, pp. 891–908.
20. Metzinger, C.N.; Chow, R.; Baker, J.P.; Cooperman, A.M.; Van Dam, C. Experimental and computational investigation of blunt trailing-edge airfoils with splitter plates. *AIAA J.* **2018**, *56*, 3229–3239. [CrossRef]
21. Kooijman, H.; Lindenburg, C.; Winkelaar, D.; Van der Hooft, E. DOWEC 6 MW Pre-Design: Aero-Elastic Modeling of the DOWEC 6 MW Pre-Design in PHATASDOWEC Dutch Offshore Wind Energy Converter 1997–2003 Public Reports. 2003. Available online: https://citeseerx.ist.psu.edu/document?repid=rep1&type=pdf&doi=b331d7f80401ca1058058f4d130c3246843a1989 (accessed on 11 August 2024).
22. Blaylock, M.L.; Maniaci, D.C.; Resor, B.R. Numerical simulations of subscale wind turbine rotor inboard airfoils at low Reynolds number. In Proceedings of the 33rd Wind Energy Symposium, Kissimmee, FL, USA, 5–9 January 2015; p. 0493. [CrossRef]
23. Blaylock, M.L.; Maniaci, D.C.; Resor, B.R. *Numerical Simulations of Subscale Wind Turbine Rotor Inboard Airfoils at Low Reynolds Number*; Technical Report; Sandia National Laboratories (SNL): Albuquerque, NM, USA, 2015. [CrossRef]
24. Available online: https://github.com/old-NWTC/FAST/tree/master/CertTest/5MW_Baseline/Airfoils (accessed on 11 August 2024).
25. Kim, S.H.; Bang, H.J.; Shin, H.K.; Jang, M.S. Composite structural analysis of flat-back shaped blade for multi-MW class wind turbine. *Appl. Compos. Mater.* **2014**, *21*, 525–539. [CrossRef]
26. Jonkman, J.; Butterfield, S.; Musial, W.; Scott, G. *Definition of a 5-MW Reference Wind Turbine for Offshore System Development*; Technical Report; National Renewable Energy Lab. (NREL): Golden, CO, USA, 2009. [CrossRef]
27. Burton, T.; Jenkins, N.; Sharpe, D.; Bossanyi, E. *Wind Energy Handbook*; John Wiley & Sons: Hoboken, NJ, USA, 2011.
28. Griffith, D.; Richards, P.W. *The SNL100-03 Blade: Design Studies with Flatback Airfoils for the Sandia 100-meter Blade*; Technical Report; Sandia National Lab.(SNL-NM): Albuquerque, NM, USA, 2014. [CrossRef]

eng

MDPI

Article

WebTraceSense—A Framework for the Visualization of User Log Interactions

Dennis Paulino [1,2,*], André Thiago Netto [1,2], Walkir A. T. Brito [1] and Hugo Paredes [1,2]

1 School of Science and Technology, University of Trás-os-Montes e Alto Douro, 5000-801 Vila Real, Portugal; andre.netto@inesctec.pt (A.T.N.); walkirbrito@utad.pt (W.A.T.B.); hparedes@utad.pt (H.P.)
2 INESC TEC, 4200-465 Porto, Portugal
* Correspondence: dpaulino@utad.pt

Abstract: The current surge in the deployment of web applications underscores the need to consider users' individual preferences in order to enhance their experience. In response to this, an innovative approach is emerging that focuses on the detailed analysis of interaction data captured by web browsers. These data, which includes metrics such as the number of mouse clicks, keystrokes, and navigation patterns, offer insights into user behavior and preferences. By leveraging this information, developers can achieve a higher degree of personalization in web applications, particularly in the context of interactive elements such as online games. This paper presents the WebTraceSense project, which aims to pioneer this approach by developing a framework that encompasses a backend and frontend, advanced visualization modules, a DevOps cycle, and the integration of AI and statistical methods. The backend of this framework will be responsible for securely collecting, storing, and processing vast amounts of interaction data from various websites. The frontend will provide a user-friendly interface that allows developers to easily access and utilize the platform's capabilities. One of the key components of this framework is the visualization modules, which will enable developers to monitor, analyze, and interpret user interactions in real time, facilitating more informed decisions about user interface design and functionality. Furthermore, the WebTraceSense framework incorporates a DevOps cycle to ensure continuous integration and delivery, thereby promoting agile development practices and enhancing the overall efficiency of the development process. Moreover, the integration of AI methods and statistical techniques will be a cornerstone of this framework. By applying machine learning algorithms and statistical analysis, the platform will not only personalize user experiences based on historical interaction data but also infer new user behaviors and predict future preferences. In order to validate the proposed components, a case study was conducted which demonstrated the usefulness of the WebTraceSense framework in the creation of visualizations based on an existing dataset.

Keywords: user log; behavior identification; web platform; DevOps; statistical analysis

Citation: Paulino, D.; Netto, A.T.; Brito, W.A.T.; Paredes, H. WebTraceSense—A Framework for the Visualization of User Log Interactions. *Eng* **2024**, *5*, 2206–2222. https://doi.org/10.3390/eng5030115

Academic Editor: Antonio Gil Bravo

Received: 31 July 2024
Revised: 20 August 2024
Accepted: 23 August 2024
Published: 5 September 2024

1. Introduction

E-commerce websites, web games, and other digital platforms will significantly benefit from the implementation of personalization strategies. The utilization of user data to tailor content and experiences has the potential to enhance user satisfaction and engagement on these platforms. For instance, the implementation of personalized recommendations in e-commerce has the potential to enhance sales and foster customer loyalty by enabling users to access products that align with their preferences [1]. Personalizing digital experiences is a complex process [2]. Moreover, research has demonstrated that the adaptation and personalization of game content significantly enhance player satisfaction and engagement, thereby contributing positively to gamified education and rehabilitation [3]. By adapting the gaming experience to align with the preferences of individual players, developers can enhance user engagement and satisfaction [4]. In the context of web games, gamification plays a pivotal role in the creation of immersive and interactive experiences for players. The utilization of gamification techniques and personalization facilitates player retention and encourages repeated gameplay, which contributes to the success of web games in the digital landscape [5].

Similarly, personalization can be beneficial in other contexts, such as web platforms designed for use in healthcare settings. For example, modifying the interface to accommodate users with limited digital literacy can markedly enhance the user experience and accessibility [6]. The research on understanding, developing, and assessing personalization systems is distributed across multiple disciplines and draws upon methodologies and findings from a variety of research fields and traditions, including artificial intelligence (AI), machine learning (ML), human–computer interaction (HCI), and user modeling based on social and cognitive psychology. The focus of AI and ML is the optimization of personalization applications through the implementation of accurate algorithmic decision-making and prediction models. The fields of human–computer interaction (HCI) and information systems examine the phenomena surrounding the use and interaction with personalization systems. Meanwhile, cognitive science provides the theoretical foundations for these observed effects [7].

Digital phenotyping is an approach that employs data gathered from digital devices to evaluate and comprehend user behaviors and characteristics. The term "digital phenotyping" is defined as the moment-by-moment quantification of the individual-level human phenotype in situ, utilizing data from personal digital devices [8]. A related concept is task fingerprinting, which refers to the distinctive patterns and behaviors exhibited by users when engaged with digital tasks. By analyzing these patterns, developers are able to create highly personalized experiences that are tailored to the specific behaviors and preferences of users [9]. These techniques are of great value in the enhancement of the personalization of web applications, including games, through the provision of a detailed understanding of user engagement. A similar trajectory may be envisaged for digital phenotyping, whereby traditional statistical tools may be combined with machine learning to translate smartphone sensor and usage data into biomedical and clinical insights [10]. In order to identify behavioral patterns and inform personalization strategies, it is essential to create a web platform for the visualization of user log interaction data. Such a platform can provide developers with intuitive tools for the analysis of interaction data, including mouse clicks and keystrokes, thereby enabling the identification of trends and preferences. The visualization of these data enables informed decision-making by developers regarding game design and content delivery, which in turn enhances the user experience [11].

Some articles have reported the development of platforms designed to facilitate the visualization and insight generation of user logs. In the work of Rzeszotarski and Kittur [12], the techniques employed for the visualization of user logs comprise a combination of raw event logs, aggregate behavioral features, and advanced visualization methods, which are used to provide insights into the behavior and outputs of crowd workers. The raw event logs record detailed data such as mouse movements, clicks, keypresses, and focus changes, which are then summarized into visual timelines to represent the behavior of the worker. In the context of learning analytics, an article discussed the techniques used for user log visualization [13]. The study underscores the importance of integrating data across multiple learning platforms, which is represented visually using a bespoke dashboard. The visualization component of the dashboard was designed with the objective of aggregating data from disparate sources, including standalone applications, and creating visual representations based on user input. The architectural design enables the generation of a range of visual reports that facilitate a deeper comprehension of learner behavior and performance across diverse digital platforms. In the context of digital phenotyping, a study employed a range of sophisticated visualization techniques to contextualize and interpret low-level sensor data collected from smartphones [14]. The visualizations permit analysts to explore and interpret intricate, context-rich datasets, thereby facilitating the discovery of behavioral patterns, or "phone-o-types", that can be predictive of health outcomes. This approach is particularly effective in addressing the challenges inherent to low-level data, such as the absence of contextual information and the necessity for data anonymization, which are prevalent in smartphone-sensed datasets.

The primary objective of this article is to present the utilization of the WebTraceSense project to analyze user interaction data from a dataset derived from another study involving 134 distinct crowd workers, employing the task fingerprinting technique to examine the user interaction logs [15]. The objective of our research is to investigate the potential of WebTraceSense in enhancing the personalization of web applications by leveraging these logs. The following research questions (RQs) were formulated:

- RQ1: How can the visualization of user interaction logs enhance the personalization of web applications, including e-commerce websites, web games, and other digital platforms?
- RQ2: What are the most effective statistical and machine learning techniques for analyzing user interaction data to identify and predict user behavioral patterns in web applications?

In order to respond to these queries, we generated visualizations and applied statistical analyses with a view to assessing the potential for personalized visualizations to enhance user experience and engagement on digital platforms. The generation of detailed visual representations of user interactions facilitates the uncovering of patterns and insights that can inform the customization of web content and features. Furthermore, the optimal statistical and machine learning methodologies for analyzing the interaction data gathered through the WebTraceSense project were identified. The objective is to ascertain which techniques are most efficacious in identifying and predicting user behavioral patterns, thereby informing the development of personalized web content. The behavioral traces of crowd workers were analyzed from a performance standpoint using task fingerprinting, with data from cognitive ability tests being employed to complement this analysis. The WebTraceSense platform distinguishes itself from existing systems described earlier by offering a comprehensive and advanced approach to user interaction analysis and personalization. In contrast to the aforementioned platforms, which concentrate on particular domains, such as crowd worker behavior visualization [12], learning analytics [13], or digital phenotyping [14], WebTraceSense integrates a robust backend and frontend with AI-driven insights and statistical methods to provide interactive visualizations across various digital platforms. Its unique features, including the dynamic player for detailed interaction representation and a DevOps cycle for continuous integration, enable developers to predict, and enhance user experiences with unprecedented precision. This holistic approach makes WebTraceSense a suitable tool for optimizing user interfaces and tailoring web applications to individual preferences, setting it apart from other platforms that does not focus on integration and adaptability.

The investigation of WebTraceSense in the customization of web application interaction data has the potential to yield significant outcomes for the web development industry. A more profound comprehension of user behaviors and preferences enables developers to refine the design, functionality, and content of their applications in a manner that aligns with the specific requirements and interests of their target audience.

2. Background

User log analysis serves as a versatile tool across various domains, particularly in enhancing healthcare diagnostics and treatments by integrating traditional and digital biomarkers. Termed digital phenotyping, this method improves "point-of-care" diagnostics by enabling swift and real-time feedback, thus improving patient care by offering immediate insights into patient conditions [16–18]. A study utilizing machine learning to analyze keystroke dynamics and metadata from smartphone interactions identified significant correlations with depressive symptoms, measured by PHQ-9 scores [19], highlighting the method's accuracy and its utility for routine mental health monitoring [20]. Additionally, physiological assessments, such as those analyzing arm movement functions, are seamlessly integrated into web platforms, offering quantitative and automated evaluations that facilitate the personalization and adaptation of user interfaces for enhanced interaction [21].

Evaluating performance on microtasks, commonly used in crowdsourcing, can extend beyond simple outcome measures to include interaction log analysis, offering a detailed view of worker engagement and efficiency [22,23]. These logs, often collected via simple JavaScript libraries, reveal insights into user behavior, distinguishing between high- and low-effort interactions which can inform task allocation and design. Game analytics studies underscore the importance of leveraging user data to refine game mechanics and narratives to suit individual player profiles [24].

The methodologies used in web interaction analysis are varied, including descriptive statistics, machine learning, knowledge inference, deep learning, and sequence mining. Each methodology provides unique insights into data patterns and user behaviors, aiding in the development of targeted and effective web applications. For instance, deep learning models are particularly adept at analyzing complex patterns within large datasets, providing predictive insights that enhance the identification of user behaviors [25–28]. Although significant advancements have been made in user log analysis, particularly by researchers like Mastoras and colleagues [20], there remains a research gap in the strategic application of user log analysis through advanced statistical and computational techniques, such as generating interactive visualizations, to create tailored web experiences. This approach not only boosts user satisfaction and engagement but also supports the ongoing enhancement of digital platforms.

3. WebTraceSense Platform—A Framework for the Visualization of User Log Interactions

The WebTraceSense platform aims to provide a comprehensive solution for analyzing web interaction data. The project involves the development of a backend using .NET (available online: https://dotnet.microsoft.com, accessed on 22 August 2024) (a framework based on C#) for the CRUD (Create, Read, Update, Delete) operations on the visualization data, and a Django API (available online: https://django-rest-framework.org, accessed on 22 August 2024) (a framework based on Python) for the application of statistical and machine learning methods on the user interaction log data. This backend is complemented by the establishment of a database designed to store interaction log files in JSON format, ensuring efficient data management and retrieval. The authentication for accessing the backend is made with Auth0 (available online: https://auth0.com, accessed on 22 August 2024), which handles OAuth authentication. On the frontend, the development focuses on creating a web application using React (available online: https://react.dev, accessed on 22 August 2024), to facilitate the visualization of interaction data. This involves not only the creation and editing of HTML/CSS pages but also the implementation of a step-by-step web form, commonly referred to as a "Wizard," which helps users customize the visualization of their web interaction records. Additionally, registration and login pages are developed, adhering to web usability heuristics to ensure a smooth and intuitive user experience. A significant aspect of the WebTraceSense platform is the development of visualization modules using JavaScript libraries such as D3.js (available online: https://d3js.org, accessed on 22 August 2024). These modules are essential for illustrating interaction data in a dynamic and interactive manner, providing users with clear and insightful visual representations of their data. To ensure that the platform operates efficiently in a production environment, a robust DevOps [29] and automation process is implemented. This includes the deployment of backend and frontend components into production mode, conducting thorough integration tests to validate both components, and creating containerized solutions using Docker (available online: https://www.docker.com, accessed on 22 August 2024). These steps are crucial for maintaining the reliability and scalability of the platform.

In order to address the challenges inherent in the management and analysis of large volumes of interaction data, a NoSQL database (specifically, MongoDB, available online: https://www.mongodb.com, accessed on 22 August 2024) was implemented in the .NET backend of the WebTraceSense framework. This approach was selected due to its scalability and flexibility in handling unstructured data, which is of paramount importance given the diverse and extensive nature of user interaction logs. The initial design of the backend was intended for individual user analysis. However, subsequent adaptations were made to facilitate collective data visualization, enabling the simultaneous display of interaction data from multiple users. This required significant enhancements to the D3.js-based visualization module, ensuring that it could effectively represent and analyze aggregated data across different users. These technical improvements not only enhance the platform's capability to manage large datasets, but also provide deeper insights into group behaviors and trends, further extending the applicability and effectiveness of the WebTraceSense framework.

Figure 1 illustrates a component diagram of the WebTraceSense framework. The WebTraceSense framework is contingent upon researchers, web developers, or any individual interested in analyzing user interaction data having access to a website where these logs can be collected. The interaction logs are exported into a JSON file, which captures essential metrics such as click data, key presses, and the latency of the user's session. The OAuth protocol is employed to streamline the process of user authentication, thereby enabling users to log in using a third-party provider. Upon successful authentication on an external platform, users are redirected to the WebTraceSense frontend. The main webpage provides users with the ability to manage the visualizations of user logs. The generation of these visualizations is facilitated through the utilization of a wizard form, which guides users through a series of steps, including the upload of interaction log files and the selection of metrics to be analyzed. These may include mouse data, key presses, and temporal intervals, as well as the types of visualization charts to be generated. Following the submission of a visualization request, the 'Visualization Module' processes it to produce dynamic, data-driven graphics. Subsequently, the backend collates data on the visualization metrics employed, which are then subjected to statistical analysis in order to optimize the visualizations produced. The refined visualizations provide valuable insights that assist in the definition of requirements for researchers, game developers, or web developers to enhance their web platforms. This iterative process of continuous refinement entails the generation of new user interaction logs and the conduct of further analyses. Throughout its operational lifespan, the WebTraceSense project incorporates a DevOps cycle to automate the integration of statistical methods into its data-driven processes. This automation facilitates a seamless transition of backend and frontend components from the development phase to the production environment, with unit and integration tests ensuring efficient and seamless functionality.

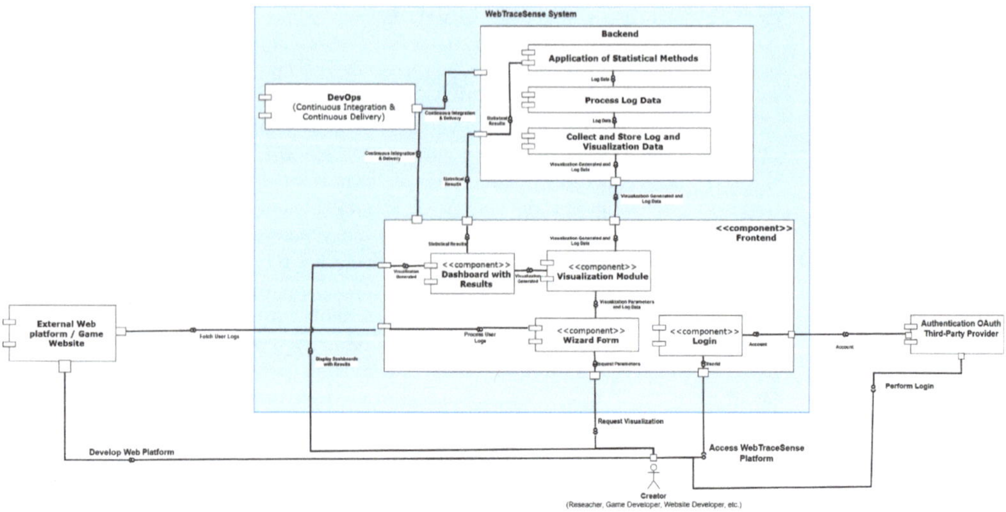

Figure 1. Unified Modeling Language (UML) component diagram for the WebTraceSense system.

The activity diagram for the visualization module in the WebTraceSense project outlines a structured process for creating data visualizations based on user interaction logs (see Figure 2). Initially, the module fetches the interaction log data from the specified web platform that the researcher wishes to analyze. The user interaction logger is capable of collecting a multitude of granular data points, which collectively provide profound insights into the manner in which users interact with web applications (e.g., [15]). The interaction logger is equipped with the capability to meticulously capture and record mouse clicks, encompassing not only the frequency but also the precise coordinates of each click within the interface. This enables a spatial analysis of user behavior, thereby identifying areas of the interface that receive the most attention or interaction. Furthermore, the logger records the context in which each click occurs, such as whether it is part of a navigation sequence, a form submission, or an interaction with a specific UI element. These contextual data are of great importance for the comprehension of the user's intent and the assessment of the effectiveness of the interface design. Another essential function is keystroke logging, whereby the logger records both the frequency and timing of key presses. These data offer insights into user typing patterns, text input behavior, and potential cognitive load. The logger can differentiate between various types of key presses, including alphanumeric input, function keys, and special characters, facilitating a comprehensive analysis of text entry tasks and user proficiency.

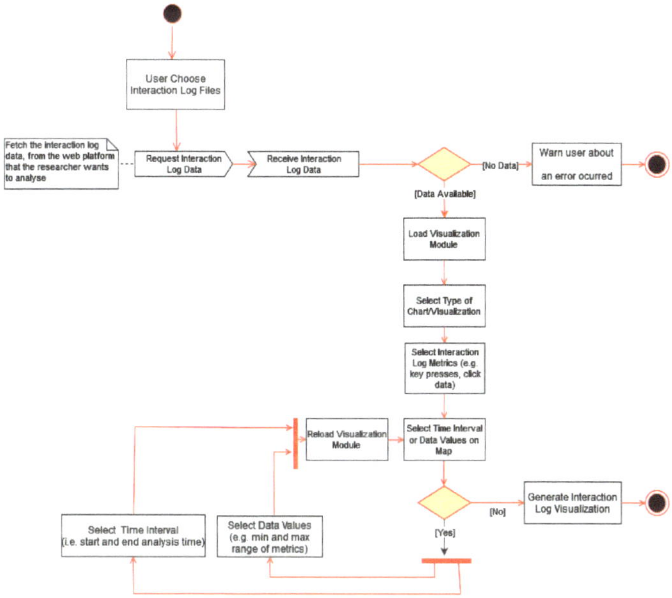

Figure 2. UML activity diagram for the visualization module.

Once the data are retrieved, the user can choose the type of chart or visualization they prefer, which sets the framework for how the data will be presented. Following this, the user selects specific interaction log metrics to focus on, such as mouse clicks or keystrokes, depending on what aspects of user behavior they are most interested in exploring. The user also specifies a time interval or selects specific data values to map, which helps in narrowing down the data analysis to relevant time frames or specific events. Finally, the module generates the visualization of the interaction log, translating the complex data into a visual format that is easier to understand and analyze. This streamlined process allows for efficient and tailored visualizations, providing valuable insights into user interactions on the web platform.

The sequence diagram of the visualization module (see Figure 3) provides a comprehensive overview of the process, which commences when a creator (researcher, web developer, or game developer) requests interaction log data from a particular web platform. This constitutes the inaugural stage in a series of data-driven activities with the objective of creating sophisticated visualizations. Once the log data have been obtained, the creator must then select the type of visualization that is required. This may range from a relatively simple chart to a more complex graphical representation, such as a heatmap or a dynamic visualization of player interactions. The usage of line charts enables the illustration of trends over time, such as the frequency of user interactions or the progression of task completion. This provides a clear, chronological view of user behavior. Heatmaps represent another indispensable visualization instrument within the system, employed to highlight areas of elevated interaction density on the user interface, thereby effectively identifying those locations where users frequently click or direct their attention. This is particularly useful for identifying user interface (UI) elements that are most engaging or, conversely, those that may be causing user frustration. Furthermore, the dynamic player represents an additional interactive visualization option, enabling users to replay sequences of interactions, such as mouse movements and keypresses, in real time or at adjusted speeds. This tool provides an immersive method of analyzing the flow and timing of user actions, offering insights into the user's navigation patterns and decision-making processes. The aforementioned visualizations, which are powered by D3.js, not only present the interaction data in an accessible and engaging format, but also enable users to interact with the data through filtering, zooming, and panning. This allows for a more in-depth and comprehensive analysis to be conducted.

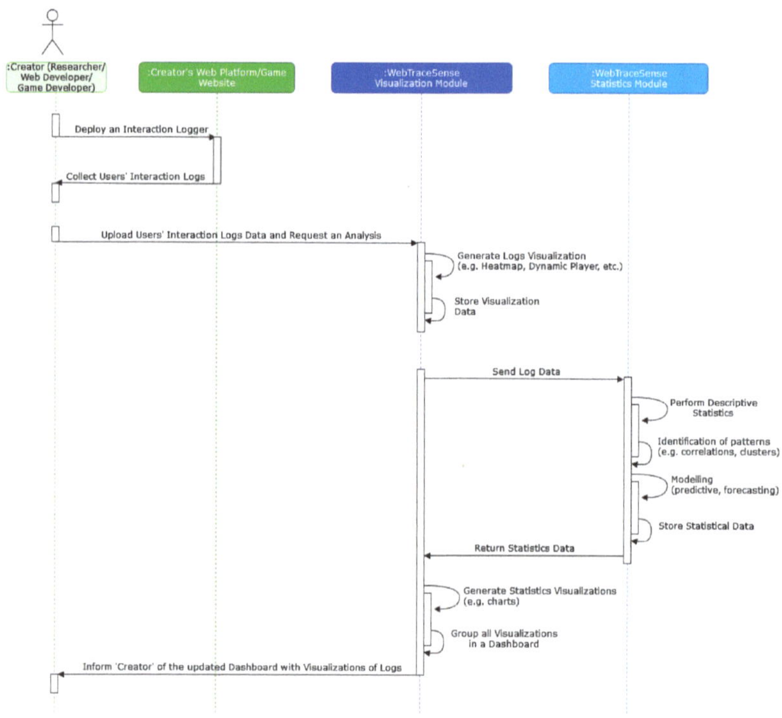

Figure 3. UML sequence diagram for the interaction between the visualization and statistics modules.

Subsequently, the creator must select the specific interaction log metrics that will be subjected to analysis. Such data points may include click rates, navigation paths, or time spent on specific tasks. Additionally, the creator establishes the temporal scope of the data, or alternatively, selects specific data values for mapping, thereby facilitating a concentrated examination of specific behaviors or occurrences. At this juncture, the visualization module deploys statistical techniques to conduct a comprehensive analysis of the specified metrics. Descriptive statistics may be employed to summarize the characteristics of the data, while inferential statistics may be used to make predictions and test hypotheses about user behaviors. It is imperative that statistical analysis is conducted on the raw data in order to accurately interpret it and transform it into actionable insights. Once the data have been statistically processed, the visualization module generates the visual representation. This final visualization integrates both the statistical analysis and the initial user specifications to produce a comprehensive, easily comprehensible display of user interactions. These visualizations serve as powerful tools for the creator, providing a visual summary of complex datasets and highlighting trends and patterns that may not be immediately apparent from raw data alone.

Figure 4 presents a comprehensive visual guide to the WebTraceSense platform, illustrating the key components of its user interface that facilitate efficient interaction analysis and visualization. The image in the top left-hand corner depicts the Home Screen, which represents the initial point of interaction for users and provides access to the various functionalities of the platform. The top right screenshot illustrates the login screen integrated with Auth0. The bottom left image depicts the CRUD visualization table, which plays a pivotal role in the management and editing of user-generated visualizations, enabling real-time modifications and updates. The bottom right screenshot presents the Wizard form, which guides users through the process of generating custom visualizations. The generated visualizations can have multiple options, and the main types of charts are the line chart (see Figure 5), bar chart, pie chart, and the dynamic player (presented in Figure 6).

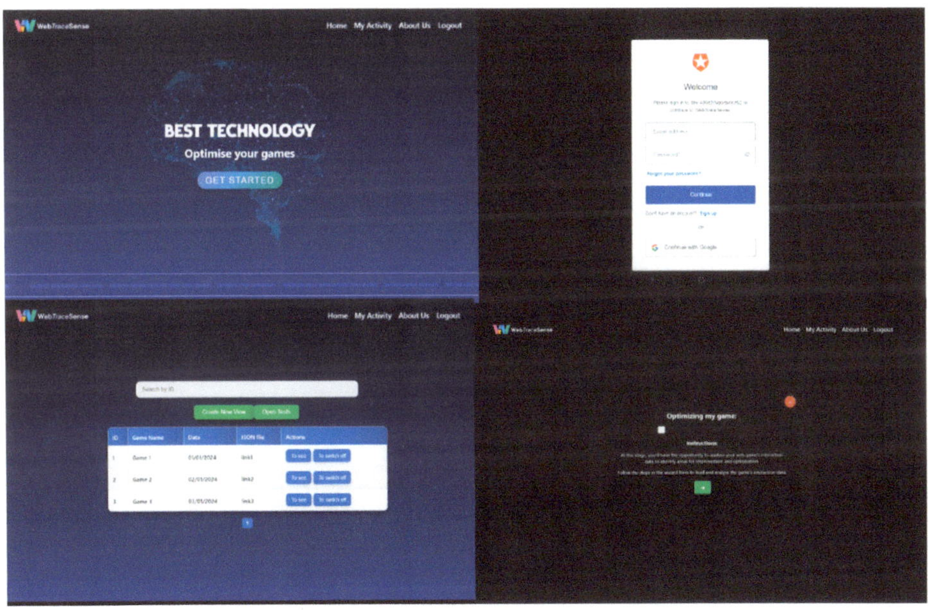

Figure 4. Overview of the WebTraceSense (**Top left**: Home Screen; **Top right**: Login screen with Auth0; **Bottom left**: CRUD visualization table; **Bottom right**: Wizard form for generating visualizations).

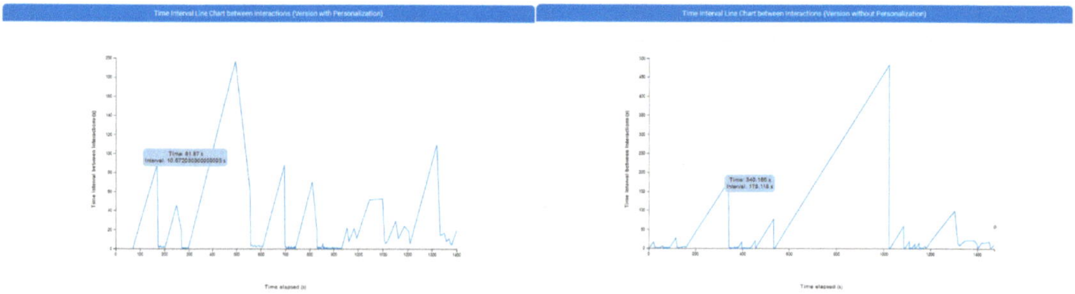

Figure 5. A line chart showing one user log for the version with personalization (**left**) and another for the version without personalization (**right**). The user log is from the Paulino et al. [12] dataset.

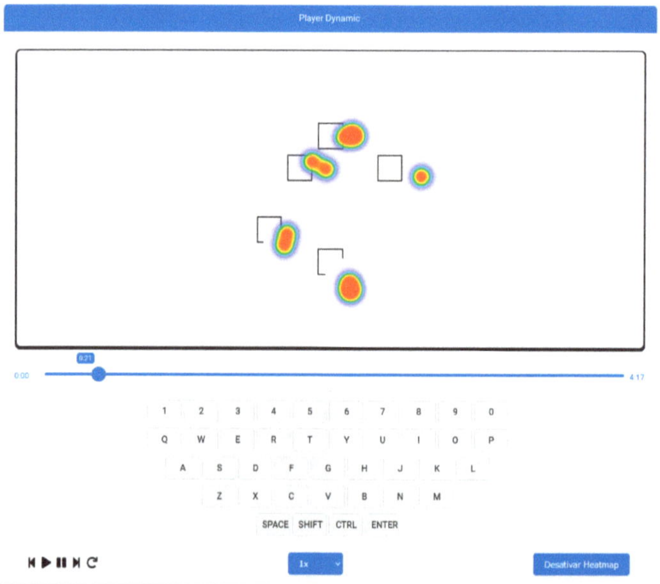

Figure 6. Dynamic player with the heatmap option for the visualization of a user log based on the scenario of Paulino et al. [12].

4. Case Study of Analyzing the User Interaction Logs in a Crowdsourcing Context

The principal aim of this article is to present the utilization of the WebTraceSense project to analyze user interaction data from a dataset derived from another study involving 134 distinct crowd workers [15]. This was achieved by employing the task fingerprinting technique to examine the user interaction logs. This study made use of the advanced capabilities of the WebTraceSense platform to gain insight into the patterns and nuances of user behavior by reconstructing the sequence of interactions from extensive log data. The application of task fingerprinting enables the differentiation of disparate user behaviors and the identification of notable trends and anomalies within the dataset. This approach not only enhances our understanding of user engagement but also provides insights that can be acted upon to optimize user interfaces and improve the overall user experience.

The dataset was initially assembled for the purpose of investigating the cognitive personalization of microtask design. It encompasses the contributions of 134 crowd workers engaged in a range of Human Intelligence Tasks (HITs). The principal metrics gathered encompass task accuracy, response times, mouse clicks, key presses, and other forms of interaction. These metrics were gathered using a combination of cognitive tests and task fingerprinting methods, thereby providing a robust foundation for the subsequent analysis of user behavior. The objective of the data collection was to identify the impact of different cognitive abilities on task performance and to ascertain how tasks could be personalized in order to enhance worker efficiency and satisfaction. This comprehensive dataset is a crucial input for the WebTraceSense framework, facilitating the generation of bespoke visualizations and the application of sophisticated statistical techniques to extract meaningful insights into user behavior.

In the context of the present study, this dataset is of pivotal importance in validating the capabilities of the WebTraceSense framework. By leveraging the rich interaction data, the framework is able to generate visualizations that reflect the specific behaviors and interactions of individual users with web-based platforms in a dynamic manner. To illustrate, the incorporation of comprehensive click and keystroke data into the dataset enables the framework to generate highly detailed heatmaps and dynamic player visualizations, which are essential for elucidating user engagement patterns and identifying potential areas for improvement in the user interface and user experience. Moreover, the response times and task accuracy metrics are of great importance in evaluating the efficacy of cognitive personalization techniques when integrated into web applications.

4.1. Visualizations Generated for the Analysis of Some Logs of the Dataset

Figure 5 presents a line chart comparing user interaction logs from two versions of a web application, one with personalization and the other without, as documented by Paulino et al. [15]. The chart effectively illustrates the temporal dynamics between user interactions. In the personalized version (left side of the chart), there is a noticeable reduction in the time intervals between interactions, suggesting a smoother and potentially more intuitive user experience. Conversely, the version without personalization (right side of the chart) exhibits longer time intervals between interactions, implying a more challenging user interface that may require more time for users to navigate and complete tasks. This visual representation underscores the efficiency gains that can be achieved through personalized interfaces, aligning with expectations that personalized systems can enhance user engagement and task performance by tailoring the experience to individual user needs.

The dynamic player shows user behavior more realistically than other visualization options. It draws the corresponding interaction (mouse or key press) in screenshots of the user's login interface. It also shows where the user clicked most. Figure 6 shows an example of the cognitive test "Pointing", analyzed in the same scenario as an experiment on Paulino et al. [15], which collected user interaction data.

4.2. Statistical Analysis

The results of the statistical analysis, conducted with the objective of evaluating the potential of personalized visualizations to optimize user experience and engagement on digital platforms, are presented in this section. In the context of the WebTraceSense system, the Python packages Pandas [30] and NumPy [31] were employed to facilitate data manipulation and numerical computations, which were essential for preparing the datasets. The Scikit-learn package [32] was employed for the application of statistical modelling techniques, such as clustering, with the objective of identifying behavioral patterns. The SciPy package [33] provided additional statistical tests that were necessary for deeper analysis. To facilitate the visualization of these insights, Matplotlib [34] and Seaborn [35] were employed to create clear, informative charts that depict the relationships and trends within the data. This enabled a thorough understanding of different user groups based on their interaction logs. The aforementioned packages integrated seamlessly, facilitating the analysis and presentation of data in a meaningful manner. The microtasks subjected to analysis were as follows: classification accuracy, counting accuracy, transcription accuracy, and sentiment analysis accuracy. These were identified as dependent variables and correlated with the results of the cognitive tests, which were considered to correspond to the independent variables.

The analysis indicates that personalization is associated with enhanced performance metrics, particularly with regard to accuracy and efficiency (see Table 1). In order to gain a more nuanced understanding and identify patterns, we employed a range of analytical techniques, including the use of correlations, trends, and potential clusters. The following steps were then undertaken: A correlation analysis was conducted to evaluate the relationships between different metrics and identify significant correlations that might indicate how certain factors influence performance. To examine trends in the data, such as changes in performance metrics across different tasks or groups, a trend analysis was conducted. In order to identify distinct groups within the analyzed data that may exhibit similar characteristics or behaviors, a cluster analysis was conducted using a variety of clustering techniques. The results of this analysis are presented in Figure 7.

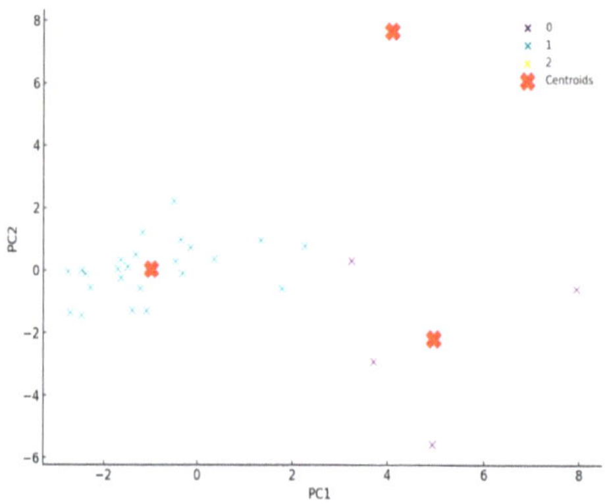

Figure 7. Clusters of microtask performance metrics with centroids (Interquartile Range Method [36]).

Table 1. Comparative analysis of personalization data.

Task	Metric	Group with Personalization	Group without Personalization
Counting Task	Accuracy Response Time Click and Action Counts	Higher average accuracy Lower response times Lower hesitant actions, low counts of hurry and special actions	Lower average accuracy Higher response times Higher hesitant actions, low counts of hurry and special actions
Transcriptions and Classification Tasks	---	Less prominent data in initial extract, might not have distinct columns or might be combined	---
Sentiment Analysis	Key Confident Actions	More consistent performance with fewer high deviations	Greater variation in key confident actions
	Special Actions	Similar low occurrences in both groups, slightly higher averages in the group without personalization	Similar low occurrences

The results demonstrated that the independent variables exhibited the most significant correlations with each of the dependent variables when personalization was applied. The limit of correlation was employed to determinate the significance, with a value of correlation coefficient greater than 0.3 or less than −0.3 with each dependent variable [37]. This approach is illustrated in Figure 8.

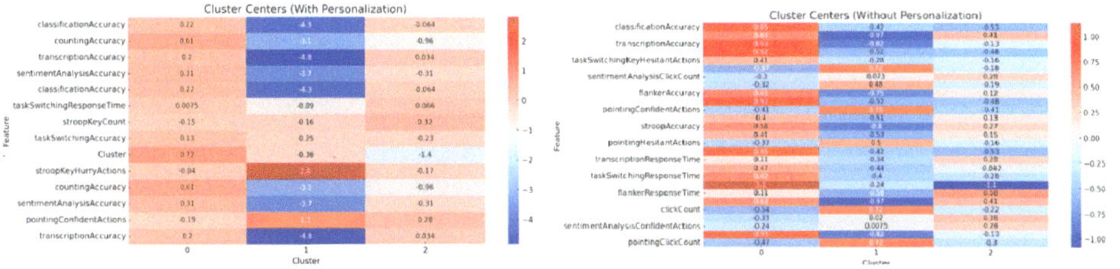

Figure 8. Clusters centers. (**Left**) Version with personalization; (**Right**) version without personalization).

This analysis aimed to identify and compare the performance metrics of different clusters of participants based on their performance in various microtasks as showed. Furthermore, we were interested in identifying the factors that were most strongly associated with counting accuracy within each cluster. This analysis is vital for elucidating the manner in which disparate factors influence task performance showing in this way the design of personalized interventions. The analysis of clusters revealed the following behavioral distinct tendencies:

- **Cluster 0**: This cluster, encompassing 4 participants, has a mean score of 0.62, a median score of 0.69, a standard deviation of 0.37, with scores ranging from a minimum of 0.11 to a maximum of 1.0. This cluster suggests a group of participants who are likely to excel across multiple task types, potentially due to higher overall engagement or superior cognitive strategies.
- **Cluster 1**: With 24 participants, this cluster shows a mean score of 0.58, a median score of 0.67, and a standard deviation of 0.35, with scores ranging from 0.0 to 1.0. It reflects a more varied group in which some participants excel in certain tasks while others struggle, indicating diverse strategies or varying levels of task-specific expertise.
- **Cluster 2**: This cluster represents a singular participant with constant scores (mean, median, min, and max all at 0.25) and no recorded standard deviation, suggesting either uniform performance across a limited set of tasks or incomplete data. This cluster represents participants who generally perform poorly or have incomplete data, indicating the need for deeper investigation or targeted support.

Finally, we analyzed the pair plot graph, as presented in Figure 9, to explore the relationships between a selection of key variables countingAccuracy, classificationAccuracy, and transcriptAccuracy.

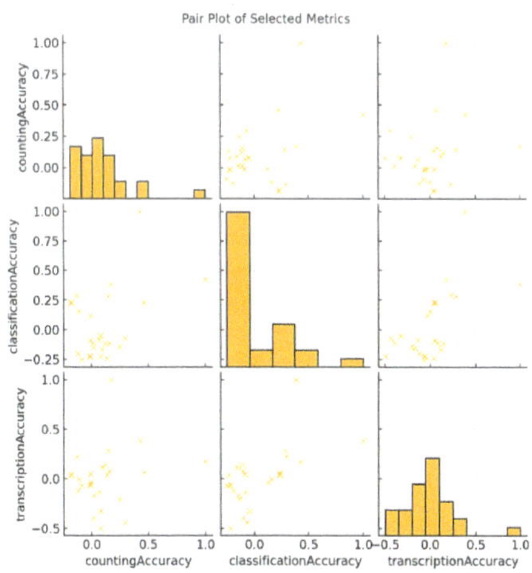

Figure 9. Pair plot of selected variables (accuracy of classification, counting, transcription, and sentiment analysis microtasks).

The presented statistics until this moment served to elucidate divergences in performance across clusters, thereby indicating that Cluster 0 and Cluster 1 exhibit a more extensive and variability in counting accuracy, whereas Cluster 2 has limited representation. This information serves to assist in the personalization of interventions and training programs to align with the specific needs of individual participants, thus enhancing overall task performance and facilitating an understanding of the diversity in selected task.

5. Discussion

5.1. RQ1: How Can the Visualization of User Interaction Logs Enhance the Personalization of Web Applications, Including E-Commerce Websites, Web Games, and Other Digital Platforms?

The visualization of user interaction logs is of significant importance in the personalization of web applications across a range of platforms, including e-commerce sites, web games, and other digital interfaces. By mapping user interactions within these environments, developers and designers can gain a profound comprehension of user behavioral patterns, preferences, and areas of difficulty. To illustrate, in the context of e-commerce platforms, the visualization of user pathways can assist in the identification of products that elicit greater interest and those sections of the site that give rise to confusion or abandonment. These insights facilitate the implementation of targeted modifications to the user interface and user experience (UI/UX) design, thereby promoting a more seamless shopping experience that can ultimately lead to increased sales and customer retention.

The capacity for visualization inherent to the WebTraceSense platform is pivotal to its functionality, facilitating a comprehensive and interactive representation of user interaction logs. The employment of D3.js for the purpose of visualization modules enables the platform to represent interaction data in a dynamic manner, thereby facilitating the intuitive visualization of complex datasets for users. This is particularly advantageous in a personalized setting, as evidenced by the comparative analysis of user logs from personalized versus non-personalized versions of web applications. The data clearly demonstrates that the implementation of personalized interfaces results in a notable reduction in the time between interactions, which suggests a more streamlined and user-friendly experience. The aforementioned visualizations facilitate the immediate acquisition of insights into user behavior, thereby guiding subsequent enhancements to the web application based on real-time user feedback.

Furthermore, the incorporation of robust backend technologies with advanced frontend visualization tools enables WebTraceSense to provide a comprehensive solution for web developers and researchers seeking to enhance web platforms through data-driven insights. The application of statistical and machine learning methods via the Django API to the interaction data facilitates the generation of highly tailored visualizations through the further refinement of the customization process. These capabilities render WebTraceSense an invaluable tool for enhancing the user experience across a range of digital platforms, whereby detailed interaction logs inform and drive the personalization process. This approach guarantees that each user interaction is not only recorded and examined, but also effectively employed to enhance the user's navigational experience, thus promoting higher engagement and satisfaction. The visualization of behavioral traces is a well-documented phenomenon in the scientific literature. For example, the work of Rzeszotarski and Kittur [12,22] demonstrated the potential of visualizing interaction data to uncover user behaviors. Nevertheless, despite these developments, there remains a notable deficiency in the availability of tools that facilitate the seamless identification of behavioral traces through ad-hoc visualizations, which are tailored to the specific requirements of researchers, game developers and website programmers. WebTraceSense addresses this gap by providing a robust platform that supports the creation of bespoke visualizations. The incorporation of sophisticated statistical and AI modules serves to augment the platform's capacity to furnish profound insights into user behavior. Such modules are capable of applying sophisticated techniques to interaction data, thereby revealing subtle patterns that may not be immediately apparent through basic visualization methods

5.2. RQ2: What Are the Most Effective Statistical and Machine Learning Techniques for Analyzing User Interaction Data to Identify and Predict User Behavioral Patterns in Web Applications?

The analysis of user interaction data for the purpose of identifying and predicting behavioral patterns in web applications requires the application of a range of statistical and machine learning techniques. The principal technique employed was correlation analysis, facilitated by Python packages such as Pandas and NumPy, which permitted the identification of significant relationships between disparate interaction metrics and user behaviors. For example, the analysis demonstrated that enhanced accuracy in counting tasks was associated with reduced response times and a lower incidence of hesitant actions. This suggests that users who demonstrated proficiency in these tasks exhibited more confident and efficient behavior. This finding was further corroborated by the identification of discrete clusters within the user data. The clusters demonstrated that participants who exhibited higher overall engagement or superior cognitive strategies demonstrated consistent superior performance across a range of microtasks. Furthermore, the analysis indicated that personalization initiatives, such as task difficulty customization based on user performance, resulted in enhanced accuracy and consistency in task completion. These findings emphasize the significance of employing correlation analysis and descriptive statistics to adapt web applications to the distinctive requirements and actions of individual users, thereby optimizing their overall experience.

While not the focus of this case study, clustering techniques, including K-Means clustering [38] and Principal Component Analysis (PCA), are highly effective for identifying distinct user groups based on interaction data. K-Means clustering is an effective method for grouping users with similar interaction patterns, thereby identifying distinct behavioral segments. In contrast, PCA reduces the dimensionality of the data, thereby facilitating the visualization and interpretation of these clusters. The combination of these methods enables researchers to identify underlying patterns and segments within the user base, which are crucial for the development of targeted interventions and the delivery of personalized experiences in web applications.

In the context of predictive modelling, machine learning algorithms such as Random Forest and Gradient Boosting have been demonstrated to be highly effective [39]. These techniques are robust in handling both linear and non-linear relationships and can manage complex interactions between variables. For example, Random Forest provides insights into the importance of various features in predicting user behavior, while Gradient Boosting can enhance prediction accuracy by iteratively improving the model. Collectively, these methods facilitate the development of models that not only explain user behavior but also predict future actions, thereby enabling web applications to adapt dynamically to user needs and preferences.

6. Final Remarks

The WebTraceSense platform represents a comprehensive methodology for the analysis and visualization of web interaction data, thereby significantly enhancing the personalization capabilities of web applications, e-commerce sites, and digital games. By leveraging user interaction logs, the platform facilitates a more profound comprehension of user behaviors, thereby enabling developers and researchers to tailor user experiences with greater precision. The integration of a four-step wizard form allows for the upload and meticulous analysis of these logs, where users can select specific interaction metrics to visualize, such as mouse clicks and keystrokes, along with choosing from various visualization options, including line charts, bar charts, and pie charts. Of particular note is the dynamic player feature, which provides a dynamic visual representation of user interactions over time on actual interface screenshots, including heatmaps that highlight frequent interaction points. Furthermore, the comprehensive architecture of the WebTraceSense platform, which encompasses robust backend and frontend components, is designed to facilitate the streamlining of operational workflows from data collection to visualization. The platform employs advanced statistical and machine learning techniques to process interaction data, which significantly aids in the identification and prediction of user behavior patterns. Techniques such as correlation analysis and clustering facilitate the discernment of relationships between different user interactions and the grouping of users with similar behaviors, thereby enabling the implementation of targeted interventions. Additionally, predictive models are employed to predict user actions, offering insights that are crucial for the dynamic adaptation of web applications to enhance user satisfaction and engagement. This integration of sophisticated analytical tools ensures that WebTraceSense can provide continuous improvements to web platforms, leading to more refined and user-centered applications. A case study was conducted to evaluate the utility of the proposed platform for the generation of visualizations based on an existing dataset. The results demonstrated the effectiveness of the proposed approach.

In the future, it is anticipated that the platform will undergo further refinement, including the development of additional visualizations. Moreover, the dynamic player will be enhanced through the generation of logs of the interface based on user interactions, thereby eliminating the necessity for screenshots for these visualizations. Additionally, improvements to the visualizations of the collective tendencies of the entire dataset are required in order to more effectively identify behavioral traces of users in the user interaction log. Another limitation has been identified, namely the necessity for broader validation through additional case studies across diverse domains. This represents a crucial future direction. Furthermore, the study referenced in [40] has been identified as a valuable source for future work, as it offers a formalized method for applying visualization taxonomies to real-world interaction data. In regards to the performance metrics, it was stipulated that a comprehensive examination of pivotal performance metrics, including the processing time for visualization generation and the efficacy of loading user interaction log data, shall be conducted in future work.

Author Contributions: Conceptualization, D.P., A.T.N., and H.P.; methodology, D.P., A.T.N., and H.P; software, D.P. and A.T.N.; validation, D.P. and A.T.N.; formal analysis, D.P., A.T.N., and W.A.T.B.; investigation, D.P. and A.T.N.; resources, D.P.; data curation, D.P. and A.T.N.; writing—original draft preparation, D.P.; writing—review and editing, D.P. and A.T.N.; visualization, D.P., A.T.N., and W.A.T.B.; supervision, D.P., A.T.N., and H.P.; project administration, D.P. and H.P.; funding acquisition, H.P. All authors have read and agreed to the published version of the manuscript.

Funding: This work was co-financed by Component 5—Capitalization and Business Innovation, integrated in the Resilience Dimension of the Recovery and Resilience Plan within the scope of the Recovery and Resilience Mechanism (MRR) of the European Union (EU), framed in the Next Generation EU, for the period 2021–2026, within project HfPT, with reference 41.

Institutional Review Board Statement: Not applicable.

Informed Consent Statement: Not applicable.

Data Availability Statement: The original contributions presented in the study are included in the article, further inquiries can be directed to the corresponding author.

Acknowledgments: The authors would like to thank the students Bernardo Pinto, Francisco Sousa, Gonçalo Silva, José Marinho, Marcelo Apolinário, and Aman Kumar for their assistance in the development of several components for the WebTraceSense system, which are described in this article.

Conflicts of Interest: The authors declare no conflicts of interest.

References

1. Smith, A.D.; Shock, J.R.; Beaves, R.G. Customer relationship management and the impact of e-coupons on B2C retail markets. *Int. J. Bus. Inf. Syst.* **2019**, *30*, 203–231. [CrossRef]
2. Nguyen, T.; Hsu, P.-F. More Personalized, More Useful? Reinvestigating Recommendation Mechanisms in E-Commerce. *Int. J. Electron. Commer.* **2022**, *26*, 90–122. [CrossRef]
3. Paraschos, P.D.; Koulouriotis, D.E. Game Difficulty Adaptation and Experience Personalization: A Literature Review. *Int. J. Human–Computer Interact.* **2022**, *39*, 1–22. [CrossRef]

4. Aria, R.; Archer, N.; Khanlari, M.; Shah, B. Influential Factors in the Design and Development of a Sustainable Web3/Metaverse and Its Applications. *Future Internet* **2023**, *15*, 131. [CrossRef]
5. Lamprinou, D.; Fotini, P. Gamification design framework based on SDT for student motivation. In Proceedings of the 2015 International Conference on Interactive Mobile Communication Technologies and Learning (IMCL), Thessaloniki, Greece, 19–20 November 2015; pp. 406–410. [CrossRef]
6. Paredes, H.; Paulino, D.; Barroso, J.; Abrantes, C.; Machado, I.; Silva, I. Supervised physical exercise therapy of peripheral artery disease patients: M-health challenges and opportunities. In Proceedings of the 54th Hawaii International Conference on System Sciences, HICSS 2021, Kauai, HI, USA, 5–8 January 2021; pp. 1–10. [CrossRef]
7. Zanker, M.; Rook, L.; Jannach, D. Measuring the impact of online personalisation: Past, present and future. *Int. J. Human-Computer Stud.* **2019**, *131*, 160–168. [CrossRef]
8. Torous, J.; Kiang, M.V.; Lorme, J.; Onnela, J.-P. New tools for new research in psychiatry: A scalable and customizable platform to empower data-driven smartphone research. *JMIR Ment. Health* **2016**, *3*, e5165. [CrossRef]
9. Onnela, J.-P.; Rauch, S.L. Harnessing Smartphone-Based Digital Phenotyping to Enhance Behavioral and Mental Health. *Neuropsychopharmacology* **2016**, *41*, 1691–1696. [CrossRef] [PubMed]
10. Mohr, D.C.; Zhang, M.; Schueller, S.M. Personal sensing: Understanding mental health using ubiquitous sensors and machine learning. *Annu. Rev. Clin. Psychol.* **2017**, *13*, 23–47. [CrossRef] [PubMed]
11. MacCormick, D.; Zaman, L. Echo: Analyzing gameplay sessions by reconstructing them from recorded data. In Proceedings of the Annual Symposium on Computer-Human Interaction in Play, Amsterdam, The Netherlands, 15–18 October 2020; pp. 281–293. [CrossRef]
12. Rzeszotarski, J.; Kittur, A. CrowdScape: Interactively visualizing user behavior and output. In Proceedings of the 25th Annual ACM Symposium on User Interface Software and Technology, Cambridge, MA, USA, 7–10 October 2012; pp. 55–62. [CrossRef]
13. Mangaroska, K.; Vesin, B.; Giannakos, M. Cross-platform analytics: A step towards personalization and adaptation in education. In Proceedings of the 9th International Conference on Learning Analytics & Knowledge, Tempe, AZ, USA, 4–8 March 2019; pp. 71–75. [CrossRef]
14. Mansoor, H.; Gerych, W.; Alajaji, A.; Buquicchio, L.; Chandrasekaran, K.; Agu, E.; Rundensteiner, E.; Rodriguez, A.I. INPHOVIS: Interactive visual analytics for smartphone-based digital phenotyping. *Vis. Inform.* **2023**, *7*, 13–29. [CrossRef]
15. Paulino, D.; Guimarães, D.; Correia, A.; Ribeiro, J.; Barroso, J.; Paredes, H. A Model for Cognitive Personalization of Microtask Design. *Sensors* **2023**, *23*, 3571. [CrossRef]
16. Arya, S.S.; Dias, S.B.; Jelinek, H.F.; Hadjileontiadis, L.J.; Pappa, A.-M. The Convergence of Traditional and Digital Biomarkers through AI-Assisted Biosensing: A New Era in Translational Diagnostics? *Biosens. Bioelectron.* **2023**, *235*, 115387. [CrossRef] [PubMed]
17. Insel, T.R. Digital phenotyping: A global tool for psychiatry. *World Psychiatry* **2018**, *17*, 276. [CrossRef] [PubMed]
18. Bufano, P.; Laurino, M.; Said, S.; Tognetti, A.; Menicucci, D. Digital phenotyping for monitoring mental disorders: Systematic review. *J. Med Internet Res.* **2023**, *25*, e46778. [CrossRef] [PubMed]
19. Kroenke, K.; Spitzer, R.L.; Williams, J.B. The PHQ-9: Validity of a brief depression severity measure. *J. Gen. Intern. Med.* **2001**, *16*, 606–613. [CrossRef] [PubMed]
20. Mastoras, R.-E.; Iakovakis, D.; Hadjidimitriou, S.; Charisis, V.; Kassie, S.; Alsaadi, T.; Khandoker, A.; Hadjileontiadis, L.J. Touchscreen typing pattern analysis for remote detection of the depressive tendency. *Sci. Rep.* **2019**, *9*, 13414. [CrossRef]
21. Rodríguez-de Pablo, C.; Savić, A.; Keller, T. Game-based assessment in upper-limb post-stroke telerehabilitation. *Biosyst. Biorobotics* **2017**, *15*, 413–417. [CrossRef]
22. Rzeszotarski, J.M.; Kittur, A. Instrumenting the crowd: Using implicit behavioral measures to predict task performance. In Proceedings of the 24th Annual ACM Symposium on User Interface Software and Technology, Santa Barbara, CA, USA, 16–19 October 2011; pp. 13–22. [CrossRef]
23. Paulino, D.; Correia, A.; Barroso, J.; Paredes, H. Cognitive personalization for online microtask labor platforms: A systematic literature review. *User Model. User-Adapted Interact.* **2023**, 1–42. [CrossRef]
24. Gomez, M.J.; Ruipérez-Valiente, J.A.; Clemente, F.J.G. A Systematic Literature Review of Game-Based Assessment Studies: Trends and Challenges. *IEEE Trans. Learn. Technol.* **2022**, *16*, 500–515. [CrossRef]
25. Gomez, M.J.; Ruipérez-Valiente, J.A.; Martinez, P.A.; Kim, Y.J. Exploring the affordances of sequence mining in educational games. In Proceedings of the 8th International Conference on Technological Ecosystems for Enhancing Multiculturality, Salamanca, Spain, 21–23 October 2020; pp. 648–654.
26. De Cnudde, S.; Ramon, Y.; Martens, D.; Provost, F. Deep learning on big, sparse, behavioral data. *Big Data* **2019**, *7*, 286–307. [CrossRef]
27. Ussath, M.; Jaeger, D.; Cheng, F.; Meinel, C. Identifying suspicious user behavior with neural networks. In Proceedings of the 2017 IEEE 4th International Conference on Cyber Security and Cloud Computing (CSCloud), New York, NY, USA, 26–28 June 2017; pp. 255–263. [CrossRef]
28. Koehn, D.; Lessmann, S.; Schaal, M. Predicting online shopping behaviour from clickstream data using deep learning. *Expert Syst. Appl.* **2020**, *150*, 113342. [CrossRef]
29. Faustino, J.; Adriano, D.; Amaro, R.; Pereira, R.; da Silva, M.M. DevOps benefits: A systematic literature review. *Software Pract. Exp.* **2022**, *52*, 1905–1926. [CrossRef]

30. McKinney, W. Data Structures for Statistical Computing in Python. In Proceedings of the 9th Python in Science Conference, Austin, TX, USA, 28 June–3 July 2010; pp. 51–56. [CrossRef]
31. Harris, C.R.; Millman, K.J.; van der Walt, S.J.; Gommers, R.; Virtanen, P.; Cournapeau, D.; Wieser, E.; Taylor, J.; Berg, S.; Smith, N.J.; et al. Array programming with NumPy. *Nature* **2020**, *585*, 357–362. [CrossRef] [PubMed]
32. Pedregosa, F.; Varoquaux, G.; Gramfort, A.; Michel, V.; Thirion, B.; Grisel, O.; Blondel, M.; Prettenhofer, P.; Weiss, R.; Dubourg, V.; et al. Scikit-learn: Machine Learning in Python. *J. Mach. Learn. Res.* **2011**, *12*, 2825–2830.
33. Virtanen, P.; Gommers, R.; Oliphant, T.E.; Haberland, M.; Reddy, T.; Cournapeau, D.; Burovski, E.; Peterson, P.; Weckesser, W.; Bright, J.; et al. SciPy 1.0: Fundamental Algorithms for Scientific Computing in Python. *Nat. Methods* **2020**, *17*, 261–272. [CrossRef]
34. Hunter, J.D. Matplotlib: A 2D graphics environment. *Comput. Sci. Eng.* **2007**, *9*, 90–95. [CrossRef]
35. Waskom, M.L. seaborn: Statistical data visualization. *J. Open Source Softw.* **2021**, *6*, 3021. [CrossRef]
36. Wan, X.; Wang, W.; Liu, J.; Tong, T. Estimating the sample mean and standard deviation from the sample size, median, range and/or interquartile range. *BMC Med. Res. Methodol.* **2014**, *14*, 135. [CrossRef]
37. Hemphill, J.F. Interpreting the magnitudes of correlation coefficients. *Am. Psychol.* **2003**, *58*, 78–79. [CrossRef] [PubMed]
38. Jain, A.K. Data clustering: 50 years beyond K-means. *Pattern Recognit. Lett.* **2010**, *31*, 651–666. [CrossRef]
39. Chen, T.; Guestrin, C. XGBoost: A scalable tree boosting system. In Proceedings of the 22nd ACM SIGKDD International Conference on Knowledge Discovery and Data Mining, San Francisco, CA, USA, 13–17 August 2016; pp. 785–794. [CrossRef]
40. Gathani, S.; Monadjemi, S.; Ottley, A.; Battle, L. A Grammar-Based Approach for Applying Visualization Taxonomies to Interaction Logs. *Comput. Graph. Forum* **2022**, *41*, 489–500. [CrossRef]

eng

MDPI

Article

Balancing the Scale: Data Augmentation Techniques for Improved Supervised Learning in Cyberattack Detection

Kateryna Medvedieva , Tommaso Tosi, Enrico Barbierato and Alice Gatti *

Department of Mathematics and Physics, Catholic University of the Sacred Heart, 25121 Brescia, Italy; kateryna.medvedieva01@icatt.it (K.M.); tommaso.tosi@unicatt.it (T.T.); enrico.barbierato@unicatt.it (E.B.)
* Correspondence: alice.gatti@unicatt.it

Abstract: The increasing sophistication of cyberattacks necessitates the development of advanced detection systems capable of accurately identifying and mitigating potential threats. This research addresses the critical challenge of cyberattack detection by employing a comprehensive approach that includes generating a realistic yet imbalanced dataset simulating various types of cyberattacks. Recognizing the inherent limitations posed by imbalanced data, we explored multiple data augmentation techniques to enhance the model's learning effectiveness and ensure robust performance across different attack scenarios. Firstly, we constructed a detailed dataset reflecting real-world conditions of network intrusions by simulating a range of cyberattack types, ensuring it embodies the typical imbalances observed in genuine cybersecurity threats. Subsequently, we applied several data augmentation techniques, including SMOTE and ADASYN, to address the skew in class distribution, thereby providing a more balanced dataset for training supervised machine learning models. Our evaluation of these techniques across various models, such as Random Forests and Neural Networks, demonstrates significant improvements in detection capabilities. Moreover, the analysis also extends to the investigation of feature importance, providing critical insights into which attributes most significantly influence the predictive outcomes of the models. This not only enhances the interpretability of the models but also aids in refining feature engineering and selection processes to optimize performance.

Keywords: data augmentation; supervised learning; cybersecurity

Citation: Medvedieva, K.; Tosi, T; Barbierato, E.; Gatti, A. Balancing the Scale: Data Augmentation Techniques for Improved Supervised Learning in Cyberattack Detection. *Eng* **2024**, *5*, 2170–2205. https://doi.org/10.3390/eng5030114

Academic Editors: Antonio Gil Bravo and Alessandro Polo

Received: 8 July 2024
Revised: 13 August 2024
Accepted: 30 August 2024
Published: 4 September 2024

1. Introduction

The National Institute of Standards and Technology (NIST www.nist.gov/cybersecurity accessed on 1 June 2024) provides multiple definitions of cybersecurity, which may be collectively synthesized into a singular principle: cybersecurity entails the implementation of protective measures and controls designed to prevent harm and secure information stored in computer systems or transmitted through communication networks. This practice is paramount to ensure the availability, integrity, and confidentiality of information.

A report by the European Union Agency for Cybersecurity (ENISA www.enisa.europa.eu/publications/enisa-threat-landscape-2022 accessed on 15 June 2024) underscores a significant augmentation in cyberattacks toward the end of 2022 and into the initial half of 2023. The ongoing Russian invasion of Ukraine is pinpointed as a primary catalyst for this uptick. Additionally, the report identifies eight major threat groups, highlighting the dynamic and evolving nature of cyber threats.

Threats against data are categorized into data leaks and data breaches, which differ primarily in the intent behind the exposure. A data leak is usually an unintentional exposure due to human error or system vulnerabilities, whereas a data breach is a deliberate attack aimed at stealing information.

Threats against availability include Denial of Service (DoS) attacks, which disrupt normal operations by overwhelming systems or networks with excessive traffic, typically originating from multiple sources. Additionally, internet threats comprise a range of deliberate or accidental disruptions to electronic communications, leading to outages, blackouts, shutdowns, or censorship caused by various factors including government actions, natural disasters, or cyberattacks. Information manipulation involves actions that influence values, procedures, and political processes, often with a manipulative agenda. While not always illegal, these actions pose significant threats by potentially undermining democratic principles, destabilizing societies, and eroding institutional trust. Supply chain attacks represent sophisticated cyberattacks targeting the interconnected relationships between organizations and their suppliers. These attacks can involve malicious code in software updates or hardware components, compromised supplier credentials, or exploited third-party service or product vulnerabilities.

Machine learning (ML) plays a crucial role in classifying and detecting cyberattacks by analyzing vast amounts of data and identifying patterns indicative of malicious activity. One significant advantage of ML is its ability to detect anomalies, which can indicate potential threats such as zero-day attacks and advanced persistent threats (APTs). Moreover, ML automates the threat identification process, reducing the time needed to detect and respond to threats, which helps mitigate damage and minimize downtime. ML's pattern recognition capabilities are particularly useful in identifying specific types of cyberattacks, thereby improving threat detection accuracy. Additionally, ML systems are scalable and can handle large-scale data, making them suitable for enterprises with extensive networks. Furthermore, ML models can continuously learn from new data, enhancing their detection capabilities over time. This adaptive learning is essential for staying ahead of emerging threats. Another benefit of ML in cybersecurity is its ability to reduce false positives in threat detection, allowing security teams to focus on genuine threats and avoid unnecessary alerts. By automating many aspects of threat detection and response, ML also reduces the operational costs associated with manual security monitoring and incident response.

ML models are trained using datasets to classify various types of cyberattacks. However, these datasets are frequently imbalanced, meaning that some types of attacks are significantly underrepresented compared to others. This imbalance poses a critical problem because the models trained on such datasets tend to become biased towards the more common attack types, resulting in poor detection rates for the rarer, yet potentially more dangerous, attacks. For example, if a dataset contains a vast majority of data on phishing attacks but very few instances of zero-day exploits, the model will likely excel at identifying phishing but fail to recognize the zero-day exploits. To address imbalance issues, cybersecurity researchers and practitioners use techniques such as data augmentation, synthetic data generation, and advanced ML algorithms designed to handle imbalanced data. These methods help ensure that the models remain robust and capable of accurately identifying both common and rare cyber threats, thereby providing a more secure defense against a wide range of attacks.

However, these efforts alone are not particularly compelling unless the ML models are also employed to assign weights to dataset features. This weighting process is crucial as it helps identify which features have the greatest influence on the model's responses. By understanding feature importance, researchers and practitioners can gain deeper insights into the factors driving cyber threats, leading to more effective detection and prevention strategies. Without this, the models remain black boxes, offering limited value beyond mere classification.

1.1. Aims of the Research

The primary objective of this research was to explore and evaluate various data augmentation techniques to enhance the effectiveness of supervised learning models in detecting cyberattacks. By generating a realistic dataset that simulates different types of cyberattacks, the study aimed to reflect the typical imbalances observed in genuine cybersecurity threats. Subsequently, the application of multiple data augmentation techniques, including SMOTE and ADASYN, sought to correct the skewed class distribution, providing a more balanced dataset for training supervised machine learning models.

Furthermore, the study hypothesized that the implementation of these data augmentation techniques would lead to significant improvements in the detection capabilities of various models, such as Random Forests and Neural Networks. Another key hypothesis was that the analysis of feature importance would offer critical insights into which attributes most significantly influence the predictive outcomes of the models, thereby enhancing their interpretability and aiding in the refinement of feature engineering and selection processes.

1.2. Contribution of This Work

The contribution of this research consists of the following:

- A novel, realistic dataset that simulates various types of cyberattacks, mirroring the complex and imbalanced nature of real-world cybersecurity scenarios;
- Furthermore, a comprehensive application of several data augmentation techniques, including SMOTE (Synthetic Minority Over-sampling Technique) and ADASYN (Adaptive Synthetic Sampling), is adapted specifically for enhancing the dataset within the cybersecurity context;
- Additionally, the research includes a systematic evaluation of how different machine learning models perform following data augmentation, identifying those most effective in recognizing diverse cyber threats;
- Finally, this exploration enhances the understanding of predictive features and contributes significantly to the models' transparency and explainability, essential for operational trust and effective deployment in cybersecurity environments.

The work is organized as per Figure 1. Specifically, Section 2 reviews the related work. Section 3 provides the background of this work, presenting some of the data augmentation techniques and machine learning methods used. Section 4 details the experiments run on a generated dataset, while Section 5 comments on the results. Section 6 draws some conclusions and provides a few comments on future work. Finally, a list of abbreviations used and more detailed results are given in Appendix A.

2. Related Work

Supervised learning is widely used in Intrusion Detection Systems (IDSs) due to its effectiveness in employing labeled datasets to train predictive models.

Apruzzese et al. [1] provide a comprehensive review of the deployment and integration of ML in cybersecurity. Key contributions include highlighting the benefits of ML over traditional human-driven detection methods and identifying additional cybersecurity tasks that can be enhanced by Supervised Learning. The study discusses intrinsic problems such as concept drift, adversarial settings, and data confidentiality that affect real-world ML deployments in cybersecurity. Limitations of the approach include the slow pace of integrating ML into production environments and the need for continuous updates to handle evolving threats. The article also presents case studies demonstrating industrial applications of ML in cybersecurity, emphasizing the necessity of collaborative efforts among stakeholders to advance ML's role in this field.

Mijwil et al. [2] explain how supervised and unsupervised learning methods, such as logistic regression and clustering, are utilized for intrusion and anomaly detection, while DL techniques like convolutional neural networks (CNNs) and recurrent neural networks (RNNs) effectively identify malware and cyber threats with high accuracy. Despite their potential, ML and DL face challenges, including the need for large datasets, high false positive rates, and the continuous evolution of cyber threats, necessitating regular updates and human oversight. The paper calls for ongoing research, better datasets, and integrated AI techniques to stay ahead of cybercriminals. Finally, it emphasizes the importance of further investigations into AI applications in cybersecurity, encouraging collaboration and the development of advanced techniques to protect digital environments.

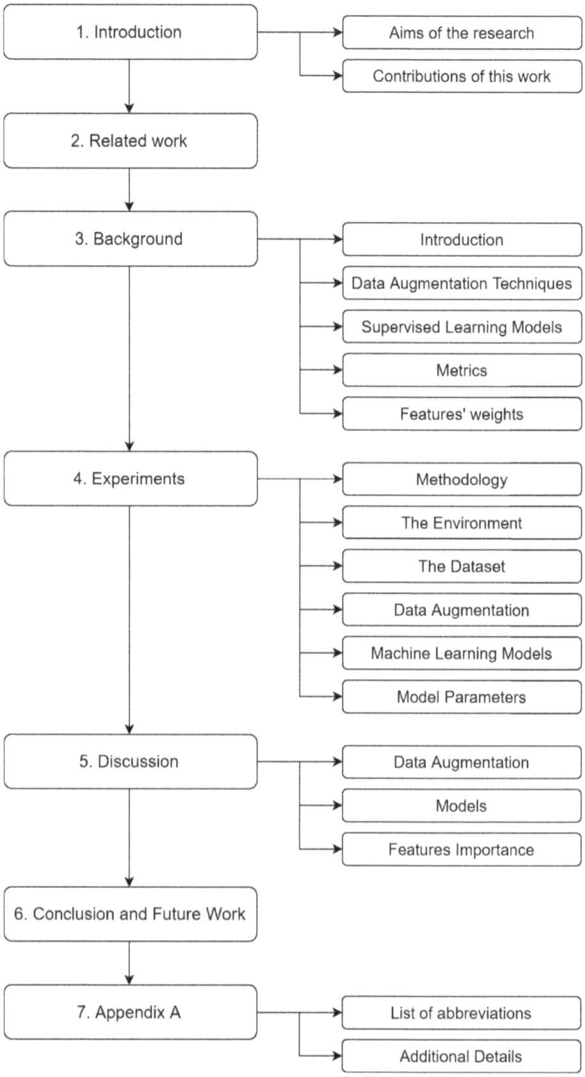

Figure 1. Visual representation of the work's structure.

A significant study by Bagui et al. [3] analyzed network connection logs using various supervised learning models, including Logistic Regression, Decision Trees (DTs), Random Forests (RFs), SVM, Naïve Bayes, and gradient-boosting trees. They introduced a new dataset, UWF-ZeekData22, which is publicly accessible through the University of West Florida's site. This dataset is labeled according to the MITRE ATT&CK framework, although the labeling process is not clearly documented. It includes 18 attributes, with the top six features identified based on information gain being the history of connection, transport layer protocol, application layer protocol, number of payload bytes the originator sent, destination IP, and number of packets the originator sent. The UWF-ZeekData22 dataset primarily covers two tactics: reconnaissance and discovery, with a significant imbalance between the number of observations for each tactic—2087 for discovery and 504,576 for reconnaissance. The focus of the research is more on classifying adversary tactics rather than specific techniques of attack. The strengths of the study by Bagui et al. include a detailed presentation of a novel approach to preprocessing log data. The methodology involves binning numerical values (except for originator and destination ports) using a moving mean approach. Nominal features are converted into numerical labels, IP addresses are categorized based on the standard network classification of the first octet, and ports are grouped by ranges. A noted limitation in the study is its reliance on binary classification to assess model performance. The findings highlight that tree-based methods—specifically Decision Trees, gradient boosting trees, and Random Forests—excelled, achieving over 99% accuracy along with high precision, recall, f-measure, and AUROC scores.

Tufan et al. [4] explored a supervised ML model trained and tested on two distinct datasets. The first dataset consists of real-world private network data collected by the authors, and the second is the UNSW-NB15 dataset, developed by the Australian Center for Cyber Security. The data from the organizational environment required significant preparation, including the conversion of private IP addresses to public ones within captured packets. A noteworthy feature extraction technique involved analyzing packets with various TCP headers, such as ICMP, SYN, SYN-ACK, NULL, FIN, XMAS (PSH-URG-FIN), and FIN-ACK, with measurements taken every two seconds for each source IP. The preprocessing steps included the removal of redundant columns, those with empty values, and those with little variation. For categorical features, one-hot encoding was employed. This approach to handling missing values and detailing the labeling process, using open-source tools like Snort and Suricata to generate alerts from packet data, marks a comprehensive method where these alerts serve as labels for training. Tufan et al. utilized both filter and wrapper methods for feature selection, aiming to effectively refine the feature set. In contrast, the study by Bagui et al. [3]. performed attribute reduction using information gain to assess the importance and relevance of features. They compared two supervised models: an ensemble model consisting of Bayesian classifiers, KNN, Logistic Regression, and SVMs, against CNNs. A significant finding from their research was that CNNs outperformed the ensemble model in both datasets, highlighting the effectiveness of deep learning approaches in handling complex data structures and patterns.

In their study, Ravi et al. [5] introduced a deep-learning ensemble approach to enhance the performance of IDSs. The effectiveness of the proposed model was evaluated using several datasets, including SDN-IoT, KDD-Cup-1999, UNSW-NB15, WSN-DS, and CICIDS-2017, although only samples from the last four were considered. These datasets were preprocessed to ensure normalization. The models implemented for classification and feature extraction were RNNs, LSTMs, and Gated Recurrent Units (GRUs). These neural networks demonstrated a high efficacy in detecting attacks across all datasets. The next phase of the proposed methodology involved dimensionality reduction using Kernel Principal Component Analysis (KPCA) to improve the manageability and effectiveness of the data. During the feature fusion process, features extracted from different models were concatenated to form a comprehensive feature set. This combined feature set was then processed by base-level classifiers, specifically SVMs and Random Forests. The outputs from these classifiers were subsequently analyzed by a meta-level classifier, Logistic Regression, to finalize the detection process. The performance metrics from this study highlighted significant success in attack classification, achieving over 98% accuracy on all datasets, except the KDD-Cup-1999 dataset, which recorded an accuracy of 89%.

Further exploration into unsupervised ML models was conducted by Verkerken et al. [6], who utilized datasets developed by the Canadian Institute for Cybersecurity. In this phase, they removed redundant features and eliminated samples with missing or infinite values, as well as duplicated rows. Notably, unlike in the study by Bagui et al. [3], previously cited, features such as IP addresses were discarded to avoid overfitting. They applied a variety of feature scaling techniques, including StandardScaler, RobustScaler, QuantileTransformer, and MinMaxScaler from the scikit-learn library. These preprocessing steps were essential for the subsequent application of models such as Principal Component Analysis (PCA), Isolation Forest, Autoencoder, and One-Class SVM, particularly focusing on anomaly detection. However, they observed a significant drop in model performance on newer data collected in 2018, with the AUROC decreasing by an average of 30.45%, and a minimum drop of 17.85% for the One-Class SVM. The authors attributed these declines to changes in the distribution of attack labels between datasets and the inadequacy of model hyperparameters during validation.

Another innovative approach to data handling was proposed by Hwang et al. [7], aimed at addressing the challenge of high memory demand typically associated with storing traffic data. Their method focused on analyzing only the initial bytes of the first few packets in a flow, significantly reducing data storage requirements. This approach showcases a practical solution to one of the fundamental issues in network traffic analysis, highlighting the potential of unsupervised learning techniques in IDSs. Each of these studies contributes to the evolving landscape of ML applications in cybersecurity, demonstrating various strategies to enhance the effectiveness and efficiency of IDSs.

In their innovative study, Aamir and Zaidi [8] introduced a semi-supervised approach based on clustering techniques. The primary objective was to employ various clustering methods to initially label the dataset, which would then facilitate the training of supervised algorithms for classification.

A significant limitation of this study was its reliance on synthetic datasets generated through simulation. Unlike other studies, the feature set used here was relatively small, comprising variables such as traffic rate, processing delay, and server CPU utilization, which were deemed sufficient for detecting Distributed Denial of Service (DDoS) attacks. The dataset's size was also restricted, containing only 1000 observations, which may not adequately represent more complex attack scenarios. Nonetheless, the results from this dataset were compared with a subset of the CICIDS2017 dataset that specifically focused on DDoS attacks.

After the collection and normalization of the dataset, two clustering techniques —agglomerative clustering and K-means—were applied. Notably, K-means was enhanced by principal components obtained from PCA. The subsequent step involved a voting mechanism to reconcile the clustering results: if an observation was consistently labeled across both results, it was assigned a definitive class (benign or DDoS); otherwise, it was tagged as "Suspicious". This process was crucial for creating a reliably labeled dataset for subsequent supervised learning.

Supervised algorithms used included K-nearest neighbor (KNN), Support Vector Machines (SVMs), and Random Forest, with hyperparameters finely tuned during the training phase. The classification accuracy on the synthetic dataset was impressive, particularly for Random Forest, which achieved a 96.66% accuracy rate. This methodology was also validated on the aforementioned subset of CICIDS2017, achieving over 86% accuracy, underscoring the effectiveness of this semi-supervised approach.

Concerning data augmentation, Maharana et al. [9] extensively reviewed augmentation techniques, particularly focusing on their application in ML for image data. Techniques such as flipping, cropping, rotation, and color space adjustments were explored, detailing how they help in creating varied datasets from limited data sources. These methods are critical for reducing model overfitting and improving the robustness of the predictions.

Naik et al. [10] discuss the integration of AI with traditional cybersecurity strategies, noting how AI can bring a significant improvement in handling cyber threats through technologies like Big Data, Blockchain, and Behavioral Analytics. In detail, it provides an in-depth analysis of both "distributed" AI methods (such as Multi-Agent Systems, Artificial Neural Networks, Artificial Immune Systems, and Genetic Algorithms) and "compact" AI methods (including ML Systems, Expert Systems, and Fuzzy Logic). These classifications help differentiate AI techniques based on their application scope and complexity.

Table 1 outlines a comparative review of the methodologies and outcomes from the discussed studies focused on ML and data augmentation approaches to intrusion detection.

Table 1. Comparison of related work in IDS using ML and data augmentation techniques.

Paper	Year	Strengths	Limitations
Hwang et al. [7]	2020	Approach used tried to minimize memory usage.	No standard classification.
Verkerken et al. [6]	2021	Multiclass classification. Comparison between intra- and inter-dataset evaluations.	No feature selection. No standard classification.
Aamir and Zaidi [8]	2021	Hyperparameter tuning. Validation of the approach on a synthetic dataset and benchmark dataset.	Synthetic dataset. Focus only on DDoS attacks. Using a small number of variables
Tufan et al. [4]	2021	Hyperparameter tuning. Comparison of the models trained and tested on private and publicly available datasets. Feature selection.	Focus only on probing. Designed to work offline. The institutional dataset was collected from a specific environment. No standard classification.
Bagui et al. [3]	2022	Created a new dataset labeled according to MITRE framework. Unique and detailed preprocessing strategy. Feature selection based on information gain. Comparison between models also in terms of using important features (top 6, top 9, all 18)	Dataset contains only 2 tactics. Using binary classification. Absence of hyperparameter tuning. Designed to work offline. Many duplicates in the dataset.
Ravi et al. [5]	2022	Multiclass classification. Used the most common benchmark datasets.	Proposed approach is sensitive to imbalanced datasets. No standard classification.
Maharana et al. [9]	2022	Methodological diversity, educational value	Lack of empirical data, focus on ML
Naik et al. [10]	2022	Comprehensive coverage, practical applications	Lack of case studies and evaluation metrics
Apruzzese et al. [1]	2023	Deep discussion on intrinsic problems	None
Mijwil et al. [2]	2023	Analysis of current ML and DL models applied in IDS, in terms of advantages and limitations	None
Agrawal et al. [11]	2024	Synthetic data generated by GANs	Lack of realism
Mohammad et al. [12]	2024	High accuracy in intrusion detection	Persistent challenge of class imbalance and the marginal performance improvements of complex DL models

In [11], Agrawal et al. explored the application of generative adversarial networks (GANs) in creating synthetic data for cybersecurity. Key contributions included comprehensively examining GANs' capabilities in generating realistic cyberattack data and their use in enhancing IDSs (IDS). The study identified challenges such as the efficacy of GAN-generated data in accurately representing real-world attacks and the need for further investigation into the robustness of deep learning models trained on synthetic data.

Limitations included persistent concerns about the quality and realism of the synthetic data produced by GANs. The paper emphasizes the importance of synthetic data in overcoming privacy and security concerns associated with real-world data sharing

Mohammed et al. [12] presented a method to improve IDS performance by combining deep learning architectures with data augmentation techniques. Key contributions included using four prominent datasets (UNSW-NB15, 5G-NIDD, FLNET2023, and CIC-IDS-2017) to demonstrate that simple CNN-based models can achieve a high accuracy in intrusion detection. Limitations highlighted include the persistent challenge of class imbalance and the marginal performance improvements observed with more complex deep learning architectures compared to simpler models. The study emphasized the importance of data quality and augmentation in enhancing detection capabilities.

3. Background

3.1. Introduction

In the rapidly evolving field of cybersecurity, ML plays a crucial role in enhancing threat detection and mitigation. Effective data augmentation techniques, such as SMOTE and ADASYN, are essential for addressing class imbalances in cybersecurity datasets. However, simply balancing datasets and training ML models is insufficient unless these models can also assign weights to dataset features. Understanding feature importance is key to identifying which factors most influence the model's responses, thereby improving the model's interpretability and effectiveness. This section explores various data augmentation methods, supervised learning models, and the significance of feature weighting in the context of cybersecurity, offering insights into optimizing model performance and decision-making.

3.2. Data Augmentation Techniques

Advanced techniques like synthetic data generation through methods such as SMOTE (Synthetic Minority Over-sampling Technique [13]) can also be employed to enrich the dataset without losing valuable information. Given a sample x_i from the minority class, SMOTE identifies its k nearest neighbors in the feature space. Let x_{nn} denote one of these k nearest neighbors. A synthetic sample x_{new} is generated by interpolating between x_i and x_{nn} using the equation:

$$x_{\text{new}} = x_i + \lambda(x_{nn} - x_i)$$

where λ is a random number between 0 and 1. This interpolation step creates a new sample that is a linear combination of the original sample and its neighbor, thus preserving the general data distribution while expanding the minority class.

ADASYN (Adaptive Synthetic Sampling, [14]) extends SMOTE by focusing more on generating synthetic samples next to the minority class samples that are wrongly classified by a classifier. Mathematically, ADASYN calculates the number of synthetic samples to generate for each minority class sample x_i by using a density distribution:

$$r_i = \frac{\gamma_i}{\sum_{i=1}^{N} \gamma_i}$$

where γ_i is the number of majority class samples in the k nearest neighbors of x_i. The number of synthetic samples G_i to generate for each x_i is proportional to r_i.

In cases where classes overlap significantly, SMOTE and ADASYN can introduce synthetic samples in regions where the classes are not well-separated. This can lead to increased misclassification, as synthetic samples in overlapping regions may be misclassified, reducing the overall performance of the model. Moreover, introducing synthetic samples in overlapping regions can blur the decision boundaries, making it difficult for the classifier to distinguish between classes.

When synthetic samples introduce noise, both SMOTE and ADASYN can suffer from decreased performance. In particular, synthetic samples that are not representative of the actual data distribution can introduce noise, leading to poor generalization of the model. Furthermore, the presence of noisy samples can cause the model to overfit the synthetic data, reducing its ability to perform well on unseen data. Additionally, noise can adversely affect precision and recall, as the model may produce more false positives and false negatives.

However, several strategies can be employed to mitigate the issues of overlapping classes and noise. For example, the data can be preprocessed to remove noise before applying SMOTE or ADASYN. Secondly, more sophisticated techniques (such as Borderline-SMOTE or SVM-SMOTE) that focus on generating samples near the decision boundary can be used. Finally, it is possible to continuously evaluate the performance and tune the parameters of the over-sampling techniques to minimize the introduction of noise.

Borderline-SMOTE [15] specifically targets minority class samples that are close to the boundary with the majority class. It uses the same interpolation strategy as SMOTE but restricts it to those minority samples whose nearest neighbors include majority class samples. The synthetic sample generation formula remains as per the SMOTE technique.

Tomek-links data augmentation [16] is a technique used primarily to enhance the performance of classifiers on imbalanced datasets. It involves identifying pairs of instances that are nearest neighbors but belong to different classes and removing them to increase the separability of the classes.

A Tomek-link exists between a pair of instances x_i and x_j from different classes if there is no instance x_k such that $d(x_i, x_k) < d(x_i, x_j)$ or $d(x_j, x_k) < d(x_j, x_i)$, where d represents the distance metric used, often the Euclidean distance. Mathematically, it can be characterized as follows:

Let S be the set of all samples, then a pair $(x_i, x_j) \in S \times S$ forms a Tomek-link if

$$(y_i \neq y_j) \wedge (\nexists x_k \in S : (d(x_i, x_k) < d(x_i, x_j) \vee d(x_j, x_k) < d(x_j, x_i)))$$

This method is especially effective for binary classification problems and is often utilized as a data cleaning technique rather than an oversampling technique.

SMOTEENN [17] combines two approaches to address the issue of class imbalance in machine learning datasets: SMOTE (Synthetic Minority Over-sampling Technique) for over-sampling the minority class and ENN (Edited Nearest Neighbor) for cleaning the data by under-sampling both classes.

ENN removes any sample that has a majority of its k nearest neighbors belonging to a different class. For a given sample x_i, it is removed if

$$\frac{1}{k} \sum_{j=1}^{k} I(y_j \neq y_i) > 0.5$$

where I is an indicator function, y_j is the class label of the j-th nearest neighbor, and y_i is the class label of x_i.

SMOTEENN applies SMOTE to generate synthetic samples and then uses ENN to remove any generated or original samples that are misclassified by their nearest neighbors. This combination helps in refining the class boundaries further than using SMOTE alone.

Finally, SMOTE-Tomek [18] is a hybrid method that combines the SMOTE approach for over-sampling the minority class with Tomek links for cleaning overlapping samples between classes. This technique is particularly effective in improving the classification of imbalanced datasets by both augmenting the minority class and enhancing class separability. SMOTE-Tomek first applies SMOTE to generate additional synthetic samples to balance the class distribution. Subsequently, it applies the Tomek links method to remove any Tomek links identified between the synthetic and original samples. This removal process helps in reducing noise and making the classes more distinct, which is beneficial for the subsequent learning process.

It is interesting to discuss the computational costs of the data augmentation techniques and their impact on the model. The time complexity of SMOTE is $O(T \times k \times d)$, where T is the number of synthetic samples, k is the number of nearest neighbors, and d is the dimensionality of the data. This results in a moderate increase in training time due to the need to find nearest neighbors and generate synthetic samples. ADASYN has a similar time complexity to SMOTE but includes additional computations to determine the difficulty of instances, resulting in slightly higher computational costs and additional processing time compared to SMOTE. Borderline SMOTE shares the same time complexity as SMOTE but with additional steps to identify boundary samples. This leads to higher computational costs as identifying boundary samples requires extra computations. The time complexity for Tomek Links is $O(n^2 \times d)$, where n is the number of samples and d is the dimensionality. This results in significant preprocessing time due to the need to compute pairwise distances between samples, thus impacting overall model training time. SMOTEENN combines the costs of SMOTE and the Edited Nearest Neighbors (ENNs) technique, typically $O(T \times k \times d) + O(n \times k \times d)$. The high computational cost results from combining synthetic sample generation and nearest neighbor cleaning, leading to a notable increase in resource usage. SMOTE Tomek combines the costs of SMOTE and Tomek Links, typically $O(T \times k \times d) + O(n^2 \times d)$. This combination results in very high computational costs due to extensive pairwise distance computations and synthetic sample generation, significantly impacting training time and resource consumption.

The computational costs of these techniques directly impact model training time and resource usage. Techniques with higher time complexity, such as SMOTE Tomek and SMOTEENN, significantly increase preprocessing time and extend overall training time. High computational costs translate to increased CPU and memory usage, which can be limiting factors for large datasets or complex models. Techniques with quadratic time complexity (e.g., Tomek Links) may not scale well with large datasets.

Data augmentation is crucial in cybersecurity for generating more comprehensive datasets that can help in better training machine learning and deep learning models. The scientific literature proposes different surveys (see, for example, [1,19–21]) centered around machine learning models applied in cybersecurity.

3.3. *Supervised Learning Models*

This work considers some of the supervised learning models such as Naive Bayes, KNN, XGBoost (XGB), Gradient Boosting Machine (GBM), Logistic Regression, and Random Forest, as well as deep learning models like RNNs and LSTMs on cyberattack datasets.

Naive Bayes is a probabilistic classifier based on Bayes' theorem with the assumption of independence between features. Its strengths include being fast and efficient, especially with large datasets, and performing well with small amounts of training data. However, it assumes independence between features, which is rarely true in real-world data, and is not suitable for datasets with highly correlated features. KNN is a simple, non-parametric algorithm that classifies a sample based on the majority class among its k-nearest neighbors. KNN is simple to implement and understand, and it is effective for small datasets with well-defined classes. Its limitations are that it is computationally intensive with large datasets and its performance can degrade with high-dimensional data. XGB is an optimized distributed gradient boosting library designed to be highly efficient and flexible. XGB offers high performance and accuracy, and it handles missing values and large datasets well. However, it requires careful tuning of hyperparameters and can be prone to overfitting if not properly regularized. GBM builds an additive model in a forward stage-wise manner, optimizing differentiable loss functions. GBM is known for its high accuracy and robustness, and it is effective for both regression and classification tasks. The main limitations are that it is computationally intensive and slow to train, and it can be prone to overfitting without proper tuning. Logistic Regression is a linear model used for binary classification that predicts the probability of a categorical dependent variable. It is simple and interpretable, and it is efficient for binary and multinomial classification. However, it assumes a linear relationship between the features and the log-odds of the outcome, and it is not suitable for complex datasets with non-linear relationships. The Random Forest Classifier is an ensemble learning method that constructs multiple decision trees and merges them to obtain a more accurate and stable prediction. Random Forest handles large datasets with higher dimensionality well and is robust to overfitting due to its ensemble nature. However, it is computationally intensive, especially with a large number of trees, and less interpretable than single decision trees.

RNNs are a class of neural networks where connections between nodes form a directed graph along a temporal sequence, allowing them to exhibit temporal dynamic behavior. RNNs are effective for sequence prediction problems and can handle time-series data and sequential data well. However, they are prone to the vanishing gradient problem, making training difficult, and require significant computational resources. Finally, LSTMs are a special kind of RNN capable of learning long-term dependencies and mitigating the vanishing gradient problem. LSTMs are capable of learning long-term dependencies and are effective for time-series and sequential data. However, they are computationally expensive and slower to train, and they require extensive hyperparameter tuning.

3.4. Metrics

Accuracy is a widely used metric that reflects the overall correctness of a model's predictions. It is calculated as the proportion of correctly predicted instances relative to the total number of instances within the dataset as follows:

$$\text{Accuracy} = \frac{\text{Number of Correct Predictions}}{\text{Total Number of Predictions}}$$

While offering a general sense of model performance, accuracy can be misleading in scenarios with imbalanced datasets.

Precision, on the other hand, focuses on the proportion of true positive predictions among all predicted positives. It essentially measures the model's ability to accurately identify positive instances:

$$\text{Precision} = \frac{\text{True Positives}}{\text{True Positives} + \text{False Positives}}$$

A high precision value indicates a low false positive rate, signifying the model's proficiency in distinguishing between positive and negative cases.

Recall, alternatively referred to as sensitivity or true positive rate, measures the model's capacity to identify all relevant positive instances. It is calculated as the ratio of correctly predicted positive observations to the total actual positive observations within the dataset:

$$\text{Recall} = \frac{\text{True Positives}}{\text{True Positives} + \text{False Negatives}}$$

A high recall value signifies a low false negative rate, implying the model's effectiveness in capturing all pertinent positive cases.

The F1-score addresses the potential shortcomings of relying solely on precision or recall by providing a harmonic mean of both metrics. This consolidated metric offers a balanced assessment, particularly valuable in situations with imbalanced class distributions. The F1-score is calculated as follows:

$$\text{F1-Score} = 2 \times \frac{\text{Precision} \times \text{Recall}}{\text{Precision} + \text{Recall}}$$

The F1-score ranges from 0 to 1, with a value of 1 signifying the ideal scenario where both precision and recall are perfect.

Finally, support refers to the total number of actual occurrences for each class within the dataset. While not a direct measure of model performance, support provides crucial context for interpreting the other metrics. By understanding the class distribution and the number of data points per class (support), we can more effectively evaluate the significance of the calculated precision, recall, and F1-score values.

3.5. Features' Weights

CatBoost [22] is a machine learning algorithm developed by Yandex, which is part of the family of gradient boosting algorithms. The term "CatBoost" reflects its capability to handle categorical features effectively and its nature as a boosting algorithm. It is specifically designed to offer high performance with a focus on speed and accuracy, which is particularly advantageous when dealing with categorical data. This model optimizes the gradient boosting process through the use of symmetric trees and Oblivious Trees, enhancing both speed and accuracy while mitigating overfitting. This makes the model particularly robust and suitable for large datasets, with an implementation that supports GPU acceleration and multi-core processing.

A key feature of CatBoost is its capability to provide insights into the importance of features in the model. Within the domain of cybersecurity, the process of assigning weights to variables during data analysis holds paramount importance for several compelling reasons. Firstly, cybersecurity necessitates the examination of vast datasets to identify anomalies, potential threats, and existing vulnerabilities. Assigning weights to variables allows for the discernment of the most impactful factors contributing to potential security breaches. This acquired knowledge empowers cybersecurity professionals to prioritize their efforts on the most critical aspects, ultimately enhancing the efficacy of threat detection and mitigation strategies.

Secondly, the ever-evolving nature of cybersecurity threats presents a constant challenge. Attackers continuously develop novel techniques to exploit system vulnerabilities. Through the assessment of variable weights, security models can be dynamically adapted and updated to reflect the evolving threat landscape. This dynamic approach ensures the continued relevance and robustness of security measures in the face of emerging threats.

Finally, the assessment of variable weights also facilitates the explainability and interpretability of machine learning models. In the context of cybersecurity, comprehending the rationale behind a specific alert generation is crucial. This transparency not only fosters trust-building with stakeholders but also aids in forensic investigations to trace the origins and methodologies employed in cyberattacks.

4. Experiments

4.1. Methodology

The methodology employed in this study was aimed to address the challenges posed by imbalanced datasets in cyberattack detection. Initially, a realistic dataset was constructed to simulate various types of cyberattacks, reflecting the imbalances typically observed in real-world cybersecurity threats. This dataset serves as the foundation for evaluating the effectiveness of different data augmentation techniques.

To correct the skewed class distribution, the study applied several data augmentation techniques, notably SMOTE (Synthetic Minority Over-sampling Technique) and ADASYN (Adaptive Synthetic Sampling). These techniques generate synthetic samples to balance the dataset, thereby providing a more equitable distribution of classes for training supervised machine learning models.

Following the augmentation, various supervised machine learning models were trained and evaluated on the balanced dataset. The performance of these models was assessed using standard metrics such as accuracy, precision, recall, and F1-score to determine the improvements brought by the data augmentation techniques (see Appendix A for more details).

Additionally, the study analyzed feature importance to identify which attributes most significantly influence the predictive outcomes of the models. This analysis not only enhances the interpretability of the models but also provides valuable insights for refining feature engineering and selection processes, thereby optimizing model performance.

4.2. The Environment

The laboratory environment used VMware vSphere version 8.0.2.00300 for creating and managing virtual networks. This setup is crucial for developing and testing network security solutions effectively. The Security Onion server, version 2.4.60, was configured with two network interfaces. One interface was connected to the internal network, and the other was used for Docker container communications. The server was equipped with 12 CPUs, 24 GB of RAM, and a 250 GB hard disk configured as Thick Provision Lazy Zeroed. A Windows Server 2022 Standard Evaluation version (21H2) was set up with 2 CPUs, 12 GB of RAM, and a 90 GB hard disk.

The Windows 10 Pro client (Version 22H2) operated with 2 CPUs, 4 GB of RAM, and a 48 GB hard disk. Its network interface was connected to the same internal network, facilitating various network security experiments. An Ubuntu 22.04.4 LTS client was part of the network. Each system was configured with 2 CPUs, 3 GB of RAM, and a 25 GB hard disk. These clients interacted with other network components to simulate real-world traffic and attack scenarios. The main web server (APACHE01) and the reverse proxy server both ran on Ubuntu 22.04.4 LTS with similar hardware configurations as the Ubuntu clients. They play critical roles in hosting and securing web applications.

To simulate realistic network traffic, several generators were implemented. These include a generator for creating traffic from the internal network to the Internet based on the "noisy" project. This project involves collecting and visiting links from specified root URLs recursively until no more links are available or a timeout is reached.

APACHE01 is protected by a reverse proxy server (APACHE-REVERSE-PROXY), which is exposed to the Internet through a firewall, enhancing the security of hosted applications. Additionally, traffic and activities are monitored using tools like Security Onion to provide insights into network traffic and potential security threats.

4.3. The Dataset

The dataset features and their meaning are depicted in Table 2.

Table 2. Description of variables in the cyberattack dataframe.

Variable Name	Description
`resp_pkts`	Number of packets sent by the responder during the connection.
`service`	The type of service being accessed (e.g., HTTP, FTP).
`local_resp`	Indicates whether the responder is local to the network.
`protocol`	Network protocol used in the connection (e.g., TCP, UDP).
`duration`	Duration of the connection in seconds.
`conn_state`	State of the connection (e.g., established, closed).
`orig_pkts`	Number of packets sent by the originator during the connection.
`dest_port`	Destination port number of the connection.
`orig_bytes`	Number of bytes sent by the originator during the connection.
`local_orig`	Indicates whether the originator is local to the network.
`resp_bytes`	Number of bytes sent by the responder during the connection.
`src_port`	Source port number of the connection.
`techniques_mitre`	MITRE ATTACK technique(s) associated with the cyberattack.

Initially, the dataset consisted of 436,404 rows, which were reduced to 307,658 after validation (many observations were duplicated, and the features {duration, orig_bytes, resp_bytes} included 98,844 NaN values) and normalization. This preprocessing brought the total number of rows in the dataset to 208,735.

Notably, the techniques_mitre variable takes the following values:

- network_service_discovery;
- benign;
- reconnaissance_vulnerability_scanning;
- reconnaissance_wordlist_scanning;
- remote_system_discovery;
- domain_trust_discovery;
- account_discovery_domain;
- reconnaissance_scan_ip_blocks.

Network service discovery refers to the process of identifying and characterizing services running on networked devices. Adversaries employ techniques such as port scanning and service enumeration to identify open ports, listening services, and their corresponding software versions. This information aids in pinpointing potential vulnerabilities within the network and constructing a comprehensive network topology map.

It is vital to distinguish between benign activities and malicious network reconnaissance. Benign activities encompass actions inherent to normal system operations, including legitimate software updates, routine maintenance procedures, and standard user behavior. In contrast, malicious network reconnaissance, as detailed in the subsequent sections, involves deliberate attempts to exploit vulnerabilities and compromise system security.

Reconnaissance vulnerability scanning involves the systematic interrogation of target systems to identify exploitable weaknesses. Adversaries leverage this technique to detect outdated software, misconfigurations, and other security gaps. The primary objective is to amass information that can be later utilized to gain unauthorized access or execute malicious actions.

This technique involves leveraging pre-defined lists of words or phrases (wordlists) to systematically probe potential points of interest within a target environment. Adversaries utilize wordlists to conduct brute-force attacks or attempt to guess critical information such as usernames, passwords, URLs, and other sensitive data. Reconnaissance wordlist scanning frequently complements other reconnaissance activities to enhance attack efficiency and accuracy.

Remote system discovery is the process of gathering information about remote systems on a network. This can involve identifying active hosts, network shares, and accessible resources. Techniques used for remote system discovery include ping sweeps, port scanning, and querying network services. The ultimate objective is to map the network layout and pinpoint potential targets for subsequent exploitation attempts.

Within domain environments, adversaries utilize domain trust discovery to comprehend the trust relationships established between various domains. Understanding these trust relationships can provide adversaries with pathways for lateral movement and privilege escalation. This may involve identifying trusted domains, domain controllers, and any cross-domain policies that govern access control.

Account discovery (domain) is a technique employed by adversaries to enumerate user accounts within a domain environment. This involves discovering usernames, associated user groups, and corresponding permissions. The information gleaned can be utilized to plan attacks that involve credential theft, privilege escalation, and lateral movement within the compromised domain. Common methods for account discovery include querying Active Directory and leveraging built-in domain commands.

Reconnaissance scan IP blocks involve systematically scanning large ranges of IP addresses to identify active devices and services. Adversaries utilize this technique to map the target network infrastructure and pinpoint potential targets for further exploitation. This type of scanning can reveal critical information such as the number of active hosts, operating systems in use, and network devices present within the target environment.

Finally, Group Policy Discovery refers to the process of identifying and analyzing Group Policy Objects (GPOs) within a Windows domain environment. Adversaries examine GPOs to gain insights into security configurations, administrative templates, and user policies. This information can be used to identify misconfigurations, understand the deployed security controls, and pinpoint potential weaknesses that can be exploited to achieve their malicious goals.

As per Table 3, the dataset appears to be imbalanced due to a significant disproportion in the occurrences of different features, specifically within the "techniques_mitre distribution". Imbalanced datasets are commonly encountered in machine learning and statistical analysis and can lead to biased models that inadequately represent the minority classes. The feature "network_service_discovery" exhibits an overwhelming dominance with 144,279 occurrences, which is nearly 2.4 times that of the next most frequent category, "benign", which has 60,997 occurrences. This dominant feature may lead predictive models to exhibit a strong bias towards predicting this category, potentially at the expense of accuracy in other less frequent categories.

Table 3. Distribution of techniques_mitre.

Techniques_Mitre Distribution	Occurrences
network_service_discovery	144,279
benign	60,997
reconnaissance_vulnerability_scanning	1581
reconnaissance_wordlist_scanning	715
remote_system_discovery	554
domain_trust_discovery	411
account_discovery_domain	84
reconnaissance_scan_ip_blocks	80
group_policy_discovery	34

Moreover, categories such as "group_policy_discovery", "reconnaissance_scan_ip_blocks", and "account_discovery_domain" are extremely underrepresented with only 34, 80, and 84 occurrences, respectively. This sparse representation complicates the learning process for statistical models, as there are insufficient data to achieve a good generalization performance on new or unseen data falling into these categories. The vast range in feature distribution, from the most to the least frequent (144,279 occurrences vs. 34 occurrences), highlights the stark imbalance, indicating not only a skew towards certain features but also a significant under-representation of others.

Machine learning algorithms generally perform better when the numbers of instances for each class are approximately equal. An imbalanced dataset can result in models that are biased towards classes with more instances, increasing the likelihood of misclassification of minority class instances. This can severely affect the model's accuracy, particularly its ability to detect less frequent but potentially important categories.

4.4. Data Augmentation

Addressing this imbalance might involve employing techniques such as oversampling the minority classes, undersampling the majority classes, or using approaches like SMOTE, ADASYN, Borderline-SMOTE, Tomel-Links, SMOTEENN, and SMOTE Temek.

Table 4 offers a comprehensive comparison of various data augmentation techniques applied to an imbalanced dataset categorized under "techniques_mitre". These techniques include SMOTE, ADASYN, Borderline-SMOTE, Tomek Links, SMOTEENN, and a combination of SMOTE and Tomek Links, each tailored to modify the distribution of minority and majority classes through synthetic data generation or data cleaning.

The application of these augmentation methods aimed to normalize the occurrence rates across categories to a target number, approximately 115,474, for most methods, indicative of the level set to achieve class balance.

Table 4. Comparison of data augmentation techniques.

Techniques _MITRE	Original	SMOTE	ADASYN	Borderline-SMOTE	Tomek-Links	SMOTEENN	SMOTE Tomek
Benign	144,279	115,474	115,870	115,474	48,353	107,187	114,001
Account Discovery Domain	60,997	115,474	115,480	115,474	40	114,351	115,334
Domain Trust Discovery	1581	115,474	115,533	115,474	238	113,080	115,030
Group Policy Discovery	715	115,474	115,473	115,474	25	114,503	115,258
Network Service Discovery	554	115,474	115,474	115,474	115,467	115,405	115,470
Reconnaissance Scan IP Blocks	411	115,474	115,478	115,474	62	115,357	115,473
Reconnaissance Vulnerability Scanning	84	115,474	115,751	115,474	1004	111,585	114,734
Reconnaissance Wordlist Scanning	80	115,474	115,475	115,474	577	115,474	115,474
Remote System Discovery	34	115,474	115,475	115,474	431	113,998	115,324

4.5. Machine Learning Models

Machine learning models like Naïve Bayes, K-nearest neighbor (KNN), XGBoost (XGB), Gradient Boosting Machine (GBM), Logistic Regression, and Random Forest Classifier are often preferred in predictive analytics due to their diverse strengths and applicability across a wide range of problems. Each model brings a unique set of capabilities that makes it suitable for different types of data and predictive tasks. The preference for these models in various analytical scenarios stems from their ability to balance accuracy and computational efficiency while providing solutions that are easy to interpret and implement in real-world applications.

Table 5 reveals insightful trends regarding the behavior of these models under test conditions.

Table 5. Accuracy values for different classifiers and data augmentation methods.

Classifier	SMOTE	ADASYN	Borderline SMOTE	Tomek Links	SMOTEENN	SMOTE Tomek
Naïve Bayes	0.497	0.453	0.602	**0.718**	0.668	0.659
KNN	0.824	0.978	0.981	0.992	**0.993**	0.990
XGB	0.838	0.925	0.942	**0.993**	0.981	0.977
GBM	0.842	0.940	0.953	**0.989**	**0.989**	0.984
RF	0.833	0.985	0.985	0.994	**0.998**	0.996
Logistic	0.738	0.741	0.847	0.807	**0.860**	0.851
RNN	0.759	0.823	**0.888**	0.979	0.647	0.797
LSTM	0.819	0.875	0.916	**0.982**	0.945	0.944

Naïve Bayes, traditionally valued for its simplicity and efficiency in handling large datasets, shows moderate accuracy. This is expected given its assumption of feature independence, which might not always hold true in real-world datasets. KNN's performance is generally better, reflecting its capability to adapt its classification strategy based on the local data structure. However, KNN's reliance on feature scaling and the curse of dimensionality can sometimes affect its performance adversely.

XGB and GBM, both boosting models, exhibit high accuracy, underscoring their strength in dealing with complex datasets that involve non-linear relationships among features. These models build upon the errors of previous trees and, hence, can adaptively improve their predictions. The high performance is indicative of their robustness, but it also brings to light the need for careful parameter tuning to avoid fitting excessively to the training data.

Logistic Regression provides a reasonable accuracy that is useful in scenarios requiring probability estimation for binary outcomes. Its performance is generally less competitive compared to ensemble methods but offers valuable insights due to its interpretability.

Random Forest typically shows excellent accuracy due to its ability to reduce overfitting through averaging multiple decision trees. This model is effective in handling various types of data, including unbalanced datasets.

While the results suggest that ensemble methods like XGB, GBM, and Random Forest tend to provide higher accuracy, this must be balanced with the understanding that high accuracy can sometimes be a result of overfitting. Although overfitting is not the primary focus of this discussion, it is implicitly relevant when interpreting the high accuracy of complex models.

Finally, a CatBoost model was applied to the augmented dataset ("tomek_links.csv") obtained using the Tomek links approach. The result is shown in Figure 2.

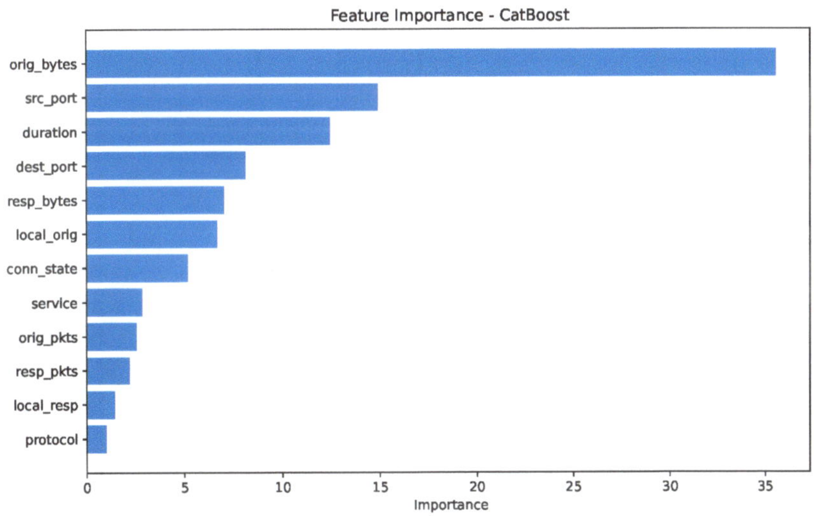

Figure 2. Features importance

4.6. Model Parameters

Table 6 shows the parameters employed by each model.

For GBM, key hyperparameters include the number of estimators, learning rate, and maximum depth. Tuning these involves the following:

- n_estimators: A higher number typically increases model complexity. Cross-validation helps find the optimal balance to avoid overfitting;
- learning_rate: Controls the contribution of each tree. Lower values typically require more trees;
- max_depth: Limits the depth of individual trees to control overfitting.

For KNN, the primary hyperparameter is the number of neighbors:

- _neighbors: A small number may lead to noisy predictions, while a large number can smooth out the prediction but may ignore local nuances. Grid search with cross-validation is commonly used to identify the optimal value.

Table 6. Models and their parameters.

Model	Parameters
GBM	n_estimators = 100, learning_rate = 0.1, max_depth = 3
KNN	n_neighbors = 5
Logistic Regression	max_iter = 10,000, class_weight = 'balanced'
LSTM	optimizer = 'adam', loss = 'categorical_crossentropy', metrics = ['accuracy']
GaussianNB	none
Random Forest	n_estimators = 100
RNN	optimizer = 'adam', loss = 'categorical_crossentropy', metrics = ['accuracy']
XGB	n_estimators = 100, learning_rate = 0.1, max_depth = 3, eval_metric = 'mlogloss'

Key hyperparameters include the maximum number of iterations and class weights:

- max_iter: Ensures convergence. Higher values allow the solver more iterations to converge, which is especially useful for complex datasets;
- class_weight: Balances the dataset by adjusting weights inversely proportional to class frequencies. It is particularly important in imbalanced datasets.

Critical hyperparameters for LSTM include the optimizer, loss function, and metrics:

- optimizer: "Adam" is commonly used for its adaptive learning rate capabilities;
- loss: "categorical_crossentropy" is used for multi-class classification problems;
- metrics: "accuracy" is a standard metric for evaluating classification performance.

GaussianNB typically requires no hyperparameter tuning as it is a straightforward probabilistic model. In contrast, important hyperparameters include the number of estimators:

- n_estimators: The number of trees in the forest. More trees generally improve performance but increase computation time.

Similar to LSTM, important hyperparameters include the optimizer, loss function, and metrics:

- optimizer: "Adam" is preferred for its efficiency and performance;
- loss: "categorical_crossentropy" for multi-class classification;
- metrics: "accuracy" for performance evaluation.

Finally, for XGBoost, key hyperparameters include the number of estimators, learning rate, maximum depth, and evaluation metric:

- n_estimators: Determines the number of boosting rounds;
- learning_rate: Lower rates require more boosting rounds;
- max_depth: Controls the depth of each tree to prevent overfitting;
- eval_metric: "mlogloss" is used for multi-class classification.

5. Discussion

5.1. Data Augmentation

The application of SMOTE and ADASYN was particularly effective in raising the number of instances in the minority classes to those in the majority classes, highlighting their capability to enhance intra-class variance through the generation of synthetic samples based on feature space similarities between existing minority samples. Similarly, Borderline-SMOTE, focusing on samples near the class borders, uniformly increased minority class counts, potentially aiding model accuracy in borderline cases.

The method involving Tomek Links, which removes pairs of closely situated opposite class samples, showed a substantial reduction in categories like "Benign", suggesting its efficacy in reducing major class sizes, thus deprioritizing the majority bias in data. SMOTEENN, which combines SMOTE's over-sampling with ENN's noise cleaning, appeared to both augment and cleanse the dataset by adding to the minorities and removing outliers or noise, respectively.

Combining SMOTE with Tomek Links resulted in a similar effect to SMOTE but with a slight reduction, indicating a cleaning effect on synthetic samples. These augmentation techniques signify a robust effort to address dataset imbalances, aiming to enhance the fairness and efficacy of predictive models.

5.2. Models

KNN achieved a high accuracy across all augmentation methods, with the highest accuracy of 0.993 using SMOTEENN. This model benefits from data augmentation, particularly with techniques like SMOTEENN and Tomek Links, which help in addressing class imbalance effectively. The accuracy of XGB also remained high across all augmentation methods, peaking at 0.993 with Tomek Links. XGB's robustness and ability to handle various types of augmented data contributed to its consistently high performance.

GBM showed a similar trend to XGB, with the highest accuracy of 0.989 using Tomek Links and SMOTEENN. The boosting approach in GBM makes it resilient to overfitting, even when augmented data are used. Random Forest (RF) achieved the highest accuracy of all models, with a maximum of 0.998 using SMOTEENN. The ensemble nature of RF allows it to generalize well across different augmented datasets.

Logistic Regressor displayed moderate performance, with the highest accuracy of 0.860 using SMOTEENN. The linear nature of this model might limit its ability to fully leverage the complex patterns introduced by some augmentation methods. RNN showed variable performance, with a peak accuracy of 0.979 using Tomek Links. The sequential nature of RNNs may benefit from Tomek Links' ability to clean noisy samples. LSTM, like RNN, showed improved performance with Tomek Links (0.982) and also benefited from other methods like SMOTEENN and SMOTETomek. LSTM's capability to capture long-term dependencies aids in leveraging augmented data effectively.

The choice of data augmentation technique has a significant impact on the performance of machine learning models. Techniques like SMOTEENN and Tomek Links generally yield higher accuracies, especially for models such as KNN, XGB, GBM, and RF. These findings highlight the importance of selecting appropriate data augmentation methods based on the machine learning model's characteristics and the dataset's nature.

Addressing the computational costs and efficiency of the data augmentation techniques and models involves several considerations. While data augmentation techniques can significantly improve the balance of the dataset and enhance model performance, they also introduce additional computational overhead. This overhead stems from the need to generate synthetic samples, which can be resource-intensive, especially for large datasets. To mitigate these costs, this study explored optimization strategies that streamline the augmentation process without compromising the quality of the generated data. This involves selecting appropriate parameters for each technique to balance the trade-off between computational efficiency and the effectiveness of the augmentation. For instance, optimizing the number of nearest neighbors in SMOTE or adjusting the density distribution in ADASYN can reduce unnecessary computations.

5.3. Features Importance

Regarding the various features commonly involved in network traffic data, which are crucial for identifying potentially malicious activities, the CatBoost model provided the following ranked features by importance:

- orig_bytes : This feature, representing the number of bytes that originated from the source, is identified as the most significant predictor. The high importance of this feature suggests that the volume of data sent from the source is a critical indicator of anomalous behavior.
- src_port and dest_port: The source and destination ports also play vital roles, indicating that particular ports may be more susceptible to exploitation or are commonly used by attackers.

- duration: The duration of the connection is another key feature, with longer connections possibly being indicative of data exfiltration activities.
- resp_bytes and resp_pkts: These features represent the response bytes and packets, respectively, highlighting the importance of the response size and frequency in detecting unusual responses that could signify a breach.

Regarding typical attack patterns, large orig_bytes values combined with extended duration are typical indicators. Effective detection rules should flag high data volume transfers, especially if they occur during off-hours or from unexpected sources. Unusual src_port and dest_port activity can signify reconnaissance efforts. Detection rules should monitor for spikes in port activity or access attempts to ports not typically used by legitimate applications within the organization. Moreover, long duration sessions should be scrutinized, especially if coupled with high resp_bytes. This pattern can indicate persistent attackers attempting to maintain access or exfiltrate data over extended periods. Based on these observations, different strategies can be considered to develop more effective detection rules. For example, it is recommended to establish thresholds for orig_bytes and duration that, when exceeded, trigger alerts. Specifically, a rule might flag any outgoing connection exceeding a certain data volume within a specific timeframe. Rules can also identify unusual port usage patterns, such as multiple access attempts to non-standard ports or a high frequency of connection attempts within a short period. This would help detect port scanning and early stages of attacks. ML models can be trained to learn normal patterns of src_port, dest_port, orig_bytes, and resp_pkts. Any significant deviations from these learned patterns can be flagged as potential threats. More sophisticated attack patterns can be detected by combining multiple features. For instance, a rule could flag connections with high orig_bytes and a long duration originating from an uncommon src_port and targeting an uncommon dest_port. Finally, user and system behaviors over time should be monitored. Sudden changes in data transfer volumes or connection durations that deviate from established behavior profiles can indicate compromised accounts or systems.

Updating models in response to evolving cyber threats requires a dynamic and continuous approach to ensure that detection systems remain effective against new and sophisticated attack patterns. One potential strategy is the implementation of a continuous learning framework. In this framework, the model is periodically retrained using recent data, which helps incorporate the latest threat patterns and anomalies observed in the network traffic. Another strategy involves the use of ensemble learning techniques. By combining multiple models, each trained on different aspects or time frames of the data, the system can achieve greater robustness against varying attack strategies. Ensemble methods can also incorporate new models trained on recent data, allowing the system to integrate fresh insights without completely discarding the knowledge from older models. Moreover, implementing a feedback loop from the security operations center (SOC) can be highly beneficial. When a potential threat is detected, the SOC can provide feedback on whether it was a true positive or a false positive. This feedback can be used to fine-tune the model, improving its accuracy over time. Data augmentation techniques play a crucial role in updating models. By continuously generating synthetic data that reflect the latest attack patterns and scenarios, the training dataset can be expanded and diversified. This approach helps in maintaining the model's effectiveness against a wide range of threats, including those that may not be prevalent in the historical data. Anomaly detection can be enhanced by integrating unsupervised learning methods alongside supervised ones. While supervised models are trained on labeled data, unsupervised models can identify new and unusual patterns without prior knowledge. By combining these approaches, the detection system can adapt to novel attack methods that deviate from known patterns.

6. Conclusions and Future Work

The adoption of advanced data augmentation techniques within supervised learning models significantly enhances the robustness and efficacy of cyberattack detection systems. This research demonstrates that by integrating SMOTE, ADASYN, and Tomek links, not only can the predictive accuracy be improved, but the generalizability of the models across diverse and evolving cyber threat landscapes can also be substantially enhanced.

Furthermore, our findings underscore the importance of leveraging a hybrid approach to data augmentation, which meticulously addresses the challenges of imbalanced datasets prevalent in cybersecurity applications. By employing these techniques, we successfully minimized the overfitting potential and improved the detection rates of cyberattacks.

As cybersecurity threats continue to evolve in complexity and subtlety, the ability of detection systems to adapt and respond with nuanced understanding becomes increasingly critical. The techniques developed and tested in this study support more sophisticated, adaptive responses to cyber threats, empowering security professionals with tools that are both reactive and preemptively adaptive.

Future work will explore the impact of the computational cost and efficiency of the augmentation methods, providing deeper insights into their practical applications. Further analysis on tailored cost-sensitive learning strategies will be pursued, where the cost of misclassifying minority classes is set higher than that of the majority classes to compel the model to pay more attention to the underrepresented classes. These measures are crucial for building robust models that perform well across all categories and are not biased toward the majority. Finally, more complex mechanisms of explanation exploiting consolidated techniques, such as LIME (Local Interpretable Model-Agnostic Explanations) and SHAP (SHapley Additive exPlanations), will be included to acquire a deeper understanding of cyberattacks.

Author Contributions: Conceptualization, K.M. and E.B.; methodology, A.G.; software, E.B.; validation, K.M., T.T. and A.G.; formal analysis, E.B.; investigation, T.T.; resources, T.T.; data curation, K.M.; writing—original draft preparation, K.M.; writing—review and editing, E.B.; visualization, E.B.; supervision, E.B. All authors have read and agreed to the published version of the manuscript.

Funding: This research received no external funding

Data Availability Statement: The original data presented in this study, including the software to reproduce the results, are openly available at https://github.com/EBarbierato/cyberattack_classification (accessed on 3 July 2024).

Conflicts of Interest: The authors declare no conflicts of interest.

Abbreviations

The following abbreviations are used in this manuscript:

MDPI	Multidisciplinary Digital Publishing Institute
DOAJ	Directory of open access journals
TLA	Three letter acronym
LD	Linear dichroism
ADASYN	Adaptive Synthetic Sampling
CNN	Convolutional Neural Networks
DL	Deep Learning
DDoS	Distributed Denial of Service
DT	Decision Trees
GBM	Gradient Boosting Machines
IDS	Intrusion Detection Systems
KNN	K-Nearest Neighbor
KPCA	Kernel Principal Component Analysis
LSTM	Long Short-Term Memory
PCA	Principal Component Analysis
RNN	Recurrent Neural Networks
RF	Random Forest
SMOTE	Synthetic Minority Over-sampling Technique
SMOTEENN	Combination of SMOTE and Edited Nearest Neighbors

SVM	Support Vector Machines
XGB	eXtreme Gradient Boost

Appendix A

Appendix A.1. Additional Details

This section details the precision achieved when predicting the values of the techniques_mitre distribution.

Appendix A.1.1. GBM

SMOTE

Table A1. Accuracy: 0.8420.

Techniques_Mitre	Precision	Recall	F1-Score	Support
account_discovery_domain	0.68	0.71	0.69	23,003
benign	0.55	0.47	0.51	23,309
domain_trust_discovery	0.54	0.54	0.54	23,141
group_policy_discovery	0.97	0.96	0.96	23,008
network_service_discovery	0.95	0.97	0.96	23,029
reconnaissance_scan_ip_blocks	0.98	0.98	0.98	23,036
reconnaissance_vulnerability_scanning	0.90	0.98	0.94	22,971
reconnaissance_wordlist_scanning	0.99	0.99	0.99	23,201
remote_system_discovery	0.99	0.99	0.99	23,156

ADASYN

Table A2. Accuracy: 0.9402.

Techniques_Mitre	Precision	Recall	F1-Score	Support
account_discovery_domain	0.96	0.99	0.98	23,170
benign	0.93	0.55	0.69	23,287
domain_trust_discovery	0.93	0.97	0.95	23,051
group_policy_discovery	0.97	0.99	0.98	22,987
network_service_discovery	1.00	1.00	1.00	22,985
reconnaissance_scan_ip_blocks	1.00	1.00	1.00	23,138
reconnaissance_vulnerability_scanning	0.75	0.97	0.85	23,251
reconnaissance_wordlist_scanning	1.00	1.00	1.00	23,209
remote_system_discovery	0.97	0.99	0.98	22,924

Borderline SMOTE

Table A3. Accuracy: 0.9539.

Techniques_Mitre	Precision	Recall	F1-Score	Support
account_discovery_domain	0.99	1.00	1.00	23,067
benign	0.96	0.62	0.75	23,253
domain_trust_discovery	0.90	0.99	0.94	22,971
group_policy_discovery	1.00	1.00	1.00	23,220
network_service_discovery	1.00	1.00	1.00	22,987
reconnaissance_scan_ip_blocks	1.00	1.00	1.00	23,023
reconnaissance_vulnerability_scanning	0.79	0.98	0.87	22,968
reconnaissance_wordlist_scanning	1.00	1.00	1.00	23,202
remote_system_discovery	1.00	1.00	1.00	23,163

Tomek Links

Table A4. Accuracy: 0.9895.

Techniques_Mitre	Precision	Recall	F1-Score	Support
account_discovery_domain	0.40	0.29	0.33	7
benign	0.97	0.99	0.98	9535
domain_trust_discovery	0.00	0.00	0.00	58
group_policy_discovery	1.00	0.25	0.40	4
network_service_discovery	1.00	1.00	1.00	23,225
reconnaissance_scan_ip_blocks	1.00	1.00	1.00	16
reconnaissance_vulnerability_scanning	0.59	0.43	0.50	192
reconnaissance_wordlist_scanning	1.00	1.00	1.00	115
remote_system_discovery	0.00	0.00	0.00	88

SMOTEENN

Table A5. Accuracy: 0.9892.

Techniques_Mitre	Precision	Recall	F1-Score	Support
account_discovery_domain	0.98	1.00	0.99	22,784
benign	0.98	0.93	0.96	21,615
domain_trust_discovery	1.00	0.99	0.99	22,536
group_policy_discovery	0.99	0.99	0.99	23,026
network_service_discovery	1.00	1.00	1.00	22,956
reconnaissance_scan_ip_blocks	1.00	1.00	1.00	23,013
reconnaissance_vulnerability_scanning	0.96	1.00	0.98	22,359
reconnaissance_wordlist_scanning	1.00	1.00	1.00	23,256
remote_system_discovery	1.00	1.00	1.00	22,643

SMOTETomek

Table A6. Accuracy: 0.9842.

Techniques_Mitre	Precision	Recall	F1-Score	Support
account_discovery_domain	0.97	1.00	0.98	23,164
benign	0.98	0.90	0.94	22,718
domain_trust_discovery	1.00	0.98	0.99	22,862
group_policy_discovery	0.99	0.99	0.99	23,048
network_service_discovery	1.00	1.00	1.00	23,213
reconnaissance_scan_ip_blocks	1.00	1.00	1.00	23,065
reconnaissance_vulnerability_scanning	0.94	1.00	0.97	22,871
reconnaissance_wordlist_scanning	1.00	1.00	1.00	23,202
remote_system_discovery	1.00	0.99	0.99	23,077

Appendix A.1.2. KNN

SMOTE

Table A7. Accuracy: 0.8245.

Techniques_Mitre	Precision	Recall	F1-Score	Support
account_discovery_domain	0.59	0.64	0.61	23,003
benign	0.46	0.39	0.42	23,309
domain_trust_discovery	0.50	0.51	0.50	23,141
group_policy_discovery	0.97	0.97	0.97	23,008
network_service_discovery	0.97	0.98	0.97	23,029
reconnaissance_scan_ip_blocks	0.98	0.98	0.98	23,036
reconnaissance_vulnerability_scanning	0.95	0.97	0.96	22,971

Table A7. *Cont.*

Techniques_Mitre	Precision	Recall	F1-Score	Support
reconnaissance_wordlist_scanning	0.99	0.99	0.99	23,201
remote_system_discovery	0.99	0.99	0.99	23,156

ADASYN

Table A8. Accuracy: 0.9782.

Techniques_Mitre	Precision	Recall	F1-Score	Support
account_discovery_domain	0.98	1.00	0.99	23,170
benign	0.96	0.85	0.90	23,287
domain_trust_discovery	0.94	0.99	0.96	23,051
group_policy_discovery	1.00	1.00	1.00	22,987
network_service_discovery	1.00	1.00	1.00	22,985
reconnaissance_scan_ip_blocks	1.00	1.00	1.00	23,138
reconnaissance_vulnerability_scanning	0.94	0.98	0.96	23,251
reconnaissance_wordlist_scanning	1.00	1.00	1.00	23,209
remote_system_discovery	0.99	1.00	1.00	22,924

Borderline SMOTE

Table A9. Accuracy: 0.9812.

Techniques_Mitre	Precision	Recall	F1-Score	Support
account_discovery_domain	0.99	1.00	1.00	23,067
benign	0.95	0.88	0.91	23,253
domain_trust_discovery	0.92	0.98	0.95	22,971
group_policy_discovery	1.00	1.00	1.00	23,220
network_service_discovery	1.00	1.00	1.00	22,987
reconnaissance_scan_ip_blocks	1.00	1.00	1.00	23,023
reconnaissance_vulnerability_scanning	0.97	0.98	0.98	22,968
reconnaissance_wordlist_scanning	1.00	1.00	1.00	23,202
remote_system_discovery	1.00	1.00	1.00	23,163

Tomek Links

Table A10. Accuracy: 0.9925.

Techniques_Mitre	Precision	Recall	F1-Score	Support
account_discovery_domain	0.75	0.43	0.55	7
benign	0.99	0.99	0.99	9535
domain_trust_discovery	0.69	0.84	0.76	58
group_policy_discovery	0.33	0.25	0.29	4
network_service_discovery	1.00	1.00	1.00	23,225
reconnaissance_scan_ip_blocks	0.93	0.88	0.90	16
reconnaissance_vulnerability_scanning	0.59	0.45	0.51	192
reconnaissance_wordlist_scanning	1.00	0.99	1.00	115
remote_system_discovery	0.75	0.92	0.83	88

SMOTEENN

Table A11. Accuracy: 0.9938.

Techniques_Mitre	Precision	Recall	F1-Score	Support
account_discovery_domain	0.98	1.00	0.99	22,784
benign	0.99	0.96	0.97	21,615

Table A11. *Cont.*

Techniques_Mitre	Precision	Recall	F1-Score	Support
domain_trust_discovery	1.00	0.99	1.00	22,536
group_policy_discovery	0.99	1.00	1.00	23,026
network_service_discovery	1.00	1.00	1.00	22,956
reconnaissance_scan_ip_blocks	1.00	1.00	1.00	23,013
reconnaissance_vulnerability_scanning	0.99	1.00	0.99	22,359
reconnaissance_wordlist_scanning	1.00	1.00	1.00	23,256
remote_system_discovery	0.99	1.00	1.00	22,643

SMOTETomek

Table A12. Accuracy: 0.9902.

Techniques_Mitre	Precision	Recall	F1-Score	Support
account_discovery_domain	0.97	0.99	0.98	23,164
benign	0.98	0.94	0.96	22,718
domain_trust_discovery	0.99	0.99	0.99	22,862
group_policy_discovery	0.99	1.00	1.00	23,048
network_service_discovery	1.00	1.00	1.00	23,213
reconnaissance_scan_ip_blocks	1.00	1.00	1.00	23,065
reconnaissance_vulnerability_scanning	0.98	1.00	0.99	22,871
reconnaissance_wordlist_scanning	1.00	1.00	1.00	23,202
remote_system_discovery	0.99	1.00	0.99	23,077

Appendix A.1.3. Logistic Regressor

SMOTE

Table A13. Accuracy: 0.7380.

Techniques_Mitre	Precision	Recall	F1-Score	Support
account_discovery_domain	0.63	0.72	0.67	23,003
benign	0.35	0.12	0.18	23,309
domain_trust_discovery	0.47	0.17	0.25	23,141
group_policy_discovery	0.92	0.90	0.91	23,008
network_service_discovery	0.62	0.89	0.73	23,029
reconnaissance_scan_ip_blocks	0.88	0.96	0.92	23,036
reconnaissance_vulnerability_scanning	0.89	0.91	0.90	22,971
reconnaissance_wordlist_scanning	0.99	0.99	0.99	23,201
remote_system_discovery	0.61	0.97	0.75	23,156

ADASYN

Table A14. Accuracy: 0.7414.

Techniques_Mitre	Precision	Recall	F1-Score	Support
account_discovery_domain	0.64	0.47	0.54	23,170
benign	0.74	0.16	0.26	23,287
domain_trust_discovery	0.88	0.85	0.87	23,051
group_policy_discovery	0.54	0.63	0.58	22,987
network_service_discovery	1.00	1.00	1.00	22,985
reconnaissance_scan_ip_blocks	0.90	0.98	0.94	23,138
reconnaissance_vulnerability_scanning	0.72	0.96	0.82	23,251
reconnaissance_wordlist_scanning	0.99	1.00	0.99	23,209
remote_system_discovery	0.42	0.64	0.51	22,924

Borderline SMOTE

Table A15. Accuracy: 0.8470.

Techniques_Mitre	Precision	Recall	F1-Score	Support
account_discovery_domain	0.84	0.69	0.76	23,067
benign	0.84	0.23	0.36	23,253
domain_trust_discovery	0.84	0.94	0.89	22,971
group_policy_discovery	0.84	0.92	0.88	23,220
network_service_discovery	1.00	1.00	1.00	22,987
reconnaissance_scan_ip_blocks	0.96	1.00	0.98	23,023
reconnaissance_vulnerability_scanning	0.75	0.96	0.84	22,968
reconnaissance_wordlist_scanning	0.99	1.00	1.00	23,202
remote_system_discovery	0.65	0.88	0.75	23,163

Tomek Links

Table A16. Accuracy: 0.8070.

Techniques_Mitre	Precision	Recall	F1-Score	Support
account_discovery_domain	0.02	0.43	0.03	7
benign	0.99	0.34	0.50	9535
domain_trust_discovery	0.20	0.86	0.32	58
group_policy_discovery	0.00	0.50	0.00	4
network_service_discovery	1.00	1.00	1.00	23,225
reconnaissance_scan_ip_blocks	0.01	1.00	0.03	16
reconnaissance_vulnerability_scanning	0.15	0.96	0.25	192
reconnaissance_wordlist_scanning	0.98	0.99	0.99	115
remote_system_discovery	0.03	0.97	0.06	88

SMOTEENN

Table A17. Accuracy: 0.8609.

Techniques_Mitre	Precision	Recall	F1-Score	Support
account_discovery_domain	0.84	0.58	0.69	22,784
benign	0.76	0.46	0.57	21,615
domain_trust_discovery	0.96	0.93	0.95	22,536
group_policy_discovery	0.75	0.83	0.79	23,026
network_service_discovery	1.00	1.00	1.00	22,956
reconnaissance_scan_ip_blocks	0.93	0.99	0.96	23,013
reconnaissance_vulnerability_scanning	0.94	0.96	0.95	22,359
reconnaissance_wordlist_scanning	1.00	1.00	1.00	23,256
remote_system_discovery	0.64	0.98	0.78	22,643

SMOTETomek

Table A18. Accuracy: 0.8512.

Techniques_Mitre	Precision	Recall	F1-Score	Support
account_discovery_domain	0.84	0.57	0.67	23,164
benign	0.74	0.42	0.54	22,718
domain_trust_discovery	0.95	0.93	0.94	22,862
group_policy_discovery	0.73	0.83	0.78	23,048
network_service_discovery	1.00	1.00	1.00	23,213
reconnaissance_scan_ip_blocks	0.92	0.99	0.95	23,065
reconnaissance_vulnerability_scanning	0.92	0.95	0.94	22,871
reconnaissance_wordlist_scanning	1.00	1.00	1.00	23,202
remote_system_discovery	0.64	0.98	0.77	23,077

Appendix A.1.4. LSTM

SMOTE

Table A19. Accuracy: 0.8194.

Techniques_Mitre	Precision	Recall	F1-Score	Support
account_discovery_domain	0.68	0.70	0.69	23,003
benign	0.51	0.42	0.46	23,309
domain_trust_discovery	0.52	0.46	0.49	23,141
group_policy_discovery	0.95	0.93	0.94	23,008
network_service_discovery	0.91	0.96	0.93	23,029
reconnaissance_scan_ip_blocks	0.97	0.98	0.97	23,036
reconnaissance_vulnerability_scanning	0.90	0.98	0.94	22,971
reconnaissance_wordlist_scanning	0.99	0.99	0.99	23,201
remote_system_discovery	0.85	0.97	0.90	23,156

ADASYN

Table A20. Accuracy: 0.8754.

Techniques_Mitre	Precision	Recall	F1-Score	Support
account_discovery_domain	0.88	0.77	0.82	23,170
benign	0.93	0.37	0.53	23,287
domain_trust_discovery	0.91	0.92	0.92	23,051
group_policy_discovery	0.92	0.93	0.92	22,987
network_service_discovery	1.00	1.00	1.00	22,985
reconnaissance_scan_ip_blocks	0.97	1.00	0.99	23,138
reconnaissance_vulnerability_scanning	0.71	0.99	0.83	23,251
reconnaissance_wordlist_scanning	1.00	0.98	0.99	23,209
remote_system_discovery	0.70	0.94	0.80	22,924

Borderline SMOTE

Table A21. Accuracy: 0.9166.

Techniques_Mitre	Precision	Recall	F1-Score	Support
account_discovery_domain	0.97	0.87	0.92	23,067
benign	0.81	0.45	0.58	23,253
domain_trust_discovery	0.86	0.98	0.92	22,971
group_policy_discovery	0.97	1.00	0.98	23,220
network_service_discovery	1.00	1.00	1.00	22,987
reconnaissance_scan_ip_blocks	0.98	0.99	0.99	23,023
reconnaissance_vulnerability_scanning	0.75	0.99	0.85	22,968
reconnaissance_wordlist_scanning	0.99	0.99	0.99	23,202
remote_system_discovery	0.93	0.99	0.96	23,163

Tomek Links

Table A22. Accuracy: 0.9825.

Techniques_Mitre	Precision	Recall	F1-Score	Support
account_discovery_domain	0.00	0.00	0.00	7
benign	0.95	0.99	0.97	9535
domain_trust_discovery	0.00	0.00	0.00	58
group_policy_discovery	0.00	0.00	0.00	4
network_service_discovery	1.00	0.99	1.00	23,225
reconnaissance_scan_ip_blocks	0.00	0.00	0.00	16
reconnaissance_vulnerability_scanning	0.00	0.00	0.00	192

Table A22. *Cont.*

Techniques_Mitre	Precision	Recall	F1-Score	Support
reconnaissance_wordlist_scanning	0.90	0.99	0.94	115
remote_system_discovery	0.00	0.00	0.00	88

SMOTEENN

Table A23. Accuracy: 0.9459.

Techniques_Mitre	Precision	Recall	F1-Score	Support
account_discovery_domain	0.89	0.85	0.87	22,784
benign	0.95	0.80	0.87	21,615
domain_trust_discovery	0.99	0.95	0.97	22,536
group_policy_discovery	0.94	0.95	0.95	23,026
network_service_discovery	1.00	1.00	1.00	22,956
reconnaissance_scan_ip_blocks	0.99	0.99	0.99	23,013
reconnaissance_vulnerability_scanning	0.95	0.99	0.97	22,359
reconnaissance_wordlist_scanning	1.00	1.00	1.00	23,256
remote_system_discovery	0.83	0.98	0.90	22,643

SMOTETomek

Table A24. Accuracy: 0.9447.

Techniques_Mitre	Precision	Recall	F1-Score	Support
account_discovery_domain	0.93	0.81	0.87	23,164
benign	0.90	0.81	0.85	22,718
domain_trust_discovery	0.98	0.96	0.97	22,862
group_policy_discovery	0.95	0.96	0.96	23,048
network_service_discovery	1.00	1.00	1.00	23,213
reconnaissance_scan_ip_blocks	0.98	1.00	0.99	23,065
reconnaissance_vulnerability_scanning	0.93	0.99	0.96	22,871
reconnaissance_wordlist_scanning	1.00	1.00	1.00	23,202
remote_system_discovery	0.84	0.98	0.90	23,077

Appendix A.1.5. Naïve Bayes

SMOTE

Table A25. Accuracy: 0.4979.

Techniques_Mitre	Precision	Recall	F1-Score	Support
account_discovery_domain	0.06	0.00	0.00	23,003
benign	0.14	0.00	0.00	23,309
domain_trust_discovery	0.23	0.02	0.03	23,141
group_policy_discovery	0.68	0.87	0.77	23,008
network_service_discovery	0.19	0.22	0.20	23,029
reconnaissance_scan_ip_blocks	0.73	0.48	0.57	23,036
reconnaissance_vulnerability_scanning	0.35	0.96	0.51	22,971
reconnaissance_wordlist_scanning	0.97	0.99	0.98	23,201
remote_system_discovery	0.46	0.94	0.62	23,156

ADASYN

Table A26. Accuracy: 0.4532.

Techniques_Mitre	Precision	Recall	F1-Score	Support
account_discovery_domain	0.35	0.11	0.17	23,170
benign	0.31	0.04	0.07	23,287
domain_trust_discovery	0.85	0.87	0.86	23,051
group_policy_discovery	0.54	0.06	0.11	22,987
network_service_discovery	0.98	1.00	0.99	22,985
reconnaissance_scan_ip_blocks	0.21	0.95	0.35	23,138
reconnaissance_vulnerability_scanning	0.69	0.23	0.34	23,251
reconnaissance_wordlist_scanning	0.73	0.61	0.66	23,209
remote_system_discovery	0.29	0.22	0.25	22,924

Borderline SMOTE

Table A27. Accuracy: 0.6020.

Techniques_Mitre	Precision	Recall	F1-Score	Support
account_discovery_domain	0.60	0.17	0.26	23,067
benign	0.28	0.04	0.07	23,253
domain_trust_discovery	0.86	0.93	0.89	22,971
group_policy_discovery	0.98	0.16	0.28	23,220
network_service_discovery	0.98	1.00	0.99	22,987
reconnaissance_scan_ip_blocks	0.32	1.00	0.49	23,023
reconnaissance_vulnerability_scanning	0.71	0.70	0.71	22,968
reconnaissance_wordlist_scanning	0.86	0.76	0.81	23,202
remote_system_discovery	0.49	0.66	0.56	23,163

Tomek Links

Table A28. Accuracy: 0.7185.

Techniques_Mitre	Precision	Recall	F1-Score	Support
account_discovery_domain	0.00	0.14	0.00	7
benign	0.98	0.03	0.06	9535
domain_trust_discovery	0.12	0.86	0.21	58
group_policy_discovery	0.06	0.25	0.10	4
network_service_discovery	0.99	1.00	0.99	23,225
reconnaissance_scan_ip_blocks	0.00	0.81	0.01	16
reconnaissance_vulnerability_scanning	0.04	0.97	0.07	192
reconnaissance_wordlist_scanning	0.93	0.99	0.96	115
remote_system_discovery	0.34	0.94	0.50	88

SMOTEENN

Table A29. Accuracy: 0.6688.

Techniques_Mitre	Precision	Recall	F1-Score	Support
account_discovery_domain	0.40	0.35	0.37	22,784
benign	0.34	0.03	0.06	21,615
domain_trust_discovery	0.90	0.90	0.90	22,536
group_policy_discovery	0.81	0.07	0.13	23,026
network_service_discovery	0.99	1.00	0.99	22,956
reconnaissance_scan_ip_blocks	0.36	0.92	0.51	23,013
reconnaissance_vulnerability_scanning	0.70	0.74	0.72	22,359
reconnaissance_wordlist_scanning	0.99	0.99	0.99	23,256
remote_system_discovery	0.79	0.98	0.88	22,643

SMOTETomek

Table A30. Accuracy: 0.6599.

Techniques_Mitre	Precision	Recall	F1-Score	Support
account_discovery_domain	0.40	0.35	0.37	23,164
benign	0.33	0.03	0.05	22,718
domain_trust_discovery	0.88	0.90	0.89	22,862
group_policy_discovery	0.81	0.08	0.14	23,048
network_service_discovery	0.99	1.00	0.99	23,213
reconnaissance_scan_ip_blocks	0.34	0.91	0.50	23,065
reconnaissance_vulnerability_scanning	0.68	0.74	0.71	22,871
reconnaissance_wordlist_scanning	0.99	0.99	0.99	23,202
remote_system_discovery	0.80	0.94	0.87	23,077

Appendix A.1.6. Random Forest

SMOTE

Table A31. Accuracy: 0.8338.

Techniques_Mitre	Precision	Recall	F1-Score	Support
account_discovery_domain	0.62	0.62	0.62	23,003
benign	0.45	0.42	0.43	23,309
domain_trust_discovery	0.58	0.61	0.59	23,141
group_policy_discovery	0.97	0.97	0.97	23,008
network_service_discovery	0.97	0.96	0.97	23,029
reconnaissance_scan_ip_blocks	0.97	0.97	0.97	23,036
reconnaissance_vulnerability_scanning	0.94	0.98	0.96	22,971
reconnaissance_wordlist_scanning	0.99	0.99	0.99	23,201
remote_system_discovery	1.00	1.00	1.00	23,156

ADASYN

Table A32. Accuracy: 0.9856.

Techniques_Mitre	Precision	Recall	F1-Score	Support
account_discovery_domain	1.00	1.00	1.00	23,170
benign	0.97	0.90	0.93	23,287
domain_trust_discovery	0.94	0.98	0.96	23,051
group_policy_discovery	1.00	1.00	1.00	22,987
network_service_discovery	1.00	1.00	1.00	22,985
reconnaissance_scan_ip_blocks	1.00	1.00	1.00	23,138
reconnaissance_vulnerability_scanning	0.96	0.99	0.98	23,251
reconnaissance_wordlist_scanning	1.00	1.00	1.00	23,209
remote_system_discovery	1.00	1.00	1.00	22,924

Borderline SMOTE

Table A33. Accuracy: 0.9856.

Techniques_Mitre	Precision	Recall	F1-Score	Support
account_discovery_domain	1.00	1.00	1.00	23,067
benign	0.98	0.89	0.93	23,253
domain_trust_discovery	0.91	0.98	0.95	22,971
group_policy_discovery	1.00	1.00	1.00	23,220
network_service_discovery	1.00	1.00	1.00	22,987
reconnaissance_scan_ip_blocks	1.00	1.00	1.00	23,023
reconnaissance_vulnerability_scanning	0.98	1.00	0.99	22,968

Table A33. *Cont.*

Techniques_Mitre	Precision	Recall	F1-Score	Support
reconnaissance_wordlist_scanning	1.00	1.00	1.00	23,202
remote_system_discovery	1.00	1.00	1.00	23,163

Tomek Links

Table A34. Accuracy: 0.9947.

Techniques_Mitre	Precision	Recall	F1-Score	Support
account_discovery_domain	0.75	0.43	0.55	7
benign	0.99	0.99	0.99	9535
domain_trust_discovery	0.74	0.74	0.74	58
group_policy_discovery	1.00	0.25	0.40	4
network_service_discovery	1.00	1.00	1.00	23,225
reconnaissance_scan_ip_blocks	1.00	1.00	1.00	16
reconnaissance_vulnerability_scanning	0.72	0.52	0.61	192
reconnaissance_wordlist_scanning	1.00	1.00	1.00	115
remote_system_discovery	1.00	0.98	0.99	88

SMOTEENN

Table A35. Accuracy: 0.9981.

Techniques_Mitre	Precision	Recall	F1-Score	Support
account_discovery_domain	1.00	1.00	1.00	22,784
benign	1.00	0.98	0.99	21,615
domain_trust_discovery	1.00	1.00	1.00	22,536
group_policy_discovery	1.00	1.00	1.00	23,026
network_service_discovery	1.00	1.00	1.00	22,956
reconnaissance_scan_ip_blocks	1.00	1.00	1.00	23,013
reconnaissance_vulnerability_scanning	0.99	1.00	0.99	22,359
reconnaissance_wordlist_scanning	1.00	1.00	1.00	23,256
remote_system_discovery	1.00	1.00	1.00	22,643

SMOTETomek

Table A36. Accuracy: 0.9966.

Techniques_Mitre	Precision	Recall	F1-Score	Support
account_discovery_domain	1.00	1.00	1.00	23,164
benign	1.00	0.97	0.98	22,718
domain_trust_discovery	1.00	1.00	1.00	22,862
group_policy_discovery	1.00	1.00	1.00	23,048
network_service_discovery	1.00	1.00	1.00	23,213
reconnaissance_scan_ip_blocks	1.00	1.00	1.00	23,065
reconnaissance_vulnerability_scanning	0.98	1.00	0.99	22,871
reconnaissance_wordlist_scanning	1.00	1.00	1.00	23,202
remote_system_discovery	1.00	1.00	1.00	23,077

Appendix A.1.7. RNN

SMOTE

Table A37. Accuracy: 0.7594.

Techniques_Mitre	Precision	Recall	F1-Score	Support
account_discovery_domain	0.69	0.68	0.69	23,003
benign	0.48	0.08	0.14	23,309
domain_trust_discovery	0.47	0.50	0.48	23,141
group_policy_discovery	0.96	0.87	0.92	23,008
network_service_discovery	0.71	0.91	0.80	23,029
reconnaissance_scan_ip_blocks	0.96	0.86	0.91	23,036
reconnaissance_vulnerability_scanning	0.80	0.97	0.88	22,971
reconnaissance_wordlist_scanning	0.99	0.99	0.99	23,201
remote_system_discovery	0.66	0.97	0.79	23,156

ADASYN

Table A38. Accuracy: 0.8231.

Techniques_Mitre	Precision	Recall	F1-Score	Support
account_discovery_domain	0.80	0.68	0.74	23,170
benign	0.92	0.29	0.44	23,287
domain_trust_discovery	0.91	0.89	0.90	23,051
group_policy_discovery	0.78	0.93	0.85	22,987
network_service_discovery	1.00	0.99	0.99	22,985
reconnaissance_scan_ip_blocks	0.96	0.85	0.90	23,138
reconnaissance_vulnerability_scanning	0.69	0.95	0.80	23,251
reconnaissance_wordlist_scanning	0.94	1.00	0.97	23,209
remote_system_discovery	0.62	0.83	0.71	22,924

Borderline SMOTE

Table A39. Accuracy: 0.8881.

Techniques_Mitre	Precision	Recall	F1-Score	Support
account_discovery_domain	0.95	0.77	0.85	23,067
benign	0.93	0.34	0.50	23,253
domain_trust_discovery	0.86	0.96	0.91	22,971
group_policy_discovery	0.97	0.98	0.97	23,220
network_service_discovery	1.00	1.00	1.00	22,987
reconnaissance_scan_ip_blocks	0.90	0.99	0.94	23,023
reconnaissance_vulnerability_scanning	0.74	0.99	0.84	22,968
reconnaissance_wordlist_scanning	0.99	0.99	0.99	23,202
remote_system_discovery	0.78	0.99	0.87	23,163

Tomek Links

Table A40. Accuracy: 0.9790.

Techniques_Mitre	Precision	Recall	F1-Score	Support
account_discovery_domain	0.00	0.00	0.00	7
benign	0.95	0.98	0.97	9535
domain_trust_discovery	0.00	0.00	0.00	58
group_policy_discovery	0.00	0.00	0.00	4
network_service_discovery	0.99	0.99	0.99	23,225
reconnaissance_scan_ip_blocks	0.00	0.00	0.00	16
reconnaissance_vulnerability_scanning	0.00	0.00	0.00	192

Table A40. *Cont.*

Techniques_Mitre	Precision	Recall	F1-Score	Support
reconnaissance_wordlist_scanning	0.80	0.98	0.88	115
remote_system_discovery	0.00	0.00	0.00	88

SMOTEENN

Table A41. Accuracy: 0.6472.

Techniques_Mitre	Precision	Recall	F1-Score	Support
account_discovery_domain	0.79	0.31	0.45	22,784
benign	0.25	0.65	0.36	21,615
domain_trust_discovery	0.89	0.94	0.91	22,536
group_policy_discovery	0.88	0.58	0.70	23,026
network_service_discovery	1.00	0.96	0.98	22,956
reconnaissance_scan_ip_blocks	0.47	0.92	0.62	23,013
reconnaissance_vulnerability_scanning	0.98	0.46	0.62	22,359
reconnaissance_wordlist_scanning	0.99	0.99	0.99	23,256
remote_system_discovery	0.00	0.00	0.00	22,643

SMOTETomek

Table A42. Accuracy: 0.7977.

Techniques_Mitre	Precision	Recall	F1-Score	Support
account_discovery_domain	0.69	0.34	0.45	23,164
benign	0.57	0.61	0.59	22,718
domain_trust_discovery	0.99	0.93	0.96	22,862
group_policy_discovery	0.84	0.59	0.69	23,048
network_service_discovery	1.00	0.99	0.99	23,213
reconnaissance_scan_ip_blocks	0.85	0.89	0.87	23,065
reconnaissance_vulnerability_scanning	0.89	0.86	0.87	22,871
reconnaissance_wordlist_scanning	0.99	0.99	0.99	23,202
remote_system_discovery	0.55	0.98	0.70	23,077

Appendix A.1.8. XGB

SMOTE

Table A43. Accuracy: 0.8387.

Techniques_Mitre	Precision	Recall	F1-Score	Support
account_discovery_domain	0.69	0.69	0.69	23,003
benign	0.55	0.49	0.51	23,309
domain_trust_discovery	0.53	0.53	0.53	23,141
group_policy_discovery	0.96	0.95	0.95	23,008
network_service_discovery	0.93	0.97	0.95	23,029
reconnaissance_scan_ip_blocks	0.98	0.98	0.98	23,036
reconnaissance_vulnerability_scanning	0.90	0.98	0.94	22,971
reconnaissance_wordlist_scanning	0.99	0.99	0.99	23,201
remote_system_discovery	0.99	0.98	0.98	23,156

ADASYN

Table A44. Accuracy: 0.9254.

Techniques_Mitre	Precision	Recall	F1-Score	Support
account_discovery_domain	0.93	0.98	0.96	23,170
benign	0.96	0.45	0.61	23,287
domain_trust_discovery	0.92	0.95	0.93	23,051
group_policy_discovery	0.96	0.98	0.97	22,987
network_service_discovery	1.00	1.00	1.00	22,985
reconnaissance_scan_ip_blocks	1.00	1.00	1.00	23,138
reconnaissance_vulnerability_scanning	0.72	0.99	0.84	23,251
reconnaissance_wordlist_scanning	1.00	1.00	1.00	23,209
remote_system_discovery	0.94	0.98	0.96	22,924

Borderline SMOTE

Table A45. Accuracy: 0.9427.

Techniques_Mitre	Precision	Recall	F1-Score	Support
account_discovery_domain	0.98	1.00	0.99	23,067
benign	0.96	0.51	0.67	23,253
domain_trust_discovery	0.87	0.99	0.92	22,971
group_policy_discovery	1.00	1.00	1.00	23,220
network_service_discovery	1.00	1.00	1.00	22,987
reconnaissance_scan_ip_blocks	1.00	1.00	1.00	23,023
reconnaissance_vulnerability_scanning	0.75	1.00	0.86	22,968
reconnaissance_wordlist_scanning	1.00	1.00	1.00	23,202
remote_system_discovery	1.00	1.00	1.00	23,163

Tomek Links

Table A46. Accuracy: 0.9932.

Techniques_Mitre	Precision	Recall	F1-Score	Support
account_discovery_domain	1.00	0.29	0.44	7
benign	0.98	0.99	0.99	9535
domain_trust_discovery	0.64	0.88	0.74	58
group_policy_discovery	0.00	0.00	0.00	4
network_service_discovery	1.00	1.00	1.00	23,225
reconnaissance_scan_ip_blocks	1.00	1.00	1.00	16
reconnaissance_vulnerability_scanning	0.63	0.33	0.44	192
reconnaissance_wordlist_scanning	1.00	1.00	1.00	115
remote_system_discovery	0.99	0.97	0.98	88

SMOTEENN

Table A47. Accuracy: 0.9819.

Techniques_Mitre	Precision	Recall	F1-Score	Support
account_discovery_domain	0.97	0.98	0.98	22,784
benign	0.96	0.90	0.93	21,615
domain_trust_discovery	1.00	0.97	0.98	22,536
group_policy_discovery	0.97	0.99	0.98	23,026
network_service_discovery	1.00	1.00	1.00	22,956
reconnaissance_scan_ip_blocks	1.00	1.00	1.00	23,013
reconnaissance_vulnerability_scanning	0.95	1.00	0.97	22,359
reconnaissance_wordlist_scanning	1.00	1.00	1.00	23,256
remote_system_discovery	0.99	0.99	0.99	22,643

SMOTETomek

Table A48. Accuracy: 0.9771.

Techniques_Mitre	Precision	Recall	F1-Score	Support
account_discovery_domain	0.95	0.99	0.97	23,164
benign	0.96	0.87	0.91	22,718
domain_trust_discovery	0.99	0.97	0.98	22,862
group_policy_discovery	0.97	0.98	0.98	23,048
network_service_discovery	1.00	1.00	1.00	23,213
reconnaissance_scan_ip_blocks	1.00	1.00	1.00	23,065
reconnaissance_vulnerability_scanning	0.93	1.00	0.96	22,871
reconnaissance_wordlist_scanning	1.00	1.00	1.00	23,202
remote_system_discovery	0.99	0.99	0.99	23,077

References

1. Apruzzese, G.; Laskov, P.; Montes de Oca, E.; Mallouli, W.; Brdalo Rapa, L.; Grammatopoulos, A.V.; Di Franco, F. The role of machine learning in cybersecurity. *Digit. Threat. Res. Pract.* **2023**, *4*, 1–38. [CrossRef]
2. Mijwil, M.; Salem, I.E.; Ismaeel, M.M. The significance of machine learning and deep learning techniques in cybersecurity: A comprehensive review. *Iraqi J. Comput. Sci. Math.* **2023**, *4*, 87–101.
3. Bagui, S.; Mink, D.; Bagui, S.; Ghosh, T.; McElroy, T.; Paredes, E.; Khasnavis, N.; Plenkers, R. Detecting reconnaissance and discovery tactics from the MITRE ATT&CK framework in Zeek conn logs using spark's machine learning in the big data framework. *Sensors* **2022**, *22*, 7999. [CrossRef] [PubMed]
4. Tufan, E.; Tezcan, C.; Acartürk, C. Anomaly-based intrusion detection by machine learning: A case study on probing attacks to an institutional network. *IEEE Access* **2021**, *9*, 50078–50092. [CrossRef]
5. Ravi, V.; Chaganti, R.; Alazab, M. Recurrent deep learning-based feature fusion ensemble meta-classifier approach for intelligent network intrusion detection system. *Comput. Electr. Eng.* **2022**, *102*, 108156. [CrossRef]
6. Verkerken, M.; D'hooge, L.; Wauters, T.; Volckaert, B.; De Turck, F. Towards model generalization for intrusion detection: Unsupervised machine learning techniques. *J. Netw. Syst. Manag.* **2022**, *30*, 1–25. [CrossRef]
7. Hwang, R.H.; Peng, M.C.; Huang, C.W.; Lin, P.C.; Nguyen, V.L. An unsupervised deep learning model for early network traffic anomaly detection. *IEEE Access* **2020**, *8*, 30387–30399. [CrossRef]
8. Aamir, M.; Zaidi, S.M.A. Clustering based semi-supervised machine learning for DDoS attack classification. *J. King Saud-Univ.-Comput. Inf. Sci.* **2021**, *33*, 436–446. [CrossRef]
9. Maharana, K.; Mondal, S.; Nemade, B. A review: Data pre-processing and data augmentation techniques. *Glob. Transitions Proc.* **2022**, *3*, 91–99. [CrossRef]
10. Naik, B.; Mehta, A.; Yagnik, H.; Shah, M. The impacts of artificial intelligence techniques in augmentation of cybersecurity: A comprehensive review. *Complex Intell. Syst.* **2022**, *8*, 1763–1780. [CrossRef]
11. Agrawal, G.; Kaur, A.; Myneni, S. A review of generative models in generating synthetic attack data for cybersecurity. *Electronics* **2024**, *13*, 322. [CrossRef]
12. Mohammad, R.; Saeed, F.; Almazroi, A.A.; Alsubaei, F.S.; Almazroi, A.A. Enhancing Intrusion Detection Systems Using a Deep Learning and Data Augmentation Approach. *Systems* **2024**, *12*, 79. [CrossRef]
13. Fernández, A.; Garcia, S.; Herrera, F.; Chawla, N.V. SMOTE for learning from imbalanced data: Progress and challenges, marking the 15-year anniversary. *J. Artif. Intell. Res.* **2018**, *61*, 863–905. [CrossRef]
14. He, H.; Bai, Y.; Garcia, E.A.; Li, S. ADASYN: Adaptive synthetic sampling approach for imbalanced learning. In Proceedings of the 2008 IEEE International Joint Conference on Neural Networks (IEEE World Congress on Computational Intelligence), Hong Kong, China, 1–8 June 2008; IEEE: Piscataway Township, NJ, USA, 2008; pp. 1322–1328.
15. Han, H.; Wang, W.Y.; Mao, B.H. Borderline-SMOTE: A new over-sampling method in imbalanced data sets learning. In Proceedings of the International Conference on Intelligent Computing, Hefei, China, 23–26 August 2005; Springer: Berlin/Heidelberg, Germany, 2005; pp. 878–887.
16. Swana, E.F.; Doorsamy, W.; Bokoro, P. Tomek link and SMOTE approaches for machine fault classification with an imbalanced dataset. *Sensors* **2022**, *22*, 3246. [CrossRef] [PubMed]
17. He, H.; Garcia, E.A. Learning from imbalanced data. *IEEE Trans. Knowl. Data Eng.* **2009**, *21*, 1263–1284.
18. Yang, H.; Li, M. Software Defect Prediction Based on SMOTE-Tomek and XGBoost. In *Bio-Inspired Computing: Theories and Applications*; Pan, L., Cui, Z., Cai, J., Li, L., Eds.; Springer: Singapore, 2022; pp. 12–31.
19. Handa, A.; Sharma, A.; Shukla, S.K. Machine learning in cybersecurity: A review. *Wiley Interdiscip. Rev. Data Min. Knowl. Discov.* **2019**, *9*, e1306. [CrossRef]
20. Dasgupta, D.; Akhtar, Z.; Sen, S. Machine learning in cybersecurity: A comprehensive survey. *J. Def. Model. Simul.* **2022**, *19*, 57–106. [CrossRef]

21. Martínez Torres, J.; Iglesias Comesaña, C.; García-Nieto, P.J. Machine learning techniques applied to cybersecurity. *Int. J. Mach. Learn. Cybern.* **2019**, *10*, 2823–2836. [CrossRef]
22. Prokhorenkova, L.; Gusev, G.; Vorobev, A.; Dorogush, A.V.; Gulin, A. CatBoost: Unbiased boosting with categorical features. *Adv. Neural Inf. Process. Syst.* **2018**, *31*, 6639–6649.

eng

MDPI

Article

Grading Evaluation of Marbling in Wagyu Beef Using Fractal Analysis

Yuya Suzuki [1] and Bao Yue [2,*]

[1] Department of Information Science and Technology, Graduate School of Science and Engineering, Tokyo City University, Setagaya Campus, Setagaya, Tokyo 158-8557, Japan; g2291402@tcu.ac.jp
[2] Department of Computer Science, Tokyo City University, Setagaya Campus, Setagaya, Tokyo 158-8557, Japan
* Correspondence: bao@tcu.ac.jp

Abstract: Wagyu beef is gaining worldwide popularity, primarily due to the fineness of its marbling. Currently, the evaluation of this marbling is performed visually by graders. This method has several issues: varying evaluation standards among graders, reduced accuracy due to long working hours and external factors causing fatigue, and fluctuations in grading standards due to the grader's mood at the time. This paper proposes the use of fractal analysis for the grading evaluation of beef marbling to achieve automatic grading without the inconsistencies caused by human factors. In the experiments, cross-sectional images of the parts used for visual judgment were taken, and fractal analysis was performed on these images to evaluate them using fractal dimensions. The results confirmed a correlation between the marbling evaluation and the fractal dimensions, demonstrating that quantitative evaluation can be achieved, moving away from qualitative visual assessments.

Keywords: fractal analysis; image analysis; meat quality grade; grading evaluation; Wagyu beef

Citation: Suzuki, Y.; Yue, B. Grading Evaluation of Marbling in Wagyu Beef Using Fractal Analysis. *Eng* **2024**, *5*, 2157–2169. https://doi.org/10.3390/eng5030113

Academic Editors: Antonio Gil Bravo and Jordan Hristov

Received: 10 July 2024
Revised: 16 August 2024
Accepted: 27 August 2024
Published: 2 September 2024

1. Introduction

The global popularity of Japanese cuisine has led to increased demand for Japanese Wagyu beef [1]. The most distinctive feature of Wagyu beef is its intricate marbling, which enhances the meat's smooth texture, tenderness, and unique flavor by providing a melt-in-the-mouth experience.

Currently, the evaluation of Wagyu marbling is performed visually by graders. Beef is classified into two grades: meat quality grade and yield grade. The meat quality grade consists of four evaluation criteria, with marbling being one of these criteria. This marbling evaluation standard is known as the Beef Marbling Standard Number (BMSNo), which ranges from 1 to 12. The evaluation is based on the extent of marbling in the longissimus thoracis, as well as the spinalis dorsi and semispinalis capitis muscles, between the sixth and seventh ribs. Higher marbling levels correspond to higher BMSNo values. The grading is conducted according to the standards set by the Japan Meat Grading Association (JMGA) [2], as shown in Figure 1. Table 1 illustrates the relationship between the 12-level BMSNo and the five-level marbling evaluation.

However, visual evaluation by graders presents several challenges: differences in evaluation criteria among graders, reduced accuracy due to long working hours and external factors causing fatigue, and fluctuations in grading standards due to the grader's mood. These factors introduce variability into the evaluation.

Hashimoto et al. [3] investigated a low-cost, objective method for estimating marbling using biopsy to measure moisture and crude fat content. However, this method is complex as it requires grinding the longissimus thoracis after cutting it at the slaughterhouse, and it necessitates specialized knowledge and equipment.

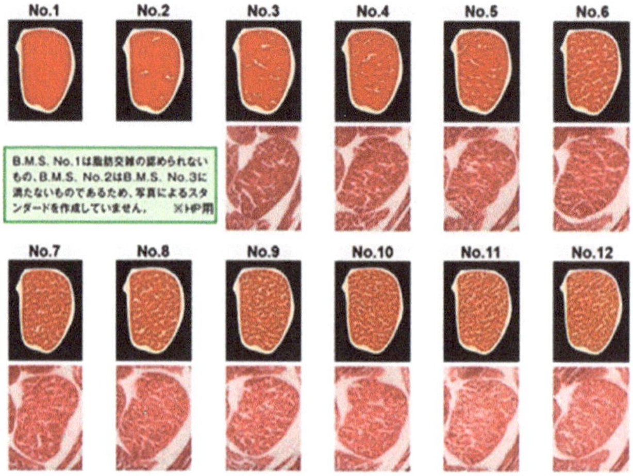

Figure 1. Beef Marbling Standards defined by the Japan Meat Grading Association (JMGA) [2]. The Japanese text within the green frame in the diagram explains that, for BMS No. 1 and 2, it does not meet the standards of BMS No. 3, and therefore, a standard based on photographs has not been created.

Table 1. Relationship between BMSNo and grading as defined by JMGA.

BMSNo	Grading
1	1
2	2
3–4	3
5–7	4
8–12	5

Kuchida et al. [4] proposed an image analysis method for estimating BMSNo from the fat area ratio using a dedicated imaging device. While this method can estimate BMSNo, it does not account for the complexity and fineness of marbling and requires specialized imaging equipment.

To address the complexity and fineness of marbling, Chen et al. [5] applied fractal analysis to the marbling images of beef from China and the United States. Fractal theory, proposed by Mandelbrot, uses the fractal dimension as a key indicator of the complexity of fractal patterns. In recent years, fractals have been utilized in various applications. Chen et al. used fractals for the evaluation of beef quality [5]. Additionally, our group has employed fractals in super-resolution processing [6], the analysis of plant growth [7], and the analysis of composite materials [8,9].

In the United States, beef quality grades are divided into eight levels, with the highest grade being "prime". The marbling grade, unlike in Japan, is based on the cross section of the longissimus thoracis between the twelfth and thirteenth ribs, classified into ten levels. The top three levels are categorized as "prime" [10,11]. Figure 2 compares images of "moderately abundant", the second-highest marbling grade in the U.S., with BMSNo. 11, the second-highest marbling grade in Japan. The complexity of Wagyu marbling is more pronounced than that of U.S. beef. Increased marbling complexity can result in coarse marbling or fine marbling, as shown in Figure 3. These characteristics are considered in Wagyu evaluation, necessitating a new method tailored to the detailed grading system used for Wagyu.

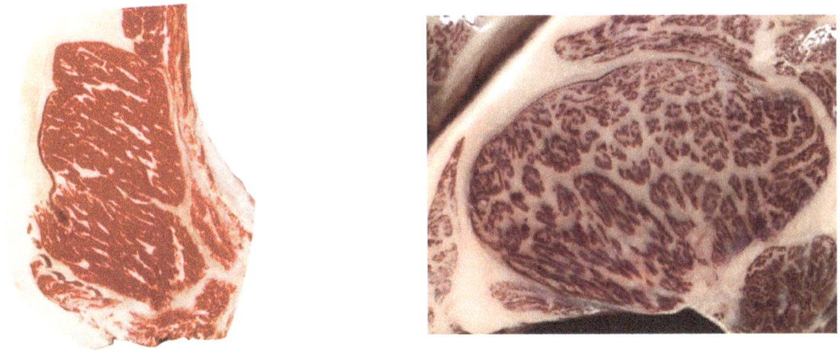

(a) (b)

Figure 2. Comparison of marbling between U.S. and Japanese beef: (**a**) moderately abundant (U.S., 2nd from top) [10]; (**b**) BMSNo.11 (JP, 2nd from top).

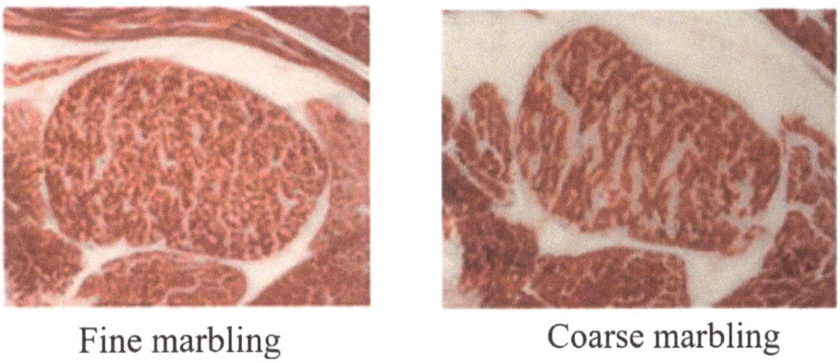

(a) (b)

Figure 3. Examples of coarse marbling and fine marbling [12]: (**a**) fine marbling; (**b**) coarse marbling.

In contrast, Xiao et al. [13] adjusted the marbling grades of Chinese beef from four to seven classes. However, even the highest grade with a marbling area of 14% is less complex compared to Wagyu. This paper proposes a fractal analysis method that considers the unique marbling characteristics of Japanese Wagyu, such as coarse marbling and fine marbling, for accurate evaluation.

2. Materials and Methods

2.1. Beef Quality Grading Evaluation

The quality grading of Wagyu beef is currently conducted through qualitative visual assessment by certified graders based on the evaluation method established by the Japan Meat Grading Association (JMGA). This evaluation encompasses four criteria: "meat color and brightness", "firmness and texture of the meat", "marbling", and "fat color and quality". Each criterion is rated on a five-point scale, where higher numbers indicate better quality.

This study focuses specifically on the evaluation of "marbling", a characteristic feature of Wagyu beef. The "marbling" criterion is assessed using the Beef Marbling Standard Number (BMSNo), which ranges from 1 to 12. This standard, developed by the Livestock Industry Technology Station of the Ministry of Agriculture, Forestry and Fisheries, measures the fineness and distribution of marbling within the beef [14].

2.2. Fractal Analysis

Fractal analysis is a technique that analyzes fractal patterns in images to calculate a parameter known as the fractal dimension, which is then used for various evaluations.

While fractal analysis can include multifractal analysis—where additional concepts like the information dimension (D_1) and correlation dimension (D_2) are used—previous studies have shown that these dimensions have low correlation with the marbling in beef used in this study [3]. Therefore, this paper uses a monofractal method rather than multifractal analysis, focusing solely on the fractal dimension (D_0). A higher D_0 value indicates a more complex pattern, which is similar to the evaluation methods used for assessing the marbling of beef.

There are several methods for performing fractal analysis to determine the fractal dimension. In this study, we use the box-counting method, the most commonly used technique in computer-based fractal analysis, to calculate the fractal dimension. The box-counting method involves covering the image with a grid of boxes of varying sizes and counting the number of boxes that contain part of the fractal pattern. The fractal dimension is then derived from the relationship between the size of the boxes and the number of boxes that contain part of the pattern.

2.3. Box-Counting Method

The box-counting method is one of the techniques used to calculate the fractal dimension. As shown in Figure 4, the box-counting method involves determining the presence probability for each box at a given box size ε, and then calculating the fractal dimension. The process involves repeatedly halving the box size ε.

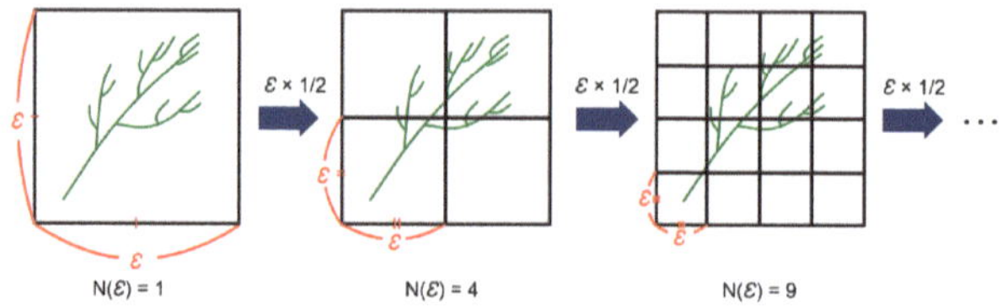

Figure 4. Schematic illustration of the box-counting method (the patterns in the figure are modeled after fractal shapes).

The fractal dimension D_q can be calculated using Equations (1) and (2).

$$D_q = \frac{1}{q-1}\lim_{\varepsilon\to 0}\frac{\log Z_q(\varepsilon)}{\log \varepsilon} \tag{1}$$

$$Z_q(\varepsilon) = \sum_{i=1}\{P_i(\varepsilon)\}^q \tag{2}$$

In Equation (1), D_q represents the generalized dimension, q is the moment order, and ε is the box size. In Equation (2), $P_i(\varepsilon)$ represents the probability of presence for each box. In this study, to use the monofractal method, the fractal dimension D_0 is employed by setting the moment order q to 0. Thus, setting q = 0 in Equation (1) leads to Equation (3).

$$D_0 = -\lim_{\varepsilon\to 0}\frac{\log Z_q\ (\varepsilon)}{\log \varepsilon} \tag{3}$$

Similarly, setting q = 0 in Equation (2) implies that the presence probability, regardless of its value (however, 0^0 is treated as 0 as an exception when no pattern exists within the box), results in a value of 1.

To summarize, $Z_q(\varepsilon)$ in Equation (3) simply counts the number of boxes that cover the pattern. Since the smallest pixel size is 1, it is impossible to determine the limit precisely. Therefore, we plot log ε on the x-axis and $\log Z_q(\varepsilon)$ on the y-axis, and then approximate the result by taking the slope of the best-fit line through these points.

2.4. Images Used for the Box-Counting Method

When using the box-counting method, it is preferable for the images to be of sizes that are powers of 2 (e.g., $2^n \times 2^n$). This ensures that when the box size ε is repeatedly halved, the entire image can be referenced without any parts being excluded or falling outside the scan area. For this study, images were prepared to meet this criterion, ensuring they fit neatly within the boxes during the analysis process. An example of how the images were prepared for use in the box-counting method is provided below.

2.4.1. Photography

The BMSNo evaluation is based on the extent of marbling in the longissimus thoracis, as well as the spinalis dorsi and semispinalis capitis muscles, between the sixth and seventh ribs. The more extensive the marbling, the higher the evaluation. In this study, we focused on the longissimus thoracis, as it is the main component used in the JMGA's Beef Marbling Standard model shown in Figure 1. Therefore, the subject of our photography was the cross section of the longissimus thoracis between the sixth and seventh ribs.

Figure 5 illustrates the photography setup. The distance of 14 cm between the camera and the meat in Figure 5 was set with the goal of capturing the entire longissimus thoracis in the image and ensuring that the actual size of one pixel is consistent across all images. The author took photographs of the longissimus thoracis at the sixth–seventh intercostal incision plane using an iPhone X in the freezer of a wholesaler with their permission. As shown in Figure 5, the camera was positioned parallel to the beef at a distance of 14 cm, with no zoom, flash, or filters used, and the focus was set to auto-adjust. The image size was 4032 × 3024 pixels, and an example of the actual photograph is shown in Figure 6.

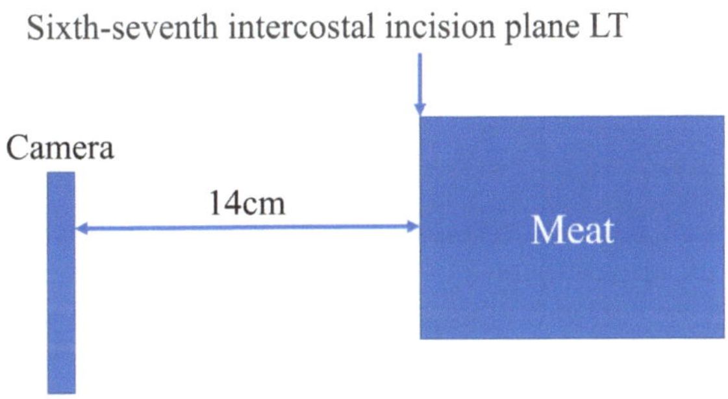

Figure 5. Schematic diagram of the photography setup (LT is longissimus thoracis).

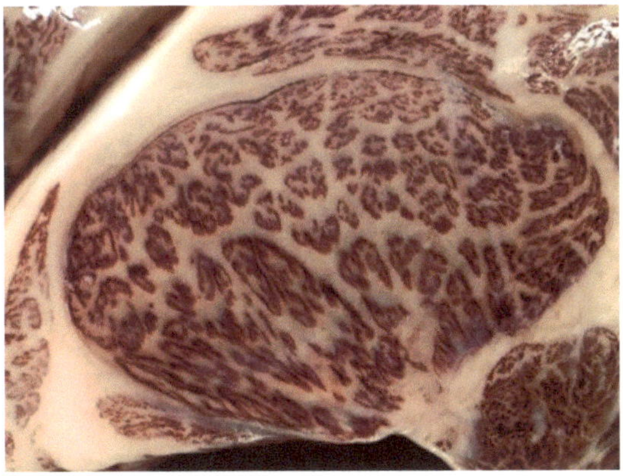

Figure 6. Actual photographs of the cross section of the longissimus thoracis between the sixth and seventh ribs (4032 × 3024 pixels).

2.4.2. Trimming

As mentioned in Section 2.4, it is preferable for the images used to be of size $2^n \times 2^n$ pixels. Therefore, the images are trimmed to a size of 1024 × 1024 pixels. However, trimming to 1024 × 1024 pixels might limit the analyzable area of the longissimus thoracis and may introduce errors if only one analysis per piece of beef is conducted. To address this, the longissimus thoracis is approximated as an ellipse, and nine areas—one central and eight surrounding—are trimmed from a single photograph, ensuring the entire muscle is covered as much as possible. This trimming is performed manually. Each of the nine trimmed images remains 1024 × 1024 pixels in size. An example of this trimming process is shown in Figure 7.

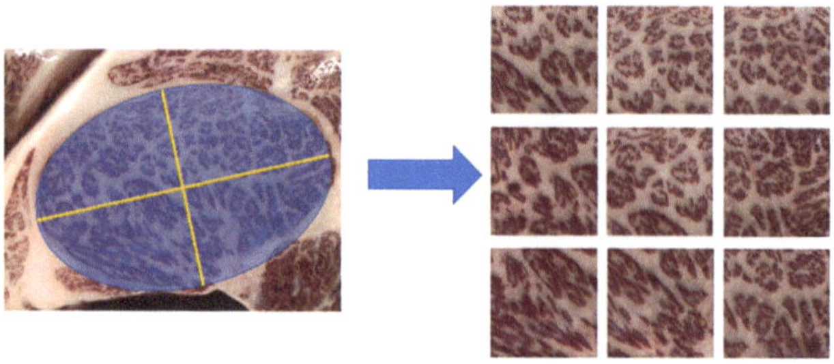

Figure 7. Example of trimming images to 1024 × 1024 pixels.

2.4.3. Grayscale Conversion and Binarization

To analyze the marbling patterns using the box-counting method, the images are first converted to grayscale and then binarized. In this study, we use a standard grayscale with 256 shades, which is typical for grayscale images. This means that each pixel can represent 256 levels of brightness, ranging from 0 (black) to 255 (white). The grayscale values are linearly related to the RGB values, as each grayscale shade corresponds to an equal value in the RGB channels. Regarding binarization, since different beef samples require different threshold values, we use Otsu's method [15] to automatically determine the optimal threshold for each image, rather than setting the threshold manually. An example of an image after these preprocessing steps is shown in Figure 8.

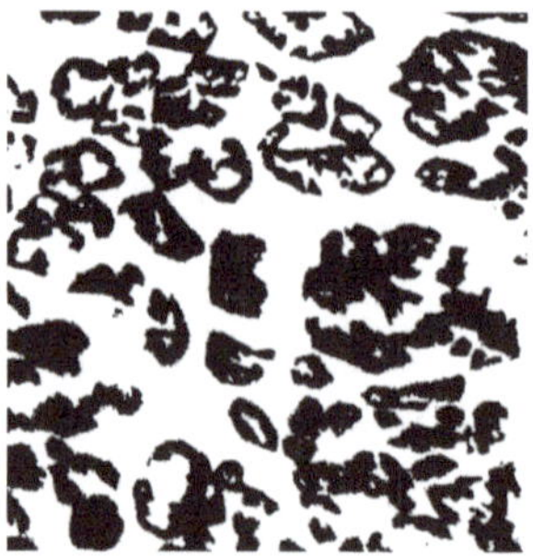

Figure 8. Example of a preprocessed image after grayscale conversion and binarization.

2.4.4. Box Size

Box-counting was performed on the binarized images. In this study, the box sizes ranged from 1 × 1 pixels to 1024 × 1024. Since the box size was doubled incrementally from 1 × 1 pixels, a total of 11 different box sizes were used.

2.5. Characteristics of Beef Marbling

The grading of beef marbling in Wagyu tends to be higher for more complex marbling patterns, as finer and more intricate marbling indicates higher quality. Within these marbling patterns, coarse marbling refers to coarser marbling, and fine marbling refers to finer marbling. Obama et al. [12,16] established roughness and fineness indices to evaluate coarse marbling and fine marbling. These indices are calculated using Equations (4) and (5).

$$R(\%) = \frac{M_a(\text{cm}^2)}{M_b(\text{cm}^2)} \tag{4}$$

$$S(\text{Amount}/\text{cm}^2) = \frac{G(\text{Amount})}{A(\text{cm}^2)} \tag{5}$$

In Equation (4), R represents the roughness index, M_a is the area of fat after thin lines are removed, and M_b is the total fat area. In Equation (5), S is the fineness index, G is the number of fine marbling particles, and A is the area of the ribeye muscle. Examples of coarse marbling and fine marbling calculated using these indices are shown in Figure 9.

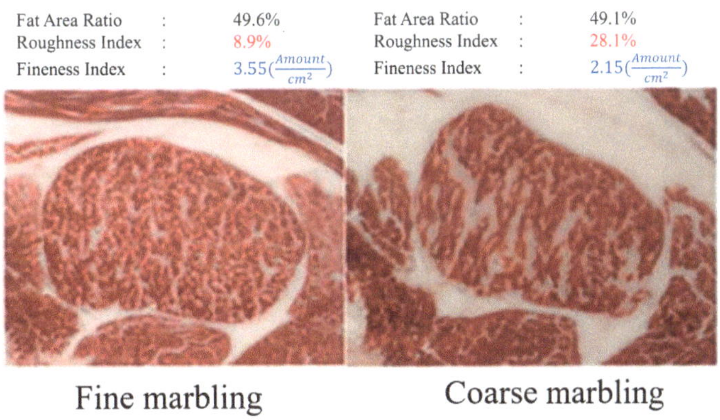

Figure 9. Examples of coarse marbling and fine marbling using roughness and fineness indices [12].

3. Results and Discussion

To demonstrate the effectiveness of the proposed method, three experiments were conducted: verifying the fractality of beef marbling, examining the correlation between BMSNo and the fractal dimension of beef marbling, and considering the characteristics of beef marbling in the experiments.

3.1. Verification of Fractality in Beef Marbling

For BC-FDA (box counting fractal dimension analysis) to be applied to a target pattern, the target must exhibit fractal properties. Although Chen et al. [5] utilized BC-FDA on beef marbling, they did not explicitly demonstrate its fractal nature. Therefore, an experiment was conducted to determine whether beef marbling exhibits fractality. Fractal patterns follow power laws, so if a log–log plot yields a coefficient of determination (r^2) close to 1, it indicates that the pattern has fractal properties [17].

After capturing the images, BC-FDA was performed according to the methods described in Section 2 (Materials and Methods).

Figure 10 shows a graph summarizing the results of applying the box-counting method to nine preprocessed images of a single piece of beef. The coefficient of determination r^2 (which indicates the degree of linearity, where a value closer to 1 suggests stronger linearity) was calculated for each of the nine images, with the results shown in Table 2. The r^2 values were obtained using the RSQ function in Excel as RSQ($\log \varepsilon$, $\log Z_q(\varepsilon)$) to determine the linearity of the graphs. From the values in Table 2, the trimmed mean was calculated by excluding the highest and lowest values, resulting in approximately 0.9996. This value is sufficiently close to 1, indicating that the log–log graph plotted in Figure 10 has strong linearity, as can be seen from the correspondence between the plotted points and the approximate straight line. This demonstrates that fractality exists in the marbling of the beef. The same process was applied to all 33 pieces of beef used in this study. The number of beef samples used in the experiment is shown in the table. For each piece of beef, the trimmed mean of the nine results was calculated by excluding the highest and lowest values, and the r^2 for each piece of beef was determined, followed by the overall result for all beef samples. For all the beef samples, the trimmed mean excluding the top and bottom 5% resulted in an r^2 of approximately 0.9994. From these results, it can be concluded that fractality exists similarly in the marbling of all the beef samples.

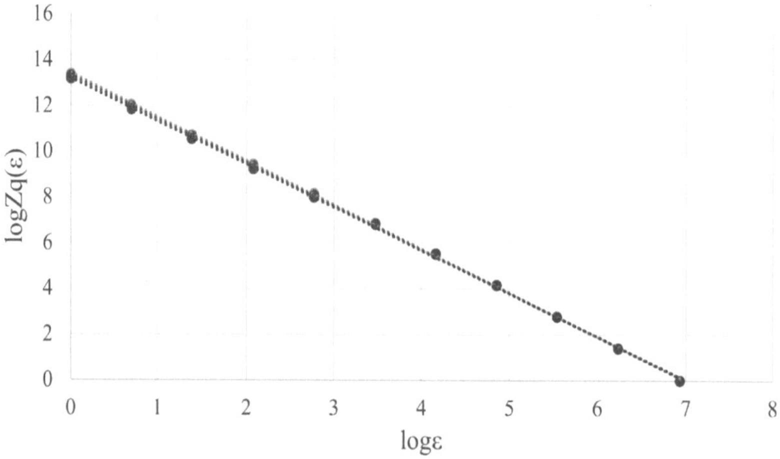

Figure 10. Results for 9 images for which the box-counting method was applied (the graphs, each with a different color, almost overlap).

Table 2. The coefficient of determination r^2 in Figure 10. Image No. refers to the number assigned to the images after they were trimmed, as shown in Figure 7. Since the order in which they are processed does not affect the results, there is no strict distinction such as 'this number corresponds to this specific trimming area'.

Image No.	r^2
a	0.9997
b	0.9997
c	0.9997
d	0.9995
e	0.9997
f	0.9996
g	0.9995
h	0.9995
i	0.9997

3.2. Experiment on the Correlation between BMSNo and the Fractal Dimentsion of Beef Marbling

BC-FDA was performed on all beef samples according to the methods described in Section 2 (Materials and Methods). Figure 11 shows the original images of all the beef samples used in the experiment. The vertical axis represents BMSNo, and the horizontal axis represents BeefNo.

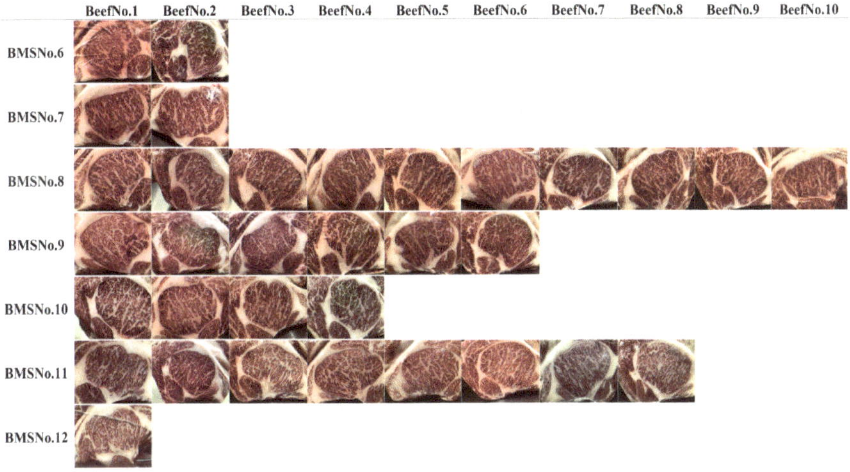

Figure 11. Original images of the beef samples used in the experiment. The vertical axis represents BMSNo., and the horizontal axis represents BeefNo.

Table 3 shows the fractal dimensions calculated for each piece of beef, and Figure 12 illustrates the correlation between the fractal dimension and BMSNo. The fractal dimensions in these tables and graphs were calculated as the trimmed mean of the nine images, excluding the maximum and minimum values. The dashed line in Figure 12 represents the trendline of the data, with a correlation coefficient of 0.7579, indicating a positive correlation. This suggests that the fractal dimension can be used to estimate BMSNo, a key indicator of Wagyu beef quality. However, some points were distant from the regression line, likely due to characteristics of marbling such as coarse marbling and fine marbling.

Table 3. Fractal dimension calculation results.

BMSNo-BeefNo.	Fractal Dimension
6-1	1.8658
6-2	1.8550
7-1	1.8680
7-2	1.8704
8-1	1.8919
8-2	1.9019
8-3	1.8620
8-4	1.8665
8-5	1.8743
8-6	1.8896
8-7	1.8840
8-8	1.8892
8-9	1.8899
8-10	1.8830
9-1	1.8946
9-2	1.8731
9-3	1.8973
9-4	1.8769
9-5	1.8891
9-6	1.8830
10-1	1.8961
10-2	1.9132
10-3	1.9012

Table 3. *Cont.*

BMSNo-BeefNo.	Fractal Dimension
10-4	1.8883
11-1	1.9055
11-2	1.8880
11-3	1.8947
11-4	1.9083
11-5	1.8931
11-6	1.8924
11-7	1.8963
11-9	1.9011
12-1	1.9205

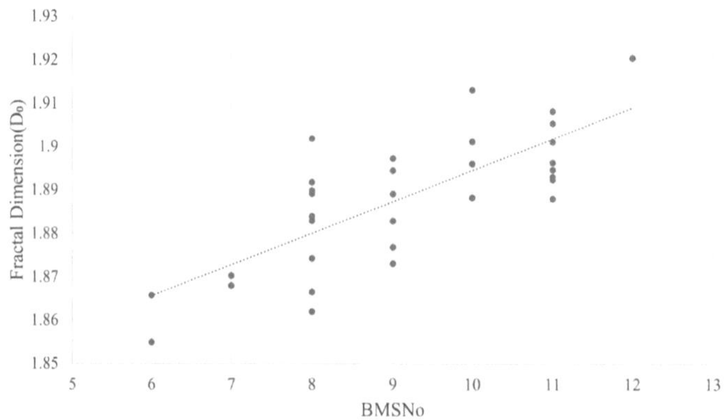

Figure 12. Graph showing the correlation between fractal dimension and BMSNo.

3.3. Experiment Considering the Characteristics of Beef Marbling

In the experiment conducted in Section 3.2, points distant from the regression line were observed due to the influence of marbling characteristics such as coarse marbling and fine marbling. Coarse marbling tends to receive a lower evaluation due to its rough appearance, but the D_0 value tends to be higher. On the other hand, fine marbling, which is finely distributed, tends to receive a higher evaluation, but the D_0 value tends to be lower. Therefore, it is necessary to consider these characteristics when performing BC-FDA. The roughness index and fineness index explained in Section 2.5 were used to account for these characteristics. In this experiment, the roughness index was calculated using Equation (4) and the fineness index using Equation (5). The area M_a and total fat area M_b in Equation (4) were determined by counting the white pixels in the binarized images and calculating the area per pixel. The area per pixel was approximately 0.000018716 cm^2, based on the measured maximum width of the longissimus thoracis.

For the number of fine marbling particles G in Equation (5), ten reduction and dilation operations were performed to isolate the fine marbling particles, and connected white pixel regions were labeled using an 8-neighbor connectivity algorithm. The number of labeled regions was treated as G. The area A of the ribeye muscle was fixed at 19.625 cm^2, calculated from the known size of the analysis region (1024 $\times$ 1024 pixels).

To verify if the results from Section 3.2 could be improved, the same beef samples were analyzed using the roughness and fineness indices. Table 4 shows the calculated roughness and fineness indices for each sample. The average roughness index was approximately 30.5, and the average fineness index was approximately 25.8. Beef samples with a roughness index of 34.3 or higher or a fineness index of 22 or lower were classified as coarse marbling, while those with a roughness index of 26.7 or lower or a fineness index of 29.6 or higher were classified as fine marbling. The remaining samples were plotted in Figure 13.

Table 4. Calculation results for roughness index and fineness index.

BMSNo-Beef No.	Roughness Index [%]	Fineness Index $\left[\frac{\text{Amount}}{\text{cm}^2}\right]$
6-1	26.6	28.5
6-2	28.8	24.0
7-1	31.4	25.8
7-2	32.1	22.6
8-1	32.0	24.8
8-2	29.7	28.0
8-3	26.8	26.8
8-4	29.1	26.7
8-5	31.2	25.7
8-6	30.2	25.9
8-7	33.0	22.7
8-8	31.2	26.4
8-9	31.2	25.3
8-10	30.5	28.5
9-1	31.5	24.9
9-2	26.2	25.4
9-3	27.0	28.2
9-4	29.8	28.2
9-5	32.4	24.2
9-6	24.0	33.7
10-1	33.4	22.6
10-2	34.4	23.2
10-3	29.2	25.4
10-4	30.5	27.1
11-1	32.0	27.5
11-2	24.7	39.0
11-3	33.6	14.5
11-4	35.6	20.8
11-5	32.6	22.8
11-6	31.4	21.9
11-7	29.8	30.3
11-8	31.0	24.3
12-1	33.3	24.6

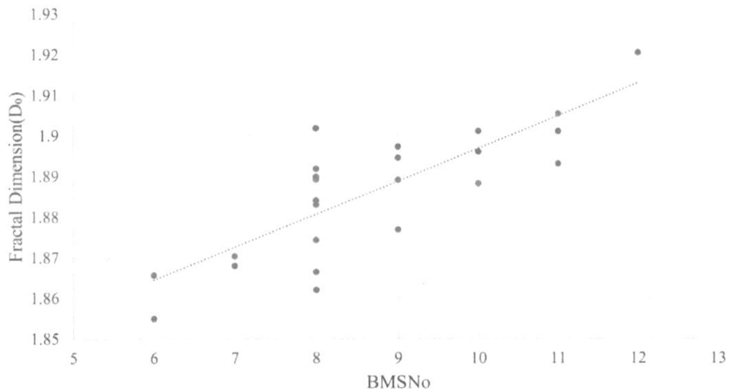

Figure 13. Graph showing correlation between fractal dimension and BMSNo excluding coarse and fine marbling.

The dashed line in Figure 13 represents the trendline of the data. The correlation coefficient is approximately 0.8019, which is higher than that in Section 3.2, indicating an improvement in accuracy. This suggests that considering the characteristics of coarse marbling and fine marbling can enhance the accuracy of BMSNo estimation. However, since BMSNo is determined by human graders, there is inherent variability, which likely increased the scatter in the results. Nonetheless, the average BMSNo values for each group in Figure 13 show correct grading, confirming that accounting for marbling characteristics such as coarse marbling and fine marbling improves BMSNo estimation accuracy.

4. Conclusions

In this study, we proposed quantifying the grading evaluation of Wagyu beef, which is traditionally performed visually, by using image processing. Three experiments were conducted to achieve this goal.

The first experiment aimed to verify the fractality of beef marbling. Through BC-FDA of beef marbling patterns, it was suggested that beef marbling indeed possesses fractal properties. This implies that BC-FDA is applicable to beef marbling.

The second experiment investigated the correlation between BMSNo and the fractal dimension of beef marbling. BC-FDA was performed on all prepared beef samples, and it was suggested that there is a positive correlation between BMSNo and the fractal dimension of beef marbling. This indicates that BMSNo can be estimated using the fractal dimension. However, the results showed some outlier points, which were suggested to be influenced by specific characteristics of beef marbling.

The third experiment considered these characteristics by incorporating the concepts of the roughness index and fineness index into the analysis. By classifying and excluding beef samples estimated to have coarse or fine marbling, the correlation coefficient between BMSNo and the fractal dimension of beef marbling improved compared to the second experiment. This suggests that classification of marbling characteristics can enhance accuracy.

In the future, it will be necessary to establish thresholds for defining coarse and fine marbling characteristics, considering their impact on BMSNo. By adjusting the BMSNo value based on these characteristics, we can improve the precision of grading evaluations.

Author Contributions: Conceptualization, Y.S. and B.Y.; methodology, Y.S.; software, Y.S.; validation, Y.S. and B.Y.; formal analysis, Y.S.; investigation, Y.S.; resources, Y.S.; data curation, Y.S.; writing—original draft preparation, Y.S.; writing—review and editing, Y.S. and B.Y.; visualization, Y.S.; supervision, B.Y.; project administration, B.Y.; funding acquisition, B.Y. All authors have read and agreed to the published version of the manuscript.

Funding: This research received no external funding.

Institutional Review Board Statement: Not applicable.

Informed Consent Statement: Not applicable.

Data Availability Statement: The datasets used or analyzed during the current study are available from the corresponding author upon reasonable request.

Conflicts of Interest: We declare that there is no conflict of interest.

References

1. Inagaki, S.; Amano, Y.; Kumazawa, K. Identification and Characterization of Volatile Components Causing the Characteristic Flavor of Wagyu Beef (Japanese Black Cattle). *Agric. Food Chem.* **2017**, *65*, 8691–8695. [CrossRef] [PubMed]
2. Japan Meat Grading Asociation. Available online: http://www.jmga.or.jp/standard/beef/ (accessed on 29 June 2022).
3. Hashimoto, D.; Iwamoto, T.; Hayata, G.; Nakanishi, Y. The Relationship Between Moisture or Crude Fat Content in Biopsy Muscle Samples and Carcass Fat Marbling in Fattening Cattle. *Warm Reg. Soc. Anim. Sci.* **2014**, *57*, 141–145. [CrossRef]
4. Kuchita, K.; Osawa, T.; Hori, T.; Kotaka, H.; Maruyama, S. Evaluation and genetics of carcass cross section of beef carcass by computer image analysis. *Anim. Genet.* **2006**, *34*, 45–52. [CrossRef] [PubMed]
5. Chen, J.; Liu, M.; Zong, L. The Fractal Dimension Research of Chinese and American Beef Marbling Standards Images. In Proceedings of the Computer and Computing Technologies in Agriculture VI: 6th IFIP WG 5.14 International Conference, CCTA 2012, Zhangjiajie, China, 19–21 October 2012; Springer: Berlin/Heidelberg, Germany, 2013; pp. 199–209. [CrossRef]
6. Mizukami, Y.; Matsuda, Y.; Bao, Y. Super-resolution processing using luminance dispersion in local area. *OSA Contin.* **2020**, *3*, 2893–2901. [CrossRef]
7. Mizukami, Y.; Sato, Y.; Bao, Y. Extraction of Leaf Features by Multifractal Analysis of Leaf Veins. In Proceedings of the International Conference on Advanced Imaging 2020, Chiba, Japan, 2–6 November 2020.

8. Munakata, F.; Takeda, M.; Nemoto, K.; Ookubo, K.; Sato, Y.; Mizukami, Y.; Koga, M.; Abe, S.; Bao, Y.; Kobayashi, R. Multifractal characteristics of the self-assembly material texture of β-Si3N4/SUS316L austenitic stainless steel composites. *J. Alloys Compd.* **2021**, *853*, 156570. [CrossRef]
9. Munakata, F.; Ookubo, K.; Takeda, M.; Sato, Y.; Mizukami, Y.; Nemoto, K.; Abe, S.; Bao, Y.; Kobayashi, R. Self-assembly process under a solid-state reaction of β-Si3N4/austenitic stainless-steel composites: Stirring conditions and material texture. *J. Compos. Mater.* **2021**, *56*, 455–466. [CrossRef]
10. Agricultural Marketing Service. U.S. Department of Agriculture. Available online: https://www.ams.usda.gov/ (accessed on 9 July 2024).
11. U.S. Meat Export Federation. Available online: https://www.americanmeat.jp/trd/database/rank/r_001.html (accessed on 9 July 2024).
12. Hyogo Prefectural Technology Center for Agriculture. Forestry and Fisheries. Available online: http://hyogo-nourinsuisangc.jp/archive/19-kenkyu/article/kenkyu_2709.html (accessed on 9 July 2024).
13. Xiao, T.; Xing, X.U.; Yong, Q.; Yuan, J.; Min, W.; Guang, Z. Improvement of marbling grade in Chinese beef grading standard. *Sci. Agric. Sin.* **2006**, *39*, 2101–2106. Available online: https://www.chinaagrisci.com/EN/Y2006/V39/I10/2101 (accessed on 13 August 2024).
14. Ministry of Agriculture, Forestry and Fisheries. Available online: https://www.maff.go.jp/index.html (accessed on 9 July 2024).
15. Otsu, N. A Threshold Selection Method from Gray-Level Histograms. *IEEE Trans. SMC* **1979**, *9*, 62–66. Available online: https://dspace.tul.cz/server/api/core/bitstreams/36abcc1c-cd72-4569-90ed-607017063124/content (accessed on 13 August 2024). [CrossRef]
16. Obama, N. Establishment of a New Meat Quality Evaluation Method (Estimation of Genetic Parameters for 'Fine Marbling' and 'Fat Quality'). *Livest. Technol. Hyogo* **2013**, *112*, 15–17. Available online: https://hyotiku.ecweb.jp/wp/wp-content/uploads/2024/03/gizyutuhyougo112.pdf (accessed on 13 August 2024).
17. Chonabayashi, H.; Negoro, H.; Ouchi, H. *Study on the Correlation Between Fractal Dimension and Environmental Cognition Using GIS and 3D Image Analysis: Considerations in Coastal Fishing Villages of the Izu and Boso Peninsulas;* College of Industrial Engineering, Nihon University Abstracts of Academic Lectures: Tokyo, Japan, 2004. Available online: https://www.cit.nihon-u.ac.jp/kouendata/No.37/4_kenchiku/4-040.pdf (accessed on 13 August 2024).

eng

MDPI

Article

Evaluating Field-Effect Separation on Rare Earth and Critical Metals

Benjamin Schroeder *, Michael Free, Prashant Sarswat, Easton Sadler, Jacob Burke and Zoe Evans

Department of Materials Science and Engineering, University of Utah, Salt Lake City, UT 84112, USA; michael.free@utah.edu (M.F.); prashant.sarswat@utah.edu (P.S.); u1191959@utah.edu (J.B.); zoeevans@arizona.edu (Z.E.)

* Correspondence: u1240350@utah.edu; Tel.: +1-(720)-266-0983

Abstract: The unique electromagnetic properties of rare earth elements (REEs) have led to rapid technological advances, creating a sharp increase in demand for these materials. The inherent challenges of separating REEs and the significant drawbacks of existing processes have driven the development of a new method known as field-effect separation (FES). This technology leverages electrical and magnetic fields to achieve separation by exploiting the differences in magnetic moments or effective charges of REEs in solution. Experiments on REEs were conducted using a microchannel based separation device, which confines fluid flow to facilitate separation within a field, with metal cations in solution being transported based on their respective electrostatic or magnetic properties. The results demonstrate that separation based on effective charge or paramagnetic properties is achievable. The confinement of fluid flow to microchannels allowed advective and osmotic forces to be suppressed sufficiently such that a reasonable separation of ions was achieved, though the impact of these forces were not completely removed. This innovative approach promises to improve the efficiency and effectiveness of REE separation, addressing both the growing demand and the limitations of current methods.

Keywords: rare earths; separation; magnetic field; electric field; microfluidics

Citation: Schroeder, B.; Free, M.; Sarswat, P.; Sadler, E.; Burke, J.; Evans, Z. Evaluating Field-Effect Separation on Rare Earth and Critical Metals. *Eng* **2024**, *5*, 2016–2032. https://doi.org/10.3390/eng5030107

Academic Editor: Antonio Gil Bravo

Received: 17 July 2024
Revised: 9 August 2024
Accepted: 15 August 2024
Published: 1 September 2024

1. Introduction

Recently, fast advances in the fields of electronics and energy storage have led to high demand for rare earth elements (REEs). The high demand for these critical materials, combined with their scarcity and the difficulty of producing them in large quantities, has led many governments globally to classify these elements as critical strategic resources for the future of technology. REEs are seen as key elements in a global renewable energy transition, and the use of heavier REEs in defense applications makes securing stable REE reserves a priority for a nation's economic viability, energy independence, and defense posture [1,2]. REEs are considered crucial for the renewable energy transition in the near future. Neodymium and other light to medium REEs, essential for strong permanent magnets, significantly contribute to the demand, primarily through their use in electric car motors and wind turbine generators [3,4]. As a result, new and more efficient methods of extracting and separating REEs for commercial use will be critical in the coming future.

REEs possess unique properties that result from their 4f valence orbital shells. However, the relatively similar physical properties of these elements as a result of each element having a similar ionic radius and their propensity to form +3 ions almost exclusively means that they all exhibit remarkably similar chemical properties that prove difficult to exploit in separation [2,5]. Generally, the REEs can be divided by atomic mass into light (La-Sm) and heavy (Eu-Lu), with some making a further distinction into a middle, medium REE class (Sm, Eu, Gd) [3,4,6]. A primary driver of recent demand for REEs has been the heavy REEs (HREEs), which have incredibly unique photo, electrical, and magnetic properties that make them the most useful of the lanthanides for advanced technologies, particularly crucial for advanced sensors used in the defense industry [6]. Currently, most industrial production of REEs uses liquid–liquid extraction, otherwise known as solvent extraction, which utilizes the small differences in chemical affinities of the REEs for a particular extractant to separate them [7].

Solvent extraction for REE separations has some significant disadvantages, including high consumption of the chemicals needed to achieve required purity, which make the process expensive and inefficient [2,8]. The process also has markedly worse performance in separating HREEs, and incomplete stripping of the solvent can lead to heavy losses of critical elements during processing [6]. Additionally, the organic solvents necessary for this process are flammable, toxic, carcinogenic, and environmentally damaging, making them a large source of problematic waste which can be challenging to safely store and dispose of [7]. Solvent extraction for rare earth production using monazite/bastnasite ore has a relatively high global warming impact per kg of produced REE due to release of greenhouse gases as well as generation of compounds that have some toxicity [9]. While actual energy consumption within a process is dependent on a number of factors that relate to the feed material such as the type and grade, the solvent extraction step can consume more than 50% of the required energy to produce one ton of rare earth metal [10]. However, until further research into alternative methods of separating these metals can be done, solvent extraction remains the most feasible process for separating bulk quantities of REEs [11]. As a result, a significant step forward in making REE extraction more economically feasible, and, more importantly, less environmentally damaging and safer for humans, will be any progress that can be made towards improving or replacing the tradition solvent extraction process. The demand for critical metals extends beyond REEs to other transition and alkali metals such as cobalt, nickel, copper, and lithium [12]. These metals are critical for a nation's energy future, and securing more environmentally friendly ways of recovering these metals is important. Additionally, as demand for these metals increase, and the quality of the supply decreases, the environmental impact of extracting these metals increases significantly [13].

The unique chemical challenge these elements present in their separation has inspired the present research on field-based separation. While the REEs may exist in identical ionic forms with similar chemical affinities, they have different electron configurations resulting in differing magnetic moment strengths. In this context, any ion with unpaired electrons in its valence orbital will have a magnetic moment [14]. It is this property which allows the 4f and 3d metals tested to respond to the magnetic fields, as well as allowing for the mobilities of these ions in the solution to be quite different depending on the number of unpaired electrons each ion has. Additionally, the REEs have been shown to exhibit differing affinities for complexing with certain organic molecules at different pH levels, which modifies their effective charge to be different enough from one another that they can be separated on the basis of charge [15,16]. This is largely due to the small differences in atomic radius of the lanthanides, as well as the difference in coordination number from the heavy to light REEs [5,17]. The aim of this research is to exploit these differences in physical properties to drive separation without the use of traditional solvent extractants. The process proposed is envisioned to either replace or supplement traditional solvent extraction; therefore, the system in question is a solution of relatively concentrated mixed REE–chloride salts with some additional valuable critical metals and contaminants. Work was carried out to understand how ions of differing magnetic moments can be separated in solution as well as how ions of differing charge can be separated electrostatically using methods similar to magnetic separation. The fundamental underlying idea behind the studied separation mechanism is that, given a fixed residence time for a dilute solution of mixed REE ions, these ions will diffuse different distances in response to a force field. The difference in this mobility can be exploited to drive separation on the basis of this generated concentration gradient if external, inhibiting forces can be sufficiently mitigated. This has already to be proven feasible; however, both the timescale and scalability of previous set-ups need to be improved to work on an industrial scale [18–20].

1.1. Theory

The concentration profile for a solution of susceptible ions subject to a force field is given by:

$$\frac{\partial C}{\partial t} + \nabla \cdot \left[\frac{D}{RT} \left(-\nabla C + C F_f \right) + C v \right] + r = 0 \quad (1)$$

The partial differential on the left is changing concentration over time, the second and third terms represent diffusion under a field and convection respectively, and the last term describes volumetric generation from a chemical reaction [21]. D is the diffusion coefficient of the species, R is the universal gas constant, T is the thermodynamic temperature, C is concentration, F_f is the external force field applied, v is the fluid velocity, and r is the volumetric generation term. Assuming no chemical reaction takes place and steady state conditions are reached, the concentration profile of the ions in solution becomes:

$$0 = -\nabla \cdot \left[\frac{D}{RT} \left(-\nabla C + C F_f \right) + C v \right] \quad (2)$$

This equation describes three sources of changing concentration which will be of significant concern to performing separations on an ionic scale. These sources of ion flux will be: (1) Migration due to external applied force, (2) diffusion due to an induced concentration gradient, and (3) convection due to fluid flow. As a result, each of these must be individually addressed to optimize ion separation.

1.1.1. Magnetic Separation

The separation of paramagnetic ions in an inhomogeneous magnetic field is done based on the significant variability in the magnetic moments of 3d transition metals and 4f lanthanides. The strength of the magnetic moments of each of the ions can be predicted from their electron structures using their orbital quantum numbers S, L, and J. The effective magnetic moment of an ion can be calculated as:

$$\mu_{eff} = \mu_b g_j \sqrt{J(J+1)} \quad (3)$$

where μ_b is the Bohr magneton ($\mu_b = \frac{e\hbar}{2m_e}$) and g_j is the Landé g-factor [18]:

$$g_j = 1 + \frac{J(J+1) + S(S+1) - L(L+1)}{2J(J+1)} \quad (4)$$

The values of μ_{eff} are calculated for the various lanthanides and a few select 3d transition metals in Table 1. When placed inside of an inhomogeneous magnetic field of strength B, the solution of paramagnetic ions will have a volumetric potential energy (E) given by:

$$E = \frac{\chi_{sol}}{2\mu_0} B^2 \quad (5)$$

where χ_{sol} is the magnetic susceptibility of the solution, which is the weighted average of the individual susceptibilities of the components of the solution, and μ_0 is the permeability of free space [18,19]. The molar magnetic susceptibility of a material is related to the strength of its magnetic moment described in Equation (3). Molar Susceptibility (χ_m) can be calculated using Curie's law, which states:

$$\chi_m = \frac{C_m}{T} = \frac{\mu_0 N_a \mu_{eff}{}^2}{3K_B T} \quad (6)$$

where N_a is Avogadro's number and K_B is the Boltzmann constant [19]. The volumetric force exerted on the solution by the magnet (F_m) is therefore:

$$F_m = \nabla E_m = \frac{\chi_{sol}}{2\mu_0} \nabla B^2 \quad (7)$$

The following volumetric force term can then be used in Equation (2) to describe the concentration gradient generated by a given magnetic field. Equation (7) shows the clear dependence of the magnetic separation on both the individual susceptibility of the ions in solution as well as the strength of the magnetic field gradient, showing that this kind of separation would not be possible in a homogenous magnetic field ($\nabla B = 0$).

Table 1. Quantum numbers and magnetic properties of a few selected metals.

Ion	Configuration	S	L	J	g_j	μ_{eff}/μ_B	Type
Ce^{3+}	4f1	0.50	3.00	2.50	0.86	2.54	LREE
Pr^{3+}	4f2	1.00	5.00	4.00	0.80	3.58	LREE
Gd^{3+}	4f7	3.50	0.00	3.50	2.00	7.94	MREE
Dy^{3+}	4f9	2.50	5.00	7.50	1.33	10.63	HREE
Er^{3+}	4f11	1.50	6.00	7.50	1.20	7.57	HREE
Li^{+}	1s2	0.00	0.00	0.00	0.00	0.00	Alkali
Fe^{3+}	3d5	2.50	0.00	2.50	2.00	5.90 *	TM
Co^{2+}	3d7	1.50	3.00	4.50	1.33	6.63 *	TM

* Experimental moments of 3d metals are complex and lower than the theoretical values by a significant amount due to orbital quenching [14].

1.1.2. Electrostatic Separation

Separation based on electrostatic (ES) attraction is relatively simple compared to magnetic separation, and the main factor impacting separation will be ionic charge of the species in solution. The force term (F_e), which is to be substituted into Equation (2), is simply Coulomb's equation for electrostatic attraction:

$$F_e = \frac{K_c q_1 q_2}{r^2} \quad (8)$$

where K_c is Coulomb's constant, q_1 and q_2 are the charges of the high voltage source and the ion respectively, and r is the distance between the two charges. Using the definition for electrostatic potential, $V = K_c \frac{q}{r}$, Equation (8) can be rewritten as:

$$F_e = \frac{V q_{ion}}{r} \quad (9)$$

This form of the force term shows the electrostatic separation to be determined primarily by the voltage of the source used for separation, the charge of the ionic species in solution, and the separation distance. The first two factors are, for the present purposes, unchanging and therefore effective ionic charge will be the primary factor impacting separation of ions. The change in total concentration generated by a potential difference between two points (ΔV) can be calculated as:

$$\Delta V = K_c Z q_e N_A \frac{\Delta C}{r} \quad (10)$$

where Z is the ionic charge, q_e is the charge of an electron (1.602×10^{-19} C/eV), N_A is Avogadro's number (6.023×10^{23}), and ΔC is the change in concentration from one point to another in mol/L. For a 100 kV source and a separation distance of 1×10^{-4} m, the predicted change in concentration is on the order of 10^{-12} mol/L. This would seemingly make separation of this kind impractical without an excessively powerful potential source or minuscule length scale. However, this fact overshadows that, for a solution that is composed of many different ions of many different valencies, only the net charge and ionic strength of the solution must remain effectively constant. Ionic strength (I) is calculated using the following equation:

$$I = \frac{1}{2}\sum_{i=1}^{n} C_i z_i^2 \tag{11}$$

Ionic strength is simply a sum of ionic charges (z) and concentrations (C), and so long as the sum of charges and concentrations remains constant, the system can exist in a number of different configurations [22]. The system will end up in a configuration which minimizes its potential energy by arranging ions in a way that results in their effective ionic charge corresponding to the gradient of the electric field. This manifests as elements with higher positive charge moving further towards the potential source and those with lower charge diffusing to maintain a net zero change in ionic strength.

1.1.3. Advection and Mixing

Though the externally applied force term has been adequately defined, Equation (2) describes a force which works against a separation of ions, advection. While the forces described above can generate a concentration gradient in a solution, realizing that separation into two distinct fractions can be difficult because of mixing that will inevitably occur as one tries to physically separate them. Designing a device which would allow fluid to flow through the field in two laminar flows, a "top" flow and a "bottom" flow, would solve this problem. Designing the flow channels such that the flow is laminar will reduce turbulent mixing while having two channels, a top and a bottom, allows ions to be deflected from one stream into another and separated.

A balance of momentum in a flow carried by either inertial or viscous forces is described using the Reynolds number. For internal flow in a pipe, the Reynolds number (Re) is:

$$Re = \frac{\rho U D_H}{\mu} \tag{12}$$

where ρ is the density of the fluid (kg/m^3), U is the velocity of the flow (m/s), D_H is the hydraulic diameter (m), and μ is dynamic viscosity (Pa s) [23]. For internal flow, the critical Reynolds number is 2300 and a $Re > 2300$ denotes a transition from a laminar flow to a turbulent flow. Therefore, Equation (10) shows that to decrease the Reynolds number of a given flow, either the hydraulic diameter of the pipe can be decreased, or the dynamic viscosity of the fluid increased. For the intended separation, the Reynolds number is always going to be below 100. Therefore, the importance of the Reynolds number is not to tell whether the flow is laminar or not, as it will be laminar for every separation device tested, but instead, it will be mostly used to describe the magnitude of inertial forces in the flow which can drive even minor amounts of unintended mixing of the top and bottom streams on the boundary where they meet.

Another dimensionless number in fluid dynamics that can be used to evaluate the ratio of advective mass transfer to molecular (mass) diffusion is the Schmidt number (Sc):

$$Sc = \frac{\mu}{\rho D} \tag{13}$$

A larger Sc indicates a flow with momentum transfer dominated by advective forces rather than the diffusive mass transfer [23]. The values for ions in water at room temperature are typically 200–1500. Table 2 shows the Sc for a few different ions in water. This table demonstrates that most of these ions will have an Sc near the upper limit typical of a liquid–liquid system, meaning that general momentum in the stream is carried more by advective mass transfer than the back diffusion created by a concentration gradient. This is the motivation behind the design of the microchannel device because minimizing advective mass transfer is key to higher separations of the target metals. Additionally, facilitating diffusion between layers in the device is critical to maintaining a balance of charges in the system.

Table 2. Diffusivities and Schmidt numbers for different metal ions.

Ion	D ($\times 10^{-10}$ m^2/s)	Sc	Reference
La^{3+}	6.19	1438	[24]
Nd^{3+}	6.16	1445	[24]
Li^{+}	10.29	865	[25]
Fe^{3+}	6.04	1474	[25]
Co^{2+}	6.78	1313	[25]

All values are for water at 25 °C.

2. Materials and Methods

To control the flow of the testing solution such that it is laminar, a device to constrain the flow into many narrow channels was designed. This device consists of two hard, plastic layers with very thin flow channels on the surface of each. The two opposing layers are angled slightly and pressed together. The channels on each layer are spaced evenly and each channel is as close to parallel as possible. The device is then placed directly in a magnetic or electrostatic field and fluid is allowed to flow through the layers. The channels guide the fluid to allow it to flow without turbulence as separation in the field is taking place. The source of the magnetic field was a 4 cm × 4 cm × 1 cm N52 NdFeB magnet placed 1 cm from the flow. An electrostatic field of 100 kV was generated using a Van de Graaf generator placed 1 cm from the flow. Figure 1a is a schematic representation of the microchannel device and demonstrates how ions are separated by deflection from the top channel into the bottom channel. Figure 1b shows an alternative view of this process, where susceptible ions are preferentially concentrated in the bottom stream, separating them from non-susceptible and mildly susceptible ions.

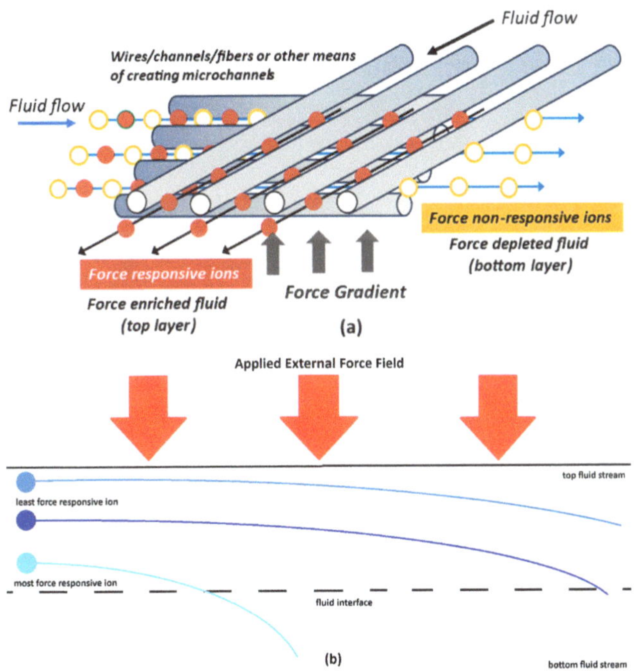

Figure 1. (**a**) Schematic diagram representing the utilization of thin microchannels to facilitate field-effect separation of dissolved metal ions. (**b**) Diagram demonstrating the principles by which the separation mechanism of FES works. Figures not to scale.

Figure 2 show the setup of the magnetic and electrostatic separations, respectively. As an ion in solution passes through the magnetic or electric field, it is deflected from one stream into the other. The two streams flow out of the device into separate containers and are collected to be analyzed. To test the relative impact of different flow velocities and channel sizes, different channel widths were compared. The two channel widths that were tested were 170 μm and 320 μm. The testing solution was composed of 50 mg/L of the following trivalent REE chloride hexahydrate salts: La, Ce, Pr, Sm, Eu, Gd, Dy, Er (Sigma-Aldrich (Burlington, MA, USA), >99%), as well as 50 mg/L of selected transition and alkali metal chloride hydrate salts: Li (Alfa-Aesar (Haverhill, MA, USA), 99%), Fe, Ni, Co, Cu (Acros Organics (Waltham, MA, USA), 98%), and a background salt of 0.25 wt% NaCl (Macron Chemicals (Radnor, PA, USA), 99%). The solution was adjusted to a pH of 0.75 using concentrated HCl (Fisher Chemicals (Waltham, MA, USA), 38.0%). The addition of transition metals to the REEs was done in these tests in order to both demonstrate the magnetic properties of 3d ions in addition to the 4f lanthanides, but also to show the viability of this system for separating metals other than REEs. Ni, Co, and Cu are metals of significant interest and Fe is a common contaminant in REE purification processes. Another set of tests were performed on the same 170 μm microchannel device with the same 50 mg/L REE+TM test solution, but the device was oriented at a variety of different angles to modify the velocity of the fluid flowing through the device. The first solution tested with electrostatic separation consisted of 100 mg/L of the same $FeCl_3$, $NiCl_2$, LiCl, and KCl (Alfa-Aesar (Haverhill, MA, USA), 99%) salts with 0.25 wt% NaCl, adjusted to a pH of 0.75 using concentrated HCl. A different ES separation was performed on a solution consisting of 100 mg/L of LiCl, and 100 mg/L of $K_2Cr_2O_7$ (J.T. Baker, 99%) adjusted to either a pH of 2.0 or 10.0 using either concentrated HCl or KOH (Sigma-Aldrich (Burlington, MA, USA), 99%), respectively.

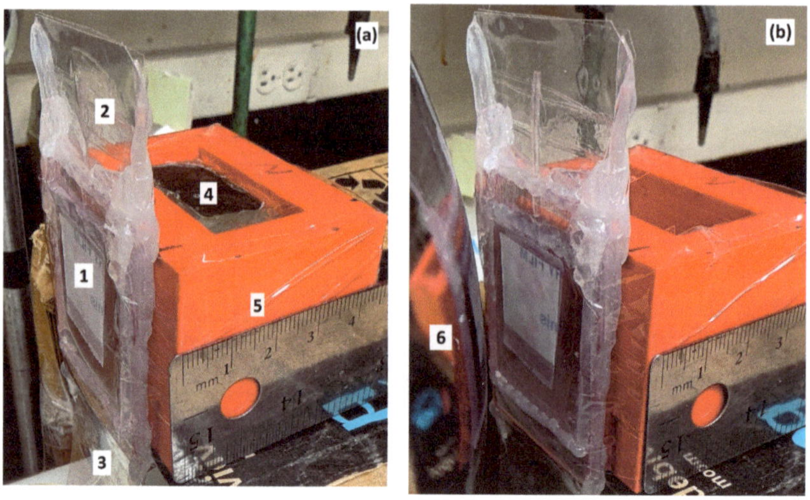

Figure 2. Experimental set up for: (**a**) magnetic separation test and (**b**) electrostatic separation test. 1: Microchannel separation device, 2: inlet funnel, 3: ramps to sample collection vessels, 4: NdFeB magnet, 5: PLA plastic device/magnet holder, 6: Van de Graaf generator.

Lastly, an ES separation was done on a solution containing 50 mg/L of the selected trivalent REE chloride salts mixed with 25 mM of ethylenediaminetriacetic acid (EDTA) (Sigma-Aldrich (Burlington, MA, USA), 99%) and adjusted with concentrated NH_4OH (Fisher Chemicals (Waltham, PA, USA), 30.0%) to a pH of 6.5 and 9.5 [17]. The test solutions were allowed to flow through the microchannel device in a magnetic or electric field, the top and bottom portions were then collected, diluted in 5% HNO_3 (Fisher Chemicals

(Waltham, PA, USA), trace metals grade), and analyzed with inductively coupled plasma optical emission spectroscopy (Agilent 5800 ICP-OES, Agilent Technologies, Santa Clara, CA, USA) to compare their various ion separations. pH was tested using a pH probe (Thermo Scientific Orion Ag/AgCl pH probe) for certain samples to gauge the magnitude of H^+ diffusion. Theoretical speciation of ions in solution was calculated with the software Visual MINTEQ 4.0.

3. Results and Discussion

3.1. Microchannel Device Analysis

The primary factor affecting separation that was tested for the microchannel device was the diameter of the channels. The increase in channel diameter will lead to a lower density of channels, greater channel spacing, and a larger internal volume. This has the impact of decreasing residence in the field as well as increasing the flow rate through the device. Therefore, a poorer separation with a larger channel diameter is predicted from analyzing the Reynolds number of the flows in the two different devices. To do so, the channel width was calculated using channel density (channels/cm) to calculate the spacing between each channel wall.

Table 3 shows the flow parameters through each device. As can be seen, the reduction in channel diameter from 320 μm to 170 μm has a significant impact on flow velocity, residence time, and, therefore, turbulence. The 170 μm channel device is about half the size of the 320 μm channel device, and this is the primary factor with contributes to the significant difference in performance between these devices. The Reynolds number for these two devices shows that while both flows should be demonstrating laminar characteristics, the 320 μm device's flow will be more turbulent along the fluid layer separating the two top and bottom streams. This is a likely factor contributing to the inherently lower performance of the larger diameter channel devices. In addition, the residence time is approximately 41% of the larger channel device, thereby leading to a much shorter (2.4×) exposure time in the applied field, resulting in less separation.

Table 3. Flow parameters of devices with two different channel diameters.

Spacing	320 μm	170 μm
# Channels (per side)	104	184
Residence Time (s)	25.9	61.1
Flow Velocity (cm/s)	0.154	0.066
Re	54.6	12.3

With a Reynolds number determined, other mass transfer relationships can be calculated. The Sherwood number is a dimensionless constant in fluid dynamics that describes the ratio of the advective mass transfer coefficient to the diffusivity of the species over its diffusion length [23]. In practice, it represents the ratio of the two rates and incorporates *Re* into the calculation using the Froessling Equation:

$$Sh = \frac{k_c L}{D} \cong 2 + 0.6\left(Re^{\frac{1}{2}} Sc^{\frac{1}{3}}\right) \quad (14)$$

where k_c is the advective mass transfer coefficient, L is the diffusion length, Re is the Reynolds number, and Sc is the Schmidt number. Table 4 shows the Sherwood numbers for a few selected ions and demonstrates that the actual magnitude of the difference between the advective and mass transfer is only roughly one order of magnitude, so both effects will be significant. Additionally, it is evident that the effects of advection are reduced significantly for the smaller microchannel device, making the impact of mass transfer due to diffusion more significant.

Table 4. Sherwood numbers for different ions in both the 170 μm and 320 μm microchannels.

	Sc	Sh_{170}	Sh_{320}
La^{3+}	1437.803	25.761	52.018
Nd^{3+}	1444.805	25.799	52.099
Li^{+}	864.917	22.058	44.224
Fe^{3+}	1473.510	25.956	52.429
Co^{2+}	1312.684	25.051	50.523

3.2. Magnetic Separation Performance

Magnetic separation was performed using the newly developed microchannel devices and the performance is summarized in Figure 3a. The magnetic nature of the concentration gradient created is confirmed by the lack of lithium mobility in the solution. Li^{+} was predicted to have no magnetic moment and be unresponsive to an inhomogeneous magnetic field, which makes it a good tracer to verify magnetic effects. Additionally, those ions with substantial magnetic moments were significantly separated from one stream to the other. Though slight, a trend in separation can be observed between each of the metals depending on the strength of their magnetic moments. LREEs with a small theoretical magnetic moment were moved from one stream to the other noticeably less than the HREEs with much larger magnetic moments. Figure 3b shows that at the low pH of 0.75, 99% of the metal in solution is present almost entirely uncomplexed with the exception of iron. This eliminates any doubt about the potential impact of differing speciation of metal ions in solution might create when analyzing the results.

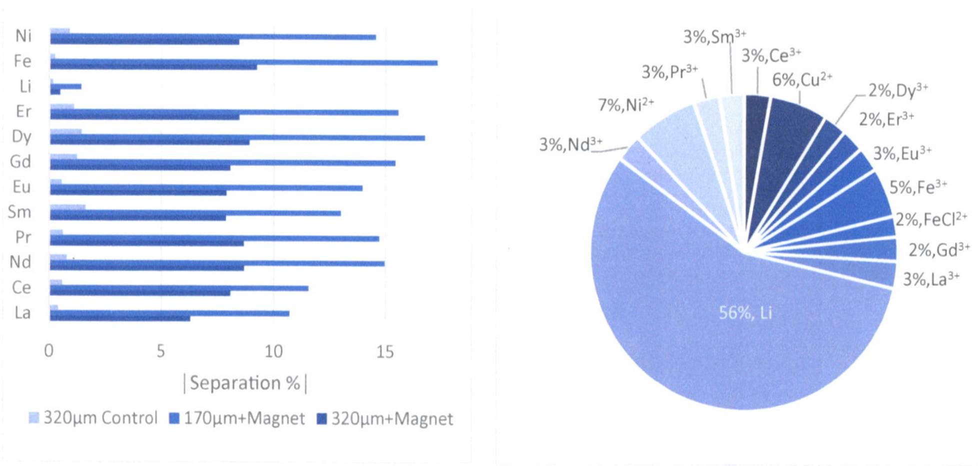

(a) (b)

Figure 3. (**a**) Separation performance of microchannel devices from top to bottom stream with different fluid flow rates in a magnetic field. (**b**) Relative speciation in mol% of the test solution as calculated from Visual MINTEQ 4.0. Note that sodium was excluded from this chart for visibility due to its extremely high concentration.

Figure 4 shows the separation of each element plotted with respect to its magnetic moment strength. The correlation between these factors is imperfect; this is a result of mixing counter to the direction of separation. This applies to both those inherent to the system (such as the back-diffusion created by the induced concentration gradient) and those that come from the imperfections that result from construction of the microchannel devices. These work to mask the impact that the magnetic moment strength has on the total separation, but a clear trend can still be observed.

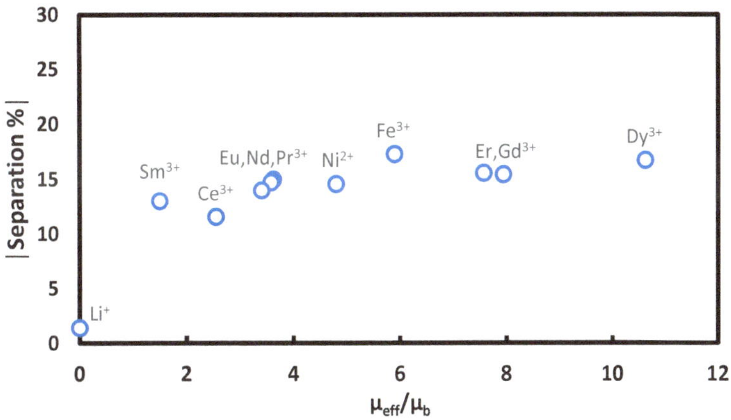

Figure 4. Separation performance of microchannel device from top to bottom stream for each element with respect to their magnetic moment strength.

The 170 μm microchannel device was modified to allow for different fluid flow rates through the device. The impact of this is twofold: (1) to modify the residence time of the fluid in the magnetic field and (2) to modify the fluid flow velocity and therefore the flow parameters (such as *Re*). Figure 5 shows the separation of the same test solution as previously with multiple different residence times. As can be observed, the effect of residence time on separation is significant and consistent. In the 170 μm microchannel device, the *Re* is <15 for any system configuration, and therefore the impact on advection as a function of changing flow velocity is, in this set of circumstances, insignificant. Table 5 shows the flow parameters of the differently oriented devices. Therefore, in the smaller microchannel device, the primary factors influencing the separation of ions from one another are the force that an ion experiences in a magnetic field and the time that ion is exposed to that magnetic field for.

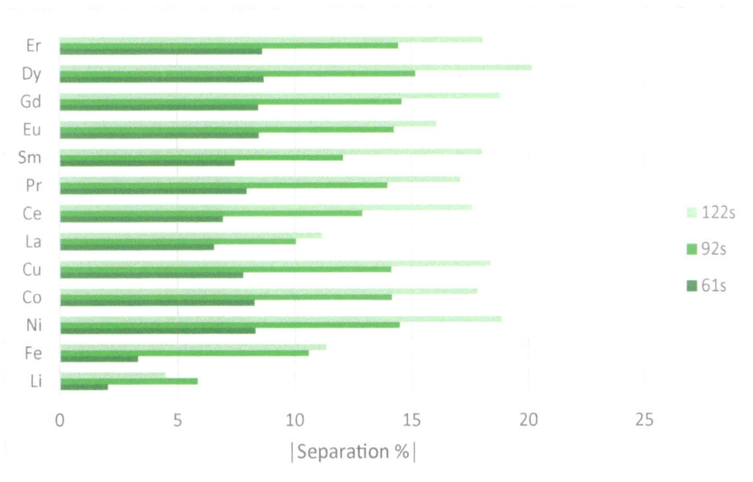

Figure 5. Separation performance of microchannel device from top to bottom stream for different fluid residence times.

Table 5. Impact of fluid velocity on residence time and Reynolds number.

Device Angle	90°	45°	30°
Pump Rate (μL/min)	150	100	75
Residence Time (s)	61.11	91.67	122.22
Re	12.31	8.21	6.16

3.3. Electrostatic Separation Performance

Electrostatic separation was performed using the same channel devices developed for the magnetic separation and the performance shown in Figure 6a. As can be seen, separation based on valence of the individual ions was achieved, with the two most significantly separated ions being the Ni^{2+} and the Fe^{3+}. Curiously, the Ni^{2+} seems to break this trend for the 320 μm device and the separation for the 170 μm device was poor. Figure 6b shows the relative percentage of each ion in solution as simulated from thermodynamics. In chloride media, it is likely that a larger portion of the iron was able to form complexes with chloride ions in solution compared to nickel. When comparing separation based on "effective" ionic charge, which is the weighted average of the charges of all the species of a given metal ion in solution, the trend is far more pronounced, as can be seen from Figure 7.

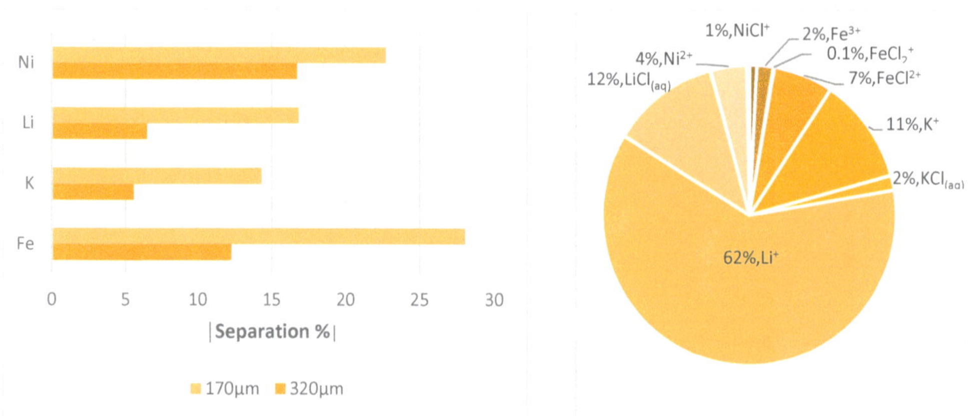

(**a**) (**b**)

Figure 6. (**a**) ES separation performance for 320 and 170 μm channel width devices. (**b**) Relative speciation in mol% of test solution calculated from thermodynamics using Visual MINTEQ 4.0. Sodium removed for clarity.

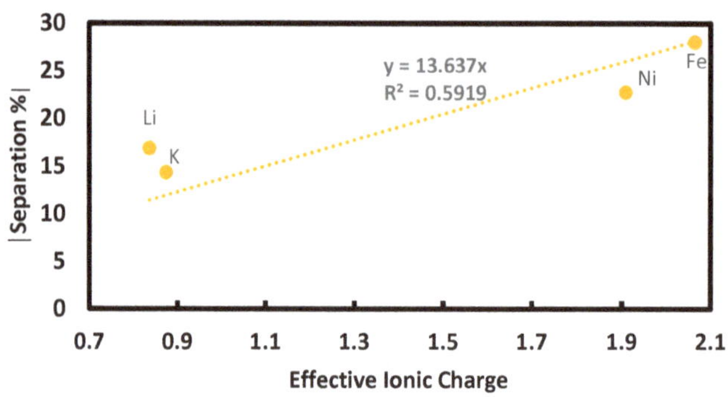

Figure 7. ES separation compared to effective ionic charge in solution for 170 μm channel test.

Further demonstration of the impact of complexation and effective ionic charge on separation is demonstrated in Figure 8. Here the same device was tested with two different solutions of LiCl and $K_2Cr_2O_7$ at a pH of 2.0 and 10.0. At a pH of 2.0, thermodynamics predicts ~95% of the chromium exists in solution as $HCrO_4^-$, while at pH = 10.0, the 95% chromium forms CrO_4^{2-}. Figure 8 demonstrates the clear disparity in separation due to this complexation effect compared to K^+ and Li^+ ions which remain relatively uncomplexed in solution relative to pH. The ultimate effect of complexation in this context is to lower the effective charge of the ion in solution. By changing the pH from highly acidic to highly basic, the effective charge of chromium was halved from CrO_4^{2-} to $HCrO_4^-$, and subsequently the separation seen between the two tests was halved as well.

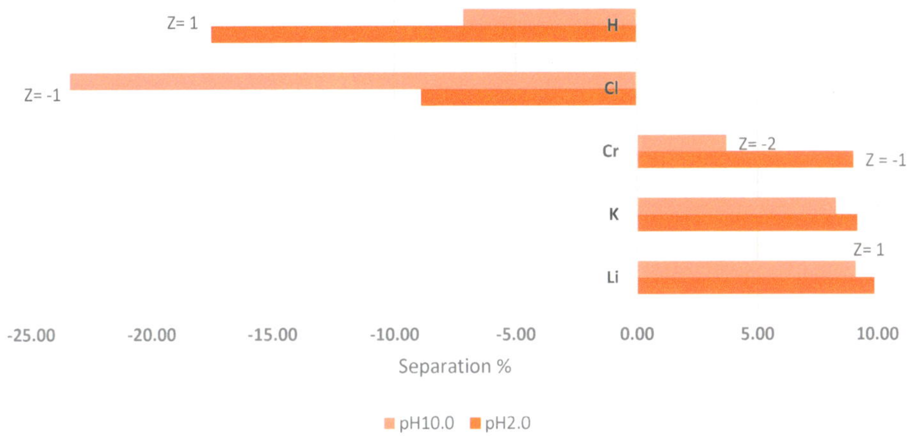

Figure 8. Effect of changing pH on separation of ions in 170 um microchannel device. Effective charges of ions are also shown.

The one aspect of this system not adequately described by chromium's effective charge is the direction of movement in the field. Despite having an opposite effective charge to the alkali metals, chromium is concentrated in the same direction. This could be indicative of the chromate anions being dragged across from one stream to another by the movement of cations such as Li^+. This is partly shown in the movement of chlorine anions in the system. A halving of the chromate movement results in a doubling of the movement of chlorine despite there being no observed impact of pH on the complexation of chlorine. It is therefore possible that the movement of chlorine is explained by the system's inherent trend towards maintaining a balance of ionic strength in the system. The counter diffusion of chlorine balances the ionic strengths of both the top and bottom streams throughout the separation. This is a critical aspect of the viability of this manner of separation, as basic calculations on the allowed separation of ions in an electric field is only 10^{-12} mol/L for this device, demonstrating that in general the amount of net charge on both sides of the separation should remain constant. Table 6 shows the total change in ionic strength of both sides of the separation at different pHs when accounting for the migration of H^+ ions in the form of pH change.

The change in ionic strength from top to bottom in the device is <1% and considered negligible. It shows the role the small, very mobile ions such as H^+ and Cl^- play in maintaining a balance of charges and concentrations between the two streams of the device. At a low pH, large amounts of H^+ can be observed moving counter to the direction of separation, and consequently a large amount of chromium in its anionic form is transported alongside its associated cations via electrostatic attraction. At a higher pH, the effective charge of the chromium was changed to −2 and part of the electrostatic attraction between it and the alkali cations moving across the boundary was partially overcome, meaning less of it was able to move across the boundary and be separated. However, its larger effective charge means that it will still have a significant impact on changing the ionic strengths of the top and bottom streams as it moves. Additionally, a higher pH means that less free H^+ can move to balance charge in the two streams. This could explain the much larger movement of Cl^- and smaller movement of H^+ at a high pH that was observed.

Table 6. Difference in ionic strength between the top and bottom streams for tests at different pHs.

		Li	K	H	Cr	Cl
pH 2.0	Zeff	1	1	1	−1	−1
	% Change	9.88	9.15	−17.50	8.98	−8.90
	% I change	−0.84%				
pH 10.0	Zeff	1	1	1	−2	−1
	% Change	9.09	8.28	−7.15	3.70	−23.36
	% I change	−0.80%				

The separation of REEs in an electric field was done by utilizing a solution of lanthanides mixed with 25 mM EDTA. The EDTA complexes with the REEs to form anions and the degree of complexation for each lanthanide can differ depending on pH. A neutral and basic pH were selected to evaluate the impact of the increasing anionic nature of an ion on its separation. Work by J.I. Alonso and R.G. Fernández provides an experimental determination of the mechanism of complexation for these ions as well as their effective charge in solution as a function of pH [17]. The main favorable characteristic of EDTA in this system is that it generally complexes differently with LREEs than it does with HREEs, and the effect can be made more dramatic by changing the pH of the solution. If Y^{4-} can be considered the conjugate base of H_4Y which represents EDTA, then at a neutral pH the system has a propensity to form a LnY^- $*(H_2O)_n$ (n = 3 for LREE, n = 2 for HREE). At a high pH of 9.5 the LREEs will form a mixture of $LnY(OH)^{2-}$ $*(H_2O)_{2\text{-}3}$ and $LnY(OH)^{3-}$ $*(H_2O)_{0\text{-}1}$ complexes, while the HREEs still preferentially form LnY^- in equilibrium with a small amount of $LnY(OH)^{2-}$ [17]. This has the impact of making all the REE complexes have nearly the same effective charge of approximately −1 at a neutral pH, and at a higher pH, the effective charges vary between approximately −1 and −2.5 depending on the atomic weight of the lanthanide. Figure 9a shows the separation of the lanthanide elements in an electric field at differing pHs. As can be observed from the figure, the exact effect described above between the two different pHs is observed. The neutral pH test shows moderate overall separation with almost no separation between elements observed. However, at a high pH a clear difference in the separation of elements can be observed. The trend of separation is that it increased from top to bottom as atomic mass increased up until Gd, at which point the separation again begins to decrease. This trend agrees very well with the experimental effective charges found by Alonso and Fernández [17]. Figure 9b shows separation plotted against the effective charge of the complex at both high and low pH, and it clearly shows a strong trend of effective charge on separation.

The results of this experiment indicate that the formation of highly anionic EDTA–hydroxyl complexes is most prevalent in MREEs (Sm-Gd) and generally falls off increasingly as atomic mass increases or decreases. Additionally, a notable discrepancy between expectation and experiment can be noted for the separation of Dy and Er, which are the two points in Figure 9b that lie furthest from the line of best fit for pH = 9.5. The larger than expected separation is likely a result of the smaller radii of these HREE complexes because these elements typically form LnY^- with few to no waters of hydration, meaning that these elements are more mobile than their LREE counterparts which generally complex with 1 to 2 hydroxyl ions and have 2–3 waters of hydration. At a neutral pH, basically every ion is present as LnY^- and because the difference in effective charge as well as the separation between each element is relatively small, the effects of differing waters of hydration on the movement of these complexes is minor. This explains why in the pH = 6.5 case the effective charge of all of the ions cluster closely around −1 and similarly all the separations also fall tightly together around ~5%.

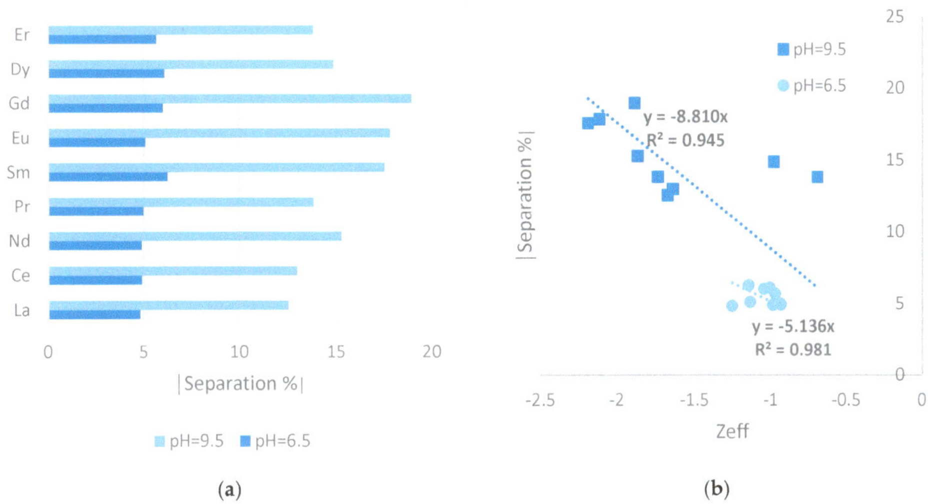

Figure 9. (**a**) Separation by element from top to bottom for REE + 25 mM EDTA test at different pHs. (**b**) Separation with respect to effective ionic charge at different pHs.

3.4. Discussion

The employment of FES was successfully demonstrated for separating metal ions in solution, in large part because the microchannel devices were effective at controlling the flow to minimize advective mass transfer while the fluid was subjected to the force field. This can be partially observed by comparing the separation of the most susceptible ions for different experiments to the Re for that particular set of testing conditions. A weak trend is present when looking at the separations presented in Figures 4a, 5 and 6a. This is because while *Re* can be used in conjunction with other dimensionless numbers such as *Sc* and *Sh* to describe the magnitude of advection relative to diffusion of ions in the force field, it is not the primary determining factor of total separation. The real effect that ties these experiments together is the correlation of total separation to residence time in the field, as might be initially expected. Figure 5 shines light on this particular effect best. In this experiment, a roughly 30 s increase in residence time is enough to increase the separation of most of the elements in solution by ~3–5% with each incremental increase. The decrease in residence time corresponds to increasing *Re*, explaining its weak correlation to separation.

A clear separation effect in the microchannel device was generated using magnetic fields generated by a strong permanent magnet. Though it leaves room for improvement, a stratification of elements on the basis of magnetic susceptibility as determined by the magnetic moment of each individual ion was clearly observed in between the top and bottom stream of the microchannel device. Between individual elements, for example Sm, Eu, and Gd, separations of 2–3% were observed. Additionally, increasing residence time maintains the observed stratification of elements even as concentration increases. While the osmotic effects of rapidly changing the concentration of solution were explored in the data presented in Figure 8 and Table 6, they were mostly neglected for these sets of experiments. This is due to the addition of a highly concentrated and non-magnetically susceptible background electrolyte (0.25 wt% NaCl). The high mobility of the salt paired with the very high ionic strength of the solution means that the magnetically susceptible ions should be free to migrate in the field without suffering the effects of back diffusion created by a chemical potential gradient.

Similarly, electrostatic separation experiments showed that ions can be separated in solution on the basis of effective charge. Figures 6 and 7 describe this particular kind of separation. While complexation seems to have little effect on magnetic properties of 4f lanthanides, anions complexing with metals will work to lower its effective charge, and therefore how strongly it will respond in a field. This explains the seemingly anomalous separation behavior of Fe and Ni in this experiment because Fe complexes more strongly with Cl^- than Ni as predicted by thermodynamics. This effect can be used to improve separation between metals where it is known that their charges can be manipulated through selective complexation. Figures 8 and 9 describe two experiments in which this effect was explored. In Figure 8, the effective charge of Cr was manipulated by changing pH which decreased its effective charge by -1 and subsequently halved its separation. Figure 9a,b describes a set of experiments wherein the pH dependent, selective complexation of REEs with EDTA was explored. It was demonstrated that EDTA can be used to create anionic complexes with REEs, and the degree of electronegativity exhibited by these complexes depended on pH, ionic radii, and coordination number. The combination of all of these factors means that medium to heavy REEs could be targeted with electrostatic separation by making them very anionic at high pHs.

4. Conclusions

This research describes the development of a new method of separation of rare earth and critical metals based on their response in magnetic or electrostatic fields, with the aid of flow-through microchannels. A device which constrained the flow of the test solution to many laminar streams and allowed the application of strong applied fields next to them to enable ion separations was designed and tested. The data demonstrated the following:

- Significant concentration changes between the top and bottom stream of the microchannel device of 10–20% can be generated using magnetic and electrostatic fields;
- Trends of ion mobility based on magnetic susceptibility and residence time in the field are observed, with individual element separations of 2–3% achieved per cycle;
- Manipulating the effective charge in solution of metal cations by changing its pH allows them to be separated electrostatically;
- The designed microchannel device sufficiently constricts flow to facilitate separation and minimize advective mixing.

While this method shows promise in separating both rare earths and critical metals in solution, in order for it to rise to meet the demand currently occupied by traditional methods of solvent extraction, further work to improve the efficiency of a single cycle of this process should be carried out to fully explore the potential of this hydrometallurgical separation technology.

5. Patents

This work has resulted in a patent application titled "Thin fluid layers and streams facilitated, force-based atom, ion, molecule, and fine particle separators and methods of using the same", patent application number 18447171.

Author Contributions: Conceptualization, B.S., M.F., and P.S.; methodology, B.S. and M.F.; formal analysis, B.S. and M.F.; investigation, B.S., E.S., J.B., and Z.E.; resources, M.F.; data curation, B.S.; writing—original draft preparation, B.S.; writing—review and editing, M.F., P.S., E.S., and J.B.; visualization, B.S.; supervision, M.F. and P.S.; project administration, M.F. and P.S.; funding acquisition, M.F. and P.S. All authors have read and agreed to the published version of the manuscript.

Funding: The research was funded through support from the Cooper-Hanson fellowship through the University of Utah as well as from DOE DE-FE0032122.

Data Availability Statement: The available data are presented in the body of the paper.

Acknowledgments: This work was supported by the University of Utah Cooper Hansen Fellowship. The authors would also like to acknowledge Xinbo Yang for providing the ICP-OES analysis used in this research.

Conflicts of Interest: The authors disclose the potential conflict of interest associated with pending patents that could benefit the authors.

References

1. Jyothi, R.K.; Thenepalli, T.; Ahn, J.W.; Parhi, P.K.; Chung, K.W.; Lee, J.-Y. Review of rare earth elements recovery from secondary resources for clean energy technologies: Grand opportunities to create wealth from waste. *J. Clean. Prod.* **2020**, *267*, 122048. [CrossRef]
2. Chen, Z.; Li, Z.; Chen, J.; Tan, H.; Wu, J.; Qiu, H. Selective Adsorption of Rare Earth Elements by Zn-BDC MOF/Graphene Oxide Nanocomposites Synthesized via In Situ Interlayer-Confined Strategy. *Ind. Eng. Chem. Res.* **2022**, *61*, 1841–1849. [CrossRef]
3. Massari, S.; Ruberti, M. Rare earth elements as critical raw materials: Focus on international markets and future strategies. *Resour. Policy* **2013**, *38*, 36–43. [CrossRef]
4. Golev, A.; Scott, M.; Erskine, P.D.; Ali, S.H.; Ballantyne, G.R. Rare earths supply chains: Current status, constraints and opportunities. *Resour. Policy* **2014**, *41*, 52–59. [CrossRef]
5. Florek, J.; Larivière, D.; Kählig, H.; Fiorilli, S.L.; Onida, B.; Fontaine, F.-G.; Kleitz, F. Understanding Selectivity of Mesoporous Silica-Grafted Diglycolamide-Type Ligands in the Solid-Phase Extraction of Rare Earths. *ACS Appl. Mater. Interfaces* **2020**, *12*, 57003–57016. [CrossRef]
6. Liu, T.; Chen, J. Extraction and separation of heavy rare earth elements: A review. *Sep. Purif. Technol.* **2021**, *276*, 119263. [CrossRef]
7. Pereira Neves, H.; Ferreira, G.M.D.; de Lemos, L.R.; Rodrigues, G.D.; Leão, V.A.; Mageste, A.B. Liquid-liquid extraction of rare earth elements using systems that are more environmentally friendly: Advances, challenges and perspectives. *Sep. Purif. Technol.* **2022**, *282*, 120064. [CrossRef]
8. Zheng, X.; Zhang, F.; Liu, E.; Xu, X.; Yan, Y. Efficient recovery of neodymium in acidic system by free-standing dual-template docking oriented ionic imprinted mesoporous films. *ACS Appl. Mater. Interfaces* **2017**, *9*, 730–739. [CrossRef] [PubMed]
9. Arshi, P.S.; Vahidi, E.; Zhao, F. Behind the Scenes of Clean Energy: The Environmental Footprint of Rare Earth Products. *ACS Sustain. Chem. Eng.* **2018**, *6*, 3311–3320. [CrossRef]
10. Pathapati, S.V.S.H.; Singh, R.S.; Free, M.L.; Sarswat, P.K. Exploring the REEs Energy Footprint: Interlocking AI/ML with an Empirical Approach for Analysis of Energy Consumption in REEs Production. *Processes* **2024**, *12*, 570. [CrossRef]
11. Pathapati, S.V.S.H.; Free, M.L.; Sarswat, P.K. A Comparative Study on Recent Developments for Individual Rare Earth Elements Separation. *Processes* **2023**, *11*, 2070. [CrossRef]
12. Zhang, S.; Ding, Y.; Liu, B.; Chang, C. Supply and demand of some critical metals and present status of their recycling in WEEE. *Waste Manag.* **2017**, *65*, 113–127. [CrossRef] [PubMed]
13. Kuipers, K.J.J.; van Oers, L.F.C.M.; Verboon, M.; van der Voet, E. Assessing environmental implications associated with global copper demand and supply scenarios from 2010 to 2050. *Glob. Environ. Chang.* **2018**, *49*, 106–115. [CrossRef]
14. Woodward, P.M.; Karen, P.; Evans, J.S.O.; Vogt, T. *Solid State Materials Chemistry*; Cambridge University Press: Cambridge, UK, 2021. [CrossRef]
15. Schijf, J.; Byrne, R.H. Speciation of yttrium and the rare earth elements in seawater: Review of a 20-year analytical journey. *Chem. Geol.* **2021**, *584*, 120479. [CrossRef]
16. Gritmon, T.F.; Goedken, M.P.; Choppin, G.R. The complexation of lanthanides by aminocarboxylate ligands—I: Stability constants. *J. Inorg. Nucl. Chem.* **1977**, *39*, 2021–2023. [CrossRef]
17. Fernández, R.G.; Alonso, J.I.G. Separation of rare earth elements by anion-exchange chromatography using ethylenediaminetetraacetic acid as mobile phase. *J. Chromatogr. A* **2008**, *1180*, 59–65. [CrossRef] [PubMed]
18. Fritzsche, B.; Lei, Z.; Yang, X.; Eckert, K. Localization of rare earth ions in an inhomogeneous magnetic field toward their magnetic separation. *J. Rare Earths* **2022**, *40*, 1598–1605. [CrossRef]
19. Rodrigues, I.R.; Lukina, L.; Dehaeck, S.; Colinet, P.; Binnemans, K.; Fransaer, J. Effect of Magnetic Susceptibility Gradient on the Magnetomigration of Rare-Earth Ions. *J. Phys. Chem. C* **2019**, *123*, 23131–23139. [CrossRef]
20. Fujiwara, M.; Chie, K.; Sawai, J.; Shimizu, D.; Tanimoto, Y. On the movement of paramagnetic ions in an inhomogeneous magnetic field. *J. Phys. Chem. B* **2004**, *108*, 3531–3534. [CrossRef]
21. Ji, B.; Wu, P.; Ren, H.; Zhang, S.; Rehman, A.; Wang, L. Segregation behavior of magnetic ions in continuous flowing solution under gradient magnetic field. *Chin. Phys. B* **2016**, *25*, 074704. [CrossRef]
22. Free, M.L. *Hydrometallurgy*, 2nd ed.; Wiley: Hoboken, NJ, USA, 2013. [CrossRef]
23. Sohn, H.Y. Constitutive elements of fluid-solid reactions. In *Fluid-Solid Reactions*; Elsevier: Amsterdam, The Netherlands, 2020; pp. 11–84. [CrossRef]

24. Liyananadirah, M.S.; Deraman, M.R.; Ibrahim, W.H.; Harun, N. Diffusion coefficients of trivalent rare earth/actinide ions in acid and alkaline aqueous solutions. In *IOP Conference Series: Materials Science and Engineering*; IOP Publishing Ltd.: Bristol, UK, 2019. [CrossRef]
25. Li, Z.; Merz, K.M. Systematic Evaluation of Ion Diffusion and Water Exchange. *J. Chem. Theory Comput.* **2021**, *18*, 3017–3026. [CrossRef] [PubMed]

MDPI AG
Grosspeteranlage 5
4052 Basel
Switzerland
Tel.: +41 61 683 77 34

Eng Editorial Office
E-mail: eng@mdpi.com
www.mdpi.com/journal/eng

www.ingramcontent.com/pod-product-compliance
Lightning Source LLC
Chambersburg PA
CBHW042014260926
51850CB00020B/3384